Fundamental Theories of Physics

Volume 197

The international monograph series "Fundamental Theories of Physics" aims to stretch the boundaries of mainstream physics by clarifying and developing the theoretical and conceptual framework of physics and by applying it to a wide range of interdisciplinary scientific fields. Original contributions in well-established fields such as Quantum Physics, Relativity Theory, Cosmology, Quantum Field Theory, Statistical Mechanics and Nonlinear Dynamics are welcome. The series also provides a forum for non-conventional approaches to these fields. Publications should present new and promising ideas, with prospects for their further development, and carefully show how they connect to conventional views of the topic. Although the aim of this series is to go beyond established mainstream physics, a high profile and open-minded Editorial Board will evaluate all contributions carefully to ensure a high scientific standard.

More information about this series at http://www.springer.com/series/6001

Horaţiu Năstase

Cosmology and String Theory

Springer

Horaţiu Năstase
Instituto de Física Teórica, UNESP
São Paulo, Brazil

ISSN 0168-1222 ISSN 2365-6425 (electronic)
Fundamental Theories of Physics
ISBN 978-3-030-15076-1 ISBN 978-3-030-15077-8 (eBook)
https://doi.org/10.1007/978-3-030-15077-8

Library of Congress Control Number: 2019934519

This Springer imprint is published by the registered company Springer Nature Switzerland AG
The registered company address is: Gewerbestrasse 11, 6330 Cham, Switzerland

To the memory of my mother,
who inspired me to become a physicist

Preface

This book deals with the intersection between string theory and cosmology, the methods to obtain cosmology from string theory, i.e., "string cosmology." While there have been many developments in the field within the last 15–20 years, to my knowledge, there is no comprehensive review or book dedicated to all the lines of research in the field. This is important, since string cosmology is a developing field, and it is not yet clear which line of research will eventually become standard. Since there are people in both string theory and cosmology interested in the subject, as well as graduate students that have a basic knowledge of both, the book introduces both general cosmology and string theory, before describing string cosmology proper.

The book is based on a graduate course I taught at my institute, the Institute for Theoretical Physics of UNESP, and for which the students indeed had this background: a bit of cosmology, and a bit of string theory background, but not proficient in either. I have added only 3 lectures at the end, denoted with an asterisk, which correspond to more recent (but no less important) developments in string cosmology. The book preserves otherwise the course format, with one lecture corresponding to one chapter, and having a list of "important points to remember" at the end of each chapter, as well as exercises. Researchers that want to learn about the subject are also an intended audience for this book, as it presents most approaches that have some success so far in describing cosmology, so they can use it as a reference.

There are no comparable books out there, to my knowledge. There is an excellent book on string inflation by Baumann and McAllister, *Inflation and String Theory*, but while it has a good presentation, it focuses only on inflationary aspects. In terms of the introduction to cosmology, there is information exclusively about inflation, assuming general cosmological knowledge, and the string theory introduction is very quick. The other book I am aware of, Gasperini's *String Cosmology*, deals with only some specific topics, mostly about inflation, and has only a small introduction to inflation, being meant for advanced researchers.

São Paulo, Brazil Horaţiu Năstase

Acknowledgements

I want to thank all the people who shaped me as a physicist throughout the years. First and foremost, my mother Ligia, the first physicist I knew, who showed me the wonders of physics. My high school teacher, Iosif Sever Georgescu, put me on the career path to physics and showed me that it is something I can do. My exchange student advisor at the Niels Bohr Institute, Poul Olesen, introduced me to string theory, my chosen field and the basis for this book. My Ph.D. advisor, Peter van Nieuwenhuizen, taught me, besides supergravity and string theory, how to be a complete theoretical physicist and the importance of rigor and precision. From Juan Maldacena, I learned a lot about the AdS/CFT correspondence, which forms the basis for some of the research described here, as well as a lot about string theory and cosmology.

This book is based on a course I taught at our institute, IFT at UNESP; so I thank all of the students who participated in the course and helped me with their questions and input.

I would like to thank all of my collaborations throughout the years, and in particular those with whom I worked on and discussed research related to the lines described in this book: Justin Khoury, Amanda Weltman, Rogerio Rosenfeld, Kurt Hinterbichler, Kostas Skenderis, Robert Brandenberger. I want to thank my wife Antonia for her understanding and encouragement while I worked on the book, in the evenings at home. My students and postdocs, for my having less time to work with them while I worked on this book. Finally, to my editor Angela Lahee for her encouragement and getting this book published.

Contents

Introduction

This book describes the intersection of string theory with cosmology or string cosmology. It is intended for people, either graduate students or researchers, with only a superficial understanding of one, or both of the sides (cosmology or string theory). The necessary background for the book is thus a thorough understanding of general relativity and quantum field theory, and some background in cosmology and string theory, but not too much.

In Part I of the book, I introduce regular cosmology, with the emphasis on the issues that will be explored using string theory, but trying to introduce all relevant concepts and most relevant calculations. Part II introduces the basic elements of string theory, with added emphasis on issues that will be used for cosmology. Besides supergravity and KK compactification, the basics of relativistic strings and their quantization, I will introduce D-branes, extended soliton-like objects in string theory, and T-duality, relevant in some cosmological models, as well as the general picture of supersymmetric string theory, their strong coupling version called M-theory, and their interconnection in the duality web. Finally, I will describe how one obtains particle physics from string theory, and the connection of string theory to strongly coupled field theories called the AdS/CFT correspondence, that can be used in a certain comological model of "holographic cosmology."

Part III of the book contains string cosmology per se. String theory has not managed yet to obtain a fully consistent model of string inflation; there is always a final step missing in those attempted constructions. I will thus describe the problems of constructing a string inflationary model as well as the problems with the supergravity approximation to string inflation. I describe several (related) attempts within the general category of string inflation, brane-antibrane inflation, the KKLT scenario for de Sitter backgrounds, and the KKLMMT scenario for string inflation and its generalizations, as well as the needed technical knowledge (for the previous constructions) of general braneworld cosmology and the Israel junction conditions. Next, I describe the competing models of ekpyrotic, new ekpyrotic and cyclic models, as well as the string gas and brane gas ones. Chameleon scalars, a type of scalars which are very light on the largest cosmological scales, are obtained from string theory as well as fuzzy dark matter (a kind of axionic scalar model). Finally, I treat the more modern and promising models of axion inflation and axion monodromy from string theory and of holographic cosmology (which uses the AdS/CFT correspondence).

Part I
Standard Cosmology

Chapter 1
The Expanding Universe and the Big Bang

In part I of the book, I will quickly describe the relevant parts of standard (non-stringy) cosmology.

In this first chapter, we will begin with a study of the basics of cosmology in a non-relativistic (non-general relativity) setting. Of course, the Universe is expanding, and the proper description of that is in general relativity, but we will see that with a simple non-relativistic analysis, supplemented with a few nontrivial points, we can go quite far in the description of cosmology. This can be thought of as just a useful trick in order to get some simpler physical insight. In the next chapter, we will move on to the relativistic theory.

After describing some observational facts, we will describe the cosmology in non-relativistic setting, and we will see that the equation of state of "matter" in the Universe dictates the time evolution of the scale factor of the Universe. After that, we will finally describe the most (general) relativistic component of "matter" in the Universe, the cosmological constant, which has a negative pressure, and leads to acceleration, instead of attraction and slowdown, of the expansion.

1.1 Observations

We will start with a few basic observational facts about the Universe.

Fact 1. The most important is the observation of the Hubble expansion, or *Hubble law*.

One defines the *redshift* as the relative change in frequency of light due to the (special relativistic) Doppler effect, which relates the frequency at emission ν_0 with the observed ("redshifted") frequency ν

$$\nu_0 = \nu\left(1 + \frac{v}{c}\right) , \tag{1.1}$$

H. Năstase, *Cosmology and String Theory*, Fundamental Theories of Physics 197, https://doi.org/10.1007/978-3-030-15077-8_1

so that the redshift is

$$z \equiv \frac{\nu_0 - \nu}{\nu} = \frac{v}{c}. \tag{1.2}$$

Since $\nu = c/\lambda$, we can also write it in terms of the wavelength at emission λ_0 and the observed wavelength λ,

$$z = \frac{\lambda - \lambda_0}{\lambda_0}. \tag{1.3}$$

A measure of the distance r is the *apparent visual magnitude* (more precisely, the d_L defined later is). One measures the light coming from a source, and knowing its brightness, we can see how far away from us it is. Then in the graph of the redshift z with the apparent visual magnitude, standing in for r, we see a linear relation for astrophysics objects, namely stars, galaxies, etc. This is known as the *Hubble law*, and it is

$$(c)z = v = Hr. \tag{1.4}$$

More precisely, this is correct in the non-relativistic approximation. In general, the relation is not linear, and $H = H(t)$. Note that here $v = \dot{r}$.

Finally, we should say that the Hubble law means that it is space itself that expands, and stars and galaxies only go along with the expansion, since there can be no mechanism that all stars and galaxies move away from us according to the Hubble law, and yet space is static.

Fact 2. The other important observation is the one of *isotropy* of the Universe. To infer it, we will preview the fact (to be explained in detail later in the book) that there exists a background radiation permeating all of the Universe, the Cosmic Microwave Background Radiation (CMBR), which is a remnant of the Big Bang, and this radiation has travelled to us since then. The CMBR spectrum is isotropic to an excellent precision, which means that the Universe (the initial Big Bang point, as well as everything in the path of the light coming from the Big Bang, that is to say, the whole of the observable Universe) is isotropic.

There is one detail about the CMBR that necessitates a relativistic treatment in order to understand properly. The CMBR has a dipole, i.e., there is a preferred direction in space. But this is easily understood as being due to the fact that the Earth is moving at a constant velocity, in the direction of this dipole, with respect to the CMBR. This is not clear within special relativity, where there shouldn't be a preferred frame for radiation. But within general relativity, in the expanding Universe metric $ds^2 = -c^2dt^2 + a^2(t)d\vec{x}^2$, there is a preferred frame, defined by this metric, where t is the "cosmological time". It is in this frame that the CMBR is defined, and the Earth is moving with respect to it.

Fact 3. Finally, we can observe the *homogeneity* of the Universe. Of course, on small scales, the Universe is not homogenous: we have stars and galaxies, either of which is surrounded by voids until the next structure. But on even larger scales, where galaxies and clusters of galaxies are considered as point-like, the Universe is homogenous. We can explicitly see this, from the distribution of galaxies and clusters of galaxies, which is found by measuring the direction, distance and mass of these structures. But moreover, while this method simply uses light to detect matter, there are other forms of matter that do not appear in light, called dark matter, about which we will speak later on in the book. To take that into consideration, one can measure the *peculiar velocity* fields of stars and galaxies. As we said, on the average, stars and galaxies move away from us according to the Hubble law (due to the expansion of space), but still, relative to this expansion of space, the objects can have extra random velocities, called peculiar velocities. In gravitational theory, the velocities of these many objects are influenced by the gravitational fields present, generated by the local mass distribution. Therefore, by measuring the peculiar velocities, in effect we can extract information about the mass distribution, that includes any matter not visible via light, i.e., dark matter. The final result is the same: on the very largest scales (of galaxies and clusters of galaxies), the Universe is homogenous to a very high degree.

1.2 Newtonian Approximation for the Expansion of Space

In order to find the equations that describe the expansion of space, we must use gravitational equations and the conservation of energy. The simplest model, that we will use in this chapter, is to use the Newtonian approximation for gravity. I should stress here that, strictly speaking, Newtonian gravity is not applicable to the expanding Universe, which is described by general relativity. However, the simplest model, Newtonian gravity, can take us very far, and by including the relativistic effect of pressure on the energy, can give us the correct equations for the expanding Universe. The method presented here is a rather known trick, whose origins I don't know, but which I have first seen in lectures by Igor Novikov some 25 years ago. The goal, besides ease of introduction, is to see that, despite appearances, we only need minimal input from general relativity, otherwise we can derive most things without it.

Assuming spherical symmetry (from isotropy) and uniformity, we can consider a sphere of matter of constant energy density ρ, and of radius r. Because of the expansion of space through the Hubble law, we have

$$\dot{r} = v = Hr. \tag{1.5}$$

Then the gravitational Gauss's law equates the total mass in the sphere (times $4\pi G_N$) with the flux of the gravitational field through the surface of the sphere, i.e.,

$$\rho \times \frac{4\pi}{3} r^3 \times (4\pi G_N) = g \times 4\pi r^2. \tag{1.6}$$

Moreover, g equals (since $m_i = m_g$, the inertial and gravitational masses are the same) the acceleration at the surface of the sphere, i.e.,

$$\ddot{r} = g = -\frac{4\pi G_N}{3}\rho r. \tag{1.7}$$

This is in fact the same as the general relativity result for pressureless matter! The only effect of general relativity is to imply that pressure has energy density as well, resulting in the shift

$$\rho \to \rho + \frac{3P}{c^2}, \tag{1.8}$$

to finally obtain the general relativity relation (renaming $r \to R$, the scale factor of the Universe)

$$\ddot{R} = -\frac{4\pi G_N}{3}\left(\rho + \frac{3P}{c^2}\right)R. \tag{1.9}$$

The Hubble law in general has

$$H = H(t) \equiv \frac{1}{R(t)}\frac{dR(t)}{dt}, \tag{1.10}$$

so the Hubble "constant" is not really a constant, but varies in time (it is a constant only in space). Moreover, as a function of the redshift z (the observational quantity), H also varies, since we measure everything by the light that comes to us, and something coming from redshift z corresponds (due to the constancy of the speed of light) with something that comes from a certain distance d, and thus a certain time t away.

We can also obtain another law from the Newtonian approximation. We can obtain it more rigorously, but first we will use a not quite correct shortcut. Neglect the pressure, $P \simeq 0$, and consider the conservation of energy in its rawest form, the sum of the kinetic and potential energies be a constant for a test mass m. The kinetic energy is $m\dot{R}^2/2$, and a very naive estimate for the potential energy, m times the potential, is m times the mass of the sphere, $\rho(4\pi R^3/3)$, times $1/R$, obtaining

$$\frac{m}{2}\left(\frac{dR}{dt}\right)^2 - m\left(\frac{4\pi}{3}R^3\rho\right)\frac{1}{R} = \text{const.} \tag{1.11}$$

But of course, this derivation is not quite correct, since to obtain the potential we would need to actually integrate the gravitational acceleration, $\int g(R)dR$, and then we would get a factor of 2 too much with respect to the correct result.

The correct way to derive the above result is to use the relation previously derived, (1.9), together with *local* energy conservation. Note that from now on we will put $c = 1$ as usual in theoretical physics. The latter amounts to writing the energy variation dU as $d(\rho V) = Vd\rho + \rho dV$, and then equating with the work done by pressure, $-PdV$, to obtain

$$\dot{\rho} = -(\rho + P)\frac{\dot{V}}{V} = -3(\rho + P)\frac{\dot{R}}{R}. \tag{1.12}$$

Eliminating P between this equation and (1.9), we obtain

$$\ddot{R} = +\frac{8\pi G_N}{3}\rho R + \frac{4\pi G_N}{3}\dot{\rho}\frac{R^2}{\dot{R}}. \tag{1.13}$$

Multiplying the latter by $\dot{R}$, we obtain

$$\frac{d}{dt}\left(\frac{\dot{R}^2}{2}\right) = \frac{d}{dt}\left(\frac{4\pi G_N}{3}\rho R^2\right), \tag{1.14}$$

which integrates to

$$\frac{\dot{R}^2}{2} - \frac{4\pi G_N}{3}\rho R^2 = \frac{k}{2} = \text{const.} \tag{1.15}$$

We see that this is the same equation as from the naive (and wrong) derivation, but now it is correct. This equation is known as the *Friedmann equation*. Moreover, we can define the constant through the value of the left hand side at current time, denoted by a zero index, obtaining

$$\left(\frac{dR}{dt}\right)^2 = \frac{8\pi G_N}{3}\rho R^2 - \frac{8\pi G_N}{3}R_0^2\left(\rho_0 - \frac{3H_0^2}{8\pi G_N}\right), \tag{1.16}$$

where the second term on the right hand side is the constant k.

We now define the *critical energy density* at current time

$$\frac{3H_0^2}{8\pi G_N} \equiv \rho_{c,0}, \tag{1.17}$$

and the ratio of the energy density at current time with the critical one,

$$\Omega_0 \equiv \frac{\rho_0}{\rho_{c,0}} = \frac{8\pi G_N}{3}\frac{\rho_0}{H_0^2}. \tag{1.18}$$

We can define the same quantities at arbitrary times, the critical energy density

$$\frac{3H^2}{8\pi G_N} \equiv \rho_c, \tag{1.19}$$

and the ratio of the energy density to the critical one,

$$\Omega \equiv \frac{\rho}{\rho_c} = \frac{8\pi G_N}{3}\frac{\rho}{H^2}. \tag{1.20}$$

Now the information that comes from general relativity and will be explained in the next chapter, is that $\Omega > 1$ implies that k is not an arbitrary constant, but rather it is $k = +1$, and the Universe is (spatially) *closed*, i.e., it something like a sphere, with a finite total volume. If on the other hand $\Omega < 1$, this means that $k = -1$, and the Universe is (spatially) open, a hyperbolic space of infinite total volume. But in fact, experimentally we are almost certainly in the case $\Omega = 1$, which implies $k = 0$, and the Universe is flat, a parabolic space of infinite total volume.

Finally, that means that we can rewrite the Friedmann equation as

$$\frac{k}{H^2R^2} = \frac{\rho}{3H^2/(8\pi G_N)} - 1 = \Omega - 1. \tag{1.21}$$

If H quantifies the velocity of the expansion, we need to define a parameter that describes also its acceleration. Or rather, we define the *deceleration parameter*

$$q \equiv \frac{1}{H^2}\left(-\frac{1}{R}\frac{d^2R}{dt^2}\right) \tag{1.22}$$

Then from the equation for $\ddot{R}$ (1.13), we obtain

$$q = -\frac{\ddot{R}}{RH^2} = -\Omega\left(1 + \frac{1}{2}\frac{\dot{\rho}/\rho}{\dot{R}/R}\right). \tag{1.23}$$

The parameter is defined as a deceleration parameter (as opposed to an acceleration parameter), since historically it was thought that the Universe is decelerated, though as we will see, now we know it currently is accelerated.

For pure pressureless matter (dust), $P = 0$, $\dot{\rho}/\rho = -3\dot{R}/R$, which gives

$$q = \frac{\Omega}{2}. \tag{1.24}$$

Experimentally, the Hubble parameter today H_0 is defined in terms of a dimensionless one h_0 as

$$H_0 \equiv h_0 \times 100\frac{\text{Km/s}}{\text{Mpc}}. \tag{1.25}$$

1.3 Time Evolution of the Scale Factor for a Linear Equation of State

In cosmology, one usually assumes a linear equation of state,

$$P = w\rho. \tag{1.26}$$

This is of course true for dust ($P = 0$), and for radiation $P = \rho/3$, i.e., $w = 1/3$, but is not necessarily true for any kind of matter. For instance, for a scalar field in a potential, we can have a varying w (nonlinear $P(\rho)$), and for QCD in the phase transition, we also have a nonlinear $P(\rho)$. Nevertheless, we will assume constant w in the following.

Then the continuity equation is

$$\frac{\dot{\rho}}{\rho} = -3(1+w)\frac{\dot{R}}{R}\,, \tag{1.27}$$

and we see that we don't need to eliminate P anymore, so we can integrate it directly,

$$\rho(R) \propto R^{-3(1+w)}. \tag{1.28}$$

If $\Omega = 1$, like it seems to be experimentally favoured, or if we are looking at $R \to \infty$, from (1.21) we obtain

$$\frac{\dot{R}^2}{R^2} = H^2 = \frac{8\pi G_N}{3}\rho \propto R^{-3(1+w)}\,, \tag{1.29}$$

which we can integrate to give

$$R \propto t^{\frac{2}{3(1+w)}}. \tag{1.30}$$

From this, we obtain

$$\left(\frac{\dot{R}}{R}\right)^2 = \frac{4}{9(1+w)^2 t^2}\,, \tag{1.31}$$

to be compared with $8\pi G_N \rho/3$, so that we have

$$\rho = \frac{1}{6\pi G_N t^2 (1+w)^2}. \tag{1.32}$$

From $R(t)$, written (by comparing with the relation today) as

$$R = R_0 (t/t_0)^{\frac{2}{3(1+w)}}\,, \tag{1.33}$$

we get

$$H_0 = \frac{2}{3(1+w)t_0} \Rightarrow t_0 = \frac{2}{3(1+w)} H_0^{-1}. \tag{1.34}$$

Thus the age of the Universe, at least in this simple model where a single w dominates the whole evolution of the Universe, is given roughly by H_0^{-1}.

1.4 Universe with Several Matter Components

Consider several energy components with different equations of state $P_i = w_i \rho_i$. Then each component has its own continuity equation (the energy is conserved for each component, since for each we have $dU_i = d(\rho_i V) = -P_i dV$)

$$\frac{\dot{\rho}_i}{\rho_i} = -3(1 + w_i)\frac{\dot{R}}{R}, \tag{1.35}$$

which integrates to

$$\rho_i(R) \propto R^{-3(1+w_i)}. \tag{1.36}$$

Thus each component decays with R in a different way.

On the other hand, if $\Omega = 1$ or $R \to \infty$, we have the Friedmann equation in terms of the total energy, i.e.,

$$\left(\frac{\dot{R}}{R}\right)^2 = H^2 = \frac{8\pi G_N}{3}\sum_i \rho_i. \tag{1.37}$$

If one of them, w_{dom}, dominates the energy density, then $\rho \simeq \rho_{dom}$, so we obtain

$$R \propto t^{\frac{2}{3(1+w_{dom})}}, \tag{1.38}$$

which leads to the scaling of the total energy density as

$$\rho = \frac{1}{6\pi G_N t^2 (1 + w_{dom})^2}. \tag{1.39}$$

But since $\rho_i(R)$ scale with R in a different way, their relative proportion changes, since

$$\rho_i(R) \propto R^{-3(1+w_i)} \Rightarrow \frac{\rho_i}{\rho_j} \propto R^{3(w_j - w_i)}. \tag{1.40}$$

That means that the component with the lowest w_i dominates at late times, and the one with the highest w_i dominates at early times.

Several components in the Friedmann equation for $\ddot{R}$ gives

$$\frac{\ddot{R}}{R} = -\frac{4\pi G_N}{3}\sum_i(\rho_i + 3P_i) = -\frac{4\pi G_N}{3}\sum_i \rho_i(1 + 3w_i), \tag{1.41}$$

and defining the relative Ω's,

$$\Omega_i = \frac{8\pi G_N \rho_i}{3H^2}, \tag{1.42}$$

we can write the Friedmann equation as the equation for the deceleration parameter

$$q \equiv -\frac{\ddot{R}}{RH^2} = \sum_i \frac{\Omega_i}{2}(1 + 3w_i). \tag{1.43}$$

1.5 A Quick History of Experimental Observations and Discoveries

At this point, we pause and give a quick timeline of the relevant discoveries for the standard cosmology described so far.

–In 1914, Slipher measures nebular velocities v, and finds that they are much larger than stellar velocities.

–In 1917, the same Slipher measures the average velocity as $\langle v \rangle = 500\,\text{km/s}$. He is the first to observe the galactic redshift. Notice that Einstein's general relativity was discovered in 1915, in between the two discoveries.

–In 1917: de Sitter writes his solution with repulsive forces proportional to the distance giving an expanding Universe. It is the first astrophysical implication of Einstein's general relativity. Einstein however doesn't like it, and invents "Einstein's static Universe" to counter it.

–In 1922 and 1924, Friedmann writes his equations for the expanding Universe.

–In 1923, Hermann Weyl describes galaxies in de Sitter's Universe.

–In 1927, Lemaître writes his theory.

–In 1928, Robertson writes his, and later Walker. Thus finally, we have the FLRW Universe, to be described in the next chapter.

–On 15 March 1929, E. Hubble finds his law for the expanding Universe. He measures 24 nebulae with known distances (we will see later how we can measure distance, which is nontrivial), and measuring also the redshift, finds $H \simeq 500\,\text{Km/s/Mpc}$. But note that now we know that $H < 100\,\text{Km/s/Mpc}$, so in reality his measurement was way off, and yet he managed to establish the correct expansion law.

Note that by inverting the value of H_0 of about $100\,\text{Km/s/Mpc}$, we obtain the order of magnitude of the age of the Universe,

$$t_0 \sim H_0^{-1} \sim 10^{10} yrs = 10 \; billion \; years. \tag{1.44}$$

This is so, since, if the Universe always expanded with the same H_0 (which we know not to be the case, in fact $H = H(t)$, but the variation is only polynomial), at $T_0 = 1/H_0$ in the past all points were situated at the same place.

This is roughly correct, since the exact age of the Universe measured now is about 13.8 billion years.

1.6 The Cosmological Constant

There is a relativistic "matter" component (see the next chapter) called the *cosmological constant* that has $w = -1$, which means that it has the equation of state

$$P = -\rho. \tag{1.45}$$

It can only be understood in relativistic theory, since we have negative pressure. In fact, there are so-called energy conditions that are believed to hold for any quantum matter, one of them being $P + \rho \geq 0$, so the cosmological constant saturates this inequality. In fact, usually one considers $w \in [-1, 1]$, but only the lower value is an actual bound, the upper value is not: we could consider $w \gg 1$, as we will in part III of the book.

Since $w = -1$ is the minimum value possible for w, in the end it will dominate the evolution of the Universe. This is what happens in the current age, when Λ just begins to dominate. It is also what happened during inflation, when there was an effective cosmological constant.

For the pure Λ Universe (cosmological constant dominating), from the continuity equation with $\rho + P = 0$ we obtain

$$\frac{\dot{\rho}}{\rho} = 0 \Rightarrow \rho = \text{const.} \tag{1.46}$$

So indeed the cosmological constant is an energy density that is not only constant in space, but also in time, making it a true constant.

Note that this is somewhat counterintuitive, since it means that despite the fact that the Universe is expanding, the energy density remains constant. That implies that as a sphere expands, energy is created (since the volume increases, but the energy density remains constant). But that is easy to explain, due to the negative pressure that exerts positive work. It is just counterintuitive, since in usual experience, pressure is positive, so exerts negative work, and the energy density drops as a volume expands.

If ρ =constant, the Friedmann equation gives

$$H^2 = \frac{8\pi G_N}{3}\rho = \text{const.} \Rightarrow R = R_0 e^{Ht} , \tag{1.47}$$

i.e., exponential expansion.

Next, consider a more physical case, with both cosmological constant (which we find experimentally that exists, at the very least as an approximation, in the real Universe), and matter, which is obviously present. This describes well the current age of the Universe. Then from (1.43), we obtain

$$q = \frac{\Omega_{\text{matter}}}{2} - \Omega_\Lambda. \tag{1.48}$$

Notice the sign for Ω_Λ, and since it dominates, we have *acceleration* of the Universe.

In fact, as seen from (1.43), we only need $w \leq -1/3$ to obtain acceleration.

Observing the Cosmological Constant

Observationally, the measurements that decides the existence of the cosmological constant are related to type I_a supernovae. It is not important what they are (though we will describe them later on in the book), but the important point for us is that they are (statistically speaking) standard candles. That is, we know they all have roughly the same brightness, so by measuring the flux that reaches us on Earth, we can calculate the distance. More precisely, we calculate a $d_P(z)$ that will be defined more precisely shortly.

As we said, the Hubble law is only approximate, and more precisely it is a local (in time, thus space) relation. Instead of $z = \dot{R} = HR$, we have

$$c\,dz = -H\,dr. \tag{1.49}$$

Moreover, $H = H(z)$ as we said, and r is the distance travelled by light, divided by the normalized expanding scale, R/R_0.

The measured distance is the integral of dr,

$$d_P = \int dr = c\int \frac{dz}{H(z)}. \tag{1.50}$$

But H depends on z implicitly, via the dependence on t, related to z via the propagation of light, namely (using $dz = -H\,dr/c = -Hdt/(R/R_0)$)

$$\frac{dH}{dz} = \frac{dt}{dz}\frac{dH}{dt} = -\frac{R/R_0}{H}\frac{d}{dt}\left(\frac{\dot{R}}{R}\right) = -\frac{R/R_0}{H}\left(\frac{\ddot{R}}{R} - \frac{\dot{R}^2}{R^2}\right) = +H\frac{R}{R_0}(q+1). \tag{1.51}$$

This implies the expansion

$$\frac{1}{H(z)} \simeq \frac{1}{H_0} - \frac{1}{H_0^2}\left(\frac{dH}{dz}\right)\bigg|_0 z. \tag{1.52}$$

Finally, substituting in $d_P = c\int dz/H(z)$, we get

$$d_P \simeq \frac{c}{H_0}\left(z - (1+q)\frac{z^2}{2}\right). \tag{1.53}$$

By measuring the curvature of $d_P(z)$ (deviation from a straight line), we find q. This gives a rough linear relation between Ω_Λ and Ω_m. Experimentally, we find $q < 0$, so we have an accelerated Universe, with $q < 0$, so $w_{total} < -1/3$. More precisely, combining the linear relation above with another linear relation, coming from the fact that CMBR measurements (to be described later) imply that $\Omega = \Omega_m + \Omega_\Lambda = 1$

(flat Universe) is favored, we obtain the current result at the intersection of the two lines, with $\Omega_m \simeq 0.3$ and $\Omega_\Lambda \simeq 0.7$.[1]

Important Concepts to Remember

- The Universe expands according to the Hubble law, $v = Hr$. In general, the Hubble law is $\dot{R}/R = H(t)$.
- The Universe is homogenous and isotropic.
- In the Newtonian approximation, the Universe obeys the Friedmann equation $\dot{R}^2/2 - 8\pi G\rho R^2/3 = k/2 =$ constant and the continuity equation $\dot{\rho} + 3(\rho + P)\dot{R}/R = 0$.
- We can define the ratio Ω of the energy density to the critical energy density $\rho_{\rm cr} = 3H^2/8\pi G_N$, so that the Friedmann equation is $\Omega - 1 = k/H^2R^2$.
- The deceleration parameter $q = -\ddot{R}/RH^2$ for the pressureless matter ($P = 0$) case is $q = \Omega/2$.
- Defining the equation of state $P = w\rho$, we have $\rho(R) \sim R^{-3(1+w)}$ and, if it dominates the energy density, $R \propto t^{2/(3(1+w))}$, so that $\rho \propto 1/t^2$.
- For several components, $\rho_i \propto R^{-3(1+w_i)}$ and $q = \sum_i \Omega(1 + 3w_i)/2$.
- The cosmological constant has $P = -\rho$, $\rho =$ constant in time, and if it dominates gives exponential expansion, $R = R_0 e^{Ht}$, with $H^2 = 8\pi G_N \rho/3$.
- The distance travelled by light is $d_P = \int dr = c\int dz/H(r)$.

Further reading: Most of the basic cosmology can be found in [1]. We can also see the books [2, 3].

Exercises

(1) Calculate $q(t)$ and $q(t)/\Omega(t)$ for arbitrary equation of state $P = w\rho$ and arbitrary k.
(2) Calculate the *exact* $\rho(t)$ in a system with two components, one with w_1 and $\rho_{0,1}$ at t_0, and one with w_2 and $\rho_{0,2}$ at t_0.
(3) Consider a Universe with a radiation component ($w = +1/3$), plus a cosmological constant. Calculate $q(\Omega_i)$ and the *exact* $\rho(t)$.
(4) Calculate the next term in the expansion for $d_P(z)$ beyond the quadratic one in (1.53).

[1]We should mention here that, strictly speaking, flatness does not imply an Universe that is open, as we will assume in the following. For instance, a torus is flat (has no curvature), yet has a finite volume. But there is no convincing argument for nontrivial topology of the Universe: in any case, the topology would have to be outside the observable Universe (outside the "horizon"), in order not to contradict experiments, in which case one would need indirect measurements to test for topology. Since there is no convincing way to test for this topology, I will not consider it in the following.

Chapter 2
Relativistic Theory

In this chapter, we re-derive, and extend, the analysis of the previous chapter within general relativity. We will re-derive the Friedmann equation from the relativistic treatment, of the FLRW metric, and then describe the various types of matter in detail, and the corresponding evolution of the Universe.

2.1 The FLRW Metric and the Friedmann Equation

Homogeneity and isotropy (experimentally observed, as we mentioned in the previous chapter) means first that we can write the metric in diagonal form, $-f(dt)^2 + dl^2$. But homogeneity means that f must be independent of r, and then we can redefine t to put it in the form

$$ds^2 = -cdt^2 + dl^2. \tag{2.1}$$

For the spatial part of the metric, dl^2, by homogeneity and isotropy, we need to have a factor $d\tilde{x}^2 + d\tilde{y}^2 + d\tilde{z}^2$, multiplied by a function of $r^2 = \tilde{x}^2 + \tilde{y}^2 + \tilde{z}^2$, and multiplied by a scale factor ($a(t)$ is the radius of curvature and $C = k/a^2$ is the curvature; here $k = 0, \pm 1$). This gives a 3 dimensional space of constant curvature (which is what homogeneity and isotropy implies), whose metric in comoving coordinates, i.e., coordinates that move together with the expansion of the space, quantified by $a(t)$, is uniquely given by

$$ds^2 = a^2(t)\frac{d\tilde{x}^2 + d\tilde{y}^2 + d\tilde{z}}{\left[1 + \frac{k}{4}(\tilde{x}^2 + \tilde{y}^2 + \tilde{z}^2)\right]^2}. \tag{2.2}$$

H. Năstase, *Cosmology and String Theory*, Fundamental Theories of Physics 197, https://doi.org/10.1007/978-3-030-15077-8_2

Defining spherical coordinates

$$\begin{aligned}\tilde{x} &= \tilde{r}\sin\theta\cos\phi\\ \tilde{y} &= \tilde{r}\sin\theta\sin\phi\\ \tilde{z} &= \tilde{r}\cos\theta\ ,\end{aligned} \tag{2.3}$$

we obtain

$$dl^2 = a^2(t)\frac{d\tilde{r}^2 + \tilde{r}^2(d\theta^2 + \sin^2\theta d\phi^2)}{\left[1 + \frac{k}{4}\tilde{r}^2\right]^2}. \tag{2.4}$$

Using the further transformations

$$\begin{aligned} k = +1 &: \frac{\tilde{r}}{1 + \frac{\tilde{r}^2}{4}} = \sin r\\ k = -1 &: \frac{\tilde{r}}{1 - \frac{\tilde{r}^2}{4}} = \sinh r\\ k = 0 &: \tilde{r} = r\ ,\end{aligned} \tag{2.5}$$

we obtain

$$\begin{aligned} dl^2|_{k=0} &= a^2(t)[dr^2 + r^2(d\theta^2 + \sin^2\theta d\phi^2)]\\ dl^2|_{k=+1} &= a^2(t)[dr^2 + \sin^2 r(d\theta^2 + \sin^2\theta d\phi^2)]\\ dl^2|_{k=-1} &= a^2(t)[dr^2 + \sinh^2 r(d\theta^2 + \sin^2\theta d\phi^2)].\end{aligned} \tag{2.6}$$

In a 2 dimensional analogy, if we had a metric $dl^2|_{k=0} = a^2(t)[dr^2 + r^2 d\theta^2]$, we would say that the radius of the circle parametrized by θ would be

$$l = 2\pi r a. \tag{2.7}$$

In the $k = +1$ case with $r \to \sin r$ in the metric,

$$l = 2\pi \sin r a\ , \tag{2.8}$$

and in the $k = -1$ case with $r \to \sinh r$ in the metric,

$$l = 2\pi \sinh r a. \tag{2.9}$$

Similarly now, the area of the 2-sphere parametrized by θ, ϕ is (in terms of the area 4π of the unit sphere)

$$\begin{aligned} S(k) &= 4\pi a^2 r^2 : \quad k = 0\\ &= 4\pi a^2 \sin^2 r; \quad k = +1\\ &= 4\pi a^2 \sinh^2 r; \quad k = -1.\end{aligned} \tag{2.10}$$

Finally, we can transform to Friedmann–Lemaître–Robertson–Walker (FLRW) coordinates, by

$$\begin{aligned} r' &= \sin r; \quad k = +1 \\ &= r; \quad k = 0 \\ &= \sinh r; \quad k = -1 \,, \end{aligned} \tag{2.11}$$

giving the FLRW Universe, with metric

$$dl^2 = a^2(t)\left[\frac{dr'^2}{1-kr'^2} + r'^2(d\theta^2 + \sin^2\theta d\phi^2)\right]. \tag{2.12}$$

We can substitute this metric in the Einstein equations, and find that they reduce to an equation for the acceleration of a,

$$\frac{d^2a}{dt^2} = -\frac{4\pi G_N}{3}a(\rho + 3P) + \frac{\Lambda}{3}a \,, \tag{2.13}$$

and an equation for H, the *Friedmann equation*,

$$H^2 = \left(\frac{da}{adt}\right)^2 = \frac{8\pi G_N}{3}\rho - \frac{k}{a^2}. \tag{2.14}$$

These are the same equations as obtained before, in the Newtonian analysis, except as we said there, the Friedmann equation comes from integrating the equation for the acceleration (2.13) and as such k is an arbitrary integration constant. Here $k = 0, \pm 1$ depending on whether the Universe is flat, closed or open. But as before, we can also go in the opposite direction, namely using (2.14) and the continuity equation, we obtain (2.13). Another difference is that now the continuity equation is obtained from the Einstein equation as well, as we will shortly see.

The Einstein equation is written as

$$R_{\mu\nu} - \frac{1}{2}g_{\mu\nu}R = 8\pi G_N T_{\mu\nu} - \Lambda g_{\mu\nu} \,, \tag{2.15}$$

where we have added a separate cosmological constant term in it, even though we could write it as part of the energy-momentum tensor $T_{\mu\nu}$. Note that here, and in the rest of the book, I will use the metric signature convention $(-+++)$ ("mostly plus").

For the energy-momentum tensor, most of the time in cosmology one uses the *perfect fluid* ansatz (non-perfect fluid is mostly only used in the cases of perturbations, without homogeneity and isotropy),

$$T^{\mu\nu} = \rho u^\mu u^\nu + P(g^{\mu\nu} + u^\mu u^\nu) \,, \tag{2.16}$$

where

$$u^\mu = \frac{dx^\mu}{d\tau} \tag{2.17}$$

is the comoving 4-velocity of the fluid, and the tensor structure $g^{\mu\nu} + u^\mu u^\nu$ is transverse to the 4-velocity of the fluid,

$$u_\mu(g^{\mu\nu} + u^\mu u^\nu) = u^\mu - u^\mu = 0 \,, \tag{2.18}$$

since $u^\mu u_\mu = -1$ (obvious in the frame where $u^\mu = (1, 0, 0, 0)$). Then in the same frame, with $u^\mu = (1, 0, 0, 0)$, we find that $T^\mu{}_\nu$ is diagonal, and more precisely

$$T^\mu{}_\nu = \text{diag}(-\rho, P, P, P) \,, \tag{2.19}$$

both in flat Minkowski space and in the FLRW Universe.

Adding the cosmological constant corresponds to shifting the energy-momentum tensor by

$$T_{\mu\nu} \to T_{\mu\nu} - \frac{\Lambda}{8\pi G_N} g_{\mu\nu} \,, \tag{2.20}$$

which amounts to the shift

$$\begin{aligned} \rho &\to \rho + \rho_\Lambda \\ P &\to P + P_\Lambda; \end{aligned} \tag{2.21}$$

where

$$\rho_\Lambda = \frac{\Lambda}{8\pi G_N}; \quad p_\Lambda = -\rho_\Lambda. \tag{2.22}$$

Energy conservation in general relativity corresponds to the conservation of the current corresponding to translation symmetry, i.e., the energy-momentum tensor,

$$D_\mu T^\mu{}_\nu = 0 \,, \tag{2.23}$$

which is a tensor relation, thus can describe energy loss or gain. With $T^\mu{}_\nu = \text{diag}(-\rho, P, P, P)$, it reduces to the usual relation, derived last time from $dU = d(\rho V) = -PdV$,

$$\dot{\rho} = -3(\rho + P)\frac{\dot{a}}{a}. \tag{2.24}$$

As in the previous chapter, we can rewrite the Friedmann equation as

$$\frac{k}{a^2} = \frac{8\pi G_N}{3}(\rho(t) - \rho_c(t)) + \frac{\Lambda}{3} = \frac{8\pi G_N}{3}(\rho_{\text{tot}}(t) - \rho_c(t)) \,, \tag{2.25}$$

where the critical energy density is

$$\rho_c \equiv \frac{3H^2}{8\pi G_N}, \tag{2.26}$$

in the second equality we have considered the cosmological constant as part of the *total* energy density ρ_{tot} as well, and the left hand side, sometimes called

$$\frac{k}{a^2} \equiv C_G, \tag{2.27}$$

the *curvature energy density* (and as such, can be included in the total energy density as well). Note that in this form of the Friedmann equation, we see that the sign of $\rho(t) - \rho_c(t)$ is the same as the sign of k, which is fixed for all time. That is, the relation of $\rho < \rho_c$ ($\Omega < 1$) or $\rho > \rho_c$ ($\Omega > 1$), or $\rho = \rho_c$ ($\Omega = 1$) is true *for all time*. Moreover, it says that if $\rho_{\text{tot}}(t) \to \rho_c(t)$ at some time, then (if $k \neq 0$, i.e., if $\rho \neq \rho_c$) we have $a \to \infty$.

2.2 Types of Matter and Resulting Expansion of the Universe

We first review the analysis of the case with linear equation of state, $P = w\rho$, from the previous chapter.

We found that

$$\begin{aligned} a(t) &\propto t^{\frac{2}{3(1+w)}} \\ H &= \frac{\dot{a}}{a} = \frac{1}{t} \propto a^{-\frac{3(1+w)}{2}} \\ \rho &= \frac{1}{6\pi G_N t^2 (1+w)^2} \propto a(t)^{-3(1+w)}. \end{aligned} \tag{2.28}$$

- $P = 0$ for *dust* matter (pressureless). If it dominates the expansion of the Universe, i.e., if we are in an era of *matter domination* (M.D.), we have, according to the above formulae,

$$\begin{aligned} a(t) &\propto t^{2/3} \\ H &\propto \frac{1}{a^{3/2}} \\ \rho &= \frac{1}{6\pi G_N t^2} = \frac{1}{a^3(t)}. \end{aligned} \tag{2.29}$$

  The last relation is expected, since the matter in a comoving volume is conserved during the expansion of the Universe, but the volume expands as a^3, so the density drops as $1/a^3$.

- $P = \rho/3$ ($w = 1/3$) for radiation. If it dominates the expansion of the Universe, i.e., if we are in an era of *radiation domination* (R.D.), we have, according to the general formulae,

$$\begin{aligned} a(t) &\propto t^{1/2} \\ H(t) &\propto \frac{1}{a^2} \\ \rho &= \frac{3}{32\pi G_N t^2} \propto \frac{1}{a^4(t)}. \end{aligned} \tag{2.30}$$

The last relation is again expected. The energy of a photon is $h\nu = hc/\lambda$, and wavelengths λ are comoving, i.e., they expand together with the Universe, which means that the energy within a comoving volume decays as $1/a$, and the energy density as $1/a^4$.

- $P = \rho$ ($w = +1$), which is usually called "stiff matter" gives, according to the general formulae,

$$\begin{aligned} a(t) &\propto t^{1/3} \\ H &\propto \frac{1}{a^3} \\ \rho &= \frac{1}{24\pi G_N t^2} \propto \frac{1}{a^6(t)}. \end{aligned} \tag{2.31}$$

The stiff matter is not really the maximum w one can have, though one rarely considers bigger ones. However, we will see towards the end of the book an example of $w \gg 1$. The case $w = +1$ is obtained for a (homogenous and isotropic) background scalar field $\phi = \phi(t)$, when the kinetic term dominates over the potential term (so the opposite to the case of inflation, when the potential dominates over the kinetic term), so that the action is approximately

$$S = -\frac{1}{2}\int d^4x\sqrt{-g}(\partial_\mu\phi)^2\ , \tag{2.32}$$

and the energy-momentum tensor is

$$T_{\mu\nu} = -\frac{2}{\sqrt{-g}}\frac{\delta S}{\delta g^{\mu\nu}} = \partial_\mu\phi\partial_\nu\phi - \frac{1}{2}g_{\mu\nu}(\partial_\rho\phi)^2\ , \tag{2.33}$$

meaning if $\phi = \phi(t)$ only, we obtain

$$T_{00} = \frac{1}{2}\dot{\phi}^2 = T_{ii} \Rightarrow w = +1. \tag{2.34}$$

- Finally, we review the case of cosmological constant, with $P = -\rho$ ($w = -1$). In this case, the general formulae are singular at $w = -1$, so we must consider them separately. As we saw in the previous chapter, we have

$$\begin{aligned} a(t) &\propto e^{Ht} \\ H &= \text{const.} \\ \rho &= \text{const.} \end{aligned} \tag{2.35}$$

The terms with highest w dominate at the beginning, and the one with lowest w at the end. Since we have $P + \rho \geq 0$ by the energy conditions (believed to be satisfied by all quantum matter), the order is as follows: first, stiff matter dominates, then radiation, then matter, then cosmological constant.

In order to find the expansion age of the Universe, t_0, as a function of H_0 (the Hubble constant today) and Ω_0 (the total Ω today), we first write the Friedmann equation as

$$\begin{aligned} \frac{k}{a^2} &= \frac{8\pi G_N}{3}(\rho_{\text{tot}}(t) - \rho_c(t)) = \frac{H^2(t)}{\rho_c(t)}(\rho_{\text{tot}}(t) - \rho_c(t)) \Rightarrow \\ \frac{k}{a^2} &= H^2(t)(\Omega(t) - 1). \end{aligned} \tag{2.36}$$

Finally, we write

$$\frac{1}{a(t)H(t)} = \sqrt{\frac{\Omega(t) - 1}{k}} \tag{2.37}$$

We can then calculate t_0 by writing dt as a function of H, as

$$t_0 = \int_0^{t_0} dt = \int_0^{a(t_0)} \frac{da'}{\dot{a}'} = \int_0^{a(t_0)} \frac{da'}{a'H'}. \tag{2.38}$$

2.3 Cosmological Evolution Formulas

Formulas as a Function of Redshift

We now define quantities as a function of the redshift. Since wavelengths in the expanding Universe are comoving (expand with the Universe),

$$\frac{\lambda(t)}{\lambda_0} = \frac{a(t)}{a_0}, \tag{2.39}$$

and then the redshift, which as found to be $1 + z = a_0/a(t)$, we obtain (remembering that $a(t)$ and $\lambda(t)$ correspond to the emission time for the wave)

$$1 + z = \frac{\lambda_{\text{detected}}}{\lambda_{\text{emitted}}} \equiv \frac{\lambda_0}{\lambda(t)} = \frac{a_0}{a(t)}. \tag{2.40}$$

Since frequencies satisfy $\nu = c/\lambda \propto 1/a$, we have

$$\nu a = \text{constant}. \tag{2.41}$$

The temperature in the Universe (which is in thermal equilibrium, at least as far as radiation goes) is defined by the black body radiation of the CMBR (cosmic microwave background radiation), which is a function of ν/T. The shape of the (Planck) distribution is obviously preserved as the Universe expands, which means that we also have

$$Ta = \text{const.} \Rightarrow T \propto \frac{1}{a(t)}. \tag{2.42}$$

We next express the age of the Universe as a function of z. Since $a_0/a(t) = 1 + z$, we replace in the formula for t (2.38)

$$x = \frac{a(t)}{a_0} = \frac{1}{1+z}. \tag{2.43}$$

We obtain

$$t = \int_0^{a_0/(1+z)} \frac{da'}{a'H'} = \frac{1}{H_0} \int_0^{1/(1+z)} \frac{dx}{x(H/H_0)}, \tag{2.44}$$

and we have to replace H/H_0 as a function of z, from $H = \dot{a}/a$,

$$a(t) = \frac{a_0}{1+z} = a_0(t/t_0)^{\frac{2}{3(1+w)}}. \tag{2.45}$$

Specifically, the Friedmann equation is rewritten as (putting the curvature k/a^2, giving Ω_R on the same side as H^2)

$$H^2 = H_0^2 \left[\Omega_{0,m} \frac{\rho}{\rho_0} + \Omega_{0,R} \frac{a_0^2}{a^2} + \Omega_{0,\Lambda} \right]. \tag{2.46}$$

Using

$$\begin{aligned} \frac{\rho}{\rho_0} &= (a/a_0)^{-3(1+w)} = (1+z)^{3(1+w)} \\ \frac{a_0^2}{a^2} &= (1+z)^2, \end{aligned} \tag{2.47}$$

in H^2, and replacing H/H_0 in the formula for the age t, we obtain finally for the age of the Universe

$$t_0 = \frac{1}{H_0} \int_0^{1/(1+z)} \frac{dx}{\sqrt{\Omega_{0,\Lambda} x^2 + \Omega_{0,R} + \Omega_{0,m} x^{-(1+3w)}}}. \tag{2.48}$$

In particular, we saw in the previous chapter that for $\Omega_{\text{tot}} = 1$, we have

$$t_0 = \frac{1}{H_0}\frac{2}{3(1+w)}. \tag{2.49}$$

Formulas as a Function of Conformal Time

We define the *conformal time* as the time when the metric is a time independent factor times an overall time dependent factor. Specifically,

$$d\eta = \frac{cdt}{a(t)}, \tag{2.50}$$

which means that

$$ds^2 = a^2(t)\left[-d\eta^2 + \frac{dl^2}{a^2(t)}\right]. \tag{2.51}$$

For instance, in FLRW coordinates, we have

$$ds^2 = a^2(t)\left[-d\eta^2 + \frac{dr'^2}{1-kr'^2} + r'^2(d\theta^2 + \sin^2\theta d\phi^2)\right]. \tag{2.52}$$

In terms of η, we obtain the table:

P	k	$t(\eta)$	$a(\eta)$	$H(\eta)$	$\Omega(\eta)$
	1	$c^{-1}A(\eta - \sin\eta)$	$A(1-\cos\eta)$	$A^{-1}c\sin\eta(1-\cos\eta)^{-2}$	$2(1+\cos\eta)^{-1}$
$P = 0$ (M.D.)	0	$c^{-1}A\eta^3/6$	$A\eta^2/2$	$A^{-1}4c\eta^{-3} = \frac{2}{3t}$	1
	-1	$c^{-1}A(\sinh\eta - \eta)$	$A(\cosh\eta - 1)$	$A^{-1}c\sinh\eta(\cosh\eta - 1)^{-2}$	$2(1+\cosh\eta)^{-1}$
	1	$c^{-1}A'(\eta - \cos\eta)$	$A'\sin\eta$	$A'^{-1}c\cos\eta(\sin\eta)^{-2}$	$(\cos\eta)^{-1}$
$P = \rho/3$ (R.D.)	0	$c^{-1}A'\eta^2/2$	$A'\eta$	$A'^{-1}4c\eta^{-2} = \frac{1}{2t}$	1
	-1	$c^{-1}A'(\cosh\eta - 1)$	$A'\sinh\eta$	$A'^{-1}c\cosh\eta(\sinh\eta)^{-2}$	$(\cos\eta)^{-1}$

We will verify the cases with $\Omega = 1$ ($k = 0$), in which case as we saw

$$a(t) = a_0 = a_0(t/t_0)^{\frac{2}{3(1+w)}}, \tag{2.53}$$

so that

$$\eta(t) = c\int\frac{dt}{a(t)} \propto t^{\frac{1+3w}{3(1+w)}}. \tag{2.54}$$

Then for $w = 0$ (M.D.) we find

$$\eta(t) \propto t^{1/3} \Rightarrow t \propto \eta^3, \tag{2.55}$$

and for $w = 1/3$ (R.D.), we find

$$\eta(t) \propto t^{1/2} \Rightarrow t \propto \eta^2. \tag{2.56}$$

Moreover, since now

$$a(t) \propto t^{\frac{2}{3(1+w)}} \propto \eta^{\frac{2}{1+3w}} , \tag{2.57}$$

and for $\Omega = 1$ we have (2.49), which means also

$$H_0 = \frac{2}{3(1+w)t_0} , \tag{2.58}$$

for the $w = 0$ case (M.D.) we obtain

$$a(\eta) \propto \eta^2 \Rightarrow H_0 = \frac{2}{3t_0} , \tag{2.59}$$

whereas for the $w = 1/3$ case (R.D.), we obtain

$$a(\eta) \propto \eta \Rightarrow H_0 = \frac{1}{2t_0}. \tag{2.60}$$

We leave it as an exercise to check the cases for $k = +1$ and $k = -1$.

Important Concepts to Remember

- The expanding Universe is described by the FLRW metric, $ds^2 = -c^2dt^2 + dl^2$, with $dl^2 \propto a^2(t)$. In one expression, $dl^2 = a^2(t)d\vec{r}^2/[1 + kr^2/4]^2$. Here $k = 0$ for flat Universe, $k = +1$ for closed Universe and $k = -1$ for open Universe, compared to arbitrary k in the Newtonian approximation.
- A perfect fluid has $T^{\mu}{}_{\nu} = \text{diag}(-\rho, P, P, P)$, and a cosmological constant has $\rho_\Lambda = \Lambda/(8\pi G_N)$, and $P_\Lambda = -\rho_\Lambda$.
- The Friedmann equation is the same as in the Newtonian case, just that with the k as above, and the continuity equation is the same.
- Dust (pressureless) matter has $w = 0$, radiation has $w = 1/3$, stiff matter has $w = 1$ (for instance a scalar with $\phi = \phi(t)$ only), and cosmological constant has $w = -1$.
- The expansion age of the Universe is $t_0 = \int dt = \int da/aH = H_0^{-1}2/(3(1 + w))$.
- The temperature scales inversely with the scale factor, $T \propto 1/a$, as does the redshift, $1/(1 + z) = a(t)/a_0$.
- Conformal time is $d\eta = cdt/a(t)$, so that $a^2(t)$ is an overall (conformal) factor in the metric.

Further reading: Again most of the basic cosmology can be found in [1], see also the books [2, 3].

Exercises

(1) Calculate $t(\eta)$, $a(\eta)$ and $H(t)$ for $k = +1$ and $k = -1$, in the cases $w = 0$ and $w = 1/3$.
(2) Check that the Einstein equations on the FRLW metric reduce to the Friedmann equations (2.14) and (2.13).
(3) Assuming $\Omega \neq 1$ now, and a Universe made up of only matter and radiation, calculate $\Omega(t)^{-1}$ at $t \sim 10^{-10}s$. This is the "flatness problem".
(4) Show that the energy-momentum conservation $D_\mu T^\mu{}_\nu = 0$ is implied by the Einstein's equations, and that it implies the continuity (energy conservation) equation (2.24).

Chapter 3
The Propagation of Light and Measurements of Distance, Luminosity and Mass

In this chapter I will explain in some detail how measurements are made in astrophysics and cosmology, so that we can gain some faith in the fact that we actually can determine things precisely. I will start describing the propagation of light in a cosmological setting, follow on to say how we can use these formulas for distance measurements, and then optical measurements of distance, based on brightness and luminosity. These are based on the ideas of standard candles, so to understand that we will discuss shortly stellar evolution, and the possible stellar objects we can have. We will then classify the various measurements of distance, followed by possible determinations of mass of stellar objects.

3.1 Propagation of Light

The physical distance travelled by light, *at time* $t = t_2$ is (using the fact that light travels on $ds^2 = -c^2dt^2 + a^2(t)dr^2 = 0$)

$$\Delta l_{t=t_2} = a(t_2)\Delta r = a(t_2)c\int_{t_1}^{t_2}\frac{dt}{a(t)} = a(t_2)\Delta\eta. \tag{3.1}$$

We see that the distance is measured now, i.e., in terms of $a(t_2)$, and is proportional to the interval of conformal time.

In particular, we are interested in the case that $t_1 = 0$, i.e., in light that has travelled to us from the Big Bang. We apply to $k = 0$ (flat Universe), since this is the relevant case, and $P = 0$ (M.D.), since for most of its history, the Universe was matter dominated. Then we obtain

$$\Delta l_h = a(t_2)ct_2\int_0^{t_2}\frac{dt/t_2}{a(t_2)(t/t_2)^{2/3}} = 3ct_2. \tag{3.2}$$

H. Năstase, *Cosmology and String Theory*, Fundamental Theories of Physics 197, https://doi.org/10.1007/978-3-030-15077-8_3

This is then the *horizon distance*, i.e., the maximal region of the Universe that is knowable in principle (since nothing travels faster than the speed of light, and the earliest it can start is during the Big Bang).

Note that we are talking here about the so-called *particle horizon*, which means what we can see from particles that travelled to us since the Big Bang. But there are other notions of horizon. One of them is the *(future) event horizon*, which is the largest (comoving) distance from which a particle sent now will even (in the future) reach us. Thus, in the case that the Universe goes on forever (and does not end, or collapse), this is the complement of the particle horizon,

$$R_{\text{ev.hor.}} = a(t)c \int_t^\infty \frac{dt'}{a(t')} \tag{3.3}$$

If the Universe ends at t_e, we only integrate up to t_e. There is also the notion of *Hubble horizon*, which is what we obtain in the linear Hubble law if we put the maximum velocity possible, $v = c$, so

$$d_H = \frac{c}{H}. \tag{3.4}$$

For a power law expansion, $a(t) \propto t^p$, all the horizons are of the same order, differing only by number factors, so we will not always be too careful of the distinctions between the horizons from now on.

In the case $k = 0$, but $P = \rho/3$ (R.D.), which was valid for a short while before the M.D. era, we have the (particle) horizon distance

$$\Delta l_h = ct_2 \int_0^{t_2} \frac{dt/t_2}{(t/t_2)^{1/2}} = 2ct_2. \tag{3.5}$$

This distance defines then, for all practical purposes, the size of the Universe (the Universe may be larger than that, but we wouldn't know, since this is the knowable Universe).

Note the strange fact that $l_h > ct_2$, so it seems like "the speed of light is greater than the speed of light". But this is of course due to the fact that l_h is measured *today*, and the Universe expands as light travels.

Another observation is that there is no global inertial reference frame, so we cannot use special relativity. That is as it should be, since we are not in Minkowski space, but in the FLRW Universe. In this space, we have a preferred time t that we have been using, defined by the cosmological expansion, and it defines a preferred reference frame (which is however not inertial).

Because our observations will have to do with light propagating from some distance place, at its constant speed, until now, we can use as a measurement of distance the redshift z, since z corresponds to a time, when we have a scale $a(t)$, from which light has propagated to us. More precisely, we have $1 + z = a_0/a(t)$. As we saw in the previous chapters, z was defined in terms of the frequency shift as

$$z = \frac{\nu_{\text{em}} - \nu_{\text{det}}}{\nu_{\text{det}}}\,, \tag{3.6}$$

and $\nu \propto 1/a$, as well as the temperature, $T \propto 1/a$, leading to

$$T \propto 1 + z \propto a \Rightarrow \frac{T}{T_0} = 1 + z = \frac{\nu}{\nu_0} \equiv \frac{\nu_{\text{em}}}{\nu_{\text{obs}}}. \tag{3.7}$$

3.2 Measurements of Distance by Light

In Euclidean space, the angular size θ is related to the linear size l and the distance to the object r by

$$\theta = \frac{l}{r}. \tag{3.8}$$

In non-Euclidean space however, it is not so simple. But reversely, we can *define* a distance called *angular size distance*, which is implicitly a function of the redshift z at which the source is located, by

$$\bar{r} \equiv \frac{l}{\theta} = f(z). \tag{3.9}$$

Using the coordinate system where we had the FLRW metric

$$ds^2 = -c^2dt^2 + a^2(t)[dr^2 + g_k(r)^2(d\theta^2 + \sin^2\theta d\phi^2)]\,, \tag{3.10}$$

where $g_k(r) = (r, \sinh r, \sin r)$ for $k = 0, -1, +1$, we have seen that a circle in the sky, defined by θ, has the circumference

$$S_k = 2\pi g_k(r)a(t). \tag{3.11}$$

For $\Delta\theta$ replacing 2π, S_k corresponds to the linear size $l = \Delta\theta\bar{r}$, so we have

$$\bar{r} = a_0 g_k(r). \tag{3.12}$$

As for r, it can be determined as before, from the fact that light propagates radially towards us, on $ds^2 = -c^2dt^2 + a^2(t)dr^2 = 0$, so

$$r = c\int\frac{dt}{a} = c\int\frac{da}{a\dot{a}} = \frac{c}{H_0}\int\frac{da}{a^2H/H_0(z)}\,, \tag{3.13}$$

is implicitly a function of the redshift z. Finally, we define the function $\psi(z)$ by

$$\bar{r} \equiv \frac{c}{H_0}\psi(z). \tag{3.14}$$

For pure dust matter ($P = 0$, M.D.), we obtain

$$\psi(z) = \frac{2}{\Omega_0^2(1+z)^2}\left\{\Omega_0 z + (\Omega_0 - 2)[(1+\Omega_0 z)^{1/2} - 1]\right\} , \tag{3.15}$$

which is left as an exercise to prove.

As a function of the variable

$$\Delta \equiv \frac{z}{1+z} , \tag{3.16}$$

which takes values from 0 to 1 (for $z = \infty$, at $t = 0$), and at fixed linear size l, the angular size θ *for* $\Lambda = 0$, $\Omega > 1$ (not the case with the current cosmology, of course) has a *minimum*. That is, as Δ increases, θ goes to a minimum, and then increases again close to 1, all the way to infinity!

Of course, we cannot see arbitrarily far in the past, i.e., arbitrarily close to the Big Bang (for $z \to \infty$). In particular, there is a maximum z, corresponding to the "surface of last scattering", when photons decoupled from matter, and atoms formed. Before that, the Universe is opaque (after that, it is transparent), and moreover there are no structures to be seen, not even atoms. So because of that, there are no large angle objects to be seen, but otherwise we would see large angles for things very close to the Big Bang.

3.3 Optical Measurements

We move on to optical measurements and their use for distance measurements.

In flat space, if there is no absorption, nor scattering, the brightness of a source is independent on the distance and on the existence of refraction.

In curved space, the analogous statement is different: the brightness of a source depends on z (and so, on the distance), but not on the curvature, nor on the density ρ.

The *brightness* is defined as the power P (energy per unit time), per unit solid angle and perpendicular area,

$$I = \frac{d^2 P_{\text{obs}}}{d\Omega dS_\perp}. \tag{3.17}$$

Note that here $d\Omega$ means $d\Omega_{\text{em}}$, which is the solid angle of the source viewed from the observer, whereas $dS_\perp$ means $dS_{\text{obs}}\cos\theta$, i.e., the area of the observer, perpendicular to the direction of propagation. But it doesn't matter, since we have the equality

$$B = \frac{d^2 P_{\text{obs}}}{d\Omega_{\text{obs}} dS_{\text{em}}\cos\theta} = I = \frac{d^2 P_{\text{obs}}}{d\Omega_{\text{em}} dS_{\text{obs}}\cos\theta}. \tag{3.18}$$

We can also define the *spectral brightness*, as the brightness per unit of frequency,

$$F(\omega) = \frac{d^3 P_{\rm obs}}{d\Omega dS_\perp d\omega}\,, \tag{3.19}$$

and then for blackbody radiation of radiation temperature T, the spectral brightness $F(\omega, T)$ is invariant. Moreover, as we already observed, the temperature decays with the expansion of the Universe, so

$$T_{\rm obs} = \frac{T_{\rm em}}{1+z}. \tag{3.20}$$

Then the brightness satisfies a law similar to Stefan–Boltzmann,

$$I = \text{const.} \times T^4 = \frac{I_0}{(1+z)^4}. \tag{3.21}$$

The *luminosity* L is defined as the total energy emitted by the source per unit time, in the rest frame. Since $\theta = l/\bar{r} = l/(c\psi(z)/H_0)$, for light propagation, the solid angle $d\Omega$ of the detector viewed from the source is, in terms of the area dS of the detector,

$$d\Omega = \frac{dS}{(c\psi(z)/H_0)^2}\,, \tag{3.22}$$

which means that the total area at the distance of the detector is

$$S = 4\pi\bar{r}^2 = 4\pi(c\psi(z)/H_0)^2. \tag{3.23}$$

Define the *total measured (bolometric) flux*, i.e., energy per unit time per unit surface of the detector. Then from the above it is given in terms of the luminosity of the source at emission $L_{\rm em}$ by

$$\mathcal{F}_{\rm obs} = \frac{L_{\rm em}}{4\pi\bar{r}^2(1+z)^4} = \frac{L_{\rm em}}{4\pi(c\psi(z)/H_0)^2(1+z)^4}\,, \tag{3.24}$$

where the factor $(1+z)^4$ is the same factor as in the brightness I, and comes from the variation of the luminosity from emission to detection.

We can define the *luminosity distance* d_L as the distance in Euclidean space producing the same $\mathcal{F}_{\rm obs}$ at the same $\mathcal{L}_{\rm em}$, i.e.,

$$\mathcal{F}_{\rm obs} = \frac{L_{\rm em}}{4\pi d_L^2} \Rightarrow d_L = \frac{c}{H_0}\psi(z)(1+z)^2. \tag{3.25}$$

Note that in terms of the luminosity *today*, $L = L_{\rm em}/(1+z)^2$, we have

$$\mathcal{F}_{\rm obs} = \frac{L}{4\pi \bar{r}^2 (1+z)^2}. \tag{3.26}$$

We can also define something that is directly observed, basically $\mathcal{F}_{\rm obs}$ on a strange scale, called the *apparent magnitude* of an object,

$$\begin{aligned} m &= -2.5 \log_{10} \mathcal{F}_{\rm obs} + \text{const.} \\ &= 5 \log_{10} d_L + \text{const.}' \end{aligned} \tag{3.27}$$

where in the second equality we have expressed $\mathcal{F}_{\rm obs}$ in terms of d_L, so that now $L_{\rm em}$ is absorbed in the constant.

Finally, define the *absolute magnitude* M, as the apparent magnitude m of an object situated at 10 parsecs (about 3×10^{19} cm) away. Therefore, since d_L is now fixed, the absolute magnitude M is a measure of the luminosity $L_{\rm em}$.

Substituting the form of $\psi(z)$ for the M.D. case into the formula for d_L (3.25), we find

$$H_0 d_L = \frac{2}{\Omega_0^2}\{\Omega_0 z + (\Omega_0 - 2)[(1+\Omega_0 z)^{1/2} - 1]\}. \tag{3.28}$$

Then the apparent (bolometric) magnitude is given by

$$\begin{aligned} m &= 5 \log_{10}(H_0 d_L) + C_1 \\ C_1 &= M_{\rm bol,em.} - 45.06 - 5 \log_{10} H_0. \end{aligned} \tag{3.29}$$

We intend to calculate this relation for nearby objects, for which we can expand $a(t)$ to second order, using H_0 and q_0 as the coefficients of the Taylor expansion, so

$$\frac{1}{1+z} = \frac{a(t)}{a(t_0)} = 1 + H_0(t - t_0) - \frac{1}{2} q_0 H_0^2 (t - t_0)^2 + \cdots \tag{3.30}$$

From it, we derive

$$z \simeq -H_0(t - t_0) + \left(1 + \frac{q_0}{2}\right) H_0^2 (t - t_0)^2 + \cdots, \tag{3.31}$$

which inverts as

$$t_0 - t \simeq H_0^{-1}\left[z - \left(1 + \frac{q_0}{2}\right) z^2 + \cdots\right]. \tag{3.32}$$

On the other hand, expanding the relation $r = c \int dt/a(t)$ to second order in the $k = 0$ (flat) case, we find

$$\bar{r} = a^{-1}(t_0)\left[(t_0 - t_1) + \frac{H_0}{2}(t_0 - t_1)^2 + \cdots\right] \tag{3.33}$$

and replacing in it the $t - t_0$ as a function of z, we obtain

$$\bar{r} = a^{-1}(t_0)H_0^{-1}\left[z - \frac{1}{2}(1 + q_0)z^2 + \cdots\right]. \tag{3.34}$$

Finally replacing this in $d_L = \bar{r}(1 + z)^2$, we find

$$H_0 d_L = z\left(1 + \frac{1}{2}(1 - q_0)z + \cdots\right), \tag{3.35}$$

and the bolometric apparent magnitude is

$$m \simeq 5\log_{10} z + 1.086(1 - q_0)z + C_1$$
$$C_1 = M_{bol} + 45.06 - 5\log_{10} H_0. \tag{3.36}$$

3.4 Stellar Evolution

At this moment, we describe shortly stellar evolution, since it is important in order to understand the measurements of distance and mass. About 100 years ago (about 1910), a diagram was devised that encapsulates very well stellar evolution, called the *Hertzsprung–Russel* diagram, after its discoverers, see Fig. 3.1. It maps luminosity (or absolute magnitude M, which is a measurement of it, as we saw), usually rescaled, so L_{em}/L_{Sun}, versus the temperature T, or rather, the colour. Note that the temperature is represented in reverse in the diagram (increasing from right to left), or colour from Blue to Red. The diagram is found from nearby stars, for which we can measure M, and we measure T from the blackbody radiation, but we assume it is valid for all stars.

We then observe a startling regularity: most stars (about 90%) fall onto a descending line in the Herzsprung–Russel diagram, called the *main sequence*. Above the main sequence, we have some lines representing red giants, below some lines for white dwarfs. It is found that the luminosity also increases with the mass, and we can represent the mass of stars on the diagram. Then the main sequence has about $M = 0.1M_{Sun}$ at the lower end, and about $M = 10M_{Sun}$ at the upper end, with some stars going until about $M = 60M_{Sun}$.

Red giants and white dwarfs are just later stages of stellar evolution, as we will see shortly. We have also variable stars, in particular *Cepheid variables* after the first discovered, Delta Cephei. These are stars that have a period of oscillation of days or months, for up to 30% of intensity. They are stars in later states of evolution as well, in which they collapse or expand repeatedly. The most important example is the North Star (Polaris), that oscillates up to 9% every 4 days. Cepheid variables are localized in a region above the main sequence also. Finally, we also have *novas*, which are stars that flare up every 30–50 years, by a factor of 10^6 times, over a

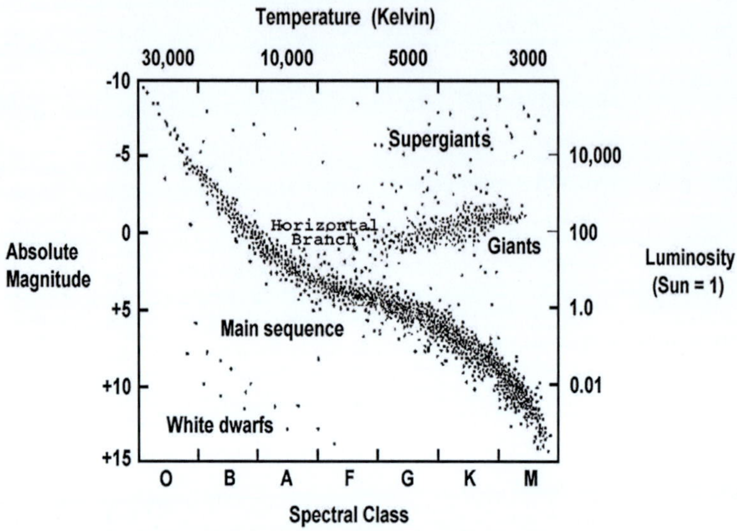

Fig. 3.1 Herzsprung–Russel diagram (Image: NASA/Chandra)

short time. They are stars in binaries, together with a white dwarf, and the accretion produces instabilities.

The evolution of stars can be described depending on their mass:

- If $M < 4M_{Sun}$, when its fuel is consumed, the star becomes a *red giant* and explodes in a "Helium flare-up", i.e., an explosion fueled by (nuclear) reacting Helium. The explosion leaves a corona (planetary nebula), and a core, which becomes a white dwarf, with a mass $< 1.4M_{Sun}$.
- If $4M_{Sun} < M < 8M_{Sun}$, the star becomes a white dwarf, with a mass $> 1.4M_{Sun}$. Then the electron degeneracy pressure (incompressibility of electrons) eventually cannot keep the star from collapsing, and it has a "Carbon flare-up", i.e., an explosion fueled by (nuclear) reacting Carbon. The explosion is a *supernova* explosion, specifically called a type I Supernova.
- If $M > 8M_{Sun}$, there is also a supernova explosion, but it has no Carbon flare-up. Instead, it produces heavier elements, and then the star collapses under gravity until the nuclei touch each other, and then explodes, creating even more heavy elements. After that, the core becomes a *neutron star*.
- A heavy enough neutron star will collapse gravitationally to a black hole (which afterwards can accrete anything nearby).

3.5 Measurements of Distance

We now turn to the various measurements of distance that we can use. The reason we can measure distances all the way to cosmological distances (sizes of the order of the Universe's) is that we have various methods that are valid *on different, yet overlapping, distance intervals*. By verifying or calibrating the method on the overlap, we can then use it at larger distances, etc.

(a) **Trigonometric Parallax**

At the smallest scales, we use a direct measure of distance, based on the fact that the Earth moves along the year around the Sun (on the ecliptic), and in doing so, the angle to a nearby star changes slightly by a $\delta\alpha$. This manifest itself in a change $\delta\alpha$ of the position on the stars on the sky (relative to very distant objects, "at infinity") as one moves on diametrically opposite positions on the ecliptic, see Fig. 3.2.

This method gives the definition of a *parsec*, as the distance at which the parallax is of one arc second. It equals about 3.26 light years. The method is used until about 100 light years.

(b) **Moving Cluster Method**

Consider a cluster of stars. Then individual stars in the cluster move at velocity v at an angle α from the direction to the star. We can then measure the radial component of the velocity, $v_r = v \cos\alpha$, through the Doppler shift. We can measure the average angle α of the stars in the cluster as follows: we can measure an angular distance on the sky of the various stars, and the angular velocities, extrapolated, (statistically) converge on a point in the sky situated at an angular distance α from the cluster, see Fig. 3.3. Then we calculate the tangent velocity $v_T = v_r \tan\alpha$. Measuring also the angular velocity on the sky μ ("proper motion"), measured over a long period, we can find the distance as

$$d = \frac{v_T}{\mu}. \tag{3.37}$$

This method is used up to about 500 light years. In particular, it is used for the Hyades cluster, situated at 120 light years, and the Scorpio–Centaurus association, situated at about 500 light years.

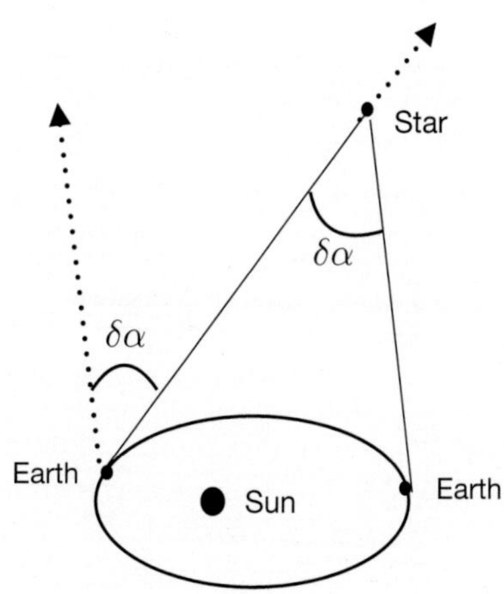

Fig. 3.2 Trigonometric parallax: the motion of the Earth around the Sun generates a corresponding motion of nearby stars on the sky

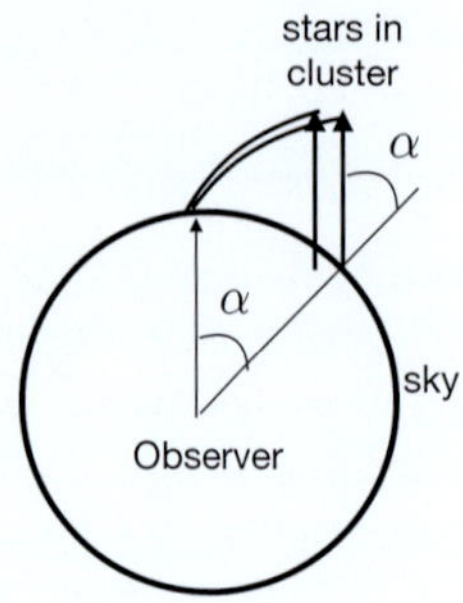

Fig. 3.3 Moving cluster method: the tangential (on the sky) velocities of the moving cluster (statistically) converge on a point in the sky, corresponding to the angle α of the velocity vector with the radial direction

(c) **Spectroscopic Parallax**

It is used for stars on the *main sequence*, for which there is collisional broadening of the spectral lines. Then one measures the apparent magnitude m, and the temperature T from its spectrum. Then, derive the absolute magnitude M from the Hertzsprung–Russel diagram. This allows us to calculate the distance from the relation between m and M,

$$d(\text{parsecs}) = 10^{\frac{m-M+5}{5}}. \tag{3.38}$$

This method is of course not very accurate, since it relies on the fact that stars lie on the Main sequence, which is not very precise (the main sequence is experimentally a quite widened line). But the advantage is that the accuracy remains constant, of about 15%, independent of the distance. This allows it to be used for distances larger than 100 light years.

(d) **Main Sequence Fitting**

It is a variant of the above. One applies the method above to *clusters of stars*, and then calibrates first by another method for nearby clusters.

(e) **Globular Clusters**

It is used for extragalactic distances (distances outside the Milky Way). One considers clusters of stars (of population II stars, on a line parallel to the Main Sequence line). We calibrate it for nearby population II stars, then use it for globular clusters in the Milky Way, then extend it to other galaxies. It is used up to about 30 million light years.

(f) **RR Lyrae Stars, or the Statistical Parallax Method**

The method is used for RR Lyrae variable stars, which are found to be *standard candles*. One measures the proper motion of the star and the peculiar motion (extra, excluding the average motion of the stars). As before, the radial component is measured by Doppler shift, and averaging over the velocities of several stars, one finds $\langle v_r \rangle = \langle v_\theta \rangle = \langle v_\phi \rangle$. Then proceed as in the moving cluster method, and find that the absolute magnitude M of RR Lyrae is always close to 0.6, i.e., the RR Lyrae are standard candles, and can be used in galaxies where they are visible.

(g) **Cepheid Variables**

The Cepheid variables are brighter than the RR Lyrae. In a similar manner, one finds that the visual absolute magnitude M is linear with the period P of the variability. That in effect acts as another type of standard candles, determining M from the easily measurable period. The method is used up to about 10 million light years.

(h) **Bright Blue Stars**

It is assumed that the brightest blue stars are blue supergiants, and have the same luminosity in all galaxies (there is a cut-off in luminosity there), $L \sim 10^6 L_{\rm Sun}$.

(i) **Size of H II Regions**

The sphere of ionized matter around bright blue stars is found to have a radius of about 150 light years. These are the H II regions, correlated to the bright blue star at the center, and have a standard width, allowing us to calculate the distance to them.

(j) **Supernovae**

It is found that the peak luminosity of a supernova explosion is about the same as the one of a bright galaxy, in all galaxies. That allows us to treat them at standard candles. The supernovae close by are spectacular: the most famous, in 1054, occurred in our own Milky Way galaxy, and was recorded in China, as being bright even during daytime. It left behind the Crab Nebula, of (still moving at relativistic speeds) ejecta from the explosion.

(k) **Galaxies**

It is found that the brightest member of a cluster of galaxies has the same luminosity L, i.e., there is a cut-off, so it can be used as a standard candle. It is a method used up to about 2 billion light years.

(l) **Redshift**

Finally, at the largest distances, we can only used the redshift for a measure of distance, assuming a cosmological model for the expansion.

3.6 Mass Determination for Distant Bodies

We finally describe methods of measurement of mass.
(a) There is an empirical law for M versus $L_{\rm em}$, from models.
(b) There is also a theoretical M versus $L_{\rm em}$ relation, from models.
(c) **Stellar spectra.**

We can measure the surface gravity of a star from the widths of spectral lines (which normally are associated with the decay of the state). The gravity at the surface of the star, where the light is emitted, influences the widths. Then the surface gravity is

$$g = \frac{MG_N}{R^2} , \tag{3.39}$$

and the luminosity is given by the Stefan–Boltzmann law,

$$L = 4\pi R^2 \sigma T^4 , \tag{3.40}$$

resulting in the relation

$$\frac{M}{L} = \frac{g}{4\pi G_N \sigma T^4}. \tag{3.41}$$

Measuring T and L (and knowing g), we can find M.

(d) **Pulsating Stars**

For the pulsating star, the period P is related to the average density $\bar{\rho}$ by the relation

$$P = 0.06\sqrt{\frac{\bar{\rho}_{\text{Sun}}}{\bar{\rho}}} \tag{3.42}$$

from which we determine $\bar{\rho}$. Determining L (for the Cepheids from the $P - L$ relation, for the others from distance), and $L = (4\pi R^2)\sigma T^4$, leading to a measurement of R, after which we have simply

$$M = \frac{4\pi}{3}\bar{\rho}R^3. \tag{3.43}$$

(e) **Gravitational Redshift**

The redshift due to the gravity at the surface of a dense object is

$$z_{GRS} \simeq \frac{2MG_N}{Rc^2}. \tag{3.44}$$

In white dwarfs, z_{GRS} can be up to $10^{-5} - 10^{-4}$, and if we measure the redshift and R, we can measure M.

Important Concepts to Remember

- The physical distance travelled by light at t_2 is $\Delta l = a(t_2)\Delta\eta = a(t_2)c \int dt/a(t) > ct_2$.
- The angular size distance is $\bar{r} = l/\theta = f(z)$, where l is linear size and θ is angular size.
- For light propagation, we have $\bar{r} \equiv c\psi(z)/H_0$.
- The luminosity distance d_L is the distance in Euclidean space producing the same observed flux $\mathcal{F}_{\text{obs}}$ for a given luminosity L_{em} at emission, so $\mathcal{F}_{\text{obs}} = L_{\text{em}}/(4\pi d_L^2)$, so $d_L = (c/H_0)\psi(z)(1+z)^2$.
- The apparent magnitude is $m = -2.5\log_{10}\mathcal{F}_{\text{obs}} + \text{const.} = 5\log_{10} d_L + \text{const.}$.

- The absolute magnitude M is the apparent magnitude of an object 10 parsecs away, and thus measures $L_{\rm em}$.
- To linear order $H_0 d_L = z\left(1 + \frac{1}{2}(1 - q_0)z + \cdots\right)$.
- Most stars are on the main sequence, for luminosity (or absolute magnitude) versus temperature (or colour).
- For $M < 4M_{\rm Sun}$, the star becomes a red giant, and explodes in a He flare-up leaving a white dwarf. For $4M_{\rm Sun} < M < 8M_{\rm Sun}$, the resulting white dwarf has a Carbon flare-up giving a supernova explosion. If $M > 8M_{\rm Sun}$, there is no Carbon flare-up, but the supernova produces heavier elements, and it creates a neutron star. A large enough neutron star collapses to a black hole.
- Measurements of distance are progressive, from small distances (where we measure them almost directly, like in the parallax method), to large distances, using overlapping domains of validity.
- There are a variety of mass measurements as well.

Further reading: Some of the things I said can be found in [1, 4], others in astrophysics textbooks.

Exercises

(1) Prove that in the M.D. case, the angular size distance is

$$\frac{c}{H_0}\psi(z) = \frac{c}{H_0}\frac{2}{\Omega_0^2(1+z)^2}\{\Omega_0 z + (\Omega_0 - 2)[(1+\Omega_0 z)^{1/2} - 1]\}. \tag{3.45}$$

(2) Write a formula for the future event horizon R_h, the maximal physical distance a fixed observer may come to observe. Calculate it approximately for the current, Λ dominated period.
(3) Calculate the particle horizon for an Universe with $\Omega = 1$, currently containing 30% matter and 70% cosmological constant, as a function of h_0, the Hubble constant today.
(4) Estimate the dimensionless gravitational potential at the surface of the Earth.

Chapter 4
Evidence for Dark Matter and the ΛCDM Model

In this chapter we will present the evidence for Dark Matter, and the resulting "Standard Model of Cosmology," the ΛCDM model (cosmological constant plus cold dark matter). Dark Matter is what the name suggests, matter that is dark, i.e., non-luminous. That is, matter that we cannot detect from emitted light. I will present the evidences, then review the possible candidates for dark matter, and finish describing the ΛCDM model.

4.1 Evidences for Dark Matter

I will present the various evidences as a list, starting with the first ones to be observed.

(1) **The Coma Cluster of Galaxies** (situated at about 20 Mpc distance)

The first evidence for dark matter dates back almost a century, to 1933, when *Zwicky*, studying the Coma Cluster, noticed an unusual thing. The Coma Cluster is made up of many galaxies, interacting gravitationally, that can be for all intents and purposes considered as point-like. This system then is no different than a gas in kinetic theory, with interactions generated by some potential.

In that case, the *virial theorem* states that for a gravitationally bound system that has virialized, i.e., "thermalized" (in the language of kinetic theory), on the statistical (ensemble) average, defined *locally*, we should have

$$2\langle T\rangle = -\langle U\rangle \Rightarrow \langle v^2\rangle \sim \frac{G_N M}{r}. \tag{4.1}$$

Here $v = v_p = v_{\text{tot}} - Hr$ is the peculiar velocity of the galaxies, i.e., the total velocity minus the Hubble expansion. But taking for M the *visible* mass, Zwicky found that

$$v^2 \gg \frac{G_N M_{\text{visible}}}{r}. \tag{4.2}$$

H. Năstase, *Cosmology and String Theory*, Fundamental Theories of Physics 197, https://doi.org/10.1007/978-3-030-15077-8_4

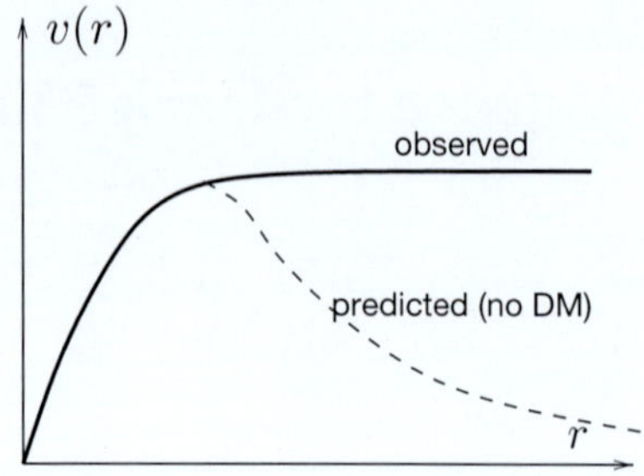

Fig. 4.1 Rotational curves of galaxies. Without the dark matter, outside the radius of the galaxy, the velocity drops as $v \sim 1/\sqrt{r}$ (dotted), but with dark matter (observed) it is approximately flat (full line)

That suggests that there should be extra matter that is dark (not visible), called *dark matter*, and that

$$M_{\rm dark} \gg M_{\rm visible}. \tag{4.3}$$

But this hypothesis was discarded untill the 1970s, assuming that it was a statistical fluke, since it was an isolated phenomenon,

(2) **The Rotational Curves of Galaxies**

In the 1970s, Ford and Rubin discovered that rotational curves of galaxies are flat. This is similar to the previous case, just that instead of clusters of galaxies, one considers a single galaxy, specifically a spiral galaxy like our own Milky Way, and stars orbiting around its center of mass. Besides the bulk of the stars, forming the characteristic spiral shape, there are outliers, that live outside the visible matter.

For such a star, orbiting the center of the galaxy and on *bound trajectories* (like the planets around the Sun, say), we have a relation that is now true independently of statistics (for each individual star, unlike the case of the cluster of galaxies), equaling the centrifugal and centripetal force,

$$F_{\rm centrifugal} = \frac{mv^2}{R} = \frac{mG_N M(r)}{r^2} = F_{\rm grav}\ , \tag{4.4}$$

from which we derive that

$$v = \sqrt{\frac{G_N M(r)}{r}}. \tag{4.5}$$

We look for stars situated at a distance greater than the location of most visible mass (outliers). In that case, we expect that $M(r) \simeq M$ =const., so $v \propto \frac{1}{\sqrt{r}}$, but instead one observes $v \simeq \text{const}(r)$, as in Fig. 4.1. That implies

$$M(r) \propto r! \tag{4.6}$$

Note that this is not what we would expect from a constant density of matter, which would give $M(r) \propto r^3$, but rather we have some diffuse matter, that is not completely bound to the galaxy, forming a *halo*, yet is still centered on the center of the galaxy. This would be *dark matter*.

The best fit for the data is then (cold) dark matter, a type of matter that doesn't interact with normal matter, except gravitationally (since gravity interacts with all matter equally). It could have some self-interaction, though that is not clear. Due to the gravitational interaction, it has the same center of mass as the galaxy, in the case of rotation curves. The fact that is cold (very low temperature) will be explained better later.

As an example, consider the galaxy NGC 6503, where the matter from the disk gives a sharp peak in the centrifugal velocity v_c at about 3 Kpc (kiloparsecs) at close to 10 km/s, then decays sharply, until about 20 Kpc. The Dark Matter halo reaches a maximum of v_c at about 10 Kpc, after which if flattens out, so that the total v_c rises quickly until about 120 km/s at about 3 Kpc, after which it stays really flat.

The limitation in distance away from the center of the galaxy is how far away can we see the light from the star, or neutral H, giving out 21 cm waves, which is about tens of Kpc.

As we said, the velocity field v_c is really flat, but that is not reproduced really perfectly from Cold Dark Matter simulations. In fact, interestingly, an alternative theory called MOND (described in an exercise) gives a better fit, though that theory cannot explain well other evidences for Dark Matter (like the Bullet Cluster below).

(3) **Hot Gas in Clusters** (for instance, in the Coma Cluster)

The principle is similar to the case of the Coma Cluster of galaxies, except we track the mass through the hot gas moving around. Its velocity (squared) v^2 is measured from the temperature of the hot virialized gas, and as before we compare with GM_{visible}/r.

(4) **The Bullet Cluster**

These are two colliding clusters of galaxies, where one goes through the other, so it is perhaps a misnomer, but it is associated with the fact that there is a lot of hot plasma (gas) remnant of the passing of one cluster through the other, like a bullet passing through some material and leaving a hot trail behind it. The interesting thing is that most visible mass is situated in the plasma, which is in between the galaxies, since it was slowed down in the collision. The center of mass of the visible mass is then in between the galaxies, but the center of mass obtained from gravitational lensing that, as we will shortly see, maps out all matter, is different. Dark matter actually traces the galaxies, so it is centered on the two galaxies. In fact, one obtains

$$\frac{M_{gas}}{M_{DM}} \sim \frac{1}{6-7}, \tag{4.7}$$

as obtained in the ΛCDM model.

This is one of the most difficult tests to reproduce outside of the Dark Matter idea (for alternatives to Dark Matter), because of its qualitatively different nature. It is usually stated that MOND has a problem in reproducing it (though in its relativistic version, TeVeS, it is sometimes claimed that the mismatch is reduced to a factor of 2). Similarly for modified gravity models.

(5) **Surveys of Visible Matter**

By taking large scale surveys of the visible matter, we can map out the distribution in the Universe. The first one was done by J. Oort in 1956, but the results have not changed much since then. One computes the ratio of mass to luminosity M/L for different galaxies, and finds that on the average

$$\left\langle \frac{M}{L} \right\rangle \sim 21 \frac{M_{\text{Sun}}}{L_{\text{Sun}}}. \tag{4.8}$$

Then one maps out the luminosities of galaxies, and from the above ratio we find the masses. Adding them up we find

$$\Omega_{\text{visible}} \sim 0.03. \tag{4.9}$$

(6) **Gravitational Lensing**

In general relativity, massive objects like stars and galaxies deflect light. Famously this is one of the first decisive tests of general relativity, when deflection of the light from the stars behind the Sun was measured during an eclipse, and found to agree with general relativity rather than special relativity (the difference between them is a factor of 2).

The angle of deflection of light (between the original and the final asymptotic lines) is

$$\delta\phi = \frac{4G_N M/c^2}{r}, \tag{4.10}$$

where r is the minimum distance at which light passes from the object as it is deflected.

The deflection means that various rays of light deflected by the object in front of them can get focused on the same observer, like in the case of a lens, hence the name *gravitational lensing*. Of course, unlike the case of a lens, the gravitational deflection is gradual (gravity being a long range force), it is not completely localized. From the point of view of the observer, he would see the light from the object behind the deflector as an arc or arcs centered on the deflector. The deflection angle observed on the sky (angular distance between the arc and the true position), i.e., the deflection angle from the point of view of the observer, is

$$\alpha = \sqrt{\frac{4G_N M}{d}}, \tag{4.11}$$

where d is the impact parameter (the minimal distance it would have passed from the deflector, were it not deflected), for small angles approximately equal to d', the smallest distance passed from the deflector, see Fig. 4.2.

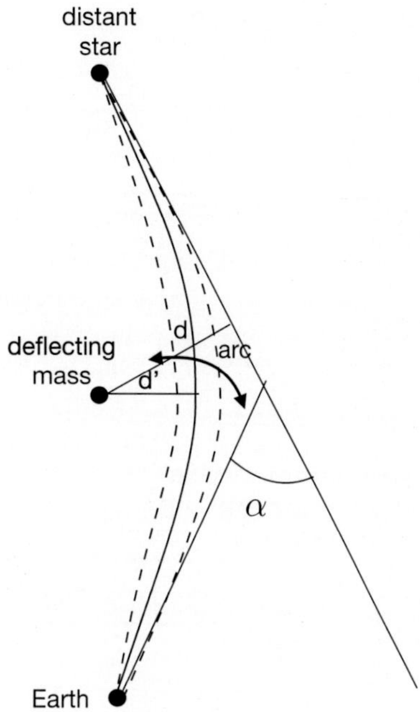

Fig. 4.2 Gravitational lensing: Light from a distant star is deflected by a deflecting mass (perhaps a local star), creating an arc on the sky, instead of a point. The angle α depends on the impact parameter d, or the smallest distance to the deflector d'

By applying this method to a cluster of galaxies (situated at tens of Mpcs), and looking at other clusters behind it (also tens of Mpcs), in the form of arcs in the field of vision, we can map out the mass distribution.

(6) **MACHOs** (MAssive Compact Halo Objects)

A kind of subcategory of gravitational lensing is MACHOs, since they are an alternative explanation for why there would be Dark Matter, and they are also related to gravitational lensing. These would be large, non-luminous objects that, as they pass in front of various sources that we see, give a very specific signal, namely *achromatic* (independent of frequency, since it is a gravitational effect) brightening. The brightening is due to the same phenomenon as the gravitational lensing, and is not possible for a non-gravitational source.

But these were not seen to the degree expected if they were supposed to in order to be an alternative to dark matter, hence they are excluded as an explanation. More on them later.

(7) **Peculiar Velocities of Galaxies in Clusters**

Considering a cluster of galaxies, one can measure the individual (peculiar) velocities

$$v_p = v_{\rm obs} - Hr \tag{4.12}$$

of galaxies, and one makes a map of the velocity flow, reconstructing the gravitational potential. A formula that is found to work relates the divergence of the peculiar velocity field to the overdensity of galaxies $\delta\rho/\rho$ by

$$\vec{\nabla} \cdot \vec{v} = -\Omega^{0.6} \frac{\delta\rho}{\rho}. \tag{4.13}$$

(8) **Big Bang Nucleosynthesis**

As we have already suggested, when we go back in time, the last moment that we can experimentally measure cosmology precisely is the Big Bang Nucleosynthesis, the period when nuclei get formed. After that, there is the moment of recombination, when atoms get formed, and the Universe becomes transparent to light, and the CMBR is formed.

During Big Bang Nucleosynthesis (BBN), the light elements are formed (the heavier ones are formed in nova and supernova explosions), mostly H, so it is used as reference. The abundances of He/H, $(D + {}^3He)/H$, Li/H created during BBN are matched against observed abundances in the Universe (which come mostly from BBN, and to a smaller degree from nuclear reactions within stars), so we have a precise test of the conditions at BBN. One of the factors that influence the abundances is $\Omega_{\rm baryons}$, and one finds a

$$\Omega_{\rm baryons} \sim 0.05. \tag{4.14}$$

A more precise description of Big Bang Nucleosynthesis (BBN) will be given in Chap. 6.

(9) **CMBR**

Finally, of course, we have the CMBR. As we already said, it fixed $\Omega_{\rm total}$ to be 1, which together with the measurement of acceleration from type I_a Supernovae means that $\Omega_{\rm matter} \sim 0.3$, implying there is dark matter through the above arguments.

But the CMBR depends also on $\Omega_{\rm baryons}$, and one finds

$$\Omega_{\rm baryons} \sim 0.05. \tag{4.15}$$

Actually, it is $\Omega_{\rm baryons} h^2$ that is the parameter that influences it ($H = h \cdot 100\,{\rm km/s/Mpc}$), but given other measurements of H, this is what we obtain.

In conclusion, we obtain

$$\Omega_{\rm total} = 1; \quad \Omega_\Lambda = 0.73; \quad \Omega_{\rm DM} = 0.23; \quad \Omega_{\rm baryons} = 0.04. \tag{4.16}$$

4.2 Candidates for Dark Matter

Having established the existence of Dark Matter, we now consider various candidates for what they can be.

Cold Dark Matter

Dark matter can be cold, i.e., at very small temperatures. This is what will happen if the mass is very large, so that the objects are almost non-relativistic. It can also happen if they are almost completely non-interacting.

1. *MACHOs*

These could be: brown dwarfs (dense like a white dwarf, except without light, i.e., no nuclear reactions happening inside them), Jupiters (would-be stars that have never ignited, like Jupiter itself, which are gaseous on the outside and not very dense), neutron stars (not emitting light), or massive black holes, etc. Of course, there are supermassive black holes at the center of every galaxy, but they represent only a small proportion of the galaxy's mass, and these are the most massive black holes there are, so they are excluded (general relativity simulations don't create overly many of them).

As we said, all of the MACHOs are excluded as a candidate for Dark Matter, both theoretically (it is hard to get enough of them to be created) and experimentally (we do not observe these MACHOs through lensing)

2. *WIMPs* (Weakly Interacting Massive Particles)

These are fundamental particles with mass that interact extremely weakly with other ones. They are the preferred explanation for Dark Matter.

3. *Axions*: more on them later

Hot Dark Matter

Dark matter can also be hot, i.e., thermalized at a temperature comparable to the CMBR's one. There is basically only one possibility, since having been in the past in thermal equilibrium with the CMBR is a condition, which pretty much restricts it to be a neutrino (other species would decouple much sooner). More on them shortly.

Modifications of Gravity

Modifications of gravity (like MOND, though that is a modification of the Newton force law, not the gravity law, or $f(R)$, or massive gravity, etc.) generally don't satisfy all of the constraints from the experimental evidences. It is very difficult to satisfy them all, since they appear at different scales (rotational curves for galaxy scale, then various at galaxy cluster scale, then CMBR scale), and a modification of gravity has generically a given distance scale at which it sets in.

Neutrino Hot Dark Matter

Neutrinos being the dark matter leads to the astrophysics constraint (see [5])

$$\sum_i m_{\nu_i} \lesssim 0.3\,\text{eV}; \quad (\text{Planck}); \quad (0.17\,\text{eV} \quad \text{combined}). \tag{4.17}$$

Direct particle physics searches for them lead to the constraints (from the Particle Data Group 2017, http://pdg.lbl.gov/2017/listings/rpp2017-list-neutrino-prop.pdf)

$$\begin{aligned} m_{\nu_e} &< 2\,\text{eV} \\ m_{\nu_\mu} &< 190\,\text{KeV} \\ m_{\nu_\tau} &< 18\,\text{MeV}. \end{aligned} \tag{4.18}$$

On the other hand, the tightest constraints come from neutrino oscillations. The observed neutrino oscillation depends only on the Δm^2 of the two neutrino species. For the species 1, 2, 3 ordered in terms of increasing mass, we find (from the Particle Data Group 2018, http://pdg.lbl.gov/2018/reviews/rpp2018-rev-neutrino-mixing.pdf)

$$\Delta m_{12}^2 = 7.4 \times 10^{-5}\,\text{eV}^2; \quad \Delta m_{31}^2 = 2.5 \times 10^{-3}\,\text{eV}^2. \tag{4.19}$$

This in turn leads to a bound on the highest neutrino mass eigenstate of

$$m_{\nu_{\text{highest}}} < 0.05\,\text{eV}\ , \tag{4.20}$$

which is the tightest constraint we have.

But the Hot Dark Matter scenario is excluded by the CMBR at low angular momentum l. Indeed, in Hot Dark Matter the fluctuation spectrum $\delta\rho/\rho$ or $\delta T/T$ at low l would drop to zero, for a linear increase with l. But even with the experimental error bars, it is clear that in reality the CMBR goes to a constant, so the main contribution to Dark Matter is Cold, though there could also be a small Hot component.

4.3 WIMP Dark Matter Candidates

Since Cold Dark Matter of the WIMP variety is the choice preferred by experiments, it is worth saying what are the main candidates.

Within supersymmetry, still one of the preferred models of high energy completion in the Standard Model, the clear Dark Matter candidate is the *Lightest Supersymmetry Particle, or LSP*. We don't need to know the details of the supersymmetric model, just that there is a particle of smallest mass that therefore cannot decay (is stable), is relatively light, so it can act as Dark Matter, and has almost no interactions with the other particles.

In supersymmetry, we have the MSSM (Minimally Supersymmetric Standard Model) or its extensions, forming the *visible sector*. We also have a *hidden sector*, which is a strongly coupled sector that breaks supersymmetry through some non-perturbative mechanism (since we know that perturbative mechanisms are excluded by experiment in the Standard Model, and there is no right nonperturbative coupling in the MSSM; in the hidden sector there can be). Finally, we have *messenger fields*, or mediators, that carry the supersymmetry breaking from the hidden sector to the visible one (MSSM). This mediation of supersymmetry breaking is usually made by gauge fields (gauge mediation) or gravity (gravity mediation).

So the only exception to the LSP being Dark Matter comes from *gravity mediated scenarios*, when the *gravitino* (superpartner of the graviton) is the LSP. This is generically inconsistent with cosmology. In this case, amplitudes for annihilation go like G_N instead of $\sqrt{G_N}$, so the annihilation rates go like

$$\Gamma \propto \mathcal{A}^2 \propto G_N^2 \, , \tag{4.21}$$

which is too small for the annihilation rate to be useful for giving gravitino decay. But in this case, the number of gravitinos is about one percent of the number of photons (which is similar to the case of the neutrinos), from which (if the gravitino is the Dark Matter) we can derive a bound on the mass of the gravitinos,

$$m_{\text{gravitino}} \lesssim 100\,\text{eV}. \tag{4.22}$$

But this is too small to make sense in the gravity mediation scenarios, so in this case the cosmology contradicts the simplest gravity mediation scenarios.

If however the gravitino is not the LSP and can decay, giving

$$\Gamma \propto \mathcal{A}^2 \propto G_N \, , \tag{4.23}$$

then one finds (like for any decaying particle) from the Big Bang Nucleosynthesis constraints that they need to decay before it, which translates into

$$m_{\text{gravitino}} > 10\,\text{TeV}. \tag{4.24}$$

This in turn means a supersymmetry breaking scale

$$M_s > 10^{11}\,\text{GeV} \, , \tag{4.25}$$

since

$$m_g \simeq \sqrt{G_N} M_s^2. \tag{4.26}$$

This is a very high scale, and it means that the gravitino is not Dark Matter in this scenario.

Axions

Another possibility is for Dark Matter to be coming from *axions*.

In the Standard Model, there is a term possible to be added to the Lagrangian of QCD, the theta term

$$\int d^4x \theta \,\text{Tr}[F_{\mu\nu}\tilde{F}^{\mu\nu}] \, , \tag{4.27}$$

where $\tilde{F}^{\mu\nu} = \frac{1}{2}\epsilon^{\mu\nu\rho\sigma}F_{\rho\sigma}$ is the Maxwell dual to the YM field.

But experimentally, it is found to be extremely small ($\theta \ll 1$), which can be enforced by the existence of a symmetry, called Peccei–Quinn symmetry, that would put $\theta = 0$. But if we spontaneously break the symmetry, the Goldstone boson for the broken Peccei–Quinn symmetry is called an *axion* and is identified with a varying $\theta(x)$. An axion could have a mass

$$m_{\text{axion}} \sim 10^{-3} - 10^{-6}\,\text{eV} \tag{4.28}$$

and would give a good candidate for Dark Matter. These axions appear easily within string theory, so they can be relevant for cosmology.

4.4 The ΛCDM Model

In conclusion, we see that the *Standard Model of Cosmology*, that best fits data, is a ΛCDM (cosmological constant plus dark matter) model, with proportions:

4% baryons, 23% cold dark matter and 73% cosmological constant.

We say cosmological constant, but really, what is sure is we have an accelerated Universe (which as we saw requires an equation of state with $w < -1/3$), but cosmological constant ($w = -1$) fits best the data. In principle it need not be a cosmological constant, but rather a *dark energy* component, which could have a time-varying w. However, there is no truly generic constraint on the time dependence of w, one has to make some assumption, and the Planck data constrains various scenarios.

Important Concepts to Remember

- The first method to observe dark matter was the Coma Cluster of galaxies, measuring statistically the virialized velocities of individual galaxies, and seeing that $v^2 \gg G_N M_{\text{visible}}/r$.
- The most classic test however is the rotational curves of galaxies, i.e., test stars rotating around the galaxy. Since $v = \sqrt{G_N M(r)/r}$ and we observe $v \simeq$ constant, it means that there is dark matter with $M(r) \propto r$ outside the visible one.
- A crucial test eliminating most alternatives to dark matter is the Bullet Cluster of galaxies, composed mostly of gas. Two clusters have collided, and the stars are ahead, but most of the gas is in between. However, the dark matter is still around the stars, so the visible and dark matter are in different regions.
- Gravitational lensing means focusing of light from a distant source by a mass distribution in front of it (with respect to us). It allows us to see dark matter distributions in front of some distant sources.
- Big Bang Nucleosynthesis is sensitive to Ω_{baryons}.
- The CMBR is sensitive to Ω_{total}, but also to Ω_{baryons}, allowing us to measure the amount of dark matter.
- Hot Dark Matter (usually neutrinos) is excluded, since the fluctuation spectrum at small l would drop to zero fast.
- MACHO cold dark matter, i.e., brown dwarfs, etc., cannot explain dark matter. We need WIMPs or axions for Cold Dark Matter to do so.

- The Lightest Supersymmetric Particle (LSP) is a good WIMP, however gravitinos are not.
- Axions are also a possibility for WIMPs.
- The standard ΛCDM model has 73% dark energy, 23% cold dark matter and 4% baryons.

Further reading: See the reviews [6, 7]. Also [8] for the Bullet Cluster.

Exercises

(1) (MOND theory) An alternative explanation for the rotational curves of galaxies is in terms of a modification of the Newton's force law at low acceleration, from $a = F/m$ to $a = \sqrt{a_0 F/m}$. Show that the latter leads to the observed $v = \text{const.}(r)$. Choose an interpolating function between the 2 regimes and find the exact $V(r)$ profile.
(2) Is it possible for the gravitational lensing effects to cancel against one another (to have false negatives: that we see nothing in the sky, yet there is lensing)? Why?
(3) Calculate the time evolution of the total energy density $\rho(t)$ in the ΛCDM model with the exact proportions Ω_i in the text.
(4) Consider noninteracting dark matter particles of mass $m \ll k_B T_{\rm DM}$, thus relativistic, and thermalized at temperature $T_{\rm DM}$ of the order of, but not equal to, the CMBR temperature of 3.75 K. If dark matter is *entirely* composed of them, calculate their number density.

Chapter 5
The Early Universe and Its Thermal History

In this chapter, I will describe the thermal history of the Early Universe, i.e., the evolution of the particle species in it. After describing the Hot Big Bang scenario, and why it is required, we will calculate the thermodynamics of various particle species in the Universe. The evolution of the particle species during the radiation dominated era will be then described, and finally we will look in detail at neutrino decoupling.

5.1 Defining the Hot Big Bang

The Universe starts off in a *hot, equilibrium* state. It is an *equilibrium* state, since at $t = 1$ s from the Big Bang, the characteristic reaction time for e^+e^- annihilation, $e^+e^- \leftrightarrow 2\gamma$, is $\tau = 10^{-17}$ s, so there is more than enough time to reach equilibrium in the annihilation reaction.

In order to characterize the Universe, we will use the quantities: entropy (per baryon)

$$S_1 \simeq \frac{N_\gamma}{N_{\text{baryon}}}\,, \tag{5.1}$$

(we will see later in the chapter that the relation is correct up to a numerical factor of order one) and normalized (electron) lepton number

$$L_e = \frac{(N_{e^-} - N_{e^+}) + (N_\nu - N_{\bar{\nu}})}{N_{\text{baryon}}}. \tag{5.2}$$

The next assumption to check is whether the Universe starts off in a hot state. We will do that historically, following the various proposals that were considered before the Hot Universe:

H. Năstase, *Cosmology and String Theory*, Fundamental Theories of Physics 197, https://doi.org/10.1007/978-3-030-15077-8_5

(1) **Only Neutrons Initially**

Looking at the Universe, it seems like it is a cold place, and mostly made up of protons and neutrons, plus electrons. It is also electrically neutral, so the simplest assumption is that it starts in a state with only neutrons n. Since n interacts with p (protons), it is hoped that we can create the necessary protons.

The initial quantum numbers of the Universe are, therefore, $S_1 = 0$ (cold Universe) and $L_e = 0$.

The neutrons (outside a nucleus) can decay, with a half time $\tau \sim 1000\,\text{s}$, and they start a reacting chain

$$\begin{aligned} n &\to p + e^- + \bar{\nu} \\ p + n &\to D + \gamma \\ D + D &\to T + p \\ &\to {}^3He + n \\ {}^3He + n &\to T + p \\ T + D &\to {}^4He + n. \end{aligned} \tag{5.3}$$

Specifically, the neutrons decay into protons, which then react with the protons to make deuterium. Two deuterium make tritium and more protons, and the tritium and deuterium make helium 4. Thus the chain starts with neutrons, and ends up with helium 4, but no protons (all the protons react with n and decay). That means we don't produce protons, i.e., hydrogen nuclei. This is bad, since the Universe is mostly made up of hydrogen.

(2) **Only Protons and Electrons in Equal Numbers**

The next possibility is to *start* with protons p's, but then in order for the Universe to be electrically neutral, we must have an equal number of electrons e^-s, $N_p = N_{e^-}$. In that case $S_1 = 0$ (cold Universe), $N_\gamma = 0$, but $L_e = 1$.

However, in this case, we start off with the reactions

$$\begin{aligned} p + e^- &\to n + \nu \\ p + n &\to D + \gamma\,, \end{aligned} \tag{5.4}$$

and from then on, the reaction chain follows in the same way, ending up at ^{4}He. So in this case as well, we consume all protons, and end up with only Helium 4. Again, this is bad, since in the end we have no hydrogen (even though we started with ionized hydrogen atoms, i.e., protons and electrons in equal amounts). But the point is that protons and neutrons interact until there is nothing left.

(3) **Protons, Electrons and Neutrinos in Equal Numbers**

Next in complexity is to assume that we have also neutrinos in equal numbers with the protons and electrons, which doesn't change the neutrality of the Universe. In this case $S_1 = 0$ and $L_e = 2$.

Now the starting point of the reaction chain,

$$p + e^- \to n + \nu \tag{5.5}$$

is forbidden because of Pauli's exclusion principle. The neutrinos are fermions, and there are already neutrinos in the state, occupying the same states.

Since the reaction chain cannot start, the protons and electrons can bind to form hydrogen atoms. So H is produced, but nothing else, so that is not good either.

All of these solutions involved a cold Universe ($S = 0$).

(4) **The Hot Universe**

But in the 1940s, George Gamow finally came up with the hypothesis of the Hot Universe, with $S_1 \gg 1$ and $L_e \sim 1$, that is, we have many photons for each baryon, and equally electrons and neutrinos, comparable with the baryons.

Now, due to the many photons present, the reaction

$$p + n \to D + \gamma \tag{5.6}$$

is not possible, since it occurs equally in the opposite direction, with the photons destroying the Deuterium atoms. That means that the reaction chain (cycle) cannot continue beyond the formation of protons (Hydrogen nuclei).

But that is fine! Since that means that there is mostly hydrogen in the Universe, and there is only 20, 30% Helium. With the drop in temperature, we can create *some* Deuterium, during Big Bang Nucleosynthesis, as will be showed in the next chapter. In terms of abundances of isotopes, the ones with more neutrons predominate, since we have free neutrons for a long time, and we have high temperature.

5.2 Equilibrium Thermodynamics

We have seen that the Universe is hot and in equilibrium, so we now describe equilibrium thermodynamics, within relativistic theory.

We denote the Fermi–Dirac or Bose–Einstein distributions by

$$f(\vec{p}) = \frac{1}{e^{\frac{E-\mu}{k_B T}} \pm 1}. \tag{5.7}$$

For a relativistic particle with degeneracy g, the number density is

$$n = \frac{g}{(2\pi)^3} \int f(\vec{p}) d^3 p. \tag{5.8}$$

The energy density is found by introducing an energy $E(\vec{p})$ inside the integral, i.e.,

$$\rho = \frac{g}{(2\pi)^3} \int E(\vec{p}) f(\vec{p}) d^3 p. \tag{5.9}$$

The pressure is found from relativistic kinetic theory,

$$P = \frac{g}{(2\pi)^3} \int \frac{|\vec{p}|^2}{3E} f(\vec{p}) d^3 p. \tag{5.10}$$

In the kinetic theory of gases, the pressure is found as momentum transfer in the x direction, perpendicular to the surface, times the velocity, i.e., $\langle p_x^2 \rangle / m = \langle p^2 \rangle / (3m)$. In the relativistic case, $m \to E$, and $|\vec{p}| = \sqrt{E^2 - m^2}$.

Writing $d^3 p = 4\pi p^2 dp$, and using $p^2 = E^2 - m^2$, we have $d^3 p = 4\pi E dE \sqrt{E^2 - m^2}$. Replacing in the integrals, we obtain

$$\begin{aligned} n &= \frac{g}{2\pi^2} \int_m^\infty \frac{(E^2 - m^2)^{1/2}}{e^{\frac{E-\mu}{k_B T}} \pm 1} E dE \\ \rho &= \frac{g}{2\pi^2} \int_m^\infty \frac{(E^2 - m^2)^{1/2} E^2}{e^{\frac{E-\mu}{k_B T}} \pm 1} dE \\ P &= \frac{g}{6\pi^2} \int_m^\infty \frac{(E^2 - m^2)^{3/2}}{e^{\frac{E-\mu}{k_B T}} \pm 1} dE. \end{aligned} \tag{5.11}$$

We are however not so interested in the intermediate region of energies described by the general formulas above, but rather in the extreme regions: ultrarelativistic and non-relativistic.

Relativistic Species

Consider now (ultra)relativistic species, i.e., $T \gg m, \mu$. In that case, we obtain

$$\begin{aligned} n &= \frac{\zeta(3)}{\pi^2} g T^3; \quad \text{Bose} \\ &= \frac{3}{4} \frac{\zeta(3)}{\pi^2} g T^3; \quad \text{Fermi.} \end{aligned} \tag{5.12}$$

Note that in the relativistic case, the integrals can be taken from 0 instead of $m \ll T$, and then by rescaling the energies by T, we obtain the scaling of n with gT^3, with a numerical prefactor given by the adimensional integral. Similarly, for the energy density we obtain

$$\begin{aligned} \rho &= \frac{\pi^2}{30} g T^4; \quad \text{Bose} \\ &= \frac{\pi^2}{30} \frac{7}{8} g T^4; \quad \text{Fermi.} \end{aligned} \tag{5.13}$$

In the case of the pressure, in both cases we obtain

$$P = \frac{\rho}{3}\,, \tag{5.14}$$

as we should. Note that the Bose case is exemplified by photons (γ), and the Fermi case is exemplified by neutrinos (ν).

We can also calculate the average energy per particle for the relativistic species

$$\begin{aligned} \langle E_1 \rangle \equiv \frac{\rho}{n} &= \frac{\pi^4}{30\zeta(3)} T \simeq 2.7T; \quad \text{Bose} \\ &= \frac{7}{6}\frac{\pi^4}{30\zeta(3)} \simeq 3.15T; \quad \text{Fermi}. \end{aligned} \tag{5.15}$$

However, in the relativistic reactions, usually one has at least two species, particles and their antiparticles, and they are in equilibrium, with $\mu_+ = -\mu_- \equiv \mu$. Substituting in the general formulas, we obtain

$$\begin{aligned} n_+ - n_- &= \frac{gT^3}{6\pi^2}\left[\pi^2 \frac{\mu}{T} + \left(\frac{\mu}{T}\right)^3\right]; \quad T \gg m \\ &= 2g\left(\frac{mk_BT}{2\pi}\right)^{3/2} \sinh \frac{\mu}{k_BT} e^{-\frac{m}{k_BT}}; \quad T \ll m. \end{aligned} \tag{5.16}$$

In the second line, we have also considered the nonrelativistic case, in which case $E \simeq m$ and the FD and BE distributions become Maxwell, which moreover (since $E \simeq m$) gets outside the integral, giving the overall $e^{-\frac{m}{T}}$ factor.

Nonrelativistic Species

In the nonrelativistic case $T \ll m, \mu$, we obtain (in the same way as for $n_+ - n_-$ above)

$$n = g\left(\frac{mk_BT}{2\pi\hbar^2}\right)^{3/2} e^{-\frac{m-\mu}{k_BT}}. \tag{5.17}$$

For the energy density, we have an extra $E \simeq m$ inside the integral, so we have

$$\rho = mn. \tag{5.18}$$

Insider the integral for the pressure, we have an extra $(E^2 - m^2)/(3E)$, which finally gives the usual ideal gas law

$$P = nk_BT \ll \rho. \tag{5.19}$$

5.3 Thermodynamics of the Universe

In order to consider what happens in the expanding Universe, we need to consider the case of several species of relativistic particles, each in equilibrium at different temperature T_i. This is possible, since we can have decoupled systems that only expand with the Universe, but keep their thermal distribution. In this case, we define

$$\rho_R = \frac{\pi^2}{30} g_* T^4$$
$$P_R = \frac{\rho_R}{3} = \frac{\pi^2}{90} g_* T^4 \,, \tag{5.20}$$

where T is the temperature of photons. This is really a definition of g_*. By summing over the various components, we have

$$g_* = \sum_{i=\text{bosons}} g_i \left(\frac{T_i}{T}\right)^4 + \sum_{i=\text{fermions}} g_i \left(\frac{T_i}{T}\right)^4 \,, \tag{5.21}$$

and this is the *number of effectively massless degrees of freedom* (or, ultrarelativistic).

The expansion of the Universe is *adiabatic*, which means that the entropy per *comoving* volume is conserved. We write

$$TdS = d(\rho V) + PdV = d[(\rho + P)V] - VdP. \tag{5.22}$$

But from the general relativistic formulas (5.20), we obtain

$$\frac{dP}{dT} = \frac{4}{3}\frac{\rho}{T} = \frac{\rho + P}{T} \,, \tag{5.23}$$

which allows us to rewrite

$$TdS = d[(\rho + P)V] - \frac{VdT}{T}(\rho + P) \Rightarrow$$
$$dS = d\left[\frac{(\rho + P)V}{T} + \text{const.}\right] \tag{5.24}$$

That means that the entropy per comoving volume is

$$s = \frac{S}{V} = \frac{\rho + P}{T} = \frac{4}{3}\frac{\rho}{T} = \frac{2\pi^2}{45} g_{*s} T^3 \,, \tag{5.25}$$

where in the last equality we have written a formula for several species. Just that we need to do that *after* dividing by T for each individual species, so that the g_* that appears is not the one in ρ_R, which was defined with factors of T^4, but rather is one with factors of T^3, namely

$$g_{*s} = \sum_{i=\text{bosons}} g_i \left(\frac{T_i}{T}\right)^3 + \sum_{i=\text{fermions}} \frac{7}{8} g_i \left(\frac{T_i}{T}\right)^3 . \tag{5.26}$$

On the other hand, specializing for photons γ, the number density is

$$n_\gamma = \frac{\zeta(3)}{\pi^2} g_\gamma T^3 , \tag{5.27}$$

where $g_\gamma = 2$, for the two photon polarizations. Combining with (5.25) and substituting the numerical values for the constants, we obtain

$$s = 1.80 g_{*s} n_\gamma , \tag{5.28}$$

or

$$S_1 = \frac{S}{N_{\text{baryons}}} = \frac{s}{n_{\text{baryons}}} = 1.80 g_{*s} \frac{N_\gamma}{N_{\text{baryons}}} , \tag{5.29}$$

justifying the formula from the beginning of the chapter.

Moreover, putting numbers, we get the formula

$$n_\gamma(\text{photons per } cm^3) \simeq 20 T^3(^0 K). \tag{5.30}$$

Physical Values

We can measure that

$$\rho_{\text{rad,CMBR}} \simeq 30 \rho_{\text{rad,all sources}} , \tag{5.31}$$

so the main part of the radiation in the Universe is in the CMBR. We measure the temperature of the CMBR as

$$T_{\text{CMBR}} \simeq 2.7\,\text{K}. \tag{5.32}$$

According to the formula above, the density of photons in the Universe is

$$n_{\gamma,0} \simeq 20 \times (2.7)^3 \simeq 410\,\text{cm}^{-3}. \tag{5.33}$$

In turn, either by using the Stefan-Boltzmann law, or by the relation between ρ_γ and n_γ from the relativistic thermodynamics, we find

$$\rho_{\gamma,0} = \sigma T^4 \simeq 4.64 \times 10^{-34}\,\text{g/cm}^3. \tag{5.34}$$

Since

$$\frac{\rho_{\text{rad},0}}{\rho_{\gamma,0}} = 1 + \frac{7}{8} 3 \left(\frac{T_\nu}{T_\gamma}\right)^3 , \tag{5.35}$$

where the factor of 3 comes from the 3 species of neutrinos, and the temperature of neutrinos will be estimated at the end of the chapter, we find

$$\rho_{\rm rad,0} \simeq 7.8 \times 10^{-34}\,{\rm g/cm^3}. \tag{5.36}$$

The number density of baryons can be estimated (how? to do in exercise) and is approximately equal to the number of protons,

$$n_{\rm baryons} \simeq n_p \simeq 10^{-6}\,{\rm cm^{-3}}\;, \tag{5.37}$$

from which we find (multiplying with the mass of the proton) the mass density of baryons,

$$\rho_{m,0} \simeq 10^{-30}\,{\rm g/cm^3}. \tag{5.38}$$

Then we have the ratio

$$\frac{\rho_{m,0}}{\rho_{\gamma,0}} \simeq 2000. \tag{5.39}$$

We also find (putting in the most precise value for the result)

$$\frac{N_\gamma}{N_{\rm baryons}} = \frac{n_\gamma}{n_{\rm baryons}} \simeq 1.66 \times 10^9. \tag{5.40}$$

But we know that the matter and radiation scale respectively as $\rho_m \propto R^{-3}$, $\rho_\gamma \propto R^{-4}$, therefore we have equality between the matter and radiation energy densities, $\rho_m \simeq \rho_\gamma$ at

$$\frac{R_0}{R} \simeq 2000\;, \tag{5.41}$$

which means at the temperature (the temperature scales inversely with the radius)

$$T \simeq 2000 T_0 \simeq 6000\,{\rm K}\;, \tag{5.42}$$

which incidentally, is about the temperature of the Sun. Also, since the scale factor R scales as $1/(1+z)$ (z is the redshift of the light coming from this time in the past), we have matter radiation equality at

$$1 + z_{eq} = \frac{R_0}{R} \Rightarrow z_{eq} \simeq 2000. \tag{5.43}$$

or, since $R \simeq t^{1/2}$ during the matter dominated (M.D.) era, we have

$$t_{eq} \simeq 1500\ {\rm years} \simeq 5 \times 10^{10}\,{\rm s}. \tag{5.44}$$

5.4 Particle Species in the Radiation Dominated (R.D). Era

Around the time of matter-radiation equality happens also recombination, the time when atoms form and the CMBR last scatters. After that, there are no more thermal events in the Universe, the only interesting events being the formation of galaxies and later stars.

So all of the thermal history of the Universe happens in the radiation dominated (R.D.) era. We have given the relation of $\rho(t)$ in the R.D. era before; writing it in numbers, we get

$$\rho_\gamma(g/cm^3) \simeq \frac{10^6}{t(s)^2}. \tag{5.45}$$

Another useful formula is the transformation between temperature (expressed in Kelvin or MeV) and time,

$$T(^0K) = \frac{10^{10}}{t(s)^{1/2}} \Rightarrow T(MeV) \simeq \frac{1\,\text{MeV}}{t(s)^{1/2}}. \tag{5.46}$$

As we said at the beginning of the chapter, at times $t < 1$ s, i.e., at temperatures $T > 10^{10}$ K $= 1$ MeV, the Universe is in equilibrium, since the reactions like $e^+e^- \leftrightarrow 2\gamma$ or $e^+e^- \leftrightarrow \nu_e\bar{\nu}_e$ occur very quickly, in a time much less than the age of the Universe t.

Before this time, there are many other particles in equilibrium, not just $e^\pm$, γ. The number of photons divided by the number of baryons was found in (5.40) to be about 10^9.

The entropy per baryon S_1 is conserved in the expansion of the Universe (since $S_1 = s/n_{\text{baryon}}$ and both the entropy per comoving volume s, as a result of adiabaticity, as we said, and the number density of baryons per comoving volume n_{baryon}, are constant). But we can rewrite (5.29) as

$$\frac{N_{\text{baryon}}}{N_\gamma} = \frac{n_{\text{baryon}}}{n_\gamma} \simeq 1.80 g_{*s} \frac{n_{\text{baryon}}}{s}, \tag{5.47}$$

and since $n_{\text{baryon}}/s = 1/S_1$ is conserved, it follows that $N_{\text{baryon}}/N_\gamma$ changes only when g_{*s} changes.

But g_{*s} changes due to particle-antiparticle annihilations ending when the temperature T drops below the rest mass of the particle E_0 (as a result, annihilations only happen with exponentially small probability, de facto stopping). In that case, the particle-antiparticle pairs disappear, they dump their entropy into the photons, increasing n_γ. The excess of particles over antiparticles becomes then nonrelativistic.

Using the relation between the energy in MeV, for the rest energy of the particle, and the temperature in Kelvin,

$$E(MeV) = k_B T \simeq 10^{-10} T(^0K), \tag{5.48}$$

(so that 10^{10} K = 1 MeV), we find the temperatures of various particles (when the annihilation stops and the remaining particles become nonrelativistic)

$$\begin{aligned} T_e &\simeq 5\times 10^9\,\text{K} \quad \text{for } m_e \simeq 0.5\,\text{MeV} \\ T_\mu &\simeq 10^{12}\,\text{K} \quad \text{for } m_\mu \simeq 100\,\text{MeV} \\ T_p &\simeq 10^{13}\,\text{K} \quad \text{for } m_p \simeq 1\,\text{GeV}. \end{aligned} \tag{5.49}$$

Thus, for $T < T_e$, electrons decouple from radiation, for $T < T_\mu$, muons do, and for $T < T_p$, protons do as well.

That also tells us how it came about that we have such high $N_\gamma/N_{\text{baryons}}$: the various baryons annihilate, and their numbers and entropies are changed into photons. Going back in time, all the N_γ would eventually correspond to $N_{\text{baryon}} = N_{\text{anti-baryon}}$. Indeed, at equilibrium for these baryons, we have $N_{\text{baryons}} \simeq \bar{N}_{\text{anti-baryons}} \simeq N_\gamma$. The (small) amount of baryons we see in the Universe today is only the excess of initial baryons over antibaryons, which therefore cannot annihilate (without it, the previous approximate equality would be exact, $\Delta N_{\text{baryons}} = \bar{N}_{\text{anti-baryons}}$. Then the initial baryon to antibaryon ratio is

$$\frac{N_{\text{baryon,in.}}}{\bar{N}_{\text{antibaryon,in.}}} \simeq \left(\frac{n_\gamma + n_{\text{baryon}}}{n_\gamma}\right)_0 \simeq \frac{10^9+1}{10^9}\,, \tag{5.50}$$

i.e., the initial baryon asymmetry was $1 : 10^9$ only.

Recombination

Recombination is the last scattering of photons, thus the time atoms form, without being dissociated by the scattering of photons. After it, the Universe becomes transparent to photons.

It occurs roughly when the mean free path of photons,

$$l_f \simeq \frac{1}{n\sigma} \simeq 10^2[t(s)]^{3/2}\,m\,, \tag{5.51}$$

equals the horizon size $l_{\text{hor}} \simeq ct$ (during the M.D. era, it is a factor of 2 larger, during the R.D. era a factor of 3, but in any case a factor of order one away from ct). Here n is the number density of photons, which during the M.D. era decays as $n = 10^{23}t^{-3/2}$, and σ is their cross section (now, $\sigma_0 \simeq 10^{-24}\,\text{cm}^{-2}$). In the last equality we have substituted these values for n and σ.

The time when the mean free path equals the horizon size (thus recombination happens, since the photons cannot interact anymore, not even as they travel the horizon distance) is

$$t^{1/2} = \frac{c}{10^2} \Rightarrow t = 10^{13}\,\text{s} \simeq 3\times 10^5\,years, \tag{5.52}$$

This corresponds to a temperature $T \simeq 4000$ K. A more precise calculation will be carried out in the next chapter.

Partial Summary

So we have $T_\gamma \simeq 4 \times 10^3$ K for recombination, after which we have $T_e \simeq 5 \times 10^9$ K, $T_\mu \simeq 10^{12}$ K and $T_p \simeq 10^{13}$ K. Using the formula $T(K) = 10^{10}/[\mathrm{t(s)}]^{1/2}$, we find also

$$t_e \simeq 4\,\mathrm{s}, \quad t_\mu \simeq 10^{-4}\,\mathrm{s}; \quad t_p \simeq 10^{-6}\,\mathrm{s}. \tag{5.53}$$

5.5 Neutrino Decoupling

The last part of the thermal history of the Universe to be analyzed in this chapter is the decoupling of neutrinos. It is found from their equilibrium reactions with the corresponding leptons.

ν_e Decoupling

The relevant reaction is

$$e^+e^- \leftrightarrow \nu_e\bar{\nu}_e. \tag{5.54}$$

The cross section for this reaction is (as it occurs through the electroweak interaction that reduces to the 4-Fermi one)

$$\sigma(e^+e^- \leftrightarrow \nu_e\bar{\nu}_e) = \frac{g_F^2 E^2}{\hbar c^4} \simeq 10^{-63} T^2\,, \tag{5.55}$$

where in the last equality we have considered ultrarelativistic particles. Then, since $n \propto T^3$, the characteristic time (relaxation time) of the reaction is

$$\tau = \frac{1}{\sigma n c} \simeq \frac{10^{51}}{T(^0K)^5} s \propto t^{5/2}. \tag{5.56}$$

As in the case of recombination, the reaction stops when the characteristic time of the reaction becomes larger than the age of the Universe

$$t \simeq \frac{10^{20}}{T^2(K)} s \tag{5.57}$$

(note that the relaxation time increases faster than the horizon, so eventually the annihilation reaction creating neutrinos stops). The temperature when it stops is

$$T \simeq 10^{31/3}\,\mathrm{K} \simeq 2 \times 10^{10}\,\mathrm{K}\,, \tag{5.58}$$

corresponding to a time of

$$t_{\nu_e} \simeq \frac{10^{20}}{10^{62/3}}\,\mathrm{s} \simeq 0.25\,\mathrm{s}. \tag{5.59}$$

Note that this is slightly before the e^+e^- annihilation (which happens at 4s).

ν_μ Decoupling

The idea is similar, except now, since $t_\mu \simeq 10^{-4}$ s, at times after that, the remaining muons are nonrelativistic. It will indeed turn out that ν_μ decoupling happens after that, so in the reactions

$$\begin{aligned} \mu^+ + \mu^- &\leftrightarrow \nu_\mu + \bar{\nu}_\mu \\ \mu^+ &\leftrightarrow e^+ + \bar{\nu}_\mu + \nu_e \\ \mu^- &\leftrightarrow e^- + \nu_\mu + \bar{\nu}_e \,, \end{aligned} \tag{5.60}$$

the muons are nonrelativistic. That means that the reaction times are exponentially amplified (the reactions happen with exponentially smaller probabilities), and one finds a relaxation time of

$$\tau = 2 \times 10^{-6} e^{\frac{m_\mu c^2}{k_B T}} = 2 \times 10^{-6} \times e^{\frac{106}{T(MeV)}} \,, \tag{5.61}$$

since $m_\mu = 106$ MeV. When the relaxation time is larger than the age of the Universe,

$$\tau > t = \frac{10^{20}}{T^2(K)} \,, \tag{5.62}$$

the reaction stops. This is a transcendental equation, which is solved by

$$T = 12\,\text{MeV} \simeq 10^{11}\,\text{K} \,, \tag{5.63}$$

corresponding to

$$t_{\nu_\mu} \simeq 0.01\,\text{s}. \tag{5.64}$$

ν_τ Decoupling

The same idea applies for the decoupling of ν_τ. Because of the exponential amplification of the relaxation time, one finds approximately the same temperature (and time) for the decoupling, despite now having $m_\tau \simeq 1778$ MeV, so

$$\tau = 2 \times 10^{-6} e^{\frac{1778}{T(MeV)}} > t = \frac{10^{20}}{T^2(K)} \,, \tag{5.65}$$

results in about the same time,

$$t_{\nu_\tau} \simeq 0.01\,\text{s}. \tag{5.66}$$

Neutrino Temperature

At $t \simeq 0.01$ s, the muons are nonrelativistic and decoupled, but γ, $e^\pm$ and all the neutrinos are not. Therefore we find the ratios

$$\rho_\gamma : \rho_{e^\pm} : \rho_{\nu_e,\bar{\nu}_e} : \rho_{\nu_\mu,\bar{\nu}_\mu} : \rho_{\nu_\tau,\bar{\nu}_\tau} : \rho_{\mu^\pm} =$$
$$\rho_1 : \frac{7}{4}\rho_1 : \frac{7}{8}\rho_1 : \frac{7}{8}\rho_1 : \frac{7}{8}\rho_1 : 10^{-4}\rho_1. \tag{5.67}$$

The ratio of photons to neutrinos is due to the fact that there are two photons for each e^+ and e^- ($e^+e^- \leftrightarrow 2\gamma$), whereas one a fermion like the electron has $7/8$ of a boson's energy density. Both of these (photons and electrons) have two polarizations. The ratio of electrons to the various neutrinos is given by the fact that the electrons have two polarizations, whereas the neutrinos have only one, and otherwise the reactions give a one to one ratio in number density. Finally, the relation to $\rho_{\mu^\pm}$ is given by the factor $e^{-\frac{106}{12}} = 10^{-4}$ for the exponential drop in energy density in the nonrelativistic case.

After neutrino decoupling, we still continue to have $T_\nu = T_\gamma$, simply since both temperatures decay in the same way with the expansion of the Universe, until e^+e^- annihilation, happening at $t = 4$ s, when the energy of the electron-positron pairs is added to the energy of the photons. After that, the two temperatures, now different, scale with the Hubble expansion in the same way until now. Since as we saw

$$S_\gamma : S_{e^\pm} = 1 : \frac{7}{4}, \tag{5.68}$$

and the photon total entropy today (the entropy is conserved in the expansion, so this is the same as after e^+e^- annihilation) is given by the sum of the two entropies,

$$S_0(\gamma) = S_1(\gamma) + S_1(e^\pm) = \frac{11}{4} S_1(\gamma). \tag{5.69}$$

Since as we saw, the entropy of relativistic species goes like T^4, T^4 increases in the same proportion, and we find that the temperature of the photons today (which scaled after e^+e^- annihilation in the same way as the temperature of the neutrinos) is

$$T_{0,\gamma} = \left(\frac{11}{4}\right)^{1/3} T_{1,\gamma}, \tag{5.70}$$

whereas $S_0(\nu) = S_1(\nu)$, so $T_{0,\nu} = T_{1,\nu} = T_{1,\gamma}$ (at the e^+e^- annihilation the electrons and neutrinos still had the same temperature),

$$T_{0,\nu} = T_{0,\gamma}\left(\frac{4}{11}\right)^{1/3} \simeq 2\,\mathrm{K}. \tag{5.71}$$

Important Concepts to Remember

- The Universe starts off in a hot Universe in thermal equilibrium, with the entropy per particle (number of photons per baryon, up to a numerical constant) $S_1 \gg 1$ and the lepton number $L_e \sim 1$.

- For ultrarelativistic species, $n \propto T^3$ and $n_{\rm Fermi} = 3/4 n_{\rm Bose}$, and $\rho \propto T^4$ and $\rho_{\rm Fermi} = 7/8 \rho_{\rm Bose}$, and $P = \rho/3$.
- For nonrelativistic species, $\rho = mn$ and $P = nk_BT \ll \rho$, and $n \propto T^{3/2} e^{-\frac{m-\mu}{k_BT}}$.
- For relativistic species in the expanding Universe, $\rho_R = (\pi^2/30) g_* T^4$, with g_* the number of effectively massless degrees of freedom, and $s = S/V = (2\pi^2/45) g_{*,s} T^3$, and $s = 1.80 g_{*,s} n_\gamma$, or $S_1 = 1.80 g_{*,s} N_\gamma / N_{\rm baryons}$.
- The temperature of the CMBR now is $T \simeq 2.75K$, and $N_\gamma / N_{\rm baryons} \sim 1.6 \times 10^9$ now.
- We have ρ_m/ρ_γ now of about 2000, so we are in the M.D. era, and equality with radiation was at $z_{\rm eq} \simeq 2000$.
- Since now $N_\gamma/N_{\rm baryons} \sim 10^9$, and photons were baryon-antibaryon pairs originally, the baryon asymmetry was originally of about 10^{-9}, i.e. $N_{\rm baryon}/N_{\rm antibaryon} \sim (10^9 + 1)/10^9$.
- Neutrinos decouple, but continue on a thermal distribution that scales as usual with $a(t)$.
- The temperature of the neutrinos now is $T_{\nu,0} \simeq T_{\gamma,0}(4/11)^{1/3} \simeq 2\,{\rm K}$.

Further reading: See Chap. 3 in [1] for more details.

Exercises

(1) Calculate g_* and $g_{*,s}$ now, from the known particles available.
(2) Estimate the total number of particles and the total number of baryons in the Universe.
(3) Calculate the total specific heat C_V in the Universe today.
(4) Given the thermal history of the Universe today, and the current matter composition, calculate the particle horizon as a function of H_0, the Hubble constant today.
(5) Calculate the number densities of ν_e, ν_μ, ν_τ in the Universe today, using the information in the text.

Chapter 6
Big Bang Nucleosynthesis and Recombination

In this chapter we will describe Big Bang Nucleosynthesis (BBN), the combination of fundamental protons and neutrons into nuclei, as well as calculate more precisely the temperature of recombination. We will start with that, since we will generalize it to the BBN case next. Then we will describe how BBN can be used to test ideas about the early Universe, since the primordial abundances, defined at BBN, depend on various cosmological parameter, and we can obtain experimental values for them.

6.1 Recombination Temperature

We consider the (chemical) reaction

$$p + e^- \leftrightarrow H + \gamma\,, \tag{6.1}$$

i.e., the recombination versus ionization of hydrogen. The reaction is in equilibrium (can occur both ways), and it has the ionization energy $I = 13.6\,\text{eV}$, the negative of the energy of the fundamental electron level in hydrogen. We will write the equation for the equilibrium, known as the *Saha equation*. It is based on the equation for the number density of a nonrelativistic species, found in the previous chapter,

$$n_i = g_i \left(\frac{2\pi m_i k_B T}{h^2}\right)^{3/2} e^{\frac{\mu_i - m_i}{k_B T}} = g_i \left(\frac{m_i k_B T}{2\pi\hbar^2}\right)^{3/2} e^{\frac{\mu_i - m_i}{k_B T}}\,. \tag{6.2}$$

Here g_i is the degeneracy factor of species i, and m_i its mass. Consider that in the reaction, the chemical potentials are the same on both sides of the equation, $\mu_p + \mu_e = \mu_H$, so by calculating the ratio $n_p n_e / n_H$, using that $m_p \simeq m_H$ ($m_p \simeq 1800 m_e$), we find

$$\frac{n_e n_p}{n_H} = \frac{g_e g_p}{g_H} \left(\frac{m_e k_B T}{2\pi\hbar^2}\right)^{3/2} e^{-\frac{I}{k_B T}}\,. \tag{6.3}$$

H. Năstase, *Cosmology and String Theory*, Fundamental Theories of Physics 197, https://doi.org/10.1007/978-3-030-15077-8_6

Since $g_e = g_p = 2$ (they are both spin 1/2 fermions) and $g_H = 4$ (there are 4 states) and putting $\hbar = 1$, we have

$$\frac{n_e n_p}{n_H} = \left(\frac{m_e k_B T}{2\pi}\right)^{3/2} e^{-\frac{I}{k_B T}}. \tag{6.4}$$

Then we define the ratio

$$X_e = \frac{n_p}{n} = \frac{n_p}{n_p + n_H} \tag{6.5}$$

and the inverse of the photon to baryon ratio,

$$\frac{n}{n\gamma} \equiv \eta\,, \tag{6.6}$$

which means that (since $n_p = n_e$ because of the neutrality of the Universe)

$$\frac{X_e^2}{1 - X_e} = \frac{n_p n_e}{(n_p + n_H) n_H} = \frac{n_p n_e}{n_H} \eta^{-1} \frac{1}{n_\gamma}. \tag{6.7}$$

Using the formula for relativistic bosons from the previous chapter, $n_\gamma = (\zeta(3)/\pi^2) g T^3 = (2\zeta/\pi^2) T^3$, since $g_\gamma = 2$, we find

$$\frac{X_e}{1 - X_e} = \frac{\sqrt{\pi}}{4\sqrt{2}\zeta(3)} \eta^{-1} \left(\frac{T}{m_e}\right)^{-3/2} e^{-\frac{I}{k_B T}}. \tag{6.8}$$

Using $n_0 = 10^{-6}\,\text{cm}^{-3}$, $T_0 = 2.75\,\text{K}$, we can find

$$\eta = 2.7 \times 10^{-8} (\Omega_B h^2)\,, \tag{6.9}$$

and replacing it in the above equation, we find

$$\frac{X_e^2}{1 - X_e} = 4 \times 10^{21} (\Omega_0 h_0^2)^{-1} T(K)^{-3/2} e^{-\frac{1.58 \times 10^5}{T(K)}}. \tag{6.10}$$

We can plot the resulting $X_e = X_e(T)$ and find that it drops to zero (we have total recombination of p's into H's) when the temperature drops through

$$T \simeq 4000\,\text{K}. \tag{6.11}$$

That corresponds to $T \simeq 0.3\,\text{eV}$, a redshift of

$$z_{\text{rec}} \simeq 1300 \tag{6.12}$$

and a time of

$$t_{\rm rec} \simeq 10^{13}\ {\rm s}. \tag{6.13}$$

6.2 Big Bang Nucleosynthesis

It means the period of formation of nuclei, that gives origin at the various nuclear species observed today. To do that, we write a generalization of the Saha equation for the nuclear statistical equilibrium of nonrelativistic species. Starting with the same relation as before, now for a nuclear species $A(Z)$ (of atomic number A and charge Z)

$$n_A = g_A \left(\frac{2\pi m_A k_N T}{h^2}\right)^{3/2} e^{\frac{\mu_A - m_A}{k_B T}}. \tag{6.14}$$

The chemical reaction of nucleosynthesis is the formation of the nucleus from Z protons and $A - Z$ neutrons,

$$Z\ \ p's + (A - Z)\ \ n's \leftrightarrow A(Z). \tag{6.15}$$

Assuming moreover, as for recombination, that the chemical potentials on the two sides match,

$$\mu_A = Z\mu_p + (A - Z)\mu_n\ , \tag{6.16}$$

we find the equivalent of the Saha equation,

$$\frac{n_A}{(n_p)^Z (n_n)^{A-Z}} = g_A A^{3/2} 2^{-A} \left(\frac{h^2}{2\pi m_N k_B T}\right)^{\frac{3(A-1)}{2}} e^{\frac{I_A}{k_B T}}. \tag{6.17}$$

Here m_N is the nucleon mass, $m_p \simeq m_n = m_N$, and $m_A \simeq A m_N$, and I_A is the activation energy (equivalent to ionization). The total number of nucleons is

$$n_N = n_n + n_p + \sum_i (A n_A)_i\ , \tag{6.18}$$

and define the ratios

$$X_A \equiv \frac{A n_A}{n_N} \tag{6.19}$$

and X_n, X_p, such that $\sum_i X_i = 1$, we find for the fraction of species $A(Z)$

$$X_A = g_A \left[\zeta(3)^{A-1} \pi^{\frac{1-A}{2}} 2^{\frac{3A-5}{2}}\right] A^{5/2} \left(\frac{T}{m_N}\right)^{\frac{3(A-1)}{2}} \eta^{A-1} (X_p)^Z (X_n)^{A-Z} e^{\frac{I_A}{k_B T}}. \tag{6.20}$$

We can use this equation to find the fractions of various species produced during BBN, but for that we must find the initial conditions (the original numbers of fractions X_p and X_n when BBN starts).

Initial Conditions for BBN

Before BBN, the balance between protons and neutrons was maintained by weak interactions, through the reactions

$$\begin{aligned} n &\leftrightarrow p + e^- + \bar{\nu}_e \\ \nu + n &\leftrightarrow p + e^- \\ e^+ + n &\leftrightarrow p + \bar{\nu}_e. \end{aligned} \tag{6.21}$$

The first reaction is the decay of the neutron outside a nucleus, that at equilibrium can occur in both directions, the second is the reaction of a neutron with a neutrino to change into a proton and electron, and is characterized by the difference in energy

$$Q = m_n c^2 - m_p c^2 = 1.293\,\text{MeV}. \tag{6.22}$$

The reaction rate at high temperatures, $T \gg Q, m_e$, for the latter reaction is

$$\Gamma_{pe\to\nu n} \to G_F^2 T^5\ , \tag{6.23}$$

leading to the reaction characteristic time mentioned the previous chapter,

$$\tau(s) = \frac{1}{\Gamma} \simeq \frac{10^{51}}{T(K)^5}. \tag{6.24}$$

The reaction *freezes out* when τ becomes larger than the age of the Universe, $t(s) \sim 10^{20}/T^2(K)$. The resulting *freeze-out time* is almost the same, but a bit before, as ν decoupling,

$$T_* \simeq 0.8\,\text{MeV}. \tag{6.25}$$

At this time, both n and p are very nonrelativistic ($T \ll m_p, m_n$), and then the formula (6.2) applies, which leads to a ratio of n to p given by Q,

$$\left(\frac{n_n}{n_p}\right)_* = e^{-\frac{Q}{k_B T}} \simeq e^{-1.5} \simeq \frac{1}{5}. \Rightarrow X_{n,*} \simeq \left(\frac{n_n}{n_n + n_p}\right)_* \simeq \frac{1}{6} \tag{6.26}$$

More precisely, $X_{n,*} \simeq 0.1609$. After freeze-out, the only thing that still happens is that the neutron naturally decays to $p + e^- + \bar{\nu}_e$. Outside the nucleus, the decay occurs with a lifetime that is known surprisingly badly,

$$\tau_n = 10.5 \pm 0.2\,\text{min}. \tag{6.27}$$

The decay of the neutron happens until a time of about $t = 1 - 3$ min, or about $T \simeq 0.3 - 0.1$ MeV, when the temperature falls below the binding energy of Deuterium $D = {}^2H$, which is the first nucleus to form. Then we can use (6.20) with given n_n and n_p to find all other X_A's. The *boundary conditions*, n_n and n_p, are, at this time (D decoupling)

$$X_{n,D} \simeq \frac{1}{7}, \tag{6.28}$$

more precisely, $X_{n,D} \simeq 0.135$.

Around BBN, the number of effectively massless degrees of freedom, g_*, changes. Before BBN and until freeze-out $T = 0.8$ MeV,

$$g_* \equiv \sum_{\text{bosons}} g_i \left(\frac{T_i}{T}\right)^4 + \frac{7}{8} \sum_{\text{fermions}} g_i \left(\frac{T_i}{T}\right)^4, \tag{6.29}$$

and all the massless particles have the same temperature. In the Standard Model, we have $e^\pm$, γ and ν that are effectively massless with the same temperature $T_i = T$, and the degeneracies are $g_\gamma = g_{e^\pm} = g_\nu = 2$ (but for the latter, considering ν and $\bar{\nu}$), which leads to

$$g_* = 2\left(1 + \frac{7}{8}(3+2)\right) = 10.75. \tag{6.30}$$

Here the 3 is from the ν_e (and $\bar{\nu}_e$), ν_μ (and $\bar{\nu}_\mu$), ν_τ (and $\bar{\nu}_\tau$), and the 2 is for e^+ and e^-.

But just before the nuclei start to form, more precisely around $T = 0.5$ MeV, when we have e^+e^- annihilation (after which e^- become norelativistic), as we saw in the previous chapter, the entropy of the $e^\pm$ is transferred to the photons γ, which increases T_γ, and leads to a relation of the ν temperature with respect to $T \equiv T_\gamma$ of $T_\nu/T = (4/11)^{1/3}$, so that (there are no more e^+ and e^-, and the neutrinos have T_ν)

$$g_* = 2 + 2\frac{7}{8}3 \times \left(\frac{4}{11}\right)^{4/3} = 3.36. \tag{6.31}$$

At $t =$ 1–3 min, deuterium D starts to form, and after that, all the other nuclei form.

Once the deuterium starts to form, it starts a reaction chain

$$\begin{aligned} p + n &\to D + \gamma \\ D + D &\to {}^3H + p \\ D + D &\to {}^3He + n \\ {}^3H + D &\to {}^4He + n \\ {}^3He + n &\to {}^3H + p. \end{aligned} \tag{6.32}$$

This chain, as explained in the first chapter already, ends with all the neutrons trapped inside ^4He, the endpoint of the chain. That means that the *mass* abundance of ^4He is given by twice the abundance of neutrons, i.e.,

$$Y = 2\left(\frac{n_n}{n_n + n_p}\right)_* \simeq 0.27. \tag{6.33}$$

For each neutron trapped inside ^4He, there is a proton as well (^4He contains two neutrons and two protons), so the proton content drops as well.

Then at the end of the chain, there is some deuterium D and some ^3He left unburnt by the chain, specifically

$$\frac{D}{H}, \frac{^3He}{H} \sim 10^{-5} - 10^{-4}. \tag{6.34}$$

Moreover, using (6.20), we can see that also some ^7Li gets synthesized, namely

$$\frac{^7Li}{H} \sim 10^{-10} - 10^{-9}. \tag{6.35}$$

So we can plot the relative abundances of protons and neutrons and ^4He. The proton and neutron abundances start (at very small time) at 0.5 each, after which the n abundance drops, as the proton abundance raises correspondingly ($X_p = 1 - X_n$), since the mass of the neutron is slightly larger (so it decays more). As freeze-out ($T = 0.8\,\text{MeV}$), $X_n : X_p = 1 : 5$. Then until BBN, X_n drops very slowly, until $X_n : X_p = 1 : 6$. Then BBN starts, and all the X_n drops to zero, whereas $X_{^4He}$ rises (all neutrons are trapped in ^4He). In the same time, X_p drops by the same amount as X_n.

6.3 Dependence of BBN Primordial Abundances on Parameters

The primordial abundances turn out to depend on several parameters:

- the lifetime of the neutron $\tau_{1/2}(n)$. As we said, surprisingly, the neutron lifetime is known only with a large error. But since the reaction time for the $n + \nu \leftrightarrow p + e^-$ reaction, $\tau \simeq \frac{1}{\Gamma} \propto \tau_{1/2}(n)$. Since moreover the freeze-out occurs when $\tau \propto 1/T^5$ becomes larger than the age of the Universe, $t \propto 1/T^2$, it follows that the freeze-out temperature depends on the lifetime of the neutron as

$$T_* \propto [\tau_{1/2}(n)]^{1/3}. \tag{6.36}$$

Since the ^4He abundance Y depends drastically on T_*, it also depends on $\tau_{1/2}(n)$.

- the number of effectively massless degrees of freedom g_*. Since the energy density $\rho \propto g_* T^4$ (from the definition of g_*), and from the Friedmann equation $\rho \propto H^2$, it means that $H \propto g_*^{1/2} T^2$. But on the other hand, the age of the Universe is, up to a numerical constant of order one, $t \sim H^{-1}$, which means $t \sim g_*^{-1/2}$. But as above, $T_* \propto (\tau \sim t)^{-1/3}$, so
$$T_* \propto (g_*)^{1/6}. \tag{6.37}$$
We will shortly see that this gives a bound on the number of neutrino species N_ν.
- the ratio of the number of baryons to number of photons η. Indeed, we saw in (6.20) that $X_A \propto \eta^{A-1}$. Also, the amount of deuterium D and ^3He left unburnt after the neutrons are trapped into ^4He depends on η as $\propto \eta^{-n}$, with $n \sim 1-2$.
- also, the abundances depend on $\Omega_B h^2$, but from CMBR (that also depends on $\Omega_B h^2$) the latter is fixed rather well.

The various equations for the nuclear reactions, including the dependences above, is included into a numerical BBN code.

6.4 Observations and Comparison with BBN

We now turn to comparison with observations. We observe a deuterium abundance is
$$\frac{D}{H} \sim 2.78 \pm 0.4 \times 10^{-5}, \tag{6.38}$$
which from the BBN code implies that
$$\eta = 5.9 \pm 0.5 \times 10^{-10}. \tag{6.39}$$
Then, the observed ^3He abundance is
$$\left.\frac{^3He}{H}\right|_{\text{obs.}} \sim 1.1 \pm 0.2 \times 10^{-5}, \tag{6.40}$$
but it is not clear that this is a *primordial* abundance, since helium is also produced in stars, which burn hydrogen and create helium (like an H bomb). For the observed value of $\Omega_B h^2 = 0.0214$ (from the CMBR), the BBN code gives $^3He/H \simeq 1.04 \pm 0.06 \times 10^{-5}$, which agrees well with the observations, so perhaps all the observed amount above is of primordial origins.

^7Li Problem

From observations also, we see a Lithium 7 abundance of
$$\left.\frac{^7Li}{H}\right|_{\text{obs}} \simeq (2.07 + 0.15/-0.04) \times 10^{-10}. \tag{6.41}$$

But this is considerably less than the value obtained from the BBN code for $\Omega_B h^2 = 0.0214$, of about 3×10^{-10}. This is the Lithium 7 problem. There are ways to fix this, but at the expense of some other observable. For instance, the graph of the abundance of Lithium 7 against η has a minimum of 2×10^{-10}, but at an $\eta \sim 3 \times 10^{-10}$, which conflicts with the value extracted from the deuterium D abundance.

Number of Neutrino Species

From the BBN code, we can also constrain the g_* dependence, and the result is

$$N_\nu \geq 4 , \tag{6.42}$$

so at most four *light* neutrino species, whereas we know there are at least 3 light neutrino species $(\nu_e, \nu_\mu, \nu_\tau)$, $N_\nu \geq 3$.

Constraint on the Number of Massless Energy Modes

BBN constrains also, similarly to the g_* dependence, energy modes present before the creation of the Standard Model matter, usually graviton and inflaton modes, though others are also constrained. The constraint is

$$f_{BBN} = \frac{\rho_{\text{graviton}} + \rho_{\text{inflaton}} + \cdots}{\rho_{\text{matter}}} \leq 0.07. \tag{6.43}$$

But this constraint is easily satisfied in (most) inflationary models, as we will see later on in the book.

6.5 Summary of the Thermal History

We now summarize the thermal history of the Universe that we learned in this chapter and the previous one, adding a few events. The most common model of cosmology is based on inflation, which is usually at some high energy scale, a bit below the string scale E_s, itself usually a bit below the Planck scale E_{Pl}. The inflationary scale needs to be smaller than the string scale, since inflation is defined as an effective field theory involving a four dimensional scalar, and in the spirit of effective field theory, it needs to be valid below the cut-off, in this case string theory. There is also the a priori possibility that an ensemble of string modes gets excited at some large scale, like in the case of the only truly stringy cosmological model, the string gas model of Bradenberger and Vafa, which we will study in part III of the book. But otherwise, the string states are not excited, and we are in effective field theory.

- At $E \sim 10^{15}$ GeV, corresponding to $T \sim 10^{27}$ K, we can have a Grand Unified Theory (GUT) phase transition. From experimental evidence, there is a very strong probability that there is such a GUT scale, though it is not always so.
- There could be scales associated with other transitions, like a susy breaking scale M_{SSB}, etc.

- At $E = 300\,\text{GeV}$, corresponding to $T \sim 3 \times 10^{15}\,\text{K}$, we have the electroweak phase transition.
- At $E = m_p \simeq 900\,\text{MeV}$, the protons become nonrelativistic (protons and antiprotons annihilate).
- At $E \simeq 200\,\text{MeV}$ or a bit below, we have the QCD phase transition.
- At $E = m_\mu \simeq 100\,\text{MeV}$, the muons become nonrelativistic (muons and antimuons annihilate).
- At $E \simeq 10\,\text{MeV}$ ($T \simeq 10^{11}\,\text{K}$), both ν_μ and ν_τ decouple.
- At $E \simeq 1\,\text{MeV}$ ($T \simeq 10^{10}\,\text{K}$), ν_e decouple.
- At $E = 0.8\,\text{MeV}$, we have the freeze-out of the neutron/proton reaction.
- At $E = m_e = 0.5\,\text{MeV}$, the electrons become nonrelativistic, i.e., e^+e^- annihilation.
- Then, the BBN occurs, with the formation of light elements.
- The end of the R.D. era happens at $T \simeq 10^5\,\text{K}$, corresponding to a redshift of $z \sim 2 \times 10^4$, or a time of $t \sim 1.5 \times 10^3$ years = 6×10^{10} s.
- After it, we have M.D., with $T \propto t^{-2/3}$ ($\rho \propto 1/t^2 \propto 1/R^3 \propto T^3$).
- Recombination and the appearance of the CMBR (last scattering) happens at a temperature $T \sim 4000\,\text{K}$, corresponding to an energy $E \sim 0.5\,\text{eV}$, or a time $t \sim 5 \times 10^{12}\,\text{s} \sim 3 \times 10^5$ years, or a redshift of $z \simeq 1300$.

Important Concepts to Remember

- The recombination temperature (for atoms) is found from the Saha equation, when $X_e(T) = n_p/n$ drops to zero. One finds $T \simeq 4000\,\text{K}$, $z_{\text{rec}} \simeq 1400$ and $t_{\text{rec}} \sim 10^{13}$ s.
- One writes down Saha equations for nucleons to form nuclear species $A(Z)$, for Big Bang Nucleosynthesis.
- Before BBN, the initial condition is as follows. The protons an neutrons are in equilibrium, with $(n_n/n_p) \simeq e^{-Q/k_B T}$ that freezes out at about 1/5. Then neutrons decay into protons a bit more, until the ratio is about 1/6.
- Just before BBN, g_* drops from 10.75 to 3.36.
- The ${}^4He$ to H ratio created in BBN is about 0.27, with D and ${}^3He$ of about $10^{-5} - 10^{-4}$, and ${}^7Li$ of about $10^{-10} - 10^{-9}$.
- Primordial abundances depend on: the neutron lifetime $\tau_{1/2}(n)$, the number of effectively massless degrees of freedom g_*, the number of photons per baryon $1/\eta$, $\Omega_B h^2$.
- We have a ${}^7Li$ problem: can't fit it together with the rest (only by itself).
- We constrain the number of neutrino species from BBN.
- We constrain the number of massless energy modes, like gravitons and inflaton modes, to $\rho_{\text{modes}}/\rho_{\text{matter}} < 0.07$.

Further reading: See Chap. 4 in [1] for more details.

Exercises

(1) Calculate the number of photons (N_γ) and entropy in our Universe at recombination and at BBN.

(2) Calculate ρ_γ/ρ_m at BBN and at the electroweak phase transition.
(3) Calculate the ratio $\frac{a_{GUT}}{a_0}$, at the GUT phase transition to today, assuming that standard cosmology follows throughout.
(4) Calculate the number g_* of effectively massless degrees of freedom just below the GUT scale, assuming that only the Standard Model particles are present (and a right-handed neutrino).
(5) Calculate the number density of baryons at recombination.

Chapter 7
The Cosmic Microwave Background Radiation (CMBR) Anisotropy

In this chapter we will describe the Cosmic Microwave Background Radiation and its anisotropy. After describing the CMBR kinematics, we will see how primordial perturbations, in particular curvature perturbations, that were already observed, are described and propagate, and how they influence the CMBR.

7.1 The CMBR and Its Anisotropy

The Cosmic Microwave Background Radiation (CMBR) is a perfect blackbody spectrum, i.e., is distributed according to the Planck distribution,

$$n(\nu)d\nu = \frac{8\pi\nu^2 d\nu}{e^{\frac{h\nu}{k_B T}} - 1} , \tag{7.1}$$

present throughout the Universe. As we already mentioned, in the expanding Universe, the Planck distribution remains valid, but the temperature is redshifted like all scales, $T = T(t)$ or rather $T = T(a)$.

The CMBR was discovered in 1965, accidentally, by Arno Penzias and Robert Woodrow Wilson, but it already had a rather long history. It was first postulated by George Gamow in the 1940s, who, as we remember from Chap. 1, had proposed the Hot Universe model, that there should be a thermal blackbody radiation throughout the Universe. He didn't give a definite temperature, but later in the 1940s gave several estimates around 50 K. Then in the late 1940s and early 1950s, Ralph Alpher and Robert Herman, gave an estimate of $T = 5$ K. Robert Dicke estimated it also in the 1950s. The first *published* paper describing it as a *detectable* phenomenon was by A.G. Doroshkevich and Igor Novikov in 1964, but in a Soviet journal, so it was not common knowledge. Jim Peebles, in 1965, had a calculation and a preprint, and based on it, Robert Dicke, Jim Peebles, Roll and David Wilkinson (the one for whom the Wilkinson Microwave Anisotropy Probe, or WMAP, was named) set out

H. Năstase, *Cosmology and String Theory*, Fundamental Theories of Physics 197, https://doi.org/10.1007/978-3-030-15077-8_7

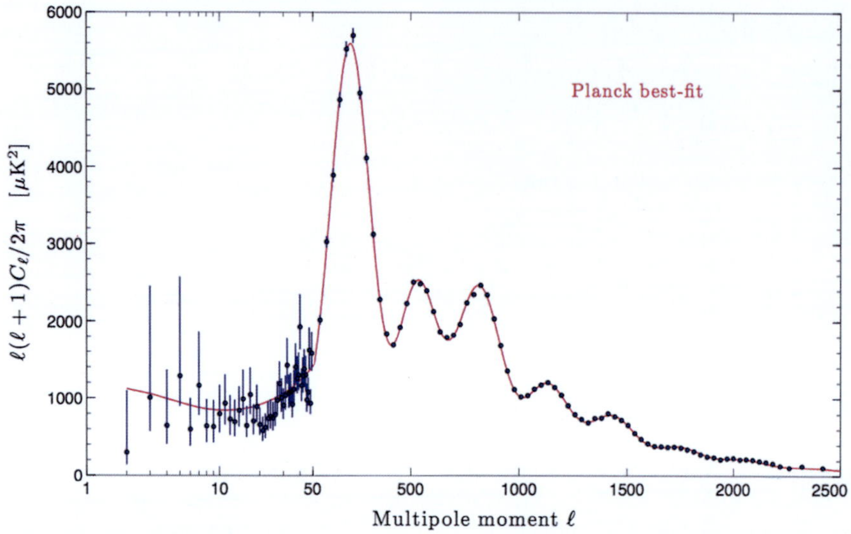

Fig. 7.1 The CMBR spectrum of fluctuations, plotting $l(l+1)C_l/(2\pi)$, quantifying correlations of $\delta T/T$ across an angular scale defined by the angular momentum l, against this l (Image: ESA)

to search for the CMBR, but before they could find it, it was accidentally discovered by Penzias and Wilson, who simply discovered this background radiation that came from extragalactic sources, but had no knowledge of what it means, being unaware of all the history described here. They later learned about the Peebles preprint and the importance of their discovery.

It is found that the CMBR is a blackbody spectrum with amazing accuracy, and the temperature $T(\vec{\Omega})$ seems to be approximately constant as a function of the direction $\vec{\Omega}$ in the sky. However, later on it was found that it is not actually constant, but there is a small *anisotropy* (variation with angle) $\delta T(\vec{\Omega})/T$ of the order of 10^{-5}. The anisotropy is found by the COBE satellite in the 1990s, which is why the absolute value of $\delta T/T$ is still known as the *COBE normalization*. Later, especially with the WMAP and Planck satellites, the CMBR anisotropy was mapped with great precision, and it was found that it exhibits a very rich structure, full of information about the Big Bang.

One plots a quantity $l(l+1)C_l/(2\pi)$, as in Fig. 7.1, roughly describing correlation functions over angular momentum l, against this angular momentum l. There is only one Universe, so the "ensemble" average described here corresponds to the fact that we have divide the sky in many angular distances $\theta \sim 1/l$, and we can correlate them. Indeed, as we will see later, at large l, we can understand l as $\sim 2\pi/\theta$, where θ is the angle. As a result, from simple statistics, the error of the "measurement" $\sim 1/\sqrt{N} \sim 1/\sqrt{l}$, so it is very large at small l. The CMBR plot starts with a plateau, the Sachs-Wolfe plateau (which at small l has huge errors), and has a small l value depending on the Integrated Sachs-Wolfe (ISW) effect. Then we have acoustic peaks, which are damped, so we have a damping tail. This is literally the same as the acoustic damped sinusoidal wave of sound propagation, as we will see later in the chapter.

7.2 Kinematics of the CMBR

The temperature anisotropy is a function of the conformal time η, the spatial position $\vec{x}$ (in our case, the Earth's position in the Universe; if we would have a fast spaceship capable of taking us to a different galaxy, maybe we could vary it, but as it is, we can't), and the angular direction $\vec{n} = \vec{e}$, the direction of the momentum of the photons incoming to us (to the detector), i.e., the angular direction in the sky. We call it

$$\Theta(\eta, \vec{x}, \vec{e}) = \frac{\delta T(\eta, \vec{x}, \vec{e})}{T}. \tag{7.2}$$

But we actually measure intensities of the waves, and because of the Stefan–Boltzmann law, we have $I \propto T^4$, so we actually measure

$$\frac{\delta I}{I} = 4\Theta. \tag{7.3}$$

We expand Θ in multipoles (angular momentum) at each point in spacetime,

$$\Theta(\eta, \vec{x}, \vec{e}) = \sum_{lm} \Theta_{lm}(\eta, \vec{x}) Y_{lm}(\vec{e}). \tag{7.4}$$

Monopole Anisotropy ($l = 0$)

The monopole, i.e., $l = 0$, Θ_{00}, is a local fluctuation (at the position $\vec{x}$ of the Earth), but independent on the direction in the sky. As such, it can't be measured, since all the detectors are at the same position $\vec{x}$ (on the Earth); only if we had a spaceship that could take us to a distant galaxy, we could. This monopole is due to a density perturbation at the current position. Since $\rho_\gamma \propto T^4$, $\Theta = \delta T/T = 1/4\delta\rho_\gamma/\rho_\gamma$, so

$$\Theta_{00}(\eta, \vec{x}) = \frac{1}{4}\frac{\delta\rho_\gamma}{\rho_\gamma}(\eta, \vec{x}). \tag{7.5}$$

Dipole Anisotropy ($l = 1$)

In the CMBR, there is a dipole anisotropy, i.e., for $l = 1$, that is associated with the motion of the observer, i.e. the motion of the Earth, relative to the photon fluid, for CMBR. Note that we are in general relativity, with FRLW expanding metric $ds^2 = -dt^2 + a^2(t)d\vec{x}^2$, instead of the Minkowski metric of special relativity, which means that all frames of reference related by boosts are not equal. There is the special frame of the CMBR, or the *cosmic frame* defined by the time t of the FRLW metric. Indeed, boosting the FRLW metric, we change it, unlike the case of the Minkowski metric.

The terms involving $l = 1$ in the expansion of Θ define the dipole,

$$\sum_m \Theta_{1m} Y_{1m}(\vec{e}) \simeq -\vec{v} \cdot \vec{e} + \cdots = -v \cos\theta + \cdots , \tag{7.6}$$

where θ is the angle of the velocity with respect to the angle of observation. Since the energy of the photon gets boosted according to the law

$$E = \gamma(1 + \beta \cos\theta) E' , \tag{7.7}$$

it means that the temperature gets changed by

$$T' = \frac{T}{\gamma(1 + \beta \cos\theta)}. \tag{7.8}$$

That gives an expansion in terms of Legendre polynomials,

$$\frac{\delta T}{T} \simeq -\beta P_1(\cos\theta) + \frac{2\beta^2}{3} P_2(\cos\theta) - \frac{\beta^2}{6} + \cdots \tag{7.9}$$

By measuring the dipole, we find the velocity of the motion of the Earth relative to the CMBR,

$$v \simeq 371\,\text{km/s} = 1.2 \times 10^{-3} c. \tag{7.10}$$

But as we saw from the above expansion, besides the dipole (defined by $P_1(\cos\theta)$, we have also contributions at higher $l \geq 2$, with $P_l(\cos\theta)$ induced by the same effect. Just that these contributions are $\mathcal{O}(10^{-6})$, much smaller than the observed 10^{-5} in all l's. That means that the CMBR anisotropy is not just from the dipole, there is really an intrinsic anisotropy of the CMBR.

CMBR Spectrum

We will define the shorthand

$$\Theta_{lm}(\eta_0, \vec{x}_0) \equiv a_{lm}. \tag{7.11}$$

We can consider the "ensemble" average $\langle\rangle$, found by averaging over the sky. We find that

$$\langle a_{lm} \rangle = 0 , \tag{7.12}$$

within experimental accuracy, from which we deduce that the Universe is isotropic. Reversely, from rotational invariance (isotropy), we find we must have $\langle a_{lm} \rangle = 0$. Then, under the same assumption of rotational invariance (isotropy), we find that the two-point function depends only on functions C_l,

$$\langle a_{lm} a_{l'm'} \rangle = \delta_{ll'} \delta_{mm'} C_l. \tag{7.13}$$

Substituting in $\Theta(\vec{e})$, we find

$$\langle\Theta(\vec{e}_1)\Theta(\vec{e}_2)\rangle = \sum_l \frac{2l+1}{4\pi} C_l P_l(\cos\theta). \tag{7.14}$$

We can define the variation of C_l,

$$(\Delta C_l)^2 \equiv \langle(|a_{lm}|^2 - C_l)^2\rangle = \langle|a_{lm}|^4\rangle - C_l^2. \tag{7.15}$$

For a *Gaussian* distribution, one obtains

$$(\Delta C_l)^2 = \frac{2C_l^2}{(2l+1)\Delta l}, \tag{7.16}$$

since in the numerator the factor of 2 comes from the real and imaginary parts of a_{lm}, and in the denominator $2l+1$ is the number of m's for a_{lm} at given l, and Δl is the binning of l (it is natural to take $\Delta l = 1$, but we can have other choices).

Meaning of Expansion in l and m

The multipole expansion at large l is like a Fourier transform, i.e., is of the type

$$\Theta(\vec{\theta}) = \frac{1}{2\pi}\int d^2\vec{l}\, a(\vec{l})e^{i\vec{\theta}\cdot\vec{l}}, \tag{7.17}$$

so the multipole (angular momentum) number l explores an angle on the sky of

$$\theta \sim \frac{1}{l}. \tag{7.18}$$

The CMBR fluctuations we will be interested in come from given tensor modes. To find the contributions of the various tensor modes, we do a Fourier expansion in the angle of rotation around the direction of the photon momentum $\vec{k}$ (or direction of observation $\vec{e}$). Indeed, the behaviour of a given tensor corresponds to a certain maximum m in the expansion of quantities

$$G(\phi) = \sum_{m\leq m_{max}} G_m e^{2\pi i m\phi}. \tag{7.19}$$

Then m_{max} corresponds to the tensor rank: $m_{max} = 0$ is a scalar, $m_{max} = 1$ is a vector, $m_{max} = 2$ is a tensor, etc.

Scalar Mode

But we actually only expect scalar, vector and tensor modes anyway for the perturbations. Moreover, we will focus on the scalar mode in this chapter, since we have a lot of experimental data on it, though there could be also vector and tensor modes, but there is no data yet on this.

Then for the scalar mode, we will only have the $m = 0$ modes, i.e., the $Y_{l0}(\vec{e})$ spherical harmonics. The expansion of Θ for scalar mode contributions only is then, when considering a Fourier expansion over $\vec{x}$ (position), to $\vec{k}$ (momentum)

$$\Theta(\eta, \vec{k}, \vec{e}) = \sum_l (-i)^l \sqrt{4\pi(2l+1)} Y_{l0}(\vec{e}) \Theta_l(\eta, \vec{k})$$
$$= \sum_l (-i)^l (2l+1) P_l[\cos(\hat{k} \cdot \vec{e})]. \tag{7.20}$$

The reverse of this expansion is

$$\Theta_{lm}(\eta, \vec{x}_0 = 0) \equiv a_{lm} = \frac{4\pi}{(2\pi)^3} i^l \int d^3\vec{k} \Theta(\eta_0, \vec{k}) Y^*_{lm}(\vec{k}) , \tag{7.21}$$

and for the scalar mode we have only $\Theta_l = \Theta_{l0} = a_{l0}$. The correlation function in this case is

$$\langle a_{l0} a_{l'0} \rangle = \delta_{ll'} C_l. \tag{7.22}$$

Gaussian Perturbations

Before we proceed, we must describe how to parametrize Gaussian perturbations in some quantity $g(\vec{x})$. Note that of course, non-Gaussian perturbations are important as well (there could be non-Gaussianities in the CMBR), but they have not been discovered yet, so we will ignore them in this chapter.

The variance of g is defined by the correlation function (ensemble average) of g^2,

$$\sigma_g^2(\vec{x}) \equiv \langle g^2(\vec{x}) \rangle , \tag{7.23}$$

and it means that the probability to find the value g for $g(\vec{x})$ is

$$P(g) = \frac{1}{\sqrt{2\pi\sigma}} e^{-\frac{g^2}{2\sigma_g^2}}. \tag{7.24}$$

But the variance is expanded in momenta, to find the spectrum of g,

$$\sigma_g^2(\vec{x}) = \frac{1}{(2\pi)^3} \int_0^\infty P_g(k) d^3k = \int_0^\infty \mathcal{P}_g(k) \frac{dk}{k} , \tag{7.25}$$

where in the last equality we have considered that if P_g is only a function of the modulus k, we can write a distribution over k,

$$\mathcal{P}_g = \frac{k^3}{2\pi^2} P_g. \tag{7.26}$$

Note that then we have, for the two-point correlation function in *momentum space* (by Fourier transform of the x space result)

$$\langle g_{\vec{k}} g_{\vec{k}'} \rangle = (2\pi)^3 \delta^3_{\vec{k}+\vec{k}'} P_g(\vec{k}). \tag{7.27}$$

Coming back to the case of Θ_l, consider the Gaussian perturbation in $g = \Theta_l = a_{l0}$, for which we have $\sigma_g^2 = \langle (a_{l0})^2 \rangle = C_l$. We obtain therefore for the spectrum

$$C_l = 4\pi \int_0^\infty \mathcal{P}_{\Theta_l}(\eta_0, k) \frac{dk}{k}. \tag{7.28}$$

Moreover, we will be interested in the effect of some primordial perturbation, to be defined shortly, called ζ_k, that is transferred into a perturbation in the CMBR, Θ_l, by the *transfer function* $T_l(k)$,

$$\Theta_l(k) = T_l(k)\zeta_k. \tag{7.29}$$

Replacing in the two-point function,

$$\begin{aligned} \langle \Theta_l^2 \rangle &= \int_0^\infty \mathcal{P}_{\Theta_l}(k) \frac{dk}{k} \\ &= (T_l(k))^2 \langle \zeta_k^2 \rangle = (T_l(k))^2 \int_0^\infty \mathcal{P}_\zeta(k) \frac{dk}{k} , \end{aligned} \tag{7.30}$$

we obtain $\mathcal{P}_{\Theta_l} = (T_l(k))^2 \mathcal{P}_\zeta$, so that

$$C_l = 4\pi \int_0^\infty (T_l(k))^2 \mathcal{P}_\zeta(k) \frac{dk}{k}. \tag{7.31}$$

7.3 Perturbation Theory for CMBR

We now move on to calculating the perturbations in the CMBR. There are in principle several types of perturbations. The one we are interested in, the most important one for its effect on the CMBR, is the *curvature perturbation* ζ, which determines the total density perturbation. There are in principle also isocurvature and tensor perturbations (though they have not been observed yet).

We must consider a gauge for general coordinate transformations, in order to define the perturbations. One considers (fixed) time slices that have uniform density ρ, and threads (fixed x_i worldlines) that are comoving. In that case, we can parametrize the *spatial* metric, perturbed by the curvature perturbation, as

$$g_{ij} = a^2(t) e^{2\zeta(\vec{x},t)} \gamma_{ij}(\vec{x}) , \tag{7.32}$$

where $\det \gamma_{ij} = 1$.

Since these curvature perturbations are determined by the density perturbations, we consider also the density perturbations of the various components,

$$\left(\frac{\delta\rho}{\rho}\right)_{\gamma,\nu,B,c} , \tag{7.33}$$

where γ is for photons, ν for neutrinos, B for baryons and c for Cold Dark Matter.

From the adiabatic condition, the various components depend only on the total density, i.e., the densities are proportional to each other, $\rho_a = \rho_a(\rho)$, and the same is true for the number densities, $n_a = n_a(\rho)$. In particular, that leads to

$$\frac{\delta(n_B/n_\gamma)}{n_B/n_\gamma} = \frac{\delta\rho_B}{\rho_B} - \frac{3}{4}\frac{\delta\rho_\gamma}{\rho_\gamma} = 0. \tag{7.34}$$

More generally, we find also

$$\frac{1}{3}\frac{\delta\rho_B}{\rho_B} = \frac{1}{3}\frac{\delta\rho_c}{\rho_c} = \frac{1}{4}\frac{\delta\rho_\gamma}{\rho_\gamma} = \frac{1}{4}\frac{\delta\rho_\nu}{\rho_\nu}. \tag{7.35}$$

For some generic gauge, transformed from the gauge above (with uniform density slices) to one with $\tilde{t} = t + \delta t(\vec{x}, t)$, we parametrize the spatial metric as

$$g_{ij} = a^2(t)e^{2\psi(\vec{x},t)}\gamma_{ij}(\vec{x}) , \tag{7.36}$$

leading to

$$\psi = \zeta - H\delta t = \zeta - H\frac{\delta\rho}{\dot{\rho}} = \psi + \frac{1}{3}\frac{\delta\rho}{\rho + P} , \tag{7.37}$$

where in the last equality we have used the conservation of energy equation.

We next consider perturbations in the fluid velocity.

- Newtonian case. The velocity field is the sum of a Hubble contribution and a peculiar velocity contribution,

$$\vec{u}(\vec{x}, t) = H(t)\vec{r} + \vec{v}(\vec{x}, t). \tag{7.38}$$

We are mostly interested in a scalar perturbation, which means that the peculiar velocity *in momentum space* satisfies

$$\vec{v}_{\vec{k}}^{\,\text{scalar}} = -i\frac{\vec{k}}{|\vec{k}|}V_{\vec{k}}. \tag{7.39}$$

Note that the $-i$ is conventional, so that the perturbation makes sense in x space. If we would consider a vector perturbation, that would be defined by $\vec{k} \cdot \vec{v}_{\vec{k}}^{\,\text{vector}} = 0$.
- General relativistic case. The background metric is $ds^2 = a^2(\eta)[-d\eta^2 + \delta_{ij}dx^i dx^j]$, and $dr^i = a\, dx^i$ is the spatial coordinate. Defining the comoving velocity as usual, by

$$u^\mu = \frac{dx^\mu}{d\tau} , \tag{7.40}$$

the normal spatial velocity is

$$v^i = \frac{dr^i}{dt} = a\, u^i. \tag{7.41}$$

Then define the general metric perturbation, independently of (before fixing) any gauge for general coordinate transformations,

$$ds^2 = a^2(\eta)\left\{-(1+2A)d\eta^2 - 2B_i d\eta dx^i + [(1+2D)\delta_{ij} + 2E_{ij}]dx^i dx^j\right\}. \tag{7.42}$$

Here E_{ij} is spatially traceless, since $(1+2D)\delta_{ij}$ is a trace, so this is indeed the most general parametrization.

For a scalar mode, as we are interested in, we would have

$$\begin{aligned}\vec{B} &= -i\frac{\vec{k}}{|\vec{k}|}B\\ E_{ij} &= \left(-\frac{k_i k_j}{k^2} + \frac{1}{3}\delta_{ij}\right)E.\end{aligned} \tag{7.43}$$

However, it is most common to consider the *conformal Newton gauge*, so called since one still has the conformal factor, but except for that, we consider the usual parametrization of small fluctuations in terms of the Newtonian potential, i.e.,

$$ds^2 = a^2(\eta)[-(1+2\Psi)d\eta^2 + (1-2\Phi)\delta_{ij}dx^i dx^j]. \tag{7.44}$$

Note that dropping the conformal factor and replacing $\Psi = \Phi = U_N$ corresponds to the form of the metric depending only on the Newtonian potential U_N, always possible in the small field approximation. Also note that $E = B$ in this gauge.
In this gauge, compared with the uniform density gauge defining ζ, we have from (7.36) expanded to first order that $\psi = -\Phi$, which leads from (7.37) (for $P = w\rho$) to

$$\frac{\delta\rho}{\rho} = 3(1+w)(\zeta + \Phi). \tag{7.45}$$

For a variety of reasons, we know that various scales k must exit the horizon, and spend some time outside the horizon. In inflation, the reason is that the scales are exponentially blown up. During this period, before re-entering the horizon (horizon entry), the perturbations at this scale will be frozen in at some value constant in time. Then, nowadays, during matter domination, the scales come back inside the horizon and then they will start again to evolve in time.
We can find during this period (before horizon entry) that $\delta\rho/\rho = -2\Psi$ (if $\dot{\Phi} = \dot{\Psi} = 0$), which means that

$$-\zeta = \Phi + \frac{2\Psi}{3(1+w)}. \tag{7.46}$$

Time Evolution of Perturbations

In order to find the time evolution of perturbations, we consider as usual the conservation of the energy-momentum tensor $D_\mu T^\mu{}_\nu = 0$, which contains the equations of motion of the fluid.

The 0 component, $D_\mu T^\mu{}_0 = 0$, gives as usual the continuity equation, whereas the i component, $D_\mu T^\mu{}_i = 0$, gives the Euler equation.

Consider a perfect fluid (with no dissipation and no anisotropy), with energy-momentum tensor

$$T^{\mu\nu} = Pg^{\mu\nu} + (P+\rho)u^\mu u^\nu. \tag{7.47}$$

Then, the continuity equation for the case of an equation of state $P = w\rho$ for the background gives

$$\dot{\left(\frac{\delta\rho}{\rho}\right)} = -(1+w)(kV - 3\dot{\Phi}) + 3aH\left(w\frac{\delta\rho}{\rho} - \frac{\delta P}{P}\right). \tag{7.48}$$

If we also have $\delta P = w\delta\rho$, which is nontrivial, since it means that fluctuations follow the same equation of state, i.e., that subsystems don't exchange energy or momentum with the exterior (even when there is a local fluctuation), then the last term in the above equation vanishes, and we get

$$\dot{\left(\frac{\delta\rho}{\rho}\right)} = -(1+w)(kV - 3\dot{\Phi}). \tag{7.49}$$

If moreover, also for any component, a subsystem doesn't exchange energy with other subsystem on a given time scale, we have the above relation valid *for any component* as well, i.e.,

$$\begin{aligned}
\dot{\left(\frac{\delta\rho_c}{\rho_c}\right)} &= -kV_c + 3\dot{\Phi}: \quad CDM\\
\dot{\left(\frac{\delta\rho_\gamma}{\rho_\gamma}\right)} &= -\frac{4}{3}kV_\gamma + 4\dot{\Phi}: \quad \text{photon}\\
\dot{\left(\frac{\delta\rho_B}{\rho_B}\right)} &= -kV_B + 3\dot{\Phi}: \quad \text{baryon}
\end{aligned} \tag{7.50}$$

The Euler equation for the case of the perfect fluid becomes now

$$\dot{V} = -aH(1-3w)V - \frac{\dot{w}}{1+w}V + k\frac{\delta P}{\rho+P} + k\Psi. \tag{7.51}$$

It is also true for subsystems (components), in the case that a subsystem doesn't exchange momentum and energy with the exterior (the other subsystems).

Consider the baryon-photon fluid, which is a tightly coupled system until recombination, meaning that the interactions between them are so strong, that they follow the same laws, depending only on the total density. We have then that $V_B = V_\gamma$, and we saw that the perturbation in the baryon energy density is 3/4 of the perturbation in the photon energy density.

Then from the Eq. (7.51) for the baryon and photon components, we can write an equation for the combined fluid, taking into account that $P_\gamma = \rho_\gamma/3$, $P_B = 0$,

$$\dot{V}_\gamma = -aH(1-3\tilde{w})V_\gamma - \frac{\dot{\tilde{w}}}{1+\tilde{w}}V_\gamma + \frac{k}{3}\frac{\rho_\gamma}{\rho_B + \frac{4}{3}\rho_\gamma}\frac{\delta\rho_\gamma}{\rho_\gamma} + k\Psi\ , \tag{7.52}$$

where the equation of state $\tilde{w}$ of the combined baryon-photon fluid is defined as

$$\tilde{w} = \frac{P_{\rm tot}}{\rho_{\rm tot}} = \frac{\rho_\gamma/3}{\rho_\gamma + \rho_B}. \tag{7.53}$$

Eliminating V_γ between (7.51) and (7.52) gives

$$\frac{1}{4}\left(\ddot{\frac{\delta\rho}{\rho}}\right)_{\gamma,k} + \frac{1}{4}\frac{\dot{R}}{1+R}\left(\dot{\frac{\delta\rho}{\rho}}\right)_{\gamma,k} + \frac{k^2}{4}c_s^2\left(\frac{\delta\rho}{\rho}\right)_{\gamma,k} = F_{\vec{k}}(\eta)\ , \tag{7.54}$$

where

$$\begin{aligned}
R(\eta) &\equiv \frac{3}{4}\frac{\rho_B}{\rho_\gamma}\\
F_{\vec{k}}(\eta) &\equiv -\frac{k^2}{3}\Psi_k(\eta) + \frac{\dot{R}(\eta)}{1+R(\eta)}\dot{\Phi}_k(\eta) + \ddot{\Phi}_k(\eta)\\
c_s^2(\eta) &\equiv \frac{\partial P}{\partial\rho} = \frac{\dot{P}_\gamma}{\dot{\rho}_\gamma + \dot{\rho}_B} = \frac{1}{3(1+R(\eta))}.
\end{aligned} \tag{7.55}$$

Here we see that $R = \delta\rho_B/\delta\rho_\gamma$ and c_s is the speed of sound of the combined fluid, $= dr/d\eta$.

Then the *sound horizon* is

$$r_s(\eta) = \int_0^\eta c_s(\eta)d\eta\ , \tag{7.56}$$

that is, the distance travelled by the sound (in the fluid) from the Big Bang until η. Note that at early times, when there are no baryons yet, $R \simeq 0$, and there is only the radiation, meaning $c_s = 1/\sqrt{3}$, the speed of sound in the photon fluid.

The Eq. (7.54) has the form of damped, forced harmonic oscillator. But in fact, the damping term (with the first derivative of the photon energy density fluctuation) turns out to be much smaller than the *Silk damping term* to be defined shortly, so we can neglect it.

Then the solution is the solution for the forced harmonic oscillator,

$$\frac{1}{4}\left(\frac{\delta\rho}{\rho}\right)_{\gamma,\vec{k}} = A_{\vec{k}}(\eta) + B_{\vec{k}}(\eta)\cos(kr_s(\eta)) + C_{\vec{k}}(\eta)\sin(kr_s(\eta)). \tag{7.57}$$

Here $A_{\vec{k}}$ is the constant (non-oscillatory) solution of (7.54),

$$A_{\vec{k}}(\eta) = -[1 + R(\eta)]\Phi_k , \tag{7.58}$$

as we can explicitly check. These are baryon acoustic oscillations.

In order to find the constant solution however, we need to use a formula that relates the Newtonian potentials Φ and Ψ to the curvature perturbation ζ_k. In the tight coupling approximation for the baryon-photon fluid, we find that $\Phi = \Psi$, which means (from (7.46)) that before horizon entry we find

$$\Phi = \Psi = -\frac{3 + 3w}{5 + 3w}\zeta. \tag{7.59}$$

If horizon entry happens close to now, i.e., during matter domination (which is what will happen for most modes observed in the CMBR), when $w = 0$, we find just before horizon entry, i.e., for modes that are on very large scales, that

$$\Phi_k = \Psi_k = -\frac{3}{5}\zeta_k. \tag{7.60}$$

But for modes that have already entered for some time inside the horizon, i.e., for modes that are at smaller scales, or larger l, we have a small modification, parametrized by a *matter transfer function* $T(k)$,

$$\Phi_k = -\frac{3}{5}T(k)\zeta_k. \tag{7.61}$$

Of course, for very large scales $T(k) \simeq 1$.

Silk Damping

The dominant damping effect in the equation for the time evolution of the density perturbation is not the damping term already present, but the so-called Silk damping, which is due to the Thomson scattering of electrons off photons, $\gamma e^- \to \gamma e^-$, which happens at a scale k_D, so gives a factor e^{-k^2/k_D^2} damping, for a solution

$$\frac{1}{4}\left(\frac{\delta\rho}{\rho}\right)_{\gamma,\vec{k}} = A_{\vec{k}}(\eta) + e^{-\frac{k^2}{k_D^2}}[B_{\vec{k}}(\eta)\cos(kr_s(\eta)) + C_{\vec{k}}(\eta)\sin(kr_s(\eta))]. \tag{7.62}$$

One can then solve more precisely the driven, damped harmonic oscillator equation (7.54) and find the coefficients as well, namely

$$\begin{aligned}\frac{1}{4}\left(\frac{\delta\rho}{\rho}\right)_{\gamma,\vec{k}} &= -(1+R)\Phi_k + e^{-\frac{k^2}{k_D^2}}\frac{\zeta_{\vec{k}}}{3}\cos(kr_s(\eta)) \Rightarrow \\ \frac{1}{4}\left(\frac{\delta\dot{\rho}}{\rho}\right)_{\gamma,\vec{k}} &= -kc_s e^{-\frac{k^2}{k_D^2}}\frac{\zeta_{\vec{k}}}{3}\sin(kr_s(\eta))\;,\end{aligned} \tag{7.63}$$

where $\Phi_k = -3T(k)\zeta_k/5$.

7.4 Effect of Density Perturbations on the CMBR

We now go back to the CMBR, to find the effect of the density perturbation above on the CMBR. Consider the *sudden decoupling approximation* for the CMBR. This means that at recombination, decoupling happens on a given time η_{ls} (ls stands for last scattering, of photons off electrons) for all the photons at the same time. Then *at last scattering*, the only anisotropy we have is due to the monopole (local density perturbation at the position we consider) plus dipole, i.e.,

$$\Theta_{ls}(\vec{e}) = \left(\frac{1}{4}\frac{\delta\rho_\gamma}{\rho_\gamma} + \vec{e}\cdot\vec{v}_\gamma\right)_{ls}. \tag{7.64}$$

Sachs-Wolfe Effect

The dependence on direction happens due to the photons travelling since last scattering until now, through the so-called Sachs-Wolfe effect, so

$$\Theta(\vec{e}) = \Theta_{ls}(\vec{e}) + \Theta_{SW}(\vec{e}). \tag{7.65}$$

Consider a 4-momentum (p^0, p^i) in a locally orthogonal (Minkowski) frame, described as (q, qu^i). On the other hand, in the conformal Newton gauge, the 4-momentum is

$$p^0 = (1+\Psi)q; \quad p^i = (1-\Phi)qu^i. \tag{7.66}$$

Then

$$0 = -(p^0)^2 + p^i p^i \simeq -(1+2\Psi)q^2 + (1-2\Phi)q^2 u^i u^i = q^\mu q^\nu g_{\mu\nu}. \tag{7.67}$$

The Sachs-Wolfe contribution comes from the redshift of nearby photons. The momentum field $q = q(\vec{x}, t)$ is written, by taking out the expanding factor, as

$$q(\vec{x}, t) = a(t)p(\vec{x}, t). \tag{7.68}$$

The redshift in the comoving frame p has therefore two contributions,

$$\frac{dp}{p} = -\frac{da}{a} + \frac{dq}{q}\,, \tag{7.69}$$

where the first term is the Hubble term (coming from the expansion of the Universe), and the second is the effect we are looking for, coming from the change in the metric of spacetime due to the perturbations,

$$\Theta_{SW} = \int_{\eta_{ls}}^{\eta_0} \frac{dq}{q} = \int_{\eta_{ls}}^{\eta_0} d\eta \frac{1}{q}\frac{dq}{d\eta}. \tag{7.70}$$

To find this expression, we consider the geodesic equation in curved space for the motion of the photon,

$$\frac{dq^\mu}{dx^0} = g^{\mu\nu}\left(\frac{1}{2}\partial_\nu g_{\alpha\beta} - \partial_\beta g_{\nu\alpha}\right)\frac{q^\alpha q^\beta}{q^0}. \tag{7.71}$$

Considering the 0 component for perturbations, and dividing by $q \equiv q^0$, and writing it in terms of η, we obtain

$$\frac{1}{q}\frac{dq}{d\eta} = \frac{\partial\Phi}{\partial\eta} - u^i \frac{d\Psi}{dx^i}\,., \tag{7.72}$$

where $u^i = dx^i/d\eta$. Since

$$\frac{d\Psi}{d\eta} = \frac{\partial\Psi}{\partial\eta} + \frac{dx^i}{d\eta}\frac{\partial\Psi}{\partial x^i}\,, \tag{7.73}$$

we obtain

$$\frac{1}{q}\frac{dq}{d\eta} = \frac{\partial(\Phi+\Psi)}{\partial\eta} - \frac{d\Psi}{d\eta}. \tag{7.74}$$

Then the Sachs-Wolfe contribution is

$$\Theta_{SW}(\vec{e}) = \Psi_{ls} - \Psi_0 + \int_{\eta_{ls}}^{\eta_0} d\eta \frac{\partial}{\partial\eta}(\Phi+\Psi). \tag{7.75}$$

In this formula, the $-\Psi_0$ term is usually neglected, since nowadays the fluctuation on very large scales in the Universe is negligible with respect to Ψ_{ls}. Then Ψ_{ls} is a contribution from the surface of last scattering, like the non-SW contribution of $1/4\delta\rho_\gamma/\rho_\gamma$, and the last contribution is an *integrated Sachs-Wolfe* (ISW) contribution.

Sachs-Wolfe Plateau

At small l, i.e., for the largest scales, for which the mode has barely left the horizon, during the matter dominated epoch, the mode has spent most of the time since decoupling to present outside the horizon, where it is constant, i.e., η-independent, i.e., no ISW contribution. Since moreover during this time $\Psi_{ls} = \Phi_{ls} = \Theta_{SW}(\vec{e})$, we find

$$\Theta(\vec{e}) = \frac{1}{4}\left(\frac{\delta\rho_\gamma}{\rho_\gamma}\right)_{ls} + \Phi_{ls}. \tag{7.76}$$

In this case, one finds that $l(l+1)C_l$, the quantity plotted in the CMBR spectrum, is approximately independent of l (flat), leading to the SW plateau, until about $l \sim 30$. Since we find $\delta\rho_\gamma/\rho_\gamma = -8\Phi/3$, we find

$$\Theta(\vec{e}) = \frac{1}{3}\Phi_{ls} = \frac{1}{5}\zeta_{ls}. \tag{7.77}$$

Integrated SW Effect

The integral of the ISW has contributions coming from radiation still present at decoupling, as well as, during late times, from the cosmological constant (dark energy).

Acoustic Peaks and Silk Damping

For l greater than about 30, we have a snapshot of acoustic oscillations at decoupling.

Substituting inside the definition

$$\Theta_l(\vec{k}) = -i^l[4\pi(2l+1)]^{-1/2}\int Y^*_{l0}(\vec{e})\Theta(\eta,\vec{k},\vec{e})d^2\vec{e} \tag{7.78}$$

the formula for Θ, we find

$$\Theta_l(\vec{k}) = \left[\frac{1}{4}\left(\frac{\delta\rho_\gamma}{\rho_\gamma}\right)_{\vec{k}} + \Phi_{\vec{k}}\right]_{ls} j_l(k\eta_0) \equiv T_l(k)\zeta_k\ , \tag{7.79}$$

where j_l are the Bessel functions. Substituting (7.63), we find

$$T_l(k) = \left[\frac{3}{5}RT(k) - \frac{1}{3}e^{-\frac{k^2}{k_D^2}}\cos kr_s\right] j_l(k\eta_0). \tag{7.80}$$

But in the formula

$$C_l = 4\pi\int_0^\infty T_l^2(k)\mathcal{P}_\zeta(k)\frac{dk}{k}\ , \tag{7.81}$$

one finds (observationally) that the spectrum $\mathcal{P}_\zeta(k)$ is approximately flat (constant), so that

$$C_l = 4\pi\mathcal{P}_\zeta\int_0^\infty\left[\frac{3}{5}RT(k) - \frac{1}{3}e^{-\frac{k^2}{k_D^2}}\cos kr_s\right]^2 j_l^2(k\eta_0)\frac{dk}{k}. \tag{7.82}$$

This shows the presence of acoustic oscillations and Silk damping. Here, from observations plus theory, we find that

$$r_s \simeq 150\,\text{Mpc}; \quad \eta_0 \simeq 14000\,\text{Mpc}; \quad k_D^{-1} \simeq 8\,\text{Mpc}. \tag{7.83}$$

Note that in this chapter we haven't described polarization modes, isocurvature or tensor modes, which could nevertheless become important, if they are observed.

Important Concepts to Remember

- The CMBR, the radiation coming from the Big Bang from every point in the Universe, is approximately homogenous and isotropic, but it has anisotropic fluctuations of $\delta T/T \sim 10^{-5}$.
- The CMBR has a monopole anisotropy at our current position in the Universe, but that can't be measured.
- The CMBR has also a dipole anisotropy due to the Earth's motion with respect to the cosmological frame (frame of the CMBR).
- The CMBR spectrum is found as an expansion over angles, leading to an expansion in multipoles l. At large l, $l \sim 1/\theta$.
- Gaussian perturbations have 2-point correlator $\langle g_{\vec{k}} g_{\vec{k}'} \rangle = (2\pi)^3 P_g(\vec{k})\delta(\vec{k} + \vec{k}')$ and $\mathcal{P}_g(k) = k^3/(2\pi^2) P_g(\vec{k})$.
- There can be scalar, vector and tensor perturbations, but we focus on the scalar ones.
- Primordial perturbations ζ_k are transferred to the CMBR as $\Theta_l(k) = T_l(k)\zeta_k$. Here ζ_k is the curvature perturbation, related to the density perturbations.
- In the conformal Newton gauge, the perturbations are defined by the Newtonian potential, $\Psi = \Phi = U_N$.
- We are interested in perturbations that have exited outside the horizon early on, so that they can come back inside during current times, as CMBR perturbations.
- We find that the density perturbation created by ζ is $1/4(\delta\rho/\rho)_{\zeta,k} = A_k(\eta) + e^{-k^2/k_D^2}[B_k(\eta)\cos(kr_s(\eta)) + C_k(\eta)\sin(kr_s(\eta))]$, giving oscillations and Silk damping.
- In the sudden decoupling approximation, the above perturbation is translated by the propagation of light into the CMBR perturbation, and one finds that an approximately flat $\mathcal{P}_\zeta(k)$ (spectrum of curvature perturbations) generates the damped oscillations of the CMBR.

Further reading: See Chaps. 5, 6, 7, 8 and 10 in [9].

Exercises

(1) Prove that for $P = w\rho$ (for the background), the continuity equation gives

$$\dot{\left(\frac{\delta\rho}{\rho}\right)} = -(1+w)(kV - 3\dot{\Phi}) + 3aHw\left(\frac{\delta\rho}{\rho} - \frac{\delta P}{P}\right). \tag{7.84}$$

(2) Prove that the Euler equation gives

$$\dot{V} = -aH(1-3w)V - \frac{\dot{w}}{1+w}V + k\frac{\delta P}{\rho + P} + \kappa\Psi. \tag{7.85}$$

(3) Eliminate V_γ between

$$\left(\frac{\dot{\delta\rho_\gamma}}{\rho_\gamma}\right) = -\frac{4}{3}kV_\gamma + 4\dot{\Phi} \tag{7.86}$$

and

$$\dot{V}_\gamma = -aH(1-3\tilde{w})V_\gamma - \frac{\dot{\tilde{w}}}{1+\tilde{w}}V_\gamma + \frac{k}{3}\frac{\rho_\gamma}{\rho_B + \frac{4}{3}\rho_\gamma}\frac{\delta\rho_\gamma}{\rho_\gamma} + k\Psi\ , \tag{7.87}$$

to find

$$\frac{1}{4}\left(\frac{\ddot{\delta\rho}}{\rho}\right)_{\gamma,k} + \frac{1}{4}\frac{\dot{R}}{1+R}\left(\frac{\dot{\delta\rho}}{\rho}\right)_{\gamma,k} + \frac{k^2}{4}c_s^2\left(\frac{\delta\rho}{\rho}\right)_{\gamma,k} = F_{\vec{k}}(\eta). \tag{7.88}$$

(4) Calculate the time evolution of the speed of sound c_s from recombination until today, and then of the sound horizon $r_s(\eta)$.

Chapter 8
Problems to Be Solved by Inflation and How They Are Solved in Inflationary Models

In this chapter we will present problems with the standard cosmology, before the advent of the inflationary paradigm, and then we will see how they are solved by inflation.

8.1 Problems with Standard Cosmology Before Inflation

1. Smoothness and horizon problem

Why is the Universe uniform and isotropic?

This problem appeared because the light that we observe in the CMBR, coming from the surface of last scattering (recombination), when considered as coming from *opposite* directions on the sky, comes from regions of the Universe that are causally disconnected, since the horizon distance is smaller than the distance travelled by light.

That sounds strange, since causality is defined also by light propagation, but it is connected with the fact that the Universe expands while light propagates, and the fact that causal contact refers *to the moment of last scattering* (so light propagation before it), whereas light propagation for the CMBR refers to the time after last scattering.

The fact that the light comes from seemingly causally disconnected regions would imply that there would be no reason for the correlation implied by the uniform and isotropic temperature and Planck distribution. Since T is constant to a high degree of precision, and moreover, there are correlations in the sky even in the fluctuations, it seems to say that the surface of last scattering in opposite directions with respect to us was causally connected. We will see that this is what happens in inflation.

H. Năstase, *Cosmology and String Theory*, Fundamental Theories of Physics 197, https://doi.org/10.1007/978-3-030-15077-8_8

We now turn to a quantitative evaluation of what we need to fix. The horizon distance at the time of last scattering t_{ls} is defined by Δx for light propagation from $t = 0$ (Big Bang) and t_{ls},

$$d_H(t_{ls}) = a(t_{ls})\Delta x = a(t_{ls}) \int_0^{t_{ls}} \frac{dt}{a(t)}. \tag{8.1}$$

This distance, translated into today's distances, by replacing $a(t_{ls})$ with $a(t_0)$, is

$$d_H(t_0) = a(t_0) \int_0^{t_{ls}} \frac{dt}{a(t)}. \tag{8.2}$$

On the other hand, the distance travelled by light from the surface of last scattering until today, in today's scales, is

$$r_H(t_0) = a(t_0) \int_{t_{ls}}^{t_0} \frac{dt}{a(t)}. \tag{8.3}$$

Light propagates during this time in a matter dominated Universe, when $a \propto t^{2/3}$, or more precisely $a(t) = a(t_0)(t/t_0)^{2/3}$, so that

$$r_H(t_0) = 3(t_0^{1/3} - t_{ls}^{1/3})t_0^{2/3}. \tag{8.4}$$

Even though on a log scale, matter domination takes a very small piece of the Universe before last scattering, in absolute terms it dominates (as we saw in Chap. 6, $t_{ls} \simeq 200 t_{eq}$), so we can also use M.D. to compute $d_H(t_{ls})$, giving

$$d_H(t_0) = 3 t_{ls}^{1/3} t_0^{2/3}, \tag{8.5}$$

so finally the ratio we want, of the distance between the opposite last scattering points vs. the horizon distance, is

$$N = \frac{2 r_H(t_0)}{d_H(t_0)} \simeq 2 \frac{t_0^{1/3} - t_{ls}^{1/3}}{t_{ls}^{1/3}} \simeq 2 \left(\frac{t_0}{t_{ls}}\right)^{1/3} = \left(\frac{a_0}{a_{ls}}\right)^{1/2} = (1 + z_{ls})^{1/2}. \tag{8.6}$$

Here we have used the fact that $t_0 \gg t_{ls}$, that $a(t) \propto t^{2/3}$ during matter domination, and that $a \propto 1 + z$, where z is the redshift. Putting in numbers, specifically that $z_{ls} \simeq 1300$, as we calculated in Chap. 6, we find

$$N = 2 \times \sqrt{1300} \simeq 72. \tag{8.7}$$

So we need to find a way to increase the horizon distance at the surface of last scattering by a factor of 72 with respect to the naive result.

2. Flatness problem

Why do we have $\Omega \simeq 1$ in the past?

We have seen that the Friedmann equation can be written in as

$$\Omega(t) - 1 = \frac{K}{a^2(t)H^2(t)} = \frac{K}{a^2(t)}\frac{3\Omega}{8\pi G\rho(t)}, \tag{8.8}$$

where in the last equality we have used $H^2 = 8\pi G\rho_{\rm cr}/3$ and $\Omega = \rho/\rho_{\rm cr}$. Then we obtain

$$\frac{\Omega(t) - 1}{\Omega(t)} = \frac{3K}{8\pi G\rho(t)a^2(t)}. \tag{8.9}$$

Since $\rho(t) \propto 1/t^2$ and $a(t) \propto t^p$, where $p = 2/3$ for M.D. and $p = 1/2$ for R.D., we find that

$$\frac{\Omega(t) - 1}{\Omega(t)} \propto t^{2(1-p)}. \tag{8.10}$$

In turn that means that even if $\Omega \neq 1$, so $K \neq 0$ (K is ± 1), considering an Ω which is of order 1, but not 1, nowadays, it will be much closer to 1 in the past, to an absurd precision. For instance, $\Omega \sim 1$ now implies $\Omega - 1 \sim 10^{-16}$ at e^+e^- annihilation, and even smaller at earlier times.

But that begs a question: why is it so unbelievably small then? Since what it is now, is a result of what it was in the past, we need to explain this unnaturally small number (saying it is what it is, via the most vanilla type of anthropic principle, the fact that the Universe is as it is, since otherwise there would be no people to see it, simply avoids any attempt at an explanation).

Of course, it could be that we have exactly $\Omega \equiv 1$, in which case $K = 0$, but we have no *experimental* evidence for that. Saying that $\Omega = 1$ would be a theoretical bias; experimentally we only know that $\Omega \simeq 1$ to about one percent, so we must assume the equality is not perfect.

But then we need to explain the absurdly small number for $\Omega - 1$ at very early times.

3. Why is the entropy so large?

We have already seen that the entropy per baryon is about $S_1 \sim 10^9$. Multiplying with the number of baryons within the current horizon, or alternatively calculating the entropy density s and multiplying by the horizon distance volume, $S = s \cdot \frac{4\pi}{3}(ct)^3$, we find the entropy $S \sim 10^{88}$ today.

But actually we need ot consider the entropy at some earlier time, when it is actually smaller, since $s \propto T^4$, but $(ct)^3 \propto (1/T^2)^3$ during the R.D. era and $(ct)^3 \propto (1/T^{3/2})^3$ during the M.D. era. For instance, at BBN, one finds $S_H(t_{BBN}) \sim 10^{63}$, which is still huge.

We need to explain the great production of entropy, since naturally we would expect again numbers of order one for the entropy within the horizon, or the entropy per baryon ratio.

4. Why do we have a (very small) baryon asymmetry?

As we have seen, we have an asymmetry

$$\frac{N_B - N_{\bar{B}}}{N_B} \sim \frac{1}{10^9}\,, \tag{8.11}$$

which needs to be explained (why it is nonzero, and why it is so small).

5. How do we generate perturbations?

This is sort of the opposite of the question at number one, in the sense that we see an Universe which is not just approximately uniform and isotropic, but also full of small fluctuations. On the largest scales, we have the fluctuations in the CMBR; on smaller scales, the fluctuations that have since then evolved to give rise to structure, namely stars and galaxies. Note that these seem to be *classical* fluctuations, and moreover on scales that were *super-horizon* in the past, i.e., they should not have been in causal contact. Indeed, scales grow as $a(t)$, but the horizon grows like ct, which during R.D. is $a(t)^2$, and during M.D. is $a(t)^{3/2}$, growing faster than the scales. Thus scales keep falling inside the horizon as time evolves: scales that were super-horizon drop to sub-horizon, where we can observe them. This is most puzzling in standard cosmology, and seems to have no explanation.

6. Monopole problem

Why so small relic densities in the Universe?

Historically, the problem referred to monopoles, which are creating during a phase transition, like a GUT phase transition, so we expect to find them in large numbers. Nowadays, we say that any kind of relics, like gravitinos, string moduli, or other topological defects (e.g., cosmic strings, domain walls, etc.), would be produced in large numbers at early times. But since we don't observe them in large numbers, there must be a mechanism to dilute them.

By the *Kibble mechanism*, we expect to create one monopole per horizon during a phase transition, as the Higgs VEV obtains a random but uniform value in a region of the order of the horizon, as the Universe cools through the phase transition, and a random set of scalar VEVs on horizons would generically create a monopole configuration for these scalars. But during the same GUT phase transition, we also expect to create one nucleon per horizon by similar arguments. Since neither decays much until now, we would expect to see about one monopole per nucleon, or one per 10^9 photons. However, direct monopole searches in materials on Earth show that there are less than 10^{-30} monopoles per nucleon, which means we should reduce the monopole density by about 10^{-30} in volume. For general relics (like for instance cosmic strings), one can calculate that in order for their mass density not to over-close the Universe, we need a reduction by a factor of about 10^{-11}, which is less stringent than the monopoles' 10^{-30}, but still important.

8.2 The Paradigm of Inflation

The paradigm that comes to solve all of the above problems is called *inflation*. It is more of a paradigm than a theory, since it is a set of very general assumptions. It was started in the West by Alan Guth in 1981, with the model now known as "old inflation", which however had some problems, to be described later. It was then refined to what is now known as "new inflation" in the independent works by Albrecht and Paul Steinhardt in 1982, and by Andrei Linde in 1982 and 1983. However, it is worth pointing out that the model that fits the current data best is the model that came before Guth, of "Starobinsky inflation", a modification of general relativity now known to be equivalent with a scalar model with a certain exponential potential. Starobinsky wrote his model in 1979 in Soviet Russia (in JETP Letters), but it didn't have an impact in the West, and was only later re-discovered.

The paradigm can be stated most generally as the need to have a (long) period of accelerated inflation, $\ddot{a} > 0$. Since the acceleration equation is

$$\frac{\ddot{a}}{a} = -\frac{4\pi G_N}{3}(\rho + P) , \tag{8.12}$$

we need $\rho + P < 0$ or, since $P = w\rho$, that $w < -1/3$. The most common way is to have also an *exponential* expansion, through an approximately constant energy density, but we don't actually need that in order to solve the problems. It is also common to assume that the approximately constant energy density is parametrized by a scalar field, but this is only the simplest case, not a necessity. In principle, we could have any sort of tensor field, e.g. vector, graviton, higher spin, that parametrizes the change in the energy density.

Then the standard incarnation of inflation involves an energy density that is approximately constant, so a dark energy component that is approximated by a cosmological constant Λ. We will review the energetics of Λ, that we already touched on before. The gravitational acceleration felt at a radius R is, from the acceleration equation,

$$\ddot{R} \equiv g_{\text{GR}} = -\frac{4\pi G}{3} R \left(\rho + \frac{3P}{c^2}\right) , \tag{8.13}$$

whereas in the Newtonian case we have the same formula, but with no P contribution,

$$\ddot{R} \equiv g_{\text{Newton}} = -\frac{4\pi G}{3} R\rho. \tag{8.14}$$

For a cosmological constant, $P_V/c^2 = -\rho_V$, which leads to the positive gravitational acceleration

$$g_V = +\frac{8\pi G}{3} R\rho_V , \tag{8.15}$$

which means a *repulsive gravity*. The reason that is possible is because the negative pressure creates positive work. Indeed, we have (for $c = 1$) $dE = -PdV$, but $E = \rho V$, which means that if $P = -\rho$, then $d\rho = 0$, so the energy density is constant

$$\rho_V = \frac{\Lambda}{8\pi G} = \text{const.} \tag{8.16}$$

as the volume expands. That means that energy is being created, by the work of the negative pressure (normally, as the volume expands, energy is lost, or at most constant, through the work of the positive pressure).

From the acceleration equation,

$$\ddot{a} = \frac{8\pi G\rho_V}{3}a\ , \tag{8.17}$$

whereas the Friedmann equation gives

$$H^2 = \frac{8\pi G}{3}\rho_{\text{tot}}\ , \tag{8.18}$$

leading to an exponential expansion,

$$a(t) = a_0 e^{H_0 t}\ , \tag{8.19}$$

where the *constant* Hubble scale H_0 is

$$H_0 = \sqrt{\frac{8\pi G\rho_V}{3}} = \sqrt{\frac{\Lambda}{3}}. \tag{8.20}$$

To finalize the description, we need to state that the exponential expansion happens near the Planck scale. The Planck scale is defined as the combination of c, $\hbar$ and G that has the right dimensions, so the Planck time and length are

$$t_P = \sqrt{\frac{\hbar G}{c^3}} \sim 10^{-43}\,\text{s}, \quad l_P = ct_P \sim 10^{-35}\,\text{m}\ , \tag{8.21}$$

whereas the Planck mass is

$$m_P = \sqrt{\frac{c\hbar}{G}} = 10^{19}\,\text{GeV} = 10^{-8}\,\text{Kg}. \tag{8.22}$$

Note however that the Planck mass m_P above is the quantity defined in particle physics, but in gravity and cosmology one uses instead a different definition, the *reduced Planck mass* M_{Pl}, defined as (for $c = \hbar = 1$)

$$M_{\rm Pl} = \frac{1}{\sqrt{8\pi G}} = \frac{m_P}{\sqrt{8\pi}}. \tag{8.23}$$

The definition is taken such that the coefficient of the Einstein–Hilbert action $\int d^4x\sqrt{-g}R$ is $\frac{M_{\rm Pl}^2}{2}$, meaning that the graviton modes have the standard canonical scalar kinetic action.

As we said, only accelerated expansion is enough for the inflation paradigm, which means that we could also have the so-called *power-law inflation*, with

$$a(t) \propto t^p, \tag{8.24}$$

where $p > 1$. Indeed, in that case, we obtain $\ddot{a} > 0$.

8.3 Inflationary Solutions to Cosmological Problems

We now turn to how to solve the problems mentioned with standard cosmology within inflation.

Flatness Problem

We saw that

$$\Omega(t) - 1 = \frac{K}{(aH)^2} \sim \left(\frac{t}{a}\right)^2 \propto t^{2(1-p)}. \tag{8.25}$$

In the case of standard cosmology, with $p < 1$ (matter domination has $p = 2/3$, radiation domination has $p = 1/2$), $\Omega(t) - 1$ grows with time. But by having a long period of accelerated expansion, i.e., inflation, with $p > 1$, we could fix it.

For instance, if we consider the case of cosmological constant (considered by Guth), with $H = H_I$ constant and $a(t) \propto e^{Ht}$,

$$\Omega(t) - 1 = \frac{K}{a^2H^2} \propto \frac{K}{a^2} \propto e^{-2H_It}, \tag{8.26}$$

which drops exponentially fast. That means that by starting with an $\Omega - 1 \sim \mathcal{O}(1)$, it drops to a very small value after some e-folds, so it can then grow again during the radiation domination and matter domination eras. Though typically, in this way we still obtain today a very small $\Omega - 1$, effectively zero. This is so since it would be very strange that after coming to an exponentially small value, it would grow back to a value of order one just now.

So the theoretical bias that we talked about is based on inflation, when it is natural to obtain an exponentially small $\Omega(t_0) - 1$ (nowadays). That is why we believe that Ω is one to an excellent accuracy now, even though experiments only show it to a small accuracy.

We now turn to a quantitative analysis, within exponential inflation, that will aim to constrain the amount of inflation needed. To do so, we we define the *number of e-foldings* N_e during inflation,

$$\frac{a(t_I)}{a(t_{bi})} = e^{2H_I t} \equiv e^{N_e} \rightarrow N_e = \ln \frac{a(t_I)}{a(t_{bi})} , \tag{8.27}$$

where t_{bi} is the the of begining of inflation, and t_I is the time of end of inflation.

Then now (at time t_0), we have

$$\Omega_0 - 1 = \frac{K}{a_0^2 H_0^2} = \frac{K}{a_I H_I^2} = \frac{K}{a_{bi}^2 H_{bi}^2} e^{-2N_e} \left(\frac{a_I H_I}{a_0 H_0}\right)^2 , \tag{8.28}$$

and if we assume $K/(a_{bi} H_{bi})^2 = \Omega(t_{bi}) - 1 \sim \mathcal{O}(1)$, then in order to solve the flatness problem, we need

$$e^{N_e} > \frac{a_I H_I}{a_0 H_0}. \tag{8.29}$$

But we consider that the time of begining of radiation domination, denoted by index 1, is just after the end of inflation, so that $a_I H_I \simeq a_1 H_1$. In reality it could be a couple of e-folds later, but it doesn't matter much for the calculation.

Since during radiation domination $H \propto 1/t \propto 1/a^2$, and the end of radiation domination corresponds to the matter-radiation equality, when $H^2 \propto \rho = \rho_m + \rho_{\rm rad} = 2\rho_{\rm rad} \propto 2H'^2_{rad}$, we obtain

$$H_1 = H' \left(\frac{a_{eq}}{a_1}\right)^2 = \frac{H_{eq}}{\sqrt{2}} \left(\frac{a_{eq}}{a_1}\right)^2 . \tag{8.30}$$

On the other hand, during radiation domination, $H \propto 1/t \propto 1/a^{3/2}$, and the end of matter domination corresponds to nowadays, when $H^2 \propto \rho_{0,\rm cr} = \rho_{m,0}/\Omega_{m,0}$, where $\Omega_{m,0} = \rho_{m,0}/\rho_{0,\rm cr}$, we similarly obtain

$$H_{eq} = \sqrt{2\Omega_{m,0}} H_0 \left(\frac{a_0}{a_{eq}}\right)^{3/2} . \tag{8.31}$$

Then the condition on the number of e-folds is

$$e^{N_e} > \frac{a_I H_I}{a_0 H_0} = \frac{a_1 H_1}{a_0 H_0} = \frac{a_1 \sqrt{H_1}}{a_0 \sqrt{H_0}} \sqrt{\frac{H_1}{H_0}} , \tag{8.32}$$

and since $a_1 \sqrt{H_1} = 2^{-1/4} a_{eq} \sqrt{H_{eq}}$, and $a_{eq}\sqrt{H_{eq}} = a_0 \sqrt{H_0} \sqrt{2\Omega_{m,0}} \sqrt{a_0/a_{eq}}$, we obtain

$$e^{N_e} > \left(\Omega_{m,0} \frac{a_{eq}}{a_0}\right)^{1/4} \sqrt{\frac{H_1}{H_0}} = \Omega_{\rm rad,0}^{1/4} \sqrt{\frac{H_1}{H_0}} = \left(\Omega_{\rm rad,0} \frac{\rho_1}{\rho_{0,\rm cr}}\right)^{1/4} , \tag{8.33}$$

where we used $H^2 = (8\pi G/3)\rho$ if $K/(a^2H^2) \to 0$ (like in the case of ρ_1, the density at the begining of the RD era), or $H^2 = (8\pi G/3)\rho_{0,\mathrm{cr}}$. We have also used the fact that $\Omega_m/\Omega_{\mathrm{rad}} = \rho_m/\rho_{\mathrm{rad}} \propto a$, so $\Omega_{m,0}/\Omega_{\mathrm{rad},0} = \Omega_{\mathrm{m,eq}}/\Omega_{\mathrm{rad,eq}} a_0/a_{eq} = a_0/a_{eq}$.

We then substitute numbers, giving

$$\rho_{\mathrm{cr},0} = [3 \times 10^{-3}\,\mathrm{eV}]^4 h^2; \quad \text{and} \quad \Omega_{\mathrm{rad},0} \simeq \frac{\Omega_{m,0}}{20000} \sim \frac{0.3}{20000}, \tag{8.34}$$

so that the bound is ($h \sim 0.7$)

$$e^{N_e} > \frac{[\rho_1]^{1/4}}{0.037\,\mathrm{heV}} \sim e^{56} \frac{[\rho_1]^{1/4}}{5 \times 10^{13}\,\mathrm{GeV}}. \tag{8.35}$$

On the other hand, the scale $[\rho_1]^{1/4}$, approximately the scale of inflation, is bounded from below by the Big Bang Nucleosynthesis scale (since we know everything that happens from BBN onwards, and inflation has no place in it), so $[\rho_1]^{1/4} > \rho_{BBN}^{1/4} \sim 1\,\mathrm{MeV}$. For this minimal value, the bound would be

$$e^{N_e} > e^{56} \times 2 \times 10^{-17} \simeq e^{17}. \tag{8.36}$$

The scale is also bounded from above by the Planck scale, $[\rho_1]^{1/4} < [\rho_{\mathrm{Pl}}]^{1/4} \sim 10^{19}\,\mathrm{GeV}$. For this maximal value, the bound would be

$$e^{N_e} > e^{56} \times 2 \times 10^{7} \simeq e^{68}. \tag{8.37}$$

In fact, the value preferred by experimental constraints on generic models is $[\rho_1]^{1/4} \sim 2 \times 10^{16}\,\mathrm{GeV}$, so that the bound in this case is

$$e^{N_e} > e^{56} \times 10^{3} \simeq e^{62}. \tag{8.38}$$

Later on in the book, we will see a more precise formula, which is

$$N_e > 56 - \frac{2}{3} \ln \frac{10^{16}\,\mathrm{GeV}}{\rho_*^{1/4}} - \frac{1}{3} \ln \frac{10^{19}\mathrm{GeV}}{T_R}, \tag{8.39}$$

where T_R is the reheat temperature, and ρ_* is the density at the end of inflation. Then for the case $T_R = \rho_*^{1/4}$, we obtain

$$N_e > 56 - \ln \frac{10^{13.6}\mathrm{GeV}}{\rho_*^{1/4}} = 56 + \ln \frac{\rho_*^{1/4}}{5 \times 10^{13}\mathrm{GeV}}, \tag{8.40}$$

which is the previous constraint for $\rho_* \to \rho_1$.

Smoothness and Horizon Problem

We have seen that during the radiation dominated era, with horizon distance $H^{-1} \sim t \propto a^2$, and during the matter dominated era, with horizon distance $H^{-1} \sim t \propto a^{3/2}$, scales enter within the horizon, since scales go like a.

The solution to the smoothness and horizon problem is that during the period of exponential expansion with $H^{-1} =$ constant (or even just accelerated expansion $a(t) \sim t^p$, $p > 1$, with $H^{-1} \propto t \propto a^{1/p}$), scales $\sim a$ get outside the horizon.

So the picture is that a small patch of space is blown up exponentially by the expansion, while the horizon distance remains constant, and in so doing, the scale gets "frozen-in", losing causal contact. Then during either the R.D. or M.D. eras, the scales falls back within the horizon, and regains causal contact, starting to react and evolve again. Then the fact that there was an apparent lack of causal contact on the largest scales is simply because the scales were frozen-in, but they actually used to be in causal contact before getting outside the horizon.

To describe the horizon problem in quantitative terms, we define first, as before, the horizon distance at time t_{ls},

$$d_H(t_{ls}) = a(t_{ls}) \int_{t_{bi}}^{t_{ls}} \frac{dt}{a(t)} , \tag{8.41}$$

which is dominated by the period of inflation begining at t_{bi}. That sounds counterintuitive, since we just argued that simply a factor of 200 in time during matter domination means that the latter period dominates the integral, but we will see that the magic of exponential expansion means that the early times actually dominate the integral. During inflation, for $t < t_I$,

$$a(t) = a_I e^{-H_I(t_I - t)} , \tag{8.42}$$

and the number of e-folds during inflation is

$$N_e = H_I(t_I - t_{bi}) , \tag{8.43}$$

leading to

$$d_H(t_{ls}) \simeq \frac{a(t_{ls})}{a_I} \int_{t_{bi}}^{t_I} dt e^{H_I(t_I - t)} = \frac{a(t_{ls})}{a_I H_I}(e^{N_e} - 1) \simeq \frac{a(t_{ls})}{a_I H_I} e^{N_e}. \tag{8.44}$$

This needs to be compared to the distance travelled by light from t_{ls} to now (t_0), measured at t_{ls}, i.e. ($H_0 = 2/(3t_0)$ during matter domination)

$$r_H(t_{ls}) = a(t_{ls}) \int_{t_{ls}}^{t_0} \frac{dt}{a(t)} = \frac{a(t_{ls})}{a_0} 3t_0 = \frac{2a(t_{ls})}{a_0 H_0}. \tag{8.45}$$

So the condition

$$\frac{d_H(t_{ls})}{2r_H(t_{ls})} > 1 \tag{8.46}$$

amounts to the same condition as obtained for the flatness problem,

$$e^{N_e} > \frac{a_I H_I}{a_0 H_0}. \tag{8.47}$$

Monopole Problem

As we said, the problem is that unlike other species, monopoles don't annihilate, and would remain until now. In a phase transition like the GUT phase transition, we expect to create about one monopole per nucleon. But from direct searches in the Universe, we know that there are less than 10^{-30} monopoles per nucleon. So there are too many monopoles (or in general, relics), and we need to dilute them somehow. It is very important that this dilution happens *after* the phase transition, so inflation needs to happen after the transition, or at most simultaneously with it.

The way the dilution happens is by the exponential expansion of a small causal patch. Since we have created one monopole per nucleon, we need expand the size of the Universe, defined by $a(t)$, by at least 10^{10}, so that the volume of the Universe grows by 10^{30}. Monopoles get diluted this way, and of course also do nucleons. But nucleons are created in much bigger numbers at the end of inflation, during reheating. So by diluting the monopoles (and other relics as well), we solve the problem. We need at last

$$N_e > ln 10^{10} = 23 \tag{8.48}$$

e-folds, which is a weaker bound than we found before. Of course, it actually says that the phase transition creating the relics needs to happen at least 23 e-folds before the end of inflation.

Why is the Entropy Large?

The reason the entropy per baryon is large is related to the presence of quantum fluctuations, and with reheating, that will populate the Universe with photons. This will be studied later on in the book. Since we have an exponential expansion, the size of the Universe will increase, leading to a large *total* number of photons.

Why Do We Have Perturbations?

We will explain this later, but it is basically because of quantum fluctuations, that will be exponentially blown up until they become classical and freeze out outside the horizon.

Why Do We Have a (Very Small) Baryon Asymmetry?

The baryon asymmetry itself is due to CP violation, but the conditions given by Sakharov (baryon number violation, CP violation, and interactions out of equilibrium) for the creation of baryons (baryogenesis) include the existence of interactions out of equilibrium, satisfied by a fastly expanding Universe like the case of inflation. Moreover, the issue is also why there is such a *very small* baryon asymmetry. The answer is that we have a large entropy per baryon ($\sim 10^9$), that gets converted into baryon-antibaryon pairs, and leads to $(N_B - N_{\bar{B}})/N_B \sim 10^{-9}$.

8.4 Inflation with a Scalar Field

Finally, we describe the most popular incarnation of the inflationary paradigm, using an approximately constant energy density for a scalar field moving down a potential. The scalar field is called an inflaton, and its equations of motion in the FRLW Universe become the Friedmann equation

$$\left(\frac{\dot{a}}{a}\right)^2 + \frac{K}{a^2} = \frac{1}{3M_{\rm Pl}^2}\left[\frac{\dot{\phi}^2}{2} + V(\phi)\right], \tag{8.49}$$

and the Klein–Gordon equation,

$$\ddot{\phi} + 3\frac{\dot{a}}{a}\dot{\phi} = -V'(\phi). \tag{8.50}$$

Note that the Hubble expansion term (the one with $\dot{a}/a$) is a friction term, that moves against the acceleration of the scalar.

There is no a priori need to have a description in terms of a scalar field, it is just the simplest possibility, but logically, it can be also in terms of a vector, or tensor field, etc.

Old Inflation

The first model, by Alan Guth, called "old inflation", was based around the idea of a phase transition, with the idea in mind to solve the monopole problem, by having inflation in the same time, and continuing slightly after, as the phase transition.

A Higgs ("Mexican hat") potential, with a minimum at some $\phi_0 \neq 0$, and a nonzero value at $\phi = 0$ gets corrected by temperature effects in quantum field theory, so that for $T > T_c$, the true vacuum is at $\phi = 0$, for $V(0) > 0$, and ϕ_0 is lifted above it, as in Fig. 8.1. At $T = T_c$, the two vacua at 0 and ϕ_0 coincide. And at $T < T_c$, but not too small, there is a false vacuum at $\phi = 0$, separated from the true vacuum at ϕ_0 by a potential barrier.

Then the Universe starts in a hot state, at the (then) minimum of $\phi = 0$. As it cools off below T_c, it still remains there, now because of the potential barrier. Since we are at a constant $V(0) > 0$, the Universe undergoes a period of exponential inflation. But then it tunnels quantum mechanically (or jumps over due to thermal fluctuations) through the barrier, creating a small bubble.

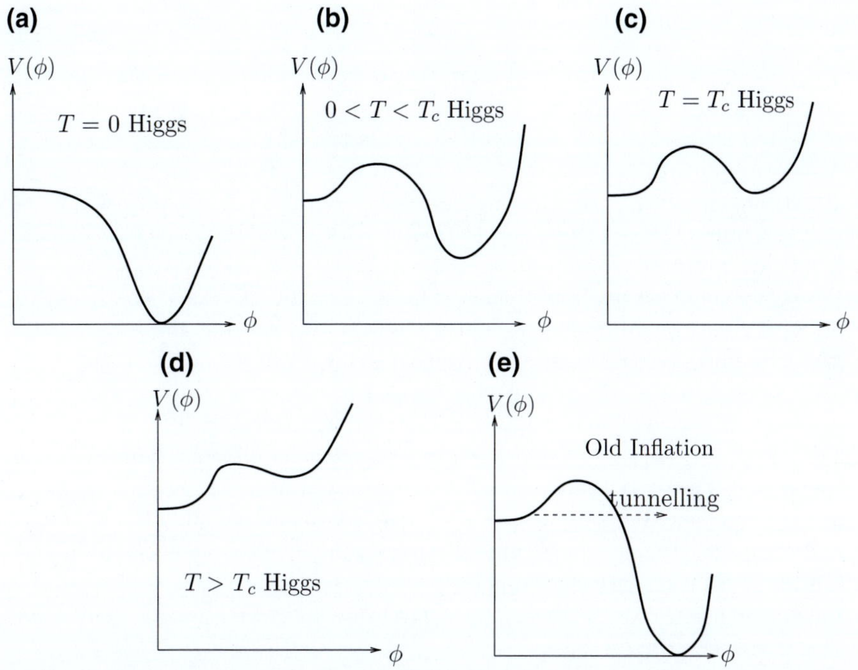

Fig. 8.1 **a** Higgs potential at $T = 0$. **b** Higgs potential at $0 < T < T_c$. **c** Higgs potential at $T = T_c$. **d** Higgs potential at $T > T_c$. **e** Old inflation potential is like case **b**, but has the true minimum at $V = 0$ (so is a combination of **a** and **b**)

The bubble, with a wall having a surface tension defined by $V(0)$, starts expanding, like a bubble of water in undercooled water vapor (or of vapor in overheated water), because it destabilizes the neighbouring regions into collapsing. This mechanism has been described by Coleman and DiLuccia. So it would seem that this bubble nucleation would give an end to inflation, since various bubbles would nucleate, and then eventually collide. This is what happens in water. But unlike water, in the inflationary Universe, the Universe in between the bubbles is still in the false vacuum at $\phi = 0$, so it still inflates, i.e., expands exponentially. That means that despite the fact that the walls of bubbles move at the speed of light, neighbouring bubbles never meet, since the space in between them expands exponentially. This is known as the "graceful exit problem" of old inflation.

Because of it, the "old inflation" model was dropped, and the "new inflation" of Albrecht, Steinhardt and Linde, supplanted it. In it, the scalar potential is almost flat, a plateau, but with a small slope, followed by a drop. Since Hubble expansion acts as a friction term, the movement on the plateau will be slow. This is the "slow-roll" paradigm, that will be described in the next chapter.

Important Concepts to Remember

- The smoothness and horizon problem is why is the Universe smooth and isotropic? Since the ratio of the distance travelled by light from last scattering until today, in today's scales, divided by the horizon at last scattering, in today's scales, is $N = 2r_H(t_0)/d_H(t_0) \sim 72$.
- The flatness problem is why has the Universe $\Omega \simeq 1$ to an absurd precision in the past (assuming $\Omega \sim 1$ today leads to $\Omega - 1 \sim 10^{-16}$ at electron-positron annihilation, for instance).
- Why is the total entropy in the Universe today of about 10^{88} (very large), why do we have a very small baryon asymmetry, how do we generate the perturbations that grow into structure (stars and galaxies); are also issues.
- The monopole (or relic) problem: what dilutes relics appearing in phase transitions, say (such that naively, we have as many baryons as relics).
- The inflationary paradigm requires most generally a period of accelerated expansion $\ddot{a} > 0$. The most common way is through a constant energy density, which leads to exponential expansion, $a \propto e^{Ht}$ (though power law inflation $a \propto t^p, p > 1$ is also fine), and usually it is assumed that it is parametrized by a scalar field, though it is likely not necessary, only simpler.
- Inflation happens near the Planck scale, but below it, so we can have an effective field theory description.
- The flatness problem is solved since $\Omega - 1 \propto t^{2(1-p)}$ or $\propto e^{-2Ht}$ drops very fast during inflation.
- Assuming that both now, and at the beginning of inflation, we have $\Omega \sim 1$, we find that we need at least $N_e > 56 + \ln \frac{[\rho_1]^{1/4}}{5\times 10^{13}\text{GeV}}$ e-folds of inflation.
- The smoothness and horizon problem is solved since during inflation, scales get blown up exponentially, while the horizon stays fixed, so scales get outside the horizon fast, and come back inside the horizon during standard cosmology. We obtain the same constraint as above on N_e.
- The monopole problem is solved if the phase transition occurs before or during inflation, so the generated relics of order one per baryon (in GUTs) are diluted away to less than 10^{-30} today, so need to be diluted by a factor of $e^{3N_e} > 10^{30}$, or $N_e > 23$ e-folds.
- Old inflation was inflation in a false vacuum $\phi = 0$, created as the temperature drops, whereas the true vacuum shifts to $\phi_0 \neq 0$. Then tunneling through the barrier ends inflation, but we have no "graceful exit", since bubbles of true vacuum never meet.
- New inflation is rolling down a plateau in order to inflate, after which there is a drop to a minimum.

Further reading: See [1, 10, 11].

Exercises

(1) If inflation occurs at temperature scale T_i and has 60 e-folds, calculate when the scale that got outside the horizon at the begining of inflation will cross inside it.
(2) Calculate the monopole/nucleon ratio that would require the same constraint on the number of e-folds as the one from reheating.
(3) Show that the equations of motion of the Einstein+scalar action reduce to

$$\left(\frac{\dot{a}}{a}\right)^2 + \frac{K}{a^2} = \frac{1}{3M_{\rm Pl}^2}\left(\frac{\dot{\phi}^2}{2} + V(\phi)\right)$$
$$\ddot{\phi} + 3\frac{\dot{a}}{a}\dot{\phi} = -V'(\phi). \tag{8.51}$$

(4) What are the equations of motion for the gravity field coupled to an antisymmetric tensor field $A_{\mu\nu}$ with uniform (space independent) field strength H_{ijk} (instead of a scalar) in the FLRW Universe?
(5) If inflation occurs at the GUT scale, and now $1 - \Omega \simeq 10^{-3}$, how many e-folds of inflation do we need in order to obtain the current Ω from a generic one at the beginning of inflation?

Chapter 9
Slow-Roll Inflation

In this chapter we will explore the simplest incarnation of inflation, namely the inflation with a scalar field that gives an approximate cosmological constant. We will see that we can quantify the condition into what is known as "slow-roll inflation". We will not address other types of inflation in this book, though we will explain what they can be.

As we said at the end of last chapter, after Guth's "old inflation" model, where the scalar sits at the false vacuum $\phi = 0$, with $V(0) > 0$ and then tunnels out of the barrier to the true vacuum at $\phi = \phi_0$, came the "new inflation" model. In "new inflation", there is a plateau in the potential, so $V(\phi)$ is large, but flat, after which there is a drop, as in Fig. 9.1. But this paradigm was much generalized afterwards, to the point that even the simplest thing we can think of, a polynomial potential $V = \lambda\phi^p$, fits into it, but we need the scalar to *slowly roll* down the potential. In the case of the polynomial potential, and other similar ones, we can have a condition on initial conditions (specifically, for the polynomial, that we start at large field ϕ). In fact, even slow-roll is not necessary for *inflation*, it is only necessary for *(almost) exponential inflation*. But in this chapter we will focus on this slow-roll inflation.

Variations of the new inflation with a plateau are "hilltop inflation", where we just start near the maximum of a potential (fine-tuned initial conditions); and "natural inflation", where we have a periodic, "axion-like" potential. Indeed, the axion of QCD is the one example we know of, of a periodic potential. In string theory, we have more "axions" with periodic potentials.

As we saw in the previous chapter, the equations of motion of the gravity+scalar system are: the Friedmann equation

$$H^2 + \frac{K}{a^2} \equiv \left(\frac{\dot{a}}{a}\right)^2 + \frac{K}{a^2} = \frac{1}{3M_{\rm Pl}^2}\left[\frac{\dot{\phi}^2}{2} + V\right] ; \tag{9.1}$$

H. Năstase, *Cosmology and String Theory*, Fundamental Theories of Physics 197, https://doi.org/10.1007/978-3-030-15077-8_9

Fig. 9.1 The "new inflation" potential

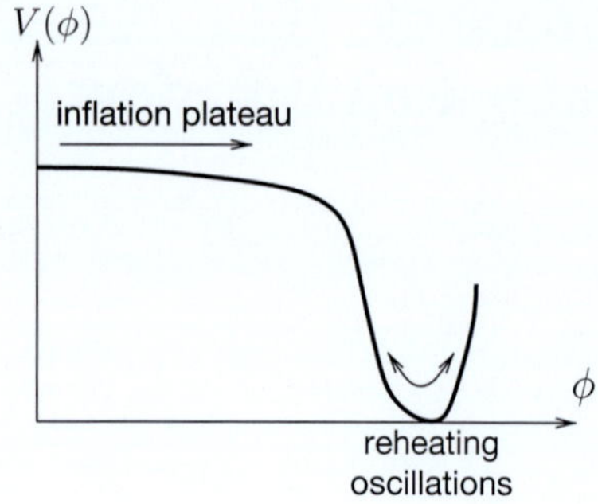

note that here $\frac{\dot{\phi}^2}{2} + V(\phi) = \rho_\phi$ is the density of the scalar field, whereas the pressure is $P_\phi = \frac{\dot{\phi}^2}{2} - V(\phi)$. We will shortly neglect k/a^2, since this goes quickly to zero (in a couple of e-folds), due to the exponential inflation. The second equation is the KG equation, but it can also be obtained from the continuity equation, i.e., the conservation of energy, $\dot{\rho} = -3H(\rho + P)$, by substituting $\rho = \rho_\phi$ and $P = P_\phi$. In both cases, the result is

$$\ddot{\phi} + 3H\dot{\phi} + V'(\phi) = 0. \tag{9.2}$$

9.1 Slow-Roll Analysis

We start the analysis with the assumption that slow roll means that the Hubble parameter $H(t)$ has a relative variation in the time = horizon time $t_H = H^{-1}$ that is much less than one,

$$\frac{|\dot{H}|}{H^2} = -\frac{\dot{H}}{H^2} \equiv \epsilon_H \ll 1. \tag{9.3}$$

This defines the first slow-roll parameter ϵ_H. Taking the time derivative of (9.1), we get

$$2H\dot{H} = \frac{\dot{\phi}}{3M_{\text{Pl}}^2}[\ddot{\phi} + V'] = -\frac{H\dot{\phi}^2}{M_{\text{Pl}}^2}, \tag{9.4}$$

where in the second equality we have used (9.2). Then finally

$$\dot{H} = -\frac{\dot{\phi}^2}{2M_{\text{Pl}}^2}. \tag{9.5}$$

Then the slow-roll condition $|\dot{H}| \ll H^2 = (\dot{\phi}^2/2 + V)/(3M_{\text{Pl}}^2)$ becomes

$$\dot{\phi}^2 \ll |V(\phi)|, \tag{9.6}$$

i.e., that the kinetic energy of the scalar is much smaller than its potential energy. This is what we would have expected anyway for a slow-roll condition. In this case, we also obtain that $\rho \simeq -P \simeq V(\phi)$, which means that we have an *approximate cosmological constant*. Then as we saw in the previous chapter, the Hubble constant is

$$H \simeq \sqrt{\frac{V}{3M_{\rm Pl}^2}}. \tag{9.7}$$

The next condition for slow roll is similar to the first, but refers to the scalar field instead of H, namely that the relative variation of $\dot{\phi}$ during the horizon time $t_H = H^{-1}$ is small,

$$\left|\frac{\ddot{\phi}}{\dot{\phi}}\right| \frac{1}{H} \ll 1. \tag{9.8}$$

Since $|\ddot{\phi}| \ll |\dot{\phi}H|$, we can neglect the first term in the KG equation (9.2), so that it becomes

$$3H\dot{\phi} \simeq -V'(\phi)\ , \tag{9.9}$$

meaning that we can write

$$\dot{\phi} \simeq -\frac{V'}{3H} \simeq -\frac{V'}{\sqrt{3V}} M_{\rm Pl}\ , \tag{9.10}$$

where in the last equality we have used the Friedmann equation in the slow-roll form (9.7). Then from (9.5) and the above formula for $\dot{\phi}$, we have the condition

$$\epsilon_H = \frac{|\dot{H}|}{H^2} = \frac{3\dot{\phi}^2}{2V} \simeq \frac{M_{\rm Pl}^2}{2}\frac{V'^2}{V^2} \equiv \epsilon \ll 1\ , \tag{9.11}$$

where the equality $\epsilon_H \simeq \epsilon$ is only valid in the case both slow-roll conditions on ϕ are satisfied ($\dot{\phi}^2 \lll |V|$ and $|\ddot{\phi}| \ll |\dot{\phi}H|$). Here we have defined the parameter (depending only on the potential)

$$\epsilon \equiv \frac{M_{\rm Pl}^2}{2}\left(\frac{V'}{V}\right)^2. \tag{9.12}$$

Thus in the case of slow roll, $\epsilon \simeq \epsilon_H \ll 1$.

To find the second condition for slow-roll in terms of the potential only, we start from the KG equation when neglecting its the first term, $\dot{\phi} = -V'/(3H)$, and we take a time derivative, obtaining

$$\ddot{\phi} = -\frac{V''\dot{\phi}}{3H} + \frac{V'\dot{H}}{3H^2} = \frac{V''V'}{9H^2} - \frac{V'^3 M_{\rm Pl}^2}{6V^2}\ , \tag{9.13}$$

where in the second equality, we have used again $\dot{\phi} = -V'/(3H)$, as well as (9.5). If $\epsilon \ll 1$, from its definition it means that $V'^3 M_{\rm Pl}^2/V^2 \ll V'$, but since we want to neglect the first, acceleration, term ($\ddot{\phi}$) in the KG equation with respect to the second, Hubble friction, term, and then the equation is $3H\dot{\phi} = -V'$, it means that we want $\ddot{\phi} \ll V$, and then (9.13) implies we need also

$$|V''| \ll 9H^2 = \frac{3|V|}{M_{\rm Pl}^2}. \tag{9.14}$$

Finally, that gives the condition

$$|\eta| \ll 1\,, \tag{9.15}$$

where we have defined the second slow-roll parameter,

$$\eta \equiv M_{\rm Pl}^2 \frac{V''}{V} \simeq \frac{V}{3H^2}. \tag{9.16}$$

In conclusion, the slow-roll conditions can be reduced to conditions on the potential itself, $\epsilon \ll 1$ and $\eta \ll 1$.

Variation of Slow Roll Quantities with the Number of e-Folds

We define the number of e-folds between two times t_1 and t_2 as before,

$$N = \ln \frac{a(t_2)}{a(t_1)}\,, \tag{9.17}$$

so that

$$dN = \frac{da}{a} = -H dt. \tag{9.18}$$

Then the first relation for the variation with respect to the number of e-folds is for $\ln H$,

$$-\frac{d(\ln H)}{dN} = \frac{dH}{H^2 dt} = -\epsilon_H \simeq -\epsilon. \tag{9.19}$$

Similarly, we find the variation of $\ln \epsilon = 2\ln(V'/V) + \ln M_{\rm Pl}^2/2$,

$$-\frac{d(\ln \epsilon)}{dN} = \frac{d\epsilon}{\epsilon H dt} \simeq \frac{2\dot{\phi}}{H}\left(\frac{V''}{V'} - \frac{V'}{V}\right) = -\frac{2V''}{V} M_{\rm Pl}^2 + \frac{2V'^2}{V^2} M_{\rm Pl}^2\,, \tag{9.20}$$

where in the last relation we have used the KG equation $3H\dot{\phi} \simeq -V'$ and the Friedmann equation $H^2 = V/(3M_{\rm Pl}^2)$, so that finally

$$-\frac{d(\ln \epsilon)}{dN} = 4\epsilon - 2\eta. \tag{9.21}$$

The last variation equation is for the variation of η with the number of e-folds. From the definition of η and N,

$$-\frac{d\eta}{dN} = M_{\rm Pl}^2 \frac{d}{Hdt}\left(\frac{V''}{V}\right) = \frac{M_{\rm Pl}^2\dot{\phi}}{H}\left(\frac{V'''}{V} - \frac{V''V'}{V^2}\right) = -\frac{M_{\rm Pl}^2}{V/M_{\rm Pl}^2}V'\left(\frac{V'''}{V} - \frac{V''V'}{V^2}\right)$$
$$= -\left(\frac{M_{\rm Pl}^4 V'''V'}{V^2}\right) + 2\left(\frac{M_{\rm Pl}^2 V''}{V}\right)\left(\frac{M_{\rm Pl}^2}{2}\frac{V'^2}{V^2}\right), \quad (9.22)$$

where in the third equality we used the Friedmann equation $\dot{\phi} = -V/(3H)$. Defining

$$\xi^2 \equiv \frac{M_{\rm Pl}^4 V'''V'}{V^2}, \quad (9.23)$$

we have finally

$$-\frac{d\eta}{dN} = -\xi^2 + 2\epsilon\eta. \quad (9.24)$$

9.2 General (Non-slow Roll) Formulas

We now derive formulas that are always valid, not just in the slow-roll case (so they would be valid even for a "fast roll", or rather non-slow roll, inflation).

First, we have seen already that we have the exact relation (9.5), derived from the Friedmann (9.1) and KG (9.2) equations, so $\dot{\phi}^2 = -2\dot{H}M_{\rm Pl}^2$. Consider next that we have the implicit relation $H = H(\phi)$, defined *on the trajectory solving the equations of motion*, via $H = H(t)$ and $\phi = \phi(t)$, so $\dot{H} = H'(\phi)\dot{\phi}$ ($H'(\phi) = \dot{H}/\dot{\phi}$). Then we find

$$\dot{\phi} = -2M_{\rm Pl}^2 H'(\phi). \quad (9.25)$$

Replacing this $\dot{\phi}$ in the Friedmann equation (9.1), we find

$$H^2(\phi) = \frac{1}{3M_{\rm Pl}^2}[2M_{\rm Pl}^4 H'^2(\phi) + V(\phi)] \Rightarrow$$
$$H'^2(\phi) - \frac{3}{2M_{\rm Pl}^2}H^2(\phi) = -\frac{V(\phi)}{2M_{\rm Pl}^4}. \quad (9.26)$$

This takes the forma of a *Hamilton–Jacobi equation for* $H(\phi)$ (understood like similar to the action function $S(\phi)$ in the classical mechanics formulation of the Hamilton–Jacobi equation).

Using the formula for $\dot{H}$ in (9.5) and the Friedmann equation (9.1), we find

$$\epsilon_H \equiv -\frac{\dot{H}}{H^2} = 3\frac{\dot{\phi}^2/2}{V + \dot{\phi}^2/2} = 3\frac{2M_{\rm Pl}^4 H'^2}{V + 2M_{\rm Pl}^4 H'^2} = 2M_{\rm Pl}^2\left(\frac{H'(\phi)}{H(\phi)}\right)^2, \quad (9.27)$$

where in the third equality we have used (9.25) in the numerator and denominator, and in the fourth equality we have used (9.26) in the denominator. The end result is similar to the definition of ϵ in terms of $V(\phi)$, just that now we have ϵ_H in terms of $H(\phi)$.

By analogy with the case of ϵ in terms of $V(\phi)$, we can define a quantity analogous to η from $V(\phi)$, namely

$$\eta_H \equiv 2M_{\rm Pl}^2 \frac{H''(\phi)}{H(\phi)}. \tag{9.28}$$

Then from (9.25), taking a time derivative, we find

$$\ddot{\phi} = -2M_{\rm Pl}^2 \dot{\phi} H''(\phi) , \tag{9.29}$$

so that

$$\eta_H = \frac{2M_{\rm Pl}^2 H''(\phi)}{H(\phi)} = -\frac{\ddot{\phi}}{\dot{\phi}H}. \tag{9.30}$$

On the other hand, from (9.25), we have

$$\epsilon_H = 2M_{\rm Pl}^2 \frac{H'^2}{H^2} = \frac{\dot{\phi}^2}{2M_{\rm Pl}^2 H^2}. \tag{9.31}$$

Taking a time derivative on this relation, we find

$$\dot{\epsilon}_H = \frac{\ddot{\phi}\dot{\phi}}{M_{\rm Pl}H^2} - \frac{\dot{\phi}^2 \dot{H}}{M_{\rm Pl}^2 H^3}. \tag{9.32}$$

In the second term, we use $\dot{H} = \dot{\phi}H' = -2M_{\rm Pl}^2 H'^2$ (from (9.25)), to obtain

$$\dot{\epsilon}_H = -\frac{\ddot{\phi}}{H\dot{\phi}}\left(-\frac{\dot{\phi}^2}{M_{\rm Pl}^2 H}\right) + \frac{\dot{\phi}^2}{HM_{\rm Pl}^2}\frac{2M_{\rm Pl}^2 H'^2}{H^2} = -2H\eta_H\epsilon_H + 2H\epsilon_H^2. \tag{9.33}$$

Solving for η_H, we find

$$\eta_H = \epsilon_H - \frac{\dot{\epsilon}_H}{2H\epsilon_H}. \tag{9.34}$$

From $\epsilon_H = -\dot{H}/H^2$, we obtain

$$\eta_H = -\frac{\dot{H}}{H^2} - \frac{1}{2H}\left(\frac{\ddot{H}}{\dot{H}} - 2\frac{\dot{H}}{H}\right) = -\frac{\ddot{H}}{2H\dot{H}}. \tag{9.35}$$

Finally, in the slow-roll case, as we saw, the Friedmann equation is $\dot{\phi} = -V'/(3H)$, so

$$\eta_H = -\frac{\ddot{\phi}}{\dot{\phi}H} = \frac{1}{H\dot{\phi}}\left(\frac{V''\dot{\phi}}{3H} + \frac{V'\dot{\phi}^2}{6M_{\rm Pl}^2 H^2}\right) = \frac{V''}{3H^2} + \frac{V'\dot{\phi}}{6H^3 M_{\rm Pl}^2} = M_{\rm Pl}^2\frac{V''}{V} - \frac{V'^2}{2\cdot 9H^4 M_{\rm Pl}^2}, \tag{9.36}$$

where in the second equality we have used (9.5) and in the fourth equality we have used again $\dot{\phi} = -V'/(3H)$. Therefore, in the slow roll case, we obtain

$$\eta_H \simeq M_{\rm Pl}^2\frac{V''}{V} - \frac{M_{\rm Pl}^2}{2}\frac{V'^2}{V^2} = \eta - \epsilon. \tag{9.37}$$

Finally, we compute a formula for the number of e-folds between time t and the end of inflation $t_{\rm end}$,

$$N \equiv \ln\frac{a(t_{\rm end})}{a(t)} = \int_t^{t_{\rm end}} Hdt = -\frac{1}{2M_{\rm Pl}^2}\int_\phi^{\phi_{\rm end}} d\phi\frac{H}{H'(\phi)} = -\frac{1}{M_{\rm Pl}}\int_\phi^{\phi_{\rm end}} d\phi\frac{1}{\sqrt{2\epsilon_H}}, \tag{9.38}$$

where we have used (9.25) in the form $dt = -d\phi/(2M_{\rm Pl}^2 H'(\phi))$.

Note that this is an *exact* formula, independent of slow-roll. In the slow roll case, $\epsilon_H \simeq \epsilon$, so

$$N \simeq -\int_\phi^{\phi_{\rm end}} \frac{d\phi}{M_{\rm Pl}}\frac{1}{\sqrt{2\epsilon}}. \tag{9.39}$$

9.3 Slow Roll Conditions

The slow-roll conditions can be conditions on the parameters of the potential, like in the case of potentials with a plateau (new inflation). They can alternatively be conditions on the initial conditions for inflation. For instance, consider the power law potential

$$V = g\phi^\alpha, \tag{9.40}$$

where $g > 0$ and $\alpha > 1$ (the last condition is so that $V'' > 0$, but otherwise it can be dropped). Then

$$2\epsilon = \left(\frac{\alpha M_{\rm Pl}}{\phi}\right)^2; \quad \eta = \alpha(\alpha-1)\frac{M_{\rm Pl}^2}{\phi^2}, \tag{9.41}$$

so $\epsilon \ll 1$, $|\eta| \ll 1$ can be solved by either $\alpha \ll 1$ (a condition on the parameters, which is discarded if $\alpha > 1$), or by the initial condition $\phi_0/M_{\rm Pl} \gg 1$. But even that condition needs to be supplemented with a condition on parameters, since for the validity of the effective field theory of the inflation scalar, we need to have no quantum gravity corrections. That in turn translates into the condition that, at least, $V(\phi_0) \ll M_{\rm Pl}^4$, which means that

$$g \ll \frac{M_{\rm Pl}^4}{\phi_0^\alpha}. \tag{9.42}$$

But we really need to choose $\phi_0 \gg M_{\rm Pl}$ not only for the initial condition, but also for a large interval, since we need to have at least 60 or so e-folds of inflation. That turns out to impose that

$$g \ll 2 \times 10^{-5} \tag{9.43}$$

in the case $\alpha = 4$, that is the power most favoured by experimental data among the possible power laws.

9.4 Exact Solution for Inflation

In most cases of potentials used for inflation, we cannot find an exact solution for the time evolution. There is one notable exception of potential, that can be used for checking the various approximations used in this chapter. We will check several, leaving the others as exercises.

The potential giving an exact solution is the inverse exponential,

$$V = ge^{-\lambda\phi}. \tag{9.44}$$

Among the possible solutions to the equations of motion (Friedmann and KG), there is the attractor solution

$$\phi(t) = \frac{1}{\lambda}\ln\left[\frac{g\epsilon^2 t^2}{M_{\rm Pl}^2(3-\epsilon)}\right] \tag{9.45}$$

together with the scaling

$$a \propto t^{1/\epsilon} \tag{9.46}$$

and where

$$\epsilon \equiv \frac{\lambda^2 M_{\rm Pl}^2}{2}. \tag{9.47}$$

From these solutions, we find

$$\dot{\phi} = \frac{2}{\lambda t} \quad \text{and} \quad H = \frac{1}{\epsilon t} = \frac{2}{\lambda^2 M_{\rm Pl}^2 t}. \tag{9.48}$$

Then the Friedmann equation for $H^2 = 1/(\epsilon^2 t^2)$, is

$$H^2 = \frac{1}{3M_{\rm Pl}^2}\left(\frac{\dot{\phi}^2}{2} + V\right) = \frac{1}{3M_{\rm Pl}^2}\left(\frac{2}{\lambda^2 t^2} + g\frac{M_{\rm Pl}^2(3-\epsilon)}{g\epsilon^2 t^2}\right) = \frac{1}{\epsilon^2 t^2}, \tag{9.49}$$

so indeed it is satisfied. Also the KG equation,

$$\ddot{\phi} + 3H\dot{\phi} + V'(\phi) = -\frac{2}{\lambda t^2} + 3\frac{2}{\lambda t}\frac{1}{\epsilon t} - \frac{\lambda g M_{\rm Pl}^2(3-\epsilon)}{g\epsilon^2 t^2} = 0\,, \tag{9.50}$$

is satisfied.

The slow roll conditions give

$$\epsilon = \frac{M_{\rm Pl}^2}{2}\frac{V'^2}{V^2} = \frac{\lambda^2 M_{\rm Pl}^2}{2}; \quad \eta = M_{\rm Pl}^2 \frac{V''}{V} = \lambda^2 M_{\rm Pl}^2 = 2\epsilon. \tag{9.51}$$

We note therefore that ϵ was well named, since indeed is the slow-roll parameter for the potential, and it is constant. The number of e-folds on the solution is

$$N = \ln \frac{a(t_2)}{a(t_1)} = \frac{1}{\epsilon} \ln \frac{t_2}{t_1} = \frac{\lambda}{2\epsilon}(\phi_2 - \phi_1) = \frac{1}{\lambda M_{\rm Pl}^2}(\phi_2 - \phi_1). \tag{9.52}$$

This matches with the result for a constant ϵ,

$$N \simeq -\frac{1}{\sqrt{2\epsilon} M_{\rm Pl}^2}(\phi_2 - \phi_1) = -\frac{1}{\lambda M_{\rm Pl}^2}(\phi_2 - \phi_1). \tag{9.53}$$

The slow-roll approximation for the solution is obtained neglecting $\ddot{\phi}$ in the KG equation, so that we need to solve $3H\dot{\phi} = -V'(\phi) = +g\lambda e^{-\lambda\phi}$, solved by

$$\phi(t) = \frac{1}{\lambda} \ln \left[\frac{\epsilon^2 t^2}{3 M_{\rm Pl}^2} \right], \tag{9.54}$$

which is indeed the general solution, neglecting the ϵ in the $3 - \epsilon$ factor in the denominator. The same solution also solves the Friedmann equation in which we drop the $\dot{\phi}^2/2$ term.

9.5 Variants of Inflation

Besides the general slow-roll paradigm for inflation, we have also other ideas that became important.

Eternal Inflation

This idea, originating with Steinhardt and others, is that, if we consider probabilities to be at different values of the potential V, distributed perhaps with a Boltzmann weight (or some other probability decreasing with energy), then the probability to be at large $V(\phi)$ will be small. If the probability refers to various patches of space to have some initial condition for the field, the initial patch with large V will be very small. However, since by being at $V \neq 0$ we inflate, the volume of the small patch quickly increases, also exponentially, so it could dominate.

Naively then, whether this patch will dominate or not depends on the relative strength of the coefficient β in the probability weight $e^{-\beta V}$ and the amount of exponential expansion of the volume $\mathcal{V} \sim e^{+3Ht} \sim e^{\sqrt{3V}t}$. But of course, in reality, we

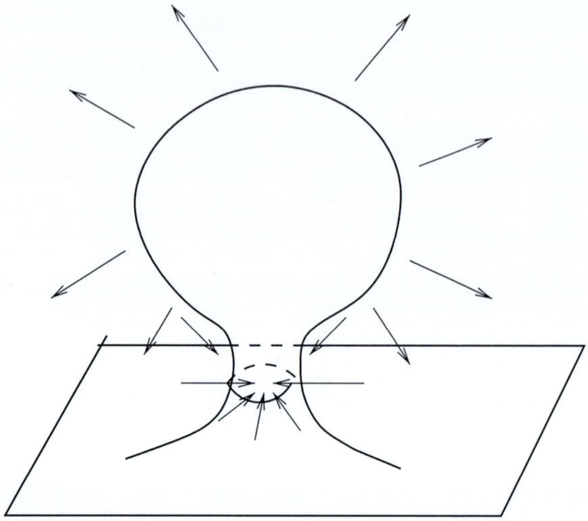

Fig. 9.2 A patch of higher V inside one of lower V creates a Baby Universe that peels off from the original one: there is a throat connecting the Baby Universe to the original one, that is collapsing. While the Baby Universe expands, viewed from the original one, the throat collapses

don't know what is the probability distribution over volumes, and moreover, it is not clear that we can even apply statistics to cosmology, where we have only one Universe (and not an ensemble of Universes; at least not to explore, there could be un-detectable ones in the many-worlds interpretation of Quantum Cosmology).

Moreover, we can consider fluctuations for the value of the potential: fluctuating upwards in V costs energy, so is less likely, but it has a higher benefit because of inflation. But again the issue of probabilities is a very tricky one.

So the issue of whether or not eternal inflation leads to an inflation defined by a large potential V seems a priori impossible to decide. What we do know is that in this idea, there will always be inflating regions, since there are always small patches of the Universe that have some high potential value, and these will eventually dominate over the rest. If they are outside the horizon, we cannot observe them, so the point is likely moot. And if they are inside the horizon, they might eventually overrun our own Universe, effectively killing it (though there is some debate about the efficiency of that scenario).

Chaotic Inflation

This idea, due to Linde, is that the initial conditions are random in space, so inflation occurs in patches. Moreover, when a patch inflates, it becomes homogenous and isotropic, smoothing out all initial inhomogeneities and anisotropies. The evolution of a patch of high $V = V_2$ inside one of lower V_1 is such that the patch inflates faster as viewed from inside it, but viewed from the outside, there is a "throat" that collapses due to its surface tension ($T = (V_2 - V_1)/Area$), as in Fig. 9.2. So eventually we have this Universe that detaches (its connection with the rest shrinks), but expands, smoothing out all inhomogeneities and vanishing an initial $\Omega - 1$.

The chaotic inflation picture is related to potentials without a plateau, but rather continuously related to zero, like the polynomial $V = g\phi^\alpha$, which is its standard example. Having an initial condition at large ϕ is associated with a (quantum or thermal) fluctuation up the potential, like in the eternal inflation idea above.

Hybrid Models

We have also considered only single-scalar models until now, but that is not necessary. The simplest generalization is in the so-called hybrid models, which contain two scalars, one (σ) with its own potential $V(\sigma)$, and the other with a potential coupling depending on σ,

$$V(\phi, \sigma) = \frac{g^2}{2}\phi^2\sigma^2 + V(\sigma). \tag{9.55}$$

Multi-scalar Models

Finally, there could be multi-scalar models (N-flation, etc.), but we will not describe them. Motion in a multi-scalar potential landscape can help with some issues, but in most cases amounts to simply powers of N in various formulas.

9.5.1 *Initial Conditions for Inflation*

The issue of initial conditions is a tricky one, and depending on the point of view, can be considered a problem (e.g., see papers by Steinhardt in 2013), or not (e.g. papers by Linde and Guth in the same year). One point is that we needed to assume $\dot{\phi}^2/2 \ll V(\phi)$ at all times, including initially. The claim is that this is not generic, and equipartition of energy would demand that initially both terms are comparable. But this assumes a lot about the initial condition defined by quantum gravity (which could appear "out of nothing" like in the gravitational instanton defining the wave function of the Universe in a "no-boundary boundary condition" of Hartle and Hawking), as well as about quantum gravity itself (is equipartition of some *effective field theory* for the scalar reasonable?)

Other issues are more obvious: as we said, homogeneity and isotropy needs not to be assumed; in the chaotic inflation scenario, it is something that comes out of the dynamics, not initial conditions.

The last point to be addressed is of the initial condition for $V(\phi)$. We need that $V(\phi_0)^{1/4} \ll M_{\rm Pl}$ (slightly lower than the quantum gravity scale), since we need that the effective field theory of the scalar is a good approximation, meaning we are below the Planck scale, yet the initial condition is defined by quantum gravity, so we are close to it.

Important Concepts to Remember

- Slow-roll inflation requires $\epsilon_H = -\dot{H}/H^2 \ll 1$ ($\dot{\phi}^2 \ll V(\phi)$) and $\eta = M_{\rm Pl}^2 V''/V \ll 1$.
- During slow-roll, $\epsilon_H \simeq \epsilon \equiv M_{\rm Pl}^2/2(V'/V)^2$.
- In a general Hamilton–Jacobi picture, we have $\epsilon_H = 2M_{\rm Pl}^2(H'(\phi)/H(\phi))^2$, and we can define $\eta_H = 2M_{\rm Pl}^2(H''(\phi)/H(\phi))$.

- During slow-roll, $\eta_H \simeq \eta - \epsilon$.
- The number of e-folds during inflation is $N_e = -\int d\phi/M_{\rm Pl}/\sqrt{2\epsilon_H}$.
- The slow-roll conditions are conditions on the parameters and the initial conditions for the fields. Imposing also the number of e-folds, usually results in conditions on the parameters only.
- $V = ge^{-\lambda\phi}$ gives an exact solution for inflation, on which we can check the various formulas.
- Eternal inflation means that patches that have higher energy inflate faster, so can dominate the ensemble.
- Chaotic inflation means that we have random values for random patches, so somewhere always inflates, but on the average we roll down.
- There is a debate over the naturalness of the initial conditions for inflation.

Further reading: See [10, 11].

Exercises

(1) Consider the potential

$$V(\phi) = V_0\left(1 - ce^{-a\frac{\phi}{M_{\rm Pl}}}\right). \tag{9.56}$$

Calculate ϵ, η and the number of e-folds and put conditions on the parameters and initial conditions such that we have slow roll inflation.
(2) For $V(\phi) = ge^{-\lambda\phi}$ and the attractor solutions, calculate $\epsilon_H, \dot{\epsilon}_H$ and η_H, and verify

$$\eta_H = \epsilon_H - \frac{\dot{\epsilon}_H}{2H\epsilon_H}. \tag{9.57}$$

Compare also η_H with $\eta - \epsilon$ and check whether we have slow-roll inflation.
(3) For $V = g\phi^p$ calculate ϵ, η, N, $d\eta/dN$ and $d(\ln\epsilon)/dN$.
Calculate $a(t)$ in the slow-roll case and from it ϵ_H and η_H.
(4) Repeat Exercise 3 for $V = V_0[1 - c(\phi_0/\phi)^p]$, where $p > 0$.
(5) Consider scalar potential given by a cosine plus a constant,

$$V = A[1 + \cos(a\phi)]\,, \tag{9.58}$$

and inflation happening near the maximum at $\phi = 0$. Calculate ϵ, η and N_e, and find over what region we have slow roll, given the constants A, a and the initial value $\phi = \phi_0$, $\dot{\phi}_0 \ll 1/a$.

Chapter 10
Reheating and Baryogenesis

In this chapter we will study reheating, which happens just after inflation, and is a way to populate the Universe with thermally distributed particles, which is followed by the standard Radiation Dominated adiabatic cosmology. Thus reheating is a way to transfer the energy of the inflaton to the Standard Model particles and to transition into standard cosmology. We will also say a few things about baryogenesis which happens around the same time.

10.1 Standard Reheating

The simplest model of reheating happens in the standard models of "new inflation", with a plateau, followed by a drop to a minimum, and then oscillations. The inflaton field falls down to the minimum, then oscillates, until the friction terms stop it. During the oscillations, the inflaton transfers its energy by decaying into particle modes. In the model studied here, ϕ decays mostly into the fermionic modes, $\phi \rightarrow \psi\psi$, with a decay constant Γ_ϕ. This decay is a friction term in the equation of motion, or otherwise a relative friction term in the kinetic energy, $\dot{\rho}_\phi = -\Gamma_\phi \rho_\phi + \dots$.

The KG equation of motion of the scalar is modified to

$$\ddot{\phi} + 3H\dot{\phi} + \Gamma_\phi \dot{\phi} + V'(\phi) = 0. \tag{10.1}$$

In this equation of motion, besides the usual Hubble friction term $3H\phi$, we have the explicit friction term due to the ϕ decay. The solution of the equations of motion corresponds to some coherent oscillations around the minimum (the field comes with some velocity, i.e., kinetic energy, and oscillates -damped- around the minimum). We multiply the modified KG equation with $\dot{\phi}$, and take the average over the oscillations (cycles), considering that the inflaton energy density,

$$\rho_\phi = \frac{\dot{\phi}^2}{2} + V\ , \tag{10.2}$$

H. Năstase, *Cosmology and String Theory*, Fundamental Theories of Physics 197, https://doi.org/10.1007/978-3-030-15077-8_10

gives on the average over a cycle (due to the equipartition of energy, kinetic = potential)

$$\langle\frac{\dot{\phi}^2}{2}\rangle + \langle V\rangle = \langle\dot{\phi}^2\rangle. \tag{10.3}$$

We obtain then

$$\dot{\rho}_\phi + 3H\rho_\phi + \Gamma_\phi\rho_\phi = 0. \tag{10.4}$$

It is easy to see that the solution of this equation in terms of the energy density at the initial time t_I, corresponding to the end of inflation, when $\rho(t_I) = V(\phi_{\text{initial}}) \equiv M^4$ (defining the mass scale M), is

$$\rho_\phi(t) = M^4\left(\frac{a_I}{a}\right)^3 e^{-\Gamma_\phi(t-t_I)}. \tag{10.5}$$

Next we write an equation for the matter (fermions) into which the inflaton decays. The conservation equation for matter in the absence of particle creation would have been the one considered in the first chapter, $\dot{\rho}_m + 3H(\rho_M + p_M) = 0$, where ρ_M is the density of matter (not inflaton), and p_M its pressure. But in fact, there is a driving term, which is minus the friction term of the inflaton, since the inflaton decays into matter. Therefore the equation of motion is

$$\dot{\rho}_m + 3H(\rho_M + p_M) = \Gamma_\phi\rho_\phi. \tag{10.6}$$

Since the mattter particles created during reheating (electrons, neutrinos, protons, etc.) are ultrarelativistic (as we saw in the thermal history of the Universe), $P_M = \rho_M/3$, and we have effectively radiation, therefore we rename $\rho_M \to \rho_R$, to obtain

$$\dot{\rho}_R + 4H\rho_R = \Gamma_\phi\rho_\phi. \tag{10.7}$$

We need to supplement this with the Friedmann equation

$$H^2 = \frac{1}{3M_{\text{Pl}}^2}(\rho_\phi + \rho_R). \tag{10.8}$$

Here we have used the fact that the total energy is composed of the energy of the inflaton and the radiation energy.

The solution of the equation of motion for ρ_R is

$$\rho_R(t) = \frac{\rho_\phi(t_I)\Gamma_\phi a^3(t_I)}{a^4(t)}\int_{t_I}^t dt'\; a(t')e^{-\Gamma_\phi(t'-t_I)}. \tag{10.9}$$

We can check that the $1/a^4(t)$ means that in $\dot{\rho}_R$ we get a $-4H\rho_R$ term, whereas the derivative acting on the integral gives

$$\frac{\rho_\phi(t_I)\Gamma_\phi a^3(t_I)}{a^4(t)}a(t)e^{-\Gamma_\phi(t'-t_I)}\;, \tag{10.10}$$

which is the same as Γ_ϕ times (10.5).

After it will thermalize, the energy density of radiation will be given by the standard relativistic thermal formula derived previously,

$$\rho_R = \frac{\pi^2}{30}g_* T^4\;, \tag{10.11}$$

which allows us to calculate T out of ρ_R.

The initial time for the begining of reheating is the end of inflation, when the time is given by the Hubble time, $t = t_I \sim H^{-1} \sim M_{\rm Pl}/M^2$ (since by the Friedmann equation $H^2 \sim V/M_{\rm Pl}^2 \sim M^4/M_{\rm Pl}^2$). From this initial time until the time constant for the decay, $t \sim \Gamma_\phi^{-1}$, the inflaton will dominate the energy density (since it has not decayed yet). Because of its large mass m_ϕ, defined by the curvature of the potential around the minimum,

$$m_\phi^2 \equiv V''(\phi_0)\;, \tag{10.12}$$

ϕ is nonrelativistic, which means that the Universe is matter dominated, meaning that $a(t) \propto t^{2/3}$.

The decay occurs quantum mechanically, but from the tree level process, we can find the usual formulas

$$\Gamma_{\phi\to\psi\psi} = \frac{g^2 m_\phi}{8\pi}\;, \tag{10.13}$$

for the decay into fermions, and for the decay into bosons χ,

$$\Gamma_{\phi\to\chi\chi} = \frac{g^2}{8\pi m_\phi}\;, \tag{10.14}$$

where g is the coupling. We assume that $t_I \sim H^{-1} \ll \Gamma_\phi^{-1}$ (so that the above interval is nonzero), which means that $\Gamma_\phi \ll H(t_I)$, and therefore $e^{-\Gamma_\phi(t-t_I)} \simeq 1$.

We then obtain

$$\begin{aligned}
\rho_R(t) &\simeq \rho_\phi(t_I)\frac{\Gamma_\phi a^3(t_I)}{a^4(t)}\int_{t_I}^t dt'\, a(t') \\
&= \frac{3}{5}\Gamma_\phi \rho_\phi(t_I) t_I \left(\frac{t_I}{t}\right)^{8/3}\left[\left(\frac{t}{t_I}\right)^{5/3} - 1\right] \\
&= \frac{3}{5}\frac{\Gamma_\phi}{t}\rho_\phi(t_I)t_I^2\left[1 - \left(\frac{t_I}{t}\right)^{5/3}\right] \\
&= \frac{3}{5}\frac{\Gamma_\phi}{t}M_{\rm Pl}^2\left[1 - \left(\frac{t_I}{t}\right)^{5/3}\right]\;,
\end{aligned} \tag{10.15}$$

where in the second equality we have used the fact that $a(t) \propto t^{2/3}$ during the period of matter domination (the massive, thus nonrelativistic, inflaton dominates the energy density), and in the last equality we have used the fact that $\rho_\phi(t_I) = M^4$ and $t_I^2 = M_{\rm Pl}^2/M^4$.

By taking the time derivative of this function, we obtain that it grows, from $\rho_R(t_I) = 0$ until a maximum, achieved at

$$t_{\max} = \left(\frac{8}{3}\right)^{3/5} t_I \ , \tag{10.16}$$

(which confirms that indeed $e^{-\Gamma(t_{\max}-t_I)} \simeq 1$ as was assumed) of value

$$\rho_{\rm R,max} = \left(\frac{3}{8}\right)^{8/5} \Gamma_\phi t_I \rho_\phi(t_I) \simeq 0.139\Gamma_\phi \frac{M_{\rm Pl}}{M^2} M^4 = 0.139\Gamma_\phi M_{\rm Pl} M^2. \tag{10.17}$$

After this maximum, $\rho_R(t)$ starts decreasing as

$$\begin{aligned} \rho_R(t) &\simeq \frac{3}{5}\Gamma_\phi M_{\rm Pl} M^2 \frac{t_I}{t}\left[1-\left(\frac{t_I}{t}\right)^{5/3}\right] = \frac{3}{5}\Gamma_\phi M_{\rm Pl} M^2 \left(\frac{a_I}{a}\right)^{3/2}\left[1-\left(\frac{a_I}{a}\right)^{5/2}\right] \\ &\sim \frac{3}{5}\Gamma_\phi M_{\rm Pl} M^2 (a_I/a)^{3/2} \ , \end{aligned} \tag{10.18}$$

where in the second equality we have used the fact that $a(t) \propto t^{3/2}$, and in the last one we have neglected the last term. We see then that, as long as we are still before $t \sim \Gamma_\phi^{-1}$, the radiation energy density drops as $a^{-3/2}$.

Finally, considering that this radiation energy density has already thermalized so that we can define a maximum temperature for the radiation, we use the relativistic relation to write

$$\rho_{\rm R,max} = \frac{\pi^2}{30} g_* T_{\max}^4 \ , \tag{10.19}$$

which gives the maximum temperature

$$T_{\max} \simeq 0.8 g_*^{-1/4} M^{1/2} (\Gamma_\phi M_{\rm Pl}\sqrt{8\pi})^{1/4}. \tag{10.20}$$

Entropy Production

In an *adiabatic* process, the total entropy $S = sa^3$ (where s is the entropy density per unit comoving volume) is conserved, so $s \propto 1/a^3$, and indeed, since in the relativistic case $s \propto T^3$ (the exact formula was $s = (2\pi^2/45)g_{*,s}T^3$), $s \propto 1/a^3$ ($T \propto 1/a$).

But in the case of reheating, we are out of equilibrium, so we are not in an adiabatic process. Entropy is in fact being generated, since the entropy density per comoving volume, $s \propto T^3$ as we saw, but also $\rho_R \propto T^4$, so $s \propto \rho_R^{3/4}$. Since $\rho_R \propto a^{-3/2}$ as we saw, we finally obtain

$$S \propto a^3 \rho_R^{3/4} \propto a^{15/8} \,, \tag{10.21}$$

so the entropy increases with the expansion of the Universe, as we said. In fact, we already said that producing the very large entropy of the Universe was one of the things that was needed of an inflationary model, and we do see that happening during reheating.

Finally, after $t \sim \Gamma_\phi^{-1}$, the inflaton ϕ has mostly decayed, and its energy has been converted into radiation, the total entropy levels off, and we enter the period of the standard adiabatic, radiation dominated cosmology.

Reheating Temperature

The most important quantity of reheating is the temperature at the end of reheating, when the standard cosmology starts, known as the *reheating temperature*. It is the temperature at $t = \Gamma_\phi^{-1}$, the last time that we can use the approximation of matter domination under which we have been doing the previous calculations, so

$$\begin{aligned} \rho_R(t = \Gamma_\phi^{-1}) &\simeq \frac{3}{5}\Gamma_\phi^2 \rho_\phi(t_I) t_I^2 = \frac{3}{5}\Gamma_\phi^2 M_{\rm Pl}^2 \\ &\equiv \frac{\pi^2}{30} g_* T_{RH}^4 \,, \end{aligned} \tag{10.22}$$

leading to

$$T_{RH} \equiv T(t = \Gamma_\phi^{-1}) \simeq 0.55 g_*^{-1/4} (\Gamma_\phi M_{\rm Pl} \sqrt{8\pi})^{1/2} \simeq g_*^{-1/4} \sqrt{\Gamma_\phi M_{\rm Pl}} . \tag{10.23}$$

On the other hand, from (10.13), for the fermionic decay $\phi \to \psi\psi$, we have $\Gamma_\phi \simeq (g^2/8\pi) m_\phi$, so

$$T_{RH} \simeq g_*^{-1/4} (g/\sqrt{8\pi}) \sqrt{m_\phi M_{\rm Pl}} . \tag{10.24}$$

10.2 New Reheating and Preheating

Nowadays however, another model of reheating is more popular. Instead of the many oscillations around a minimum, now it is assumed that the potential has a very steep drop, leading to what is known as *preheating*. This is an extremely rapid decay of modes into *bosons* (not fermions as before) via *parametric resonance*. In this case, there is no need for oscillations, or rather, oscillations stop as soon as they start (within less than a cycle). Because of the highly nonperturbative nature of this model, it is generally studied numerically, but here we will present a few salient points.

Generically, this new reheating picture is composed of 3 steps:

1. Preheating: the inflaton ϕ decays into bosons.
2. The bosons decay into everything.
3. The particles thermalize.

A simple model of the preheating step is obtained as follows. Consider the inflaton ϕ plus boson χ, with a potential composed of the mass term for ϕ (the curvature of the potential around the minimum) and a renormalizable (quartic) coupling of ϕ to χ, i.e.,

$$V \simeq \frac{1}{2}m_\phi^2\phi^2 + \frac{1}{2}g^2\phi^2\chi^2 + ... \tag{10.25}$$

Then we can effectively describe the field χ as having a *time-dependent mass* given by (the VEV of) ϕ,

$$m_\chi^2(t) \simeq g^2\phi^2(t). \tag{10.26}$$

The KG equation for the momentum modes χ_k of the boson χ becomes

$$\ddot{\chi}_k + E_k^2(t)\chi_k = 0\ , \tag{10.27}$$

where

$$E_k^2 = \left(\frac{k}{a}\right)^2 + m_\chi^2(t). \tag{10.28}$$

Assuming as before that the inflaton ϕ oscillates around its minimum, where the curvature is m_ϕ (even though as we will see we don't even need to complete one oscillation),

$$\phi = \phi_{\text{end}} \sin(m_\phi t)\ , \tag{10.29}$$

where ϕ_{end} is the value of ϕ at the end of inflation, we obtain for the χ mass

$$m_\chi^2(t) = g^2\phi_{\text{end}}^2 \sin^2(m_\phi t)\ , \tag{10.30}$$

and after defining rescaled variables and parameters as

$$z \equiv m_\phi t; \quad q \equiv \frac{g^2\phi_{\text{end}}^2}{4m_\phi^2}; \quad A(k) \equiv \frac{k^2}{m_\phi^2} + 2q\ , \tag{10.31}$$

we find the equation

$$\frac{d^2\chi_k}{dz^2} + [A(k) - 2q\cos(2z)]\chi_k = 0. \tag{10.32}$$

This is the *Mathieu equation*. In certain regions of parameter space it has a rapidly growing solution, i.e., we have parametric resonance. In particular, the largest growth is for $q \gtrsim 1$. But a growth in the boson mode χ_k means particle production. Creating these bosonic particles amounts to preheating.

Numerically we can see that we can arrange for the parameters such that we don't even need to complete a full cycle in order to transfer the energy to χ. This is called *instant preheating*.

This process can be studied numerically, however one finds that the end result for the reheating temperature (the most important quantity characterizing reheating) is not changed much with respect to the standard reheating analysis before.

10.3 Entropy Production and the Number of e-Folds

We can now perform the more careful calculation of the bound on the number of e-folds coming from entropy production that was advertised in previous chapters.

Our Universe starts off as a patch at the begining of inflation of size = horizon size, $H^{-1} \sim M_{\rm Pl}/M^2$ (since $H^2 \sim V/M_{\rm Pl}^2$ by the Friedmann equation and $V \equiv M^4$). During inflation, this patch increases by a factor of e^N, where N is the number of e-folds. But the scale factor also increases during reheating. The increase in a is found from the fact that during reheating, as we saw, $\rho \propto 1/a^3$, and moreover at the end (when it has thermalized) we can write $\rho \sim T^4$, whereas at the begining of reheating we have $\rho \sim M^4$, so that finally

$$\frac{a_{RH}}{a_I} \simeq \left(\frac{M^4}{T_{RH}^4}\right)^{1/3} . \tag{10.33}$$

At the end of reheating, the entropy density $s = S/a^3$ is thermal and relativistic, so $s \sim T_{RH}^3$, so the total entropy within our patch of Universe is

$$S_{\rm patch} = a^3 s = \left(\frac{M^4}{T_{RH}^4} e^{3N} H^{-3}\right) T_{RH}^3 = e^{3N} \frac{M_{\rm Pl}^3}{T_{RH} M^2} . \tag{10.34}$$

But we already saw that the bound that solves the flatness problem is $S_{\rm patch} \geq 10^{88}$, so now we have

$$e^{3N} \frac{M_{\rm Pl}^3}{T_{RH} M^2} \geq 10^{88} , \tag{10.35}$$

which can be rewritten (by taking the log) as

$$N \geq N_{\rm min} = 67 + \frac{1}{3} \ln \frac{M^2 T_{RH}}{M_{\rm Pl}^3} , \tag{10.36}$$

where 67 comes from $\frac{88}{3} \ln 10$, or as

$$N \geq N_{\rm min} = 56 + \frac{2}{3} \ln \frac{M}{10^{16}\,{\rm GeV}} + \frac{1}{3} \ln \frac{T_{RH}}{10^9 {\rm GeV}} . \tag{10.37}$$

This was the more precise variant of the bound on the number of e-folds that was previously advertised.

10.4 Baryogenesis

Baryogenesis, the generation of the small baryon asymmetry observed in the Universe, must happen soon after, or at most during reheating. The reason is that, like in the case of primordial relics, any initial baryon asymmetry would be diluted away by inflation, so it must happen after it, i.e., during or after inflation. But it cannot happen too late, as we will shortly see.

In order to have baryogenesis, as we hinted at before, we must have satisfied the:

Sakharov Conditions

In fact, Andrei Sakharov (who, after being crucial in the Russian nuclear program during the Cold War, advocated for peace and got the Nobel Peace prize while being imprisoned by the state for his beliefs) proved that having baryogenesis is completely equivalent to having all the following 3 conditions satisfied:

1. *B breaking.* Baryon number (B) must be nonconserved in the process. This is in fact generic in Grand Unified Theories (GUTs), that unify electroweak ($SU(2) \times U(1)$) and strong ($SU(3)$) interactions in a common gauge group. Indeed, the very definition of unification of electroweak and strong forces imply the existence of "leptoquarks", i.e., gauge bosons from the unified gauge group that mediate transitions between quarks (transforming under $SU(3)$) and leptons (transforming under $SU(2)$), $q \leftrightarrow l$. But since baryons are by definition objects containing quarks, but no leptons, the leptoquarks violate B.
   B violation is clearly needed in order to obtain B asymmetry ($B \neq 0$) from $B = 0$.
2. *Breaking of C and CP symmetries.* In fact, C is violated in weak interactions, as we know (since in weak interactions CP is approximately conserved, and P is violated: this was the Nobel prize of T.D. Lee and C.N. Yang). Moreover, CP was found to be violated as well, in various sectors.
   This is less obvious to see, but it is also required.
3. *Absence of thermal equilibrium.* Indeed, in thermal equilibrium, the chemical potentials of particle and antiparticle are the same, $\mu_X = \mu_{\bar{X}}$, so we create an equal amount of particles and antiparticles. The rapid expansion of the Universe provides the mechanism of absence of thermal equilibrium. However, during standard adiabatic cosmology we do have thermal equilibrium, so the period of reheating (and shortly thereafter) is the one we need, since as we saw, then we do have non-equilibrium.

Bound on Temperature

Baryogenesis cannot happen too much after reheating also because it needs to be at a relatively high energy. Indeed, we saw that in order for it to happen, we need B violation, usually happening through GUT leptoquark decays. But there are many experiments looking for proton p decay; if found, it would mean that there are B violating leptoquark-like channels. But p decay has not been observed (which ruled out the simplest $SU(5)$ GUT for instance), and put very stringent bounds on the proton lifetime. Among other things, these bounds constrain the mass of the particles mediating the decay to be $> 10^{10}$ GeV in generic models, though the bound can be smaller in specific (highly constrained) models.

In turn, this gives a constraint on the reheating temperature. Since baryogenesis can occur during reheating, this gives a constraint on T_{RH} that is slightly weaker,

$$T_{RH} \geq 10^9 \,\text{GeV} \tag{10.38}$$

generically, though in very particular cases it can be even smaller.

10.5 Relics

Finally, a few words about relics, like monopoles, cosmic strings, etc. As we have seen, the necessity of diluting them was one of the motivations for inflation. These relics have usually very large masses (coming from phase transitions before, or during inflation), so during the standard RD era, beginning at the end of reheating, they are nonrelativistic, $\rho_{\text{relics}} \propto 1/a^3$, whereas the total energy density is dominated by radiation, so $\rho_{\text{tot}} \propto 1/a^4$, meaning that their ratio increases,

$$\frac{\rho_{\text{relics}}}{\rho_{\text{tot}}} = \frac{\rho_{\text{non-rel}}}{\rho_{\text{rel}}} \propto a \propto \frac{1}{T}. \tag{10.39}$$

Since radiation domination ends at matter-radiation equality, when the temperature is T_{eq}, and after that we have matter domination, so the relics and the total energy density scale in the same way, thus their ratio stays constant until now, from the reheating time until now the above ratio of energy densities changes by a factor of ($T_{eq} \sim 10^5 K \sim 10$ eV)

$$\frac{T_{RH}}{T_{eq}} = 10^{18} \times \left(\frac{T_{RH}}{10^{10}\,\text{GeV}}\right). \tag{10.40}$$

That means that we need to suppress the relics by at least this factor during inflation (so relic production must happen at a number of e-folds generating at least the above factor from the end of inflation).

Important Concepts to Remember

- In standard reheating, at the end of inflation, the potential drops, and there is a minimum. There are many oscillations around the minimum. The inflaton energy is transmitted to fermionic modes, through a coupling $\phi\psi\psi$ giving decay $\Gamma_{\phi\to\psi\psi}$.
- Matter is ultrarelativistic fermions here, meaning radiation. Its density increasing to a maximum, corresponding to a maximum temperature after thermalization.
- Entropy is produced during reheating, $S \propto a^{15/8}$, but at its end we move on to adiabatic RD cosmology, with constant S.

- The reheating temperature T_{RH} is the most important quantity, which is the temperature of the radiation at $t = \Gamma_\phi^{-1}$, when the inflaton has stopped decaying.
- We obtain $T_{RH} \sim g_*^{-1/4} \sqrt{\Gamma_\phi M_{\rm Pl}}$.
- In new reheating, there is a period of preheating, which is a steep drop in the potential, during which the inflaton decays rapidly into bosons. Then the bosons decay into everything and we have thermalization. Preheating requires less than one oscillation cycle to occur.
- Preheating leads to the Mathieu equation, which has a parametric resonance in some region in parameter space.
- Correcting the number of e-folds of inflation for preheating, we find $N \geq 56 + \frac{2}{3} \ln \frac{\rho_1^{1/4}}{10^{16}\text{GeV}} + \frac{1}{3} \ln \frac{T_{RH}}{10^9\text{GeV}}$.
- The Sakharov conditions for baryogenesis are: 1. B breaking. 2. C and CP breaking. 3. Out of thermal equilibrium.
- Baryogenesis needs to occur during or not long after reheating, which gives generically (due to constraint on generic baryogenesis models) $T_{RH} > 10^9$ GeV.
- Relics need to be diluted by a factor of $10^{18} T_{RH}/10^{10}$ GeV during inflation.

Further reading: See Chap. 8.3 in [1], Chap. 5.5 in [2] and Chap. 21 in [9].

Exercises

(1) Estimate the reheat temperature in the MSSM plus scalar = inflaton of mass $m \sim 10^{13}$ GeV coupled to the fermions.
(2) Estimate the minimum number of e-folds needed for inflation at the scale 10^{16} GeV in the case at Exercise 1.
(3) Solve the Mathieu equation as a perturbation (to first order) around a given time.
(4) Consider an inflationary plateau at GUT scale, followed by a rapid drop to a minimum. Assuming that the potential energy is quickly converted completely to thermalized radiation, and that at that energy scale we have only MSSM particles being effectively massless, calculate the reheating temperature.

Chapter 11
Fluctuations in Inflation and Matching with Experimental Data

In this chapter we will show how to calculate the spectrum of fluctuations during inflation, and relate that to observables. We will quantize fluctuations during inflation, and show what power spectrum they give, and how that reflects into the CMBR fluctuations. The final point will be how to then constrain various inflationary models using experimental data.

11.1 Scales During Inflation

As we saw, during inflation, scales get out of the horizon (which is approximately constant, $H^{-1} \simeq$ const.), being exponentially blown up, and then during the standard (RD, and later, MD) cosmology, they come back inside the horizon. Moreover, as we will see, perturbations start off as quantum fluctuations, but then when they get outside the horizon, they get "frozen in" and grow with a until they become classical, and when they come back inside the horizon have therefore classical, observable, effects. When speaking about scales, we will refer to wave number modes k in coordinate (x, not $a(t)x$) space, so that the physical (momentum) scales are actually $k/(a(t))$. Therefore the horizon exit will happen when $k/a = H$, or $k = aH$. Note that the scales that leave the horizon earliest spend most time outside the horizon, thus came back latest ("first out, last in").

The largest scales that are relevant for us today are the scales of the CMBR, or the Hubble radius, of the order of $\lambda \sim 3000 Mpc$. Any bigger scales are still outside the horizon, thus unobservable, thus irrelevant. Thus this largest scale left the horizon at the earliest time, corresponding to the minimum number of e-folds before the end of inflation. We have calculated this number of e-folds in the last chapter, so we can now say that

H. Năstase, *Cosmology and String Theory*, Fundamental Theories of Physics 197, https://doi.org/10.1007/978-3-030-15077-8_11

$$N_{3000Mpc} = N_{\min} = 56 + \frac{2}{3}\ln\frac{\rho_*^{1/4}}{10^{16}\,\text{GeV}} + \frac{1}{3}\ln\frac{T_{RH}}{10^9\,\text{GeV}}. \tag{11.1}$$

On the other hand, now we can calculate the number of e-folds before the end of inflation that an arbitrary scale λ has left, considering that $\ln 3000 \simeq 8$, namely

$$N_\lambda = 48 + \ln\frac{\lambda}{Mpc} + \frac{2}{3}\ln\frac{\rho_*^{1/4}}{10^{16}\,\text{GeV}} + \frac{1}{3}\ln\frac{T_{RH}}{10^9\,\text{GeV}}. \tag{11.2}$$

11.2 Scalar Fluctuations During Inflation

The simplest model of inflation, which we analyze here, is characterized by the interaction of gravity with a scalar field. Of course, we can have other fields, but this simplest model assumes that always the leading contribution of all the fields (other than gravity) will be described by a scalar.

To calculate the scalar fluctuations during inflation, we write the KG equation for a scalar with potential $V(\phi)$ and vary it with respect to a $\delta\phi$, after which we go to momentum space, obtaining the equation for $\delta\phi_k$,

$$\delta\ddot{\phi}_k + 3H\delta\dot{\phi}_k + \left(\frac{k}{a}\right)^2 \delta\phi_k + V''(\phi)\delta\phi_k = 0. \tag{11.3}$$

Since during inflation we assume that the potential is very flat (so that we have slow-roll), we consider that $V''(\phi) \simeq 0$, and neglect it. Of course, in the previous chapter, when analyzing reheating, we have considered a very large inflaton mass, but that was given by the curvature of the potential around the minimum where reheating happens. On the plateau where inflation happens, the inflaton is nearly massless. The equation above is in the absence of backreaction, i.e., we consider that the scalar perturbation doesn't influence the gravitational background.

But in reality, there are gravitational perturbations as well. We have seen that we can choose a Newtonian gauge, where the metric perturbations are $h_{00} = -2\Psi$, $h_{0i} = 0$ and $h_{ij} = -a^2\delta_{ij}\Phi$, and $\Psi = \Phi$ for the scalar perturbations which we study here. Inputting these perturbations into the Einstein equation, $G_{\mu\nu} = 8\pi G_N T_{\mu\nu}$, and perturbing to first order, we obtain

$$\ddot{\Psi}_k + H\dot{\Psi}_k = 4\pi G_N \dot{\phi}\delta\phi_k\,, \tag{11.4}$$

since the Einstein tensor $G_{\mu\nu}$ is quadratic in derivatives. But that means that the scalar fluctuation $\delta\phi_k$ generates a Newtonian perturbation Ψ_k. That means that in reality we don't have a purely scalar perturbation, but rather a combination of the scalar and gravity modes. However, for now we will stick with the assumption of no backreaction, and later we will see that we need only modify it slightly for the real case.

We first redefine the scalar variable as

$$\varphi \equiv a\delta\phi \;, \tag{11.5}$$

and use conformal time η, defined by $d\eta = dt/a(t)$, instead of normal time t. Since $H \simeq$ constant during inflation, we have

$$\eta \simeq -\frac{1}{aH}. \tag{11.6}$$

Indeed, when differentiating, we find $d\eta = dt\dot{a}/(a^2 H) = dt/a$. Then, rewriting the KG equation (11.3) at $V''(\phi) = 0$ in these variables, we obtain

$$\frac{d^2\varphi_k(\eta)}{d\eta^2} + \omega_k^2(\eta)\varphi_k(\eta) = 0 \;, \tag{11.7}$$

that is, a free harmonic oscillator, but with a time-dependent frequency,

$$\omega_k^2(\eta) = k^2 - \frac{2}{\eta^2} = k^2 - (aH)^2. \tag{11.8}$$

We indeed see that the moment of horizon exit, $k = aH$, delimitates real from imaginary frequency of oscillation:

1. For solutions on super-horizon scales, $k \leq aH$, or rather $k \ll aH$, implying $(-k\eta) \ll 1$, i.e., at late times, the last term in (11.3) (when $V''(\phi) \simeq 0$ also) is negligible, so the equation becomes

$$\delta\ddot{\phi}_k + 3H\dot{\phi}_k = 0 \;, \tag{11.9}$$

which has the solution $\delta\phi_k =$ constant. That means that once the solution gets outside the horizon, it gets "frozen in" at a constant value, and its size (scale) expands with $a(t)$. Eventually it will become classical, and will come back inside the horizon during the standard (RD or MD) cosmology. Once it gets back inside the horizon, it will start to fluctuate again, but this time as a classical solution.

2. For solution on sub-horizon scales, $k > aH$, or more precisely $k \gg aH$, implying $(-k\eta) \gg 1$, or early times, we have a simple harmonic oscillator with constant $\omega_k^2(\eta) \simeq k^2$, so we have the usual oscillating solution,

$$\varphi_k(\eta) \simeq \frac{e^{-ik\eta}}{\sqrt{2k}} \;, \tag{11.10}$$

where we have chosen a particular normalization.

The general solution in momentum space is

$$\hat{\varphi}_k(\eta) = \frac{1}{(2\pi)^{3/2}}[\varphi_k(\eta)\hat{a}(\vec{k}) + \varphi_k^*(\eta)\hat{a}^\dagger(-\vec{k})]. \tag{11.11}$$

At the classical level, $\hat{a}(\vec{k})$, $\hat{a}^\dagger(\vec{k})$ are coefficient functions for the modes, whereas as usual, at the quantum level they will turn into annihilation and creation operators for the modes. The solution in position space is simply the momentum integral of the above,

$$\delta\varphi(\vec{x}, t) = \frac{1}{(2\pi)^{3/2}} \int d^3k[\varphi_k(\eta)e^{i\vec{k}\cdot\vec{x}}\hat{a}(\vec{k}) + \varphi_{-k}^*(\eta)e^{-i\vec{k}\cdot\vec{x}}\hat{a}^\dagger(\vec{k})] \tag{11.12}$$

We consider the solution with the initial condition (at $\eta \to -\infty$) given by the oscillating solution,

$$\varphi_k(\eta) \to \frac{e^{-ik\eta}}{\sqrt{2k}}. \tag{11.13}$$

Then we can check explicitly (by calculating $d^2\varphi_k/d\eta^2$ and $\omega_k^2\varphi_k$) that

$$\varphi_k(\eta) = \frac{e^{-ik\eta}}{\sqrt{2k}} \frac{k\eta - i}{k\eta} \tag{11.14}$$

is a solution satisfying the initial condition.

Well after horizon exit, at $(-k\eta) \ll 1$, we have

$$\varphi_k(\eta) \to \frac{-i}{\sqrt{2k}k\eta}, \tag{11.15}$$

and the physical fluctuation is

$$\delta\phi_k = \frac{\varphi_k}{a} = \frac{+i/\sqrt{2k}}{k(-a\eta)} = \frac{iH}{\sqrt{2k}k} = \text{const.}(\eta). \tag{11.16}$$

Thus indeed the solution becomes constant when reaching super-horizon scales.

11.3 Quantization of Scalar Perturbations

The solution above starts off as a quantum fluctuation, and only after horizon exit it becomes classical. That means that we need to quantize the system, done as usual by promoting $\hat{a}(\vec{k})$ and $\hat{a}^\dagger(\vec{k})$ to operators, obeying the usual $[a, a^\dagger] = 1$ algebra, or rather, with continuous $\vec{k}$,

$$[\hat{a}(\vec{k}), \hat{a}^\dagger(\vec{k}')] = (2\pi)^3\delta(\vec{k} + \vec{k}'). \tag{11.17}$$

But if we are in quantum mechanics, we need to also specify the *state* that the Universe (or rather, the patch that becomes our visible Universe) is in. There are several possible choices, but the most common one is the simplest one, called the *Bunch-Davies vacuum* $|0\rangle$, defined as the state annihilated by $\hat{a}(\vec{k})$,

$$\hat{a}(\vec{k})|0\rangle = 0, \quad \forall \vec{k}. \tag{11.18}$$

The choice of this state corresponds in the classical picture to the intial condition (11.13) for the classical solution.

Power Spectrum

Choosing the Bunch-Davies vacuum state, and with the quantum field given by (11.11), we easily obtain that

$$\begin{aligned}\langle\varphi_k\rangle &\equiv \langle 0|\hat{\varphi}_k(\eta)|\rangle = 0\\ \langle\varphi_k\varphi_{k'}\rangle &= \langle 0|\hat{\varphi}_k(\eta)\hat{\varphi}_{k'}(\eta)|0\rangle = |\varphi_k(\eta)|^2\delta^3(\vec{k}+\vec{k}')\ ,\end{aligned} \tag{11.19}$$

by using $\hat{a}(\vec{k})|0\rangle = 0 = \langle 0|\hat{a}^\dagger(\vec{k})$ and $\langle 0|\hat{a}(\vec{k})\hat{a}^\dagger(\vec{k}')|0\rangle = (2\pi)^3\delta^3(\vec{k}+\vec{k}')$.

But in general the power spectrum $\mathcal{P}_\varphi(k)$ is defined by

$$\langle\varphi_{\vec{k}}\varphi_{\vec{k}'}\rangle \equiv \frac{2\pi^2}{k^3}\mathcal{P}_\varphi(k)\delta^3(\vec{k}+\vec{k}')\ , \tag{11.20}$$

so we obtain

$$\mathcal{P}_\varphi(k,\eta) = \frac{k^3}{2\pi^2}|\varphi_k(\eta)|^2. \tag{11.21}$$

We are interested in the result after horizon exit, which then gets frozen in until today, when we can observe it in the CMBR, on the largest scales, that have just come back inside the horizon. Then we obtain

$$\mathcal{P}_\varphi(k,\eta) \to \frac{1}{(2\pi)^2\eta^2}. \tag{11.22}$$

Since $\varphi_k = a\delta\phi_k$, we have

$$\langle\varphi_k\varphi_{k'}\rangle = a^2\langle\delta\phi_k\delta\phi_{k'}\rangle\ , \tag{11.23}$$

so the power spectrum for the observable fluctuations $\delta\phi_k$ is

$$\mathcal{P}_\phi = \frac{1}{a^2}\mathcal{P}_\varphi = \left(\frac{H}{2\pi}\right)^2, \tag{11.24}$$

where we have used the fact that $a\eta = -1/H$.

In momentum space, considering that $H \simeq H_* =$ constant, and that the spectrum is approximately zero for sub-horizon scales, $k \geq aH_*$, and that the Universe is considered to have a size L, so that the minimum k is L^{-1}, is

$$\langle \delta\phi^2(\vec{x}, t)\rangle \simeq \left(\frac{H_*}{2\pi}\right)^2 \int_{L^{-1}}^{aH_*} \frac{dk}{k} = \left(\frac{H_*}{2\pi}\right)^2 \ln(LH_*a) = \left(\frac{H_*}{2\pi}\right)^2 N(t) \, , \quad (11.25)$$

where $N(t)$ is the number of e-folds after the Universe size aL leaves the horizon.

11.4 Scalar-Gravity Fluctuations

Until now we have considered only the fluctuations of the scalar, ignoring the back-reaction of gravity onto it. But as we saw, $\delta\phi$ generates a Newtonian potential fluctuation $\delta\Psi$, which in turn generates an extra contribution to $\delta\phi$ (through the perturbed metric in the $\Box$ operator), appearing on the right-hand side of the KG equation. We will not do this calculation here, but the result is (again for the case $V''(\phi) \simeq 0$)

$$\delta\ddot{\phi}_k + 3H\delta\dot{\phi}_k + \left(\frac{k}{a}\right)^2 \delta\phi_k = \left[\frac{1}{a^3}\frac{d}{dt}\left(\frac{a^3\dot{\phi}^2}{H}\right)\right]\delta\phi_k. \quad (11.26)$$

One can now also write the equation in terms of φ_k and η like in the pure scalar case, and obtain

$$\frac{d^2\varphi_k(\eta)}{d\eta^2} + \left(k^2 - \frac{1}{z}\frac{d^2z}{d\eta^2}\right)\varphi_k(\eta) = 0 \, , \quad (11.27)$$

where we have defined the variable

$$z \equiv \frac{a\dot{\phi}}{H}. \quad (11.28)$$

This is called the *Mukhanov-Sasaki equation*.

We have shown how to obtain the Mukhanov-Sasaki equation from the backreaction of gravity, in a specific gauge (the Newtonian gauge) onto the scalar. But of course, we can write the same equation in terms of a gauge-invariant variable, the curvature perturbation ζ_k introduced in Chap. 7 when discussing the CMBR. Of course, the gauge invariant formalism is more complicated, so obtaining the equation for ζ_k in it would be hard, but we will see how to use a shortcut.

The curvature perturbation can be understood at a non-rigorous level from a modification of the Friedmann equation due to the perturbation ζ,

$$H^2 = \frac{8\pi G_N}{3}\rho + \frac{2}{3}\nabla^2\zeta(\vec{x}, t) \, , \quad (11.29)$$

which when varied over a momentum scale k gives

$$2H\delta H_k = \frac{8\pi G_N}{3}\delta\rho_k - \frac{2}{3}\left(\frac{k}{a}\right)^2\zeta_k. \tag{11.30}$$

But a more rigorous definition can be used. We have seen in Chap. 7 that for a general slicing, with

$$g_{ij} = a^2(t)e^{2\psi(\vec{x},t)}\gamma_{ij} , \tag{11.31}$$

with $\det\gamma_{ij} = 1$, we have $\psi = \zeta - H\delta t(\vec{x}, t)$, where the transformation $\delta t(\vec{x}, t)$ relates this gauge to the *uniform density slicing* where one defines ζ. Consider now that the general slicing is a *flat slicing*, with $\psi = 0$ (such that there is no curvature perturbation). On such a slicing, we can define the scalar fluctuation $\delta\phi(\vec{x}, t)$, related to δt by the obvious relation

$$\delta\phi(\vec{x}, t) = -\dot{\phi}\delta t(\vec{x}, t) , \tag{11.32}$$

since in the uniform density slicing there is no $\delta\phi$ and no δt. Then moreover, $\psi = 0$ implies

$$\zeta = H\delta t = -H\frac{\delta\phi}{\dot{\phi}}. \tag{11.33}$$

Using the definition of the quantity z, we then can write

$$\zeta = -\frac{a\delta\phi}{z} = -\frac{\varphi}{z}. \tag{11.34}$$

Therefore $\varphi = -z\zeta$, which means that the correct equation for the gauge invariant quantity ζ can be obtained by replacing in the Mukhanov-Sasaki equation (11.27) $\varphi = -z\zeta$.

Doing the replacement, we see that the term with $d^2z/d\eta^2$ cancels, and we are left with the equation for ζ_k,

$$\frac{d^2\zeta_k}{d\eta^2} + \frac{2}{z}\frac{dz}{d\eta}\frac{d\zeta_k}{d\eta} + k^2\zeta_k = 0. \tag{11.35}$$

In order to write it explicitly, we need to compute $dz/zd\eta$. To do so, we first consider the definition of z, $z = a\dot{\phi}/H$, from which we obtain

$$\frac{1}{z}\frac{dz}{d\eta} = \frac{a}{z}\frac{dz}{dt} = aH + a\frac{\ddot{\phi}}{\dot{\phi}} - \frac{a\dot{H}}{H}. \tag{11.36}$$

We have seen in Chap. 9 that $\dot{H} = -\dot{\phi}^2/(2M_{\rm Pl}^2)$, so we obtain

$$\frac{1}{z}\frac{dz}{d\eta} = aH\left(1 + \frac{\ddot{H}}{2H\dot{H}} - \frac{\dot{H}}{H}\right) = aH\left(1 + \epsilon_H + \eta_H\right) , \tag{11.37}$$

where we have used the definitions of the general slow-roll parameters $\eta_H = \ddot{H}/(2H\dot{H})$ and $\epsilon_H = -\dot{H}/H^2$. In order to express aH as well, we notice that

$$\frac{d}{d\eta}\left(\frac{1}{aH}\right) = a\frac{d}{dt}\left(\frac{1}{aH}\right) = -1 - \frac{\dot{H}}{H^2} = -1 + \epsilon_H. \tag{11.38}$$

The solution of this equation is then

$$aH = -\frac{1}{(1-\epsilon_H)\eta}. \tag{11.39}$$

During slow roll, $\epsilon_H \simeq \epsilon \ll 1$ and $\eta_H \simeq \tilde{\eta} - \epsilon$, where we have used $\tilde{\eta}$ for the slow-roll parameter denoted η previously, in order not to confuse with the conformal time. Finally, we find

$$aH \simeq -\frac{1+\epsilon}{\eta}, \tag{11.40}$$

and

$$\frac{1}{z}\frac{dz}{d\eta} \simeq aH(1+\tilde{\eta}) \simeq -\frac{(1+\epsilon)(1+\tilde{\eta})}{\eta} \simeq -\frac{1+\tilde{\eta}+\epsilon}{\eta}. \tag{11.41}$$

Substituting in the Mukhanov-Sasaki equation for ζ_k, (11.35), we find, in the case of slow-roll,

$$\frac{d^2\zeta_k}{d\eta^2} - \frac{2}{\eta}(1+\epsilon+\tilde{\eta})\frac{d\zeta_k}{d\eta} + k^2\zeta_k = 0. \tag{11.42}$$

The solution of this equation is

$$\zeta_k = -\frac{\sqrt{\pi}}{2(e\pi)^{3/2}}\frac{\sqrt{\eta}}{z(\eta)}e^{\frac{i\pi\nu}{2}+\frac{i\pi}{4}}H_\nu^{(1)}(-k\eta)\,, \tag{11.43}$$

where $H_\nu^{(1)}$ is the Hankel function and

$$\nu \equiv \frac{3}{2} + 2\epsilon_H + \eta_H = \frac{3}{2} + \epsilon + \tilde{\eta}. \tag{11.44}$$

The proof of this is left as an exercise. Note that the factor $1/z(\eta)$ appears since $z(\eta)\zeta_k = \varphi_k$ is the solution of the Mukhanov-Sasaki equation (11.27).

At late times when reaching super-horizon scales, $k \ll aH$ or $(-k\eta) \ll 1$, the solution becomes

$$\zeta_k \propto k^{-\nu} \simeq \text{const.}(\eta). \tag{11.45}$$

11.5 Power Spectrum and Contact with Experimental Data

Since $\zeta \simeq -H\delta\phi/\dot{\phi}$ as we saw, it means that

$$\langle \zeta_k \zeta_{k'} \rangle \simeq \frac{H^2}{\dot{\phi}^2} \langle \delta\phi_k \delta\phi_{k'} \rangle \,, \tag{11.46}$$

which means that the power spectrum for ζ is

$$\mathcal{P}_\zeta(k) = \frac{H^2}{\dot{\phi}^2} \mathcal{P}_\phi(k) = \left(\frac{H^2}{2\pi\dot{\phi}} \right)^2 \Bigg|_{k,\text{horizon exit}} . \tag{11.47}$$

This is the power spectrum that appeared in the CMBR oscillations in Chap. 7, so it can be extracted from experimental data. But during slow-roll, as we saw in Chap. 9,

$$\dot{\phi} \simeq -\frac{V'}{3H}; \quad V \simeq 3M_{\text{Pl}}^2 H^2. \tag{11.48}$$

In fact, we can directly relate the ζ perturbation to $\delta\phi$ perturbation and the potential V as

$$\zeta(\vec{x}) = -H \frac{\delta\phi(\vec{x})}{\dot{\phi}} \simeq \frac{3H^2}{V'} \delta\phi(\vec{x}) \simeq \frac{V}{M_{\text{Pl}}^2 V'} \delta\phi(\vec{x}) = \frac{1}{\sqrt{2\epsilon}} \delta\phi(\vec{x}). \tag{11.49}$$

But more importantly, we can write now the power spectrum for ζ as a function of only parameters in the potential, as

$$\mathcal{P}_\zeta(k) = \left(\frac{H^2}{2\pi\dot{\phi}} \right)^2 \Bigg|_{k,\text{horizon exit}} = \frac{1}{M_{\text{Pl}}^2} \left(\frac{V}{M_{\text{Pl}} V'} \right)^2 \left(\frac{H}{2\pi} \right)^2 \Bigg|_{k=aH} , \tag{11.50}$$

where we have used the fact that horizon exit corresponds to $k = aH$, as we saw earlier. Finally, using the fact that $(V'/(V M_{\text{Pl}}))^2 = 2\epsilon$, we obtain

$$\mathcal{P}_\zeta(k) = \frac{1}{24\pi^2 M_{\text{Pl}}^4} \frac{V}{\epsilon} \Bigg|_{k=aH} = \frac{1}{M_{\text{Pl}}^2} \left(\frac{H_{\text{infl}}}{2\pi\sqrt{2\epsilon}} \right)^2 \Bigg|_{k=aH} . \tag{11.51}$$

From CMBR observations, one can extract the fact that $P_\zeta(k)$ is approximately constant, like we also said in Chap. 7. The first satellite that observed the CMBR fluctuations, the COBE satellite, has measured that (from the measurement of $\delta T/T$, like we said in Chap. 7)

$$P_\zeta(k) \simeq 2.4 \times 10^{-9} \,, \tag{11.52}$$

which means that

$$\frac{1}{M_{\rm Pl}^2}\left(\frac{H_{\rm infl}}{2\pi\sqrt{2\epsilon}}\right)^2\Bigg|_{k=aH} \simeq 2.9\times 10^{-9}. \tag{11.53}$$

This doesn't directly constrain the scale of inflation, but rather only its ratio to ϵ, specifically

$$\left(\frac{V}{\epsilon}\right)^{1/4}\Bigg|_{k=aH} \simeq 6.6\times 10^{16}\,\text{GeV}. \tag{11.54}$$

The power spectrum $\mathcal{P}_\zeta(k)$ is not completely flat however, it has a small *spectral tilt*, defined as

$$n_s - 1 \equiv \frac{d\ln \mathcal{P}_\zeta(k)}{d\ln k}\,, \tag{11.55}$$

such that

$$\mathcal{P}_\zeta(k) \propto k^{n_s-1}. \tag{11.56}$$

From the formula for $\mathcal{P}_\zeta(k)$, and considering that $k = aH$ (the formula is at horizon exit), we find

$$n_s - 1 = 2\frac{d\ln H}{d\ln(aH)} - \frac{d\ln\epsilon}{d\ln(aH)}. \tag{11.57}$$

But since $H \simeq$const.,

$$d(\ln aH) \simeq \frac{da}{a} \simeq H dt - dN\,, \tag{11.58}$$

where N is the number of e-folds.

Then we find

$$n_s - 1 = -2\frac{d\ln H}{dN} + \frac{d\ln\epsilon}{dN} = -2 - (4\epsilon - 2\tilde{\eta}) = (-6\epsilon + 2\tilde{\eta})|_{k=aH}\,, \tag{11.59}$$

where we have used formulas from Chap. 9 about $d(\ln H/)dN$ and $d\ln\epsilon/dN$.

Since from Planck 2018 data [77]

$$n_s = 0.965 \pm 0.004\,, \tag{11.60}$$

we can impose constraints on ϵ and $\tilde{\eta}$.

Moreover, in a similar way we can now calculate the running of the spectral tilt with scale,

$$\frac{dn}{d\ln k} = -6\frac{d\epsilon}{(-dN)} + 2\frac{d\tilde{\eta}}{(-dN)} = -6\epsilon(4\epsilon - 2\tilde{\eta}) + 2(2\epsilon\tilde{\eta} - \xi^2) = 16\epsilon\tilde{\eta} - 24\epsilon^2 - 2\xi^2. \tag{11.61}$$

11.6 Primordial Tensor Perturbations

Until now we have described the scalar perturbations, which have already been observed in the CMBR. But in principle there could be also tensor perturbations, which could be observed. In fact, the BICEP2 experiment claimed to have seen them, but then in turned out that they had not properly accounted for backgrounds.

Tensor perturbations arise as perturbations in the metric, as we saw in Chap. 7, of the type

$$ds^2 = a^2(\eta)[-d\eta^2 + (\delta_{ij} + h_{ij})dx^i dx^j] \,, \tag{11.62}$$

and by writing the Einstein equations we see that they satisfy also a quadratic equation, of the type

$$\ddot{h}_{ij} + 2aH\dot{h}_{ij} + k^2 h_{ij} = 8\pi G_N T^T_{ij} . \tag{11.63}$$

But more importantly, the tensor perurbations correspond to two propagating degrees of freedom (graviton polarizations),

$$h^i{}_j = h_+ e_+{}^i{}_j + h_\times e_\times{}^i{}_j . \tag{11.64}$$

In fact, substituting the perturbed ansatz in the Einstein-Hilbert action, we find that the action for the graviton modes $h_+, h_\times$ is just the (rescaled) action for scalar modes, with canonical scalars

$$\phi_{+,\times} = \frac{M_{\rm Pl}}{\sqrt{2}} h_{+,\times} . \tag{11.65}$$

But that means that the two-point function for $\phi_{+,\times}$ is the same as the one of ϕ_k, so

$$\langle \phi_+(\vec{k})\phi_+(\vec{k}')\rangle = \langle \phi_\times(\vec{k})\phi_\times(\vec{k}')\rangle = \frac{2\pi^2}{k^3}\left(\frac{H}{2\pi}\right)^2 \delta^3(\vec{k}+\vec{k}') . \tag{11.66}$$

On the other hand, the power spectrum for the graviton modes is defined by

$$\langle 2h_+(\vec{k}) 2h_+(\vec{k}')\rangle = \langle 2h_\times(\vec{k}) 2h_\times(\vec{k}')\rangle = \frac{2\pi^2}{k^3}\mathcal{P}_h(k)\delta^3(\vec{k}+\vec{k}') \,, \tag{11.67}$$

from which we find the power spectrum for the graviton

$$\mathcal{P}_h(k) = \frac{8}{M_{\rm Pl}^2}\left(\frac{H}{2\pi}\right)^2\Bigg|_{k=aH} . \tag{11.68}$$

Note that this power spectrum, unlike the one for ζ_k, has no ϵ in it, so it allows us to directly measure the scale of inflation H_*. The reason for this fact is that the graviton was proportional to ϕ_k, whereas ζ_k involved a factor of $H/\dot{\phi} = 1/\sqrt{2\epsilon}$ in its relation to $\delta\phi_k$.

The amplitude of the tensor perturbations is usually defined as a ratio to the amplitude of the scalar ones,

$$r = \frac{\mathcal{P}_h}{\mathcal{P}_\zeta}, \tag{11.69}$$

so we see that we have

$$r = 16\epsilon. \tag{11.70}$$

We can also define a spectral tilt for the tensor perturbation power spectrum in the same way as it was defined for the scalar perturbations,

$$n_T - 1 = \left.\frac{d \ln \mathcal{P}_h(k)}{d \ln k}\right|_{k=aH} = \frac{2d \ln H}{H dt} = \frac{2\dot{H}}{H^2} \equiv -2\epsilon_H. \tag{11.71}$$

Note that this is the same calculation as for the scalar perturbation, just with only the first term, since there is no epsilon in the graviton power spectrum. During slow-roll, $\epsilon_H \simeq \epsilon$, so

$$n_T - 1 \simeq -2\epsilon. \tag{11.72}$$

Lyth Bound

We have seen previously the relation between the number of e-folds and the variation in the canonical scalar,

$$N = \int_{\phi_i}^{\phi_f} \frac{d\phi/M_{\rm Pl}}{\sqrt{2\epsilon}}. \tag{11.73}$$

If $\epsilon \simeq$ constant, in a given region, then

$$\Delta N \simeq \frac{\Delta\phi}{\sqrt{2\epsilon} M_{\rm Pl}}. \tag{11.74}$$

Specifically, we are interested in the CMBR scales for $1 \leq l \leq 200$ or so, which are among the earliest to get outside the horizon during inflation (first one in, last one out), so that during that time slow-roll is very good, so ϵ is indeed constant to a good approximation.

Then the above relation gives a sort of limit on the total field variation, given an amount of e-folds (which also has a minimum itself),

$$\Delta\phi \geq M_{\rm Pl} \times \Delta N \sqrt{2\epsilon}. \tag{11.75}$$

But since we have just seen that $r = 16\epsilon$, we get that

$$\Delta\phi \geq M_{\rm Pl} \times \sqrt{r} \times \frac{\Delta N}{\sqrt{8}}. \tag{11.76}$$

But that in turn means that if we detect an observable amount of tensor perturbations, let's say $r \sim 0.1$, we need to have a trans-Planckian excursion for the canonical field, in this case

$$\Delta\phi \geq 5M_{\rm pl}. \tag{11.77}$$

That can be problematic, since we expect to generically have uncontrollable quantum gravity corrections in this case.

11.7 Constraints on Inflationary Models

Finally, we make a review of the constraints imposed by experimental data on the inflationary models.

–Slow-roll (which is the standard inflationary paradigm) requires $\epsilon, \tilde{\eta} \ll 1$.

–We need to have sufficient inflation, which amounts to the constraint on the number of e-folds, usually $\Delta N \geq \Delta N_{\rm min} \sim 60$ or so.

–We need to fix the amplitude of the scalar perturbation (the "COBE normalization"), $\mathcal{P}_\zeta(k) \sim 2.4 \times 10^{-9}$.

–We need to match the experimental data for $n_s - 1$, and the bounds on r, implying conditions on ϵ and $\tilde{\eta}$.

–We need to have sufficient reheating, and a reheating temperature certainly larger than the BBN temperature, $T_{RH} \geq 1\,\text{MeV}$, but usually $T_{RH} \geq 10^9\,\text{GeV}$ or so, as we saw.

–We need to get rid of unwanted relics, which means we need to suppress them by a factor of $10^{18} T_{RH}/10^{10}\,\text{GeV}$ during inflation.

–We need to suppress quantum gravity corrections due to the Lyth bound, and in general quantum corrections. We will see that this is related to the so-called η problem. We will say more about this in part III, when discussing string inflation.

–Finally, it would be ideal if the model we have is part of a good particle physics model. (though this is not always imposed by cosmologists, more by string theorists and perhaps by particle theorists).

Important Concepts to Remember

- The scales of the CMBR (about 3000 Mpc) have just got inside the horizon, so they were among the first to exit during inflation (first out, last in).
- Scalar fluctuations during inflation obey an equation that is oscillatory on highly subhorizon scales, and constant on highly superhorizon scales.
- A classical solution that was $\sim e^{-ik\eta}$ on subhorizon scales corresponds to an (initial) quantum state of the Bunch-Davies vacuum $|0\rangle$.
- Thus the initial quantum fluctuations due to the vacuum state get outside the horizon, where they grow and become classical. They return inside the horizon as classical fluctuations, that resume their oscillations.

- The power spectrum of the scalar is flat, $\mathcal{P}_\phi(k) = (H/(2\pi))^2 \simeq$constant.
- In reality, we have a scalar-gravity coupled system, and the common mode satisfies the Mukhanov-Sasaki equation. One can write down an equation for the gauge invariant curvature perturbation ζ_k, the same obtained by replacing $\varphi = a\delta\phi = z\zeta$, with $z \equiv a\dot{\phi}/H$.
- The power spectrum for ζ at horizon exit is $\mathcal{P}_\zeta(k) = (H/(2\pi\dot{\phi}))^2|_{k=aH} = (H/(2\pi \sqrt{2\epsilon}M_{\rm Pl}))^2|_{k=aH}$.
- From the CMBR observations, the power spectrum of ζ is approximately flat, of 2.4×10^{-9}, with a spectral tilt of $n_s - 1 = -6\epsilon + 2\tilde{\eta}$.
- Tensor perturbations correspond to two scalar perturbations for the graviton polarizations.
- One obtains $\mathcal{P}_h = 8(H/(2\pi M_{\rm Pl}))^2$ independent on ϵ, so $r = \mathcal{P}_h/\mathcal{P}_\zeta = 16\epsilon$.
- The Lyth bound says that $\Delta\phi/M_{\rm Pl} \geq \sqrt{r}\delta N/\sqrt{8}$, so a large observed r implies a trans-Planckian excursion for the scalar field, so we have to take care of quantum corrections (including quantum gravity ones).
- On inflationary models, we need slow-roll constraints, the constraint on the number of e-folds, on $n_s - 1$ and r, on T_{RH}, on relics. We need to take care of the Lyth bound (quantum corrections), and perhaps embed in a good particle physics model.

Further reading: See Chaps. 24, 25 in [9], Chap. 12 in [11], Chap. 10 in [12].

Exercises

(1) Consider $V(\phi) = \lambda\phi^p$. Calculate $\epsilon, \eta, n_s, r, N_e$ and $\Delta\phi$ and impose the experimental constraints on the result.
(2) Find the general solution for the fluctuation $\varphi_k(\eta)$ of the massless field, and show that it reduces to the planar wave as $\eta \to -\infty$.
(3) Show that the Mukhanov-Sasaki equation,

$$\frac{d^2\zeta_k}{d\eta^2} + \frac{2}{z}\frac{dz}{d\eta}\frac{d\zeta_k}{d\eta} + k^2\zeta_k = 0 \tag{11.78}$$

has, in the slow-roll regime, the solution

$$\zeta_k = -\frac{\sqrt{\pi}}{2(2\pi)^{3/2}}\frac{\sqrt{\eta}}{z(\eta)}e^{i\pi\frac{\nu}{2}+i\frac{\pi}{4}}H_\nu^{(1)}(-k\eta). \tag{11.79}$$

(4) Consider a cosine potential

$$V = V_0[1 + \cos(a\phi)]. \tag{11.80}$$

Given an observed tensor perturbation of $r = 0.01$, what is the constraint on a such that we have a sub-Planckian scalar field excursion? How about if we sum two of these potentials,

$$V = V_{0,1}[1 + \cos(a_1\phi_1)] + V_{0,2}[1 + \cos(a_2\phi_2)] , \tag{11.81}$$

can we improve the constraints on a_1, a_2 for any fine-tuning of the parameters (and any scalar field combination)?

Part II
Elements of String Theory

Chapter 12
Extra Dimensions and Kaluza–Klein

In this chapter we will study the idea of extra dimensions, as proposed by Kaluza and Klein.

The idea of having extra dimensions, besides our usual $3 + 1$, is an old one. Already Theodor Kaluza (in 1921) and Oskar Klein (in 1926) had considered it. The general idea of Klein was to consider that our space is a product of our four dimensional space M_4 and a small compact space K_n, i.e., $K_D = M_4 \times K_n$, with $D = 4 + n$. Then if the volume (or rather, the radii) of the extra dimensional space is really small, they would be unobservable. Perhaps the size is comparable to the Planck length $l_{\rm Pl}$, which would be natural in a quantum gravity theory, or maybe otherwise larger, yet still small. The compact space is something like a square torus $T^n = S^1 \times \cdots \times S^1$, or maybe a sphere S^n, or some more complicated thing like $\mathbb{CP}^n$, CY_n, etc.

But the reason why it is interesting is the reason that Kaluza proposed the simplest model, with S^1 space (a simple circle), namely that we can can use this Kaluza–Klein (KK) theory for unification of the fields.

In each point in our $3 + 1$ dimensional space, $(\vec{x}, t)$, we have a space K_n, that can perhaps even depend on it, $K_n(\vec{x}, t)$. This will make a general $4 + n$ dimensional spacetime K_D with points $(\vec{x}, t, \vec{y})$, but for KK theory, more precisely we want to have a product spacetime $K_D = M_4 \times K_n$.

Mathematically, we can say that functions live in the total spacetime K_D, instead of M_4, so $\phi(\vec{x}, t, \vec{y})$ instead of $\phi(\vec{x}, t)$.

In the simplest case considered by Kaluza, of $K_n = S^1$ (a circle), we know that there is the Fourier theorem, saying that any function on the total space can be expanded in Fourier modes on the circle,

$$\phi(\vec{x}, t; y) = \sum_{n \geq 0} e^{\frac{iny}{R}} \phi_n(\vec{x}, t). \tag{12.1}$$

H. Năstase, *Cosmology and String Theory*, Fundamental Theories of Physics 197, https://doi.org/10.1007/978-3-030-15077-8_12

That means that under the Fourier expansion, we can reduce a field in the larger space to an infinite, countable, set of 3+1 dimensional functions $\phi_n(\vec{x}, t)$.

The case of a general compact space K_n instead of the circle S^1 is a generalization of the above case.

12.1 KK Metrics

In the Kaluza–Klein theory, there are 3 metrics that are sometime called KK metrics.

1. **The KK Background Metric**

We need to start with a space describing the product $M_4 \times K_n$, that is a background space. It usually needs to be a solution of the equations of motion (Einstein's equations), though there are cases when one can consider also backgrounds that do not solve the equations of motion. Since the background is of product type, we have

$$g_{\Lambda\Sigma}(\vec{x}, t; \vec{y}) = \begin{pmatrix} g^{(0)}_{\mu\nu}(\vec{x}, t) & \mathbf{0} \\ \mathbf{0} & g^{(0)}_{mn}(\vec{y}) \end{pmatrix}. \tag{12.2}$$

Here $\vec{y}$ are coordinates on K_n, so $g^{(0)}_{mn}(\vec{y})$ is a metric on it.

Note that the metric itself is one of the fields of the theory, so when we write $K_D = M_4 \times K_n$ we mean the background is a product, but the full metric can include fluctuations of all types.

2. **The KK Expansion**

We next consider the full fluctuating solution for the metric and other fields, including the background and the fluctuations around it. Then the KK expansion is simply a generalization of the Fourier theorem for any compact space.

Indeed, after the Fourier case (12.1), the next one is the case of expansion on S^2, known from quantum mechanics (though there we didn't have dependence on $(\vec{x}, t)$). This is the expansion in the spherical harmonics $Y_{lm}(\theta, \phi)$ on S^2,

$$\phi(\vec{x}, t; \theta, \phi) = \sum_{lm} \phi_{lm}(\vec{x}, t) Y_{lm}(\theta, \phi). \tag{12.3}$$

Again this is an identity (it is a theorem that we can always write this expansion). In both of these cases, the functions into which we expand are eigenfunctions of the Laplancean,

$$\begin{aligned} \partial_y^2 e^{\frac{iny}{R}} &= -\left(\frac{n}{R}\right)^2 e^{\frac{iny}{R}} \\ \Delta_2 Y_{lm}(\theta, \phi) &= -\frac{l(l+1)}{R^2} Y_{lm}(\theta, \phi). \end{aligned} \tag{12.4}$$

We moreover note that n and l define the eigenvalue of the Laplacean, as well as the dimension of the $(2l+1$ dimensional) spin l representation of $SO(3) = SU(2)$, the symmetry group of the 2-sphere, whereas m is an index in the representation.

Similarly then, in the general case we write an expansion in "spherical harmonics" $Y_q^{I_q}(\vec{y})$, where q measures the eigenvalue of the Laplacean Δ_n as well as the dimension of the representation, and I_q is an index in the representation of the symmetry group of the compact space. Therefore we write

$$\phi(\vec{x}, t; \vec{y}) = \sum_{q, I_q} \phi_q^{I_q}(\vec{x}, t) Y_q^{I_q}(\vec{y}). \tag{12.5}$$

The Y_q are eigenfunctions of the Laplacean,

$$\Delta_n Y_q^{I_q}(\vec{y}) = -m_q^2 Y_q^{I_q}(\vec{y}). \tag{12.6}$$

From a 4 dimensional point of view, considering a single mode in the expansion of the field,

$$\phi(\vec{x}, t; \vec{y}) = \phi_q^{I_q}(\vec{x}, t) Y_q^{I_q}(\vec{y}) , \tag{12.7}$$

if we act with the D-dimensional D'Alembertian $\Box_D = \Box_4 + \Delta_n$ (since the space is a product, the D'Alembertian splits into the sum of a 4 dimensional D'Alembertian and an n-dimensional Laplacean), we find

$$\Box_D \phi(\vec{x}, t; \vec{y}) = (\Box_4 + \Delta_n)\phi(\vec{x}, t; \vec{y}) = (\Box_4 - m_q^2)\phi(\vec{x}, t; \vec{y}). \tag{12.8}$$

That means that if the field is D-dimensional massless,

$$\Box_D \phi(\vec{x}, t; \vec{y}) = 0 , \tag{12.9}$$

then the 4-dimensional field is massive with mass m_q,

$$[(\Box_4 - m_q^2)\phi_q^{I_q}(\vec{x}, t)] Y_q^{I_q}(\vec{y}). \tag{12.10}$$

That means that if we want to see some structure in K_n, at the level of the field $\phi_q^{I_q}$, or rather of the spherical harmonic $Y_q^{I_q}$, we need to have at least an energy of m_q to see it.

3. **The KK Reduction Ansatz**

Which brings us to the KK reduction ansatz. If we are at energies much smaller than *all* the m_q, which as we saw are $\propto 1/R$ (the scale of the compact space), we will only excite the massless mode $\phi_0^{I_0} Y_0^{I_0}$, and we will not be able to see structure in the compact space K_n, i.e., we are not able to probe it.

In effect, we have *dimensionally reduced* the D-dimensional theory to 4 dimensions. That means that we could put all the nonzero fields to zero and only keep the

"$n = 0$" representation (multiplet of fields). This is maybe "independent on $\vec{y}$", or at least has a spherical harmonic with the simplest possible dependence on $\vec{y}$, $Y_0(\vec{y})$.

Therefore the dimensional reduction ansatz is

$$\phi(\vec{x}, t; \vec{y}) = \phi_0(\vec{x}, t) Y_0(\vec{y}). \tag{12.11}$$

But note that this is an *ansatz*, which is German for "guess", i.e., it is not guaranteed to work, we need to check whether it is consistent. We will have more to say on that later.

In conclusion, we can say that: the KK background is a solution; the KK expansion is a parametrization; and the KK reduction is an ansatz.

12.2 Fields with Spin

Until now we have implicitly assumed that the fields were scalars under Lorentz transformations, but the original motivation for KK theory was to unify fields with spin into a single field with spin in higher dimensions, so we need to consider them now. Unification occurs because various components of the higher dimensional field act as different fields in 4 dimensions.

The simplest example is the one of electromagnetism with vector field A_M, and $M = (\mu, m)$, where $\mu = 0, 1, 2, 3$ and $m = 4, \ldots, 4 + n$. Then the gauge (vector) field A_M splits as

$$A_M(\vec{x}, t; \vec{y}) = (A_\mu(\vec{x}, t; \vec{y}), A_m(\vec{x}, t; \vec{y})). \tag{12.12}$$

Here A_μ is a 4 dimensional gauge (vector) field, since under an element $\Lambda_\mu{}^\nu \in SO(1, 3)$ it transforms as a vector

$$A'_\mu(\vec{x}', t'; \vec{y}) = \Lambda_\mu{}^\nu A_\nu(\vec{x}, t; \vec{y}). \tag{12.13}$$

(this happens since A_M transforms as a vector under $\Lambda_M{}^N \in SO(1, 3 + n)$, and $SO(1, 3) \subset SO(1, 3 + n)$). On the other hand, A_m is a 4 dimensional scalar, since it is invariant under the same,

$$A'_m(\vec{x}', t'; \vec{y}) = A_m(\vec{x}, t; \vec{y}). \tag{12.14}$$

Indeed, we assume that the theory in D dimensions is Lorentz invariant, i.e., under $SO(1, 3 + n)$, and that the KK expansion (and certainly the KK dimensional reduction ansatz) breaks the Lorentz group to $SO(1, 3) \times SO(n)$ (as does the KK background metric). Here $SO(n)$ only acts on A_m, as an *internal* symmetry from the point of view of $3 + 1$ dimensions,

$$A'_m(\vec{x}, t; \vec{y}') = \Lambda_m{}^n A_m(\vec{x}, t; \vec{y}). \tag{12.15}$$

This is one of the important characteristics of the KK program: the geometrization of internal symmetries. Indeed, in the Standard Model we have internal (non-spacetime) symmetries, like the local $SU(3) \times SU(2) \times U(1)$, but now these internal symmetries appear from the symmetries of the compact space (Lorentz symmetries, or isometries).

We can write the the KK expansion for the components of A_M in the usual way,

$$\begin{aligned} A_\mu(\vec{x}, t; \vec{y}) &= \sum_{q, I_q} A_\mu^{q, I_q}(\vec{x}, t) Y_q^{I_q}(\vec{y}) \\ A_m(\vec{x}, t; \vec{y}) &= \sum_{q, I_q} A^{q, I_q}(\vec{x}, t) Y_m^{q, I_q}(\vec{y}). \end{aligned} \tag{12.16}$$

Note that the internal Lorentz symmetry $SO(n)$ acts on $Y_m^{q I_q}$ only. m is an index in the (local) Lorentz symmetry of the space, the symmetry of the space tangent to K_n at that point, whereas I_q is an index in the *isometry* of the space. For an n-sphere $S^n = SO(n+1)/SO(n)$, the isometry group is $SO(n+1)$, but the local Lorentz group is $SO(n)$, so m would be in $SO(n)$, but I_q would be in $SO(n+1)$. The two groups only coincide in the simplest case of the torus T^n, when both groups are $SO(n)$, and $Y_q^{I_q} \propto$ "$\delta_m^{I_q}$". We then find that A^{q, I_q} are 4 dimensional scalars in the representation I_q of the isometry group, and A_μ^{q, I_q} are vectors in the representation of the isometry group.

The next interesting case is of gravity (symmetric traceless tensor for the metric fluctuation). Again different components have different 4 dimensional spins, and the decomposition is

$$g_{MN}(\vec{x}, t; \vec{y}) = \begin{pmatrix} g_{\mu\nu}(\vec{x}, t) & g_{\mu m}(\vec{x}, t; \vec{y}) \\ g_{m\mu}(\vec{x}, t; \vec{y}) & g_{mn}(\vec{x}, t; \vec{y}) \end{pmatrix}. \tag{12.17}$$

Then the fluctuation in $g_{\mu\nu}$ is a $SO(1, 3)$ (4 dimensional Lorentz) symmetric traceless tensor, i.e., a graviton (so the full variable is the 4 dimensional metric), $g_{\mu m} = g_{m\mu}$ is a $SO(1, 3)$ vector, i.e., gauge field, transforming as

$$g'_{\mu m}(\vec{x}', t'; \vec{y}) = \Lambda_\mu{}^\nu g_{\nu m}(\vec{x}, t; \vec{y}) , \tag{12.18}$$

and $g_{mn}(\vec{x}, t; \vec{y})$ are $SO(1, 3)$ (4 dimensional Lorentz) scalars. As before, the KK expansion (and the background metric, really) break the local Lorentz symmetry in D dimensions, $SO(1, 3+n)$, to $SO(1, 3) \times SO(n)$, 4 dimensional Lorentz times internal symmetry. Here $SO(n)$ acts on $\vec{y}$ only, through the m, n indices on the spherical harmonics, for instance $Y_{mn}^{q, I_q}(\vec{y})$ for the scalars.

12.3 The Original Kaluza–Klein Theory

The idea of Kaluza was to unify gravity ($g_{\mu\nu}$) and electromagnetism (B_μ) in a 5 dimensional metric (gravity) $g_{\Lambda\Sigma}$.

1. Writing the KK background metric for $M_4 = Mink_4 \times S^1$, we have

$$g^{(0)}_{\Lambda\Sigma}(\vec{x}, t; \vec{y}) = \begin{pmatrix} \eta_{\mu\nu} & \mathbf{0} \\ \mathbf{0} & g_{55} = 1 \end{pmatrix}. \tag{12.19}$$

2. The KK expansion (parametrization) is now simply the Fourier expansion on the circle,

$$g_{\Lambda\Sigma}(\vec{x}, t; y) = \begin{pmatrix} \eta_{\mu\nu} + \sum_{n\geq 0} h^{(n)}_{\mu\nu}(\vec{x}, t) e^{\frac{iny}{R}} & g_{\mu 5} \\ g_{\mu 5}(\vec{x}, t; y) = \sum_{n\geq 0} h^{(n)}_{\mu 5}(\vec{x}, t) e^{\frac{iny}{R}} & g_{55} = 1 + \sum_{n\geq 0} \phi^{(n)}(\vec{x}, t) e^{\frac{iny}{R}} \end{pmatrix}. \tag{12.20}$$

3. The KK reduction ansatz amounts to keeping only the $n = 0$ mode, so

$$g_{\Lambda\Sigma}(\vec{x}, t; y) = \begin{pmatrix} \eta_{\mu\nu} + h^{(0)}_{\mu\nu}(\vec{x}, t) \equiv g_{\mu\nu}(\vec{x}, t) & g_{\mu 5} \\ g_{\mu 5}(\vec{x}, t; y) = h^{(0)}_{\mu 5}(\vec{x}, t) \equiv B_\mu & g_{55} = 1 + \phi^{(0)}(\vec{x}, t) \equiv \varphi(\vec{x}, t) \end{pmatrix}. \tag{12.21}$$

But we see that we wanted to unify gravity and electromagnetism, and we seem to need to have also a scalar. Yet there is no observed scalar, until now, except the Higgs, which is extremely massive compared to the nearly massless scalar needed above. And moreover, the existence of a nearly massless scalar would mean a fifth force, which certainly contradicts experiment. So one needs to get rid of this potential scalar.

The solution proposed originally was simply to set $\varphi = 1$ ($\phi^{(0)} = 0$). However, while the original KK dimensional reduction is actually consistent, i.e., it solves the equations of motion, this further reduction is not consistent. We left the check of this statement (at the linearized level) as an exercise. We will say more on consistent truncations in the next chapter.

12.4 Generalization: The n-Torus $T^n = (S^1)^{\times n}$

The simplest generalization is obtained by considering instead of a circle, a product of circles, i.e., the square torus. Indeed, a general torus is obtained by periodic identifications. In particular, for the square torus, we have the identifications $y_i \sim y_i + 2\pi R_i$ (in general, we can have identifications at angles). The bottom line is that the torus is flat, $g^{(0)}_{mn} = \delta_{mn}$, since we have a piece of $\mathbb{R}^n$.

The *KK background metric* is then

$$g^{(0)}_{\Lambda\Sigma}(\vec{x}, t; \vec{y}) = \begin{pmatrix} g^{(0)}_{\mu\nu}(\vec{x}, t) & \mathbf{0} \\ \mathbf{0} & \delta_{mn} \end{pmatrix}. \tag{12.22}$$

The *KK expansion* is the product of Fourier expansions on the circles of radii R_i in T^n, i.e.,

$$g_{\Lambda,\Sigma}(\vec{x}, t; \vec{y}) = \begin{pmatrix} g^{(0)}_{\mu\nu}(\vec{x}, t) + \sum_{\{n_i\}} h^{\{n_i\}}_{\mu\nu}(\vec{x}, t) \prod_{i=1}^{n} e^{\frac{i n_i y_i}{R_i}}; & g_{\mu m}(\vec{x}, t; \vec{y}) \\ g_{\mu m}(\vec{x}, t; \vec{y}) = \sum_{\{n_i\}} B^{m,\{n_i\}}_{\mu}(\vec{x}, t) \prod_{i=1}^{n} e^{\frac{i n_i y_i}{R_i}}; & g_{mn} = \delta_{mn} + \sum_{\{n_i\}} h^{\{n_i\}}_{mn}(\vec{x}, t) \prod_{i=1}^{n} e^{\frac{i n_i y_i}{R_i}} \end{pmatrix}. \tag{12.23}$$

We see that here the spherical harmonics are

$$Y_{\{n_i\}}(\vec{y}) = \prod_{i=1}^{n} e^{\frac{i n_i y_i}{R_i}}. \tag{12.24}$$

The *KK reduction ansatz* amounts to taking just the zero modes, so

$$g_{\Lambda\Sigma}(\vec{x}, t; \vec{y}) = \begin{pmatrix} g^{(0)}_{\mu\nu} + h^{\{0\}}_{\mu\nu}(\vec{x}, t) \equiv g_{\mu\nu}(\vec{x}, t); & g_{\mu m}(\vec{x}, t) \\ g_{\mu m}(\vec{x}, t) = B^{m\{0\}}_{\mu}(\vec{x}, t); & g_{mn}(\vec{x}, t) = \delta_{mn} + h^{\{0\}}_{mn}(\vec{x}, t) \end{pmatrix}. \tag{12.25}$$

Then $g_{\mu\nu}(\vec{x}, t)$ is the 4 dimensional metric, $B^{m\{0\}}_{\mu}(\vec{x}, t)$ are vectors and $h^{\{0\}}_{mn}(\vec{x}, t)$ are scalars.

Important Concepts to Remember

- In Kaluza–Klein theory, the space is a product of a 4 dimensional space and a small n-dimensional compact space, $K_D = M_4 \times K_n$.
- We don't see the extra dimensions because they are small (small radii), and the fields are unified, in the sense that various components of a field with spin in the higher dimension act as various fields with spin in 4 dimensions.
- The KK theory starts with the KK background metric, usually a solution to the equations of motion (Einstein's equations), of the product form.
- We then do a KK expansion around it, in spherical harmonics, which is a generalization of the Fourier expansion. It is therefore a theorem, something that we can always do on a compact space, or another words a parametrization.
- The spherical harmonics $Y^{I_q}_q(\vec{y})$ are eigenfunctions of the Laplacean, where q defines the eigenvalue of the Laplacean, and I_q is an index in a representation of the symmetry group of K_n.
- A D-dimensional massless field is 4-dimensional massive with m_q, the eigenvalue of the Laplacean being $-m_q^2$.

- The KK reduction ansatz is an ansatz ("guess"), saying that we can truncate the series only to the first term, the "$n = 0$" representation, with "y-independent" spherical harmonic, or rather with the simplest $Y_0(\vec{y})$.
- The split of fields with spin, where the various components of the D dimensional field become various fields with spin in 4 dimensions, breaks the local Lorentz symmetry as $SO(1, 3+n) \to SO(1, 3) \times SO(n)$.
- The $SO(n)$ symmetry appears as an internal symmetry, thus the KK theory geometrizes internal symmetries.
- The indices m in $SO(n)$ appear on the spherical harmonics, like for the scalars in g_{mn} we have $Y_{mn}^{q,I_q}(\vec{y})$, whereas I_q are indices in the isometry group of K_n. For the torus, the two are identified.
- In the original KK theory, for dimensional reduction, besides $g_{\mu\nu}$ and B_μ (gravity and electromagnetism), we get a massless scalar, that cannot be put to zero (it is inconsistent to do so).
- For a torus, the spherical harmonics are the products of the Fourier modes.

Further reading: See [13] for the KK approach to supergravity, and a quick review of the KK program.

Exercises

(1) Write the dimensional reduction on S^1 of a 5 dimensional field satisfying "self-duality in odd dimensions", with action

$$S = \int d^5x \left[\epsilon^{\mu\nu\rho\sigma\tau} S_{\mu\nu}\partial_\rho S_{\sigma\tau} + \sqrt{-g} m^2 S_{\mu\nu} S^{\mu\nu}\right]. \tag{12.26}$$

(2) Write down the KK background metric and KK expansion for a compactification of a 10 dimensional metric on $M^4 \times S^2 \times S^2 \times T^2$.
(3) Show explicitly that putting $\phi = 1$ in the original KK reduction (to linearized level) is inconsistent.
(4) Dimensionally reduce the 11 dimensional 3-form gauge field A_{MNP}, with action

$$\int d^{11}x \left[-\frac{1}{2\cdot 4!} F_{MNPQ} F^{MNPQ} - \frac{1}{3}\epsilon^{M_1\dots M_{11}} A_{M_1M_2M_3} A_{M_4\dots M_7} A_{M_8\dots M_{11}}\right], \tag{12.27}$$

where $F_{MNPQ} = 4\partial_{[M} A_{NPQ]}$, on a torus T^7, down to 4 dimensions.
(5) Prove that if we split the off-diagonal metric on a compact space as $g_{\mu m}(x, y) = B_\mu^{AB}(x) V_m^{AB}(y)$, where V_m^{AB} are Killing vectors of the compact space satisfying $D_n V_m + D_n V_m = 0$, then the general coordinate transformation with $\xi_m(x, y) = \lambda^{AB}(x) V_m^{AB}(y)$ gives a nonabelian gauge transformation of B_μ^{AB}, with parameter $\lambda^{AB}(x)$ [Hint: use the fact that $V^{AB} = V^{mAB}\partial_m$ satisfies the nonabelian algebra].

Chapter 13
Electromagnetism and Gravity in Various Dimensions. Consistent Truncations

In this chapter, we will first generalize electromagnetism and gravity to general dimensions, and then we will define consistent truncations of field theories, with emphasis on the KK reduction of supergravity theories.

13.1 Review of Electromagnetism and Gravity

Electromagnetism

The Maxwell's equations, for the equations of motion

$$\partial^\mu F_{\mu\nu} = j_\nu \; , \tag{13.1}$$

and the Bianchi identities,

$$\partial_{[\mu} F_{\nu\rho]} = 0 \; , \tag{13.2}$$

solved by the existence of a gauge field A_μ,

$$F_{\mu\nu} = \partial_\mu A_\nu - \partial_\nu A_\mu \; , \tag{13.3}$$

are valid in any number of dimensions, as we can see (they do not refer to dimensionality anywhere).

For 3 spatial dimensions, we define the electric field,

$$E^i = F^{0i} = -F_{0i} \; , \tag{13.4}$$

and in terms of the gauge field

$$\vec{E} = -\partial_0 \vec{A} - \vec{\nabla}\Phi \; , \tag{13.5}$$

H. Năstase, *Cosmology and String Theory*, Fundamental Theories of Physics 197, https://doi.org/10.1007/978-3-030-15077-8_13

where $A_\mu = (-\Phi, \vec{A})$. These equations again are independent of dimension, so are valid generally. However, the definition of the magnetic field,

$$B^i = \frac{1}{2}\epsilon^{ijk} F_{jk} , \tag{13.6}$$

depends on dimensionality, so is not obviously generalizable. We can of course write a similar relation with the epsilon tensor in other dimensions, but then for instance in 2 spatial dimensions B is a scalar and in 4 spatial dimensions B^{ij} is a tensor.

The local form of Gauss's law is

$$\vec{\nabla} \cdot \vec{E} = \frac{\rho_e}{\epsilon} \Rightarrow \vec{\nabla}^2 \Phi = -\frac{\rho_e}{\epsilon}. \tag{13.7}$$

Gravity

The Einstein's equations are

$$R_{\mu\nu} - \frac{1}{2} g_{\mu\nu} R = 8\pi G_N T_{\mu\nu} , \tag{13.8}$$

and are generalized to arbitrary dimension as long as we define $G_N^{(D)}$ in a general dimension D (we see later).

The Riemann tensor in terms of the Christoffel symbol,

$$R^\mu{}_{\nu\rho\sigma} = \partial_\rho \Gamma^\mu{}_{\nu\sigma} - \partial_\sigma \Gamma^\mu{}_{\nu\rho} + \Gamma^\mu{}_{\lambda\rho}\Gamma^\lambda{}_{\nu\sigma} - \Gamma^\mu{}_{\lambda\sigma}\Gamma^\lambda{}_{\nu\rho} , \tag{13.9}$$

is also valid in general dimension, as is the Bianchi identity that encapsulates the fact that the Riemann tensor is written in terms of the Christoffel,

$$\partial_{[\lambda}(R^\mu{}_\nu)_{\rho\sigma]} = 0. \tag{13.10}$$

In the Newtonian limit of Einstein gravity, we have the Gauss's law for gravity,

$$\vec{\nabla}^2 U_{\text{Newton}} = 4\pi G_N \rho_m , \tag{13.11}$$

where the Newtonian potential is related to the perturbation of the metric by $g_{00} - 1 = \kappa_N h_{00} = -2U_{\text{Newton}}$.

13.2 Interlude: Spheres in Higher Dimensions

Before we continue, we calculate the volume of spheres in arbitrary dimensions.

A sphere in d spatial dimensions (spacetime dimension $D = d + 1$) is defined by

$$S^{d-1}(R) : \sum_{i=1}^{d} x_i^2 = R^2 \,, \tag{13.12}$$

whereas the ball is

$$B^{d-1}(R) : \sum_{i=1}^{d} x_i^2 \leq R^2. \tag{13.13}$$

The volume of the sphere is defined in terms of the volume of the sphere of unit radius as

$$vol(S^{d-1}(R)) = R^{d-1} vol(S^{d-1}). \tag{13.14}$$

For instance, $vol(S^1(R)) = 2\pi R$, $vol(S^2(R)) = 4\pi R^2$.

To find the general case for $vol(S^{d-1})$, we evaluate the d-dimensional Gaussian integral I_d in two different ways. On one hand, the integral is a product of the Gaussian integrals on each direction,

$$I_d = \int_{\mathbb{R}^d} dx_1 \ldots dx_d e^{-r^2} = \prod_{i=1}^{d} \int_{-\infty}^{+\infty} dx_i e^{-x_i^2} = (\sqrt{\pi})^d = \pi^{d/2} \,, \tag{13.15}$$

where $r^2 = x_1^2 + \cdots x_d^2$. On the other hand, we can use spherical coordinates and write

$$\begin{aligned} I_d &= \int_0^\infty dr\; vol(S^{d-1}(r)) e^{-r^2} = vol(S^{d-1}) \int_0^\infty dr r^{d-1} e^{-r^2} \\ &= vol(S^{d-1}) \int_0^\infty \frac{dt}{2} e^{-t} t^{\frac{d}{2}-1} = vol(S^{d-1}) \frac{\Gamma(d/2)}{2} \,, \end{aligned} \tag{13.16}$$

where we wrote $r^2 = t$. Equating the two ways of writing I_d, we find

$$vol(S^{d-1}) = \frac{2\pi^{d/2}}{\Gamma(d/2)}. \tag{13.17}$$

We can check that indeed, for $d = 1$, $vol(S^1) = 2\pi/\Gamma(1) = 2\pi$, and for $d = 2$, $vol(S^2) = 2\pi^{3/2}/\Gamma(3/2) = 4\pi$, since $\Gamma(1/2) = \sqrt{\pi}$ and $\Gamma(3/2) = 1/2\Gamma(1/2)$.

13.3 Gauss's Law for Electromagnetism in *d* Dimensions

3 Spatial Dimensions

The local form of Gauss's law is $\vec{\nabla} \cdot \vec{E} = \rho_e/\epsilon_0$. Integrating it over a ball $B^3(R)$, we find

$$\int_{B^3} dV\, \vec{\nabla} \cdot \vec{E} = \int_{B^3} dV\, \frac{\rho_e}{\epsilon} = \frac{Q}{\epsilon_0}. \tag{13.18}$$

On the other hand, using Stokes's theorem, the left hand side becomes

$$\int_{S^2(R)} d\vec{S}\cdot\vec{E} = vol(S^2)R^2E(R)\ , \tag{13.19}$$

where because of spherical symmetry, both the area element $d\vec{S}$ and the electric field $\vec{E}$ point radially outwards, hence their dot product equals the normal product. Finally, we find

$$E(R) = \frac{Q}{vol(S^2)\epsilon_0 R^2} = \frac{Q}{4\pi\epsilon_0 R^2}. \tag{13.20}$$

d Spatial Dimensions

Generalizing the above to d spatial dimensions, the local form of Gauss's law is the same, so we integrate it over $B^d(R)$, obtaining

$$\int dV\,\vec{\nabla}\cdot\vec{E} = \int_{B^d} dV\frac{\rho}{\epsilon_0} = \frac{Q}{\epsilon_0}. \tag{13.21}$$

Using Stokes's theorem, the left hand side becomes

$$\int_{\partial B^d(R)=S^{d-1}(R)} d\vec{S}\cdot\vec{E} = vol(S^{d-1})R^{d-1}E(R)\ , \tag{13.22}$$

so by equating the two sides we get

$$E(R) = \frac{1}{vol(S^{d-1})\epsilon_0} = \frac{\Gamma(d/2)}{2\pi^{d/2}\epsilon_0}\frac{Q}{R^{d-1}}. \tag{13.23}$$

13.4 Gauss's Law for Gravity in *d* Dimensions

3 Spatial Dimensions

The local form of Gauss's law, integrated over the ball $B^3(R)$, gives

$$\int_{B^3(R)} dV\,\vec{\nabla}^2U_N = 4\pi G_N\int_{B^3(R)} dV\,\rho_m = 4\pi G_N M. \tag{13.24}$$

Using Stokes's theorem, the left hand side becomes

$$\int_{S^2(R)} d\vec{S}\cdot\vec{\nabla}U_N = vol(S^2)R^2|\vec{g}(R)|\ , \tag{13.25}$$

where we have defined the gravitational acceleration $\vec{g} = -\vec{\nabla}U_N$. Equating the two sides, we find

$$|\vec{g}(R)| = \frac{4\pi G_N M}{vol(S^2)R^2} = \frac{G_N M}{R^2}. \tag{13.26}$$

Newton's Constant and Planck Constant in Higher Dimensions

Before we move onto higher dimensions, we need to define $G_N^{(D)}$, since only then we can have the same local form of Gauss's law for gravity. It is obtained by writing the same coefficient for the Einstein–Hilbert action in higher dimensions as in 4 dimensions,

$$S_{EH} = \frac{1}{16\pi G_N^{(D)}} \int d^D x \sqrt{-g^{(D)}} R^{(D)}. \tag{13.27}$$

But we need to KK reduce to 4 dimensions, $D = 4 + n$, with $K_D = M_4 \times K_n$,

$$g_{MN} = \begin{pmatrix} g_{\mu\nu}^{(0)} & \mathbf{0} \\ \mathbf{0} & g_{mn}^{(0)} \end{pmatrix}. \tag{13.28}$$

Then we have $\sqrt{-g^{(D)}} = \sqrt{-g^{(4)}}\sqrt{g^{(n)}}$, so by integrating over $d^n x$, the coefficient of $\int d^4x\sqrt{-g^{(4)}}R^{(D)}$ is defined as the 4 dimensional Newton's constant,

$$\frac{1}{G_N^{(D)}} \int d^n x \sqrt{g^{(n)}} = \frac{V^{(n)}}{G_N^{(D)}} \equiv \frac{1}{G_N^{(4)}}, \tag{13.29}$$

leading to

$$G_N^{(D)} = G_N^{(4)} V^{(n)}. \tag{13.30}$$

As we have already mentioned, the 4 dimensional Planck scale is constructed out of c, $\hbar$ and $G_N^{(4)}$ as $m_P = \sqrt{\hbar c / G_N^{(4)}}$, and the Planck length is $l_P = \sqrt{\hbar G_N^{(4)}/c^3}$ (so that if $\hbar = c = 1$, $m_P = (G_N^{(4)})^{-1/2}$ and $l_P = (G_N^{(4)})^{1/2}$), but in gravity and cosmology we mostly use the reduced Planck mass, $M_{\rm Pl}^{(4)} = m_P/\sqrt{8\pi}$, so that the coefficient of the Einstein–Hilbert action is $M_{\rm Pl}^2/2$.

In higher dimensions, the coefficient of the Einstein–Hilbert action has dimension $D-2$, so that the action is defined as

$$\frac{[M_{\rm Pl}^{(D)}]^{D-2}}{2} \int d^D x \sqrt{-g^{(D)}} R^{(D)}. \tag{13.31}$$

Then finally

$$\frac{1}{8\pi G_N^{(D)}} = [M_{\rm Pl}^{(D)}]^{D-2}. \tag{13.32}$$

***d* Spatial Dimensions**

With $G_N^{(D)}$ defined, the local form of the Gauss's law for gravity is the same. Integrating it over the spatial ball $B^d(R)$, we find

$$\int_{B^d(R)} dV\, \vec{\nabla}^2 U_N = 4\pi G_N^{(D)} \int dV\, \rho_m = 4\pi G_N^{(D)} M. \tag{13.33}$$

Using Stokes's theorem, the left hand side becomes

$$\int_{\partial B^d = S^{d-1}(R)} d\vec{S} \cdot \vec{\nabla} U_N = vol(S^{d-1}) R^{d-1} |\vec{g}(R)|. \tag{13.34}$$

Equating the two sides, we obtain

$$|\vec{g}(R)| = \frac{4\pi G_N^{(D)} \Gamma(d/2)}{2\pi^{d/2}} \frac{M}{R^{d-1}} = \frac{\Gamma\left(\frac{D-1}{2}\right)}{2[M_{\rm Pl}^{(D)}]^{D-2} 2\pi^{\frac{D-2}{2}}} \frac{M}{R^{D-2}}. \tag{13.35}$$

Integrating to get the Newton potential, we obtain

$$U_N(R) = -\frac{2\pi^{\frac{3-D}{2}} \Gamma\left(\frac{D-1}{2}\right)}{D-3} \frac{M G_N^{(D)}}{R^{D-3}} = -\frac{\Gamma\left(\frac{D-1}{2}\right)}{4\pi^{\frac{D-1}{2}}(D-3)} \frac{M}{[M_{\rm Pl}^{(D)}]^{D-2} R^{D-3}}. \tag{13.36}$$

For KK compactification on a square torus $T^n = (S^1_R)^{\times n}$, we find

$$G_N^{(D)} = (2\pi R)^n G_N^{(4)}. \tag{13.37}$$

Then the Gauss's law for distances much smaller than the radii, $r \ll R$, is approximately the D-dimensional law,

$$|\vec{g}(r)| = \frac{4\pi \Gamma\left(\frac{D-1}{2}\right)}{2\pi^{\frac{D-1}{2}}} \frac{G_N^{(D)} M}{r^{D-2}}. \tag{13.38}$$

For distances much larger than the radius on the other hand, we have the 4 dimensional law,

$$|\vec{g}(r)| = \frac{G_N^{(4)} M}{r^2} = \frac{G_N^{(D)} M}{(2\pi R)^{D-4} r^2}. \tag{13.39}$$

So the difference between the two is the replacement of r^{D-2} with $R^{D-4} r^2$, up to some numerical factors.

13.5 Consistent Truncations

An issue that arises when considering KK dimensional reduction is the issue of whether the truncation is consistent. Indeed, we need to put all fields with $n > 0$ to zero, but this is not guaranteed to satisfy the original equations of motion, rather it is an ansatz.

When working at the level of the action, what can go wrong is to have terms that are linear in (all the) fields $\phi_{n>0}$ (note that a term with $\phi_1\phi_2$ is quadratic in ϕ_n), which means that when we write the equations of motion for the ϕ_n, we find

$$(\Box - m^2 + \cdots)\phi_n = (\ldots)\,, \tag{13.40}$$

where the left hand side comes from terms quadratic or higher in ϕ_n's, and the right hand side is finite when you put $\phi_n = 0$, so it is inconsistent to put $\phi_n = 0$.

To see an example, consider a ϕ^3 interaction. The KK expansion, substituted in the term $\int d^D x\sqrt{-g^{(D)}}\lambda\phi^3$, leads to terms of the type (we consider the reduction to d dimensions instead of 4, for generality)

$$\lambda\int d^d x\sqrt{-\det g^{(0)}_{\mu\nu}}\phi_q^{I_q}(\vec{x},t)\phi_0^{I_0}(\vec{x},t)\phi_0^{J_0}(\vec{x},t)\times\int d^n x\sqrt{-\det g^{(0)}_{mn}}Y_q^{I_q}(\vec{y})Y_0^{I_0}(\vec{y})Y_0^{J_0}(\vec{y}). \tag{13.41}$$

If this term is nonzero, then the equation of motion of $\phi_q^{I_q}$ will have a term

$$(\Box - m_q^2)\phi_q^{I_q}(\vec{x},t) = \lambda C\phi_0^{I_0}(\vec{x},t)\phi_0^{J_0}(\vec{x},t)\,, \tag{13.42}$$

where

$$C = \int d^n x\sqrt{-\det g^{(0)}_{mn}}Y_q^{I_q}(\vec{y})Y_0^{I_0}(\vec{y})Y_0^{J_0}(\vec{y}). \tag{13.43}$$

Then $\phi_q^{I_q} = 0$ gives a contradiction, since the left-hand side is zero, but the right-hand side is not.

But if $C = 0$, i.e., if the integral of one of the Y_n's with the rest Y_0's vanishes, then the truncation is consistent. This is what happens for instance in the case of the torus T^n, when $Y_0(\vec{y}) = 1$, but then for S^1, $C_1 = \int dy Y^{I_n} = \int dy e^{\frac{iny}{R}} = 0$, and in the T^n case $C_{T^n} = \prod_n C_1 = 0$ again. Thus for the torus there is always a consistent truncation if we keep *all* the $n = 0$ fields. For instance, in the original Kaluza–Klein reduction, $\varphi = 1$ is inconsistent, since it puts one of the $n = 0$ fields to zero.

More generally, if we have invariance under a group G, and we keep *all* the singlets under G in the reduction, the reduction is consistent, since then all

$$\int Y_n Y_0 Y_0 = 0\,, \tag{13.44}$$

as the Y_n correspond by our assumption to a non-singlet, whereas the Y_0's are singlets. That means that $Y_n Y_0 Y_0$ is a non-singlet, so its integral should be also, except it is a number (which doesn't transform), which means that it must be zero.

Nonlinear Ansatz

Sometimes, when the truncation of the linearized KK expansion is inconsistent, we can write a nonlinear redefinition of the fields that makes it consistent, generically of the type

$$\begin{aligned}\phi_q' &= \phi_q + a\phi_0^2 + \cdots \\ \phi_0' &= \phi_0 + \sum_{pq,\text{incl.}0} c_{pq}\phi_p\phi_q + \cdots ,\end{aligned} \tag{13.45}$$

where we only wrote the quadratic piece of the redefinition, but there are in principle higher order terms. Or equivalently, we can write from the beginning a *nonlinear KK reduction ansatz*. Note that the linearized KK expansion is always valid (it is a theorem that it is so), but the linearized KK dimensional reduction (the truncation of this expansion to the first terms) is not guaranteed to be correct. A nonlinear ansatz for this KK dimensional reduction instead of the linearized one can sometimes work though.

The simplest example of a nonlinear ansatz is the one needed to get the correct d-dimensional Einstein–Hilbert action for the D-dimensional one. The ansatz is

$$g_{\mu\nu}(\vec{x}, t; \vec{y}) = g_{\mu\nu}(\vec{x}, t)\left[\frac{\det g_{mn}(\vec{x}, t; \vec{y})}{\det g_{mn}^{(D)}(\vec{y})}\right]^{-\frac{1}{d-2}} . \tag{13.46}$$

To fully check this result is somewhat complicated, but we can check it for the case that the square bracket is constant.

Indeed, under a constant rescaling $g_{\mu\nu} \to \lambda g_{\mu\nu}$, the Christoffel symbols $\Gamma \sim g^{-1}\partial g$ and the Ricci scalar $R_{\mu\nu} = R^\lambda{}_{\mu\lambda\nu} \sim \partial\Gamma + \Gamma\Gamma$ are invariant. In the Einstein–Hilbert action,

$$R^{(D)} = g^{\Lambda\Sigma} R^{(D)}_{\Lambda\Sigma} = g^{\mu\nu} R^{(D)}_{\mu\nu} + \cdots \tag{13.47}$$

contains the d-dimensional Ricci scalar (plus other terms). Then if we only rescale $g_{\mu\nu}$, but not g_{mn},

$$\sqrt{-g^{(D)}} g^{\mu\nu} R^{(D)}_{\mu\nu} = \sqrt{-\det g^{(d)}_{\mu\nu}}\sqrt{\det g_{mn}} g^{\mu\nu} R^{(D)}_{\mu\nu} \to \lambda^{\frac{d}{2}-1}\sqrt{-\det g^{(d)}_{\mu\nu}}\sqrt{\det g_{mn}} g^{\mu\nu} R^{(D)}_{\mu\nu} , \tag{13.48}$$

so, in order to replace the $\sqrt{\det g_{mn}}$ factor with $\sqrt{g^{(0)}_{mn}}$, so that we can obtain the correct Einstein–Hilbert action after $d^n x$ integration, we need to have

$$\lambda = \left[\frac{\det g_{mn}}{\det g^{(0)}_{mn}}\right]^{-\frac{1}{d-2}} . \tag{13.49}$$

In the case of the original KK reduction, with $d = 4$ and $D = 5$, we get $g^{(5)}_{\mu\nu} = g^{(4)}_{\mu\nu} g_{55}^{-1/2}$. Defining

$$g_{55} \equiv \Phi^{2/3} , \tag{13.50}$$

we obtain $g^{(5)}_{\mu\nu} = \Phi^{-1/3} g^{(4)}_{\mu\nu}$. In order to also find the Maxwell action for $g_{\mu 5}$, we need to redefine it also as

$$g_{\mu 5} = B_\mu(\vec{x}, t) g_{55}. \tag{13.51}$$

Finally, that means that the full nonlinear ansatz for the original KK dimensional reduction is

$$g_{\Lambda\Sigma} = \Phi^{-1/3}(\vec{x},t)\begin{pmatrix} g_{\mu\nu}(\vec{x},t) & B_\mu(\vec{x},t)\Phi(\vec{x},t) \\ B_\mu(\vec{x},t)\Phi(\vec{x},t) & \Phi(\vec{x},t)\end{pmatrix}. \tag{13.52}$$

Of course, if we want to put $\Phi = 1$, writing this nonlinear ansatz is moot, but as we said, this is not a consistent truncation.

Important Concepts to Remember

- We can easily generalize electromagnetism and gravity to higher dimensions, just defining $G_N^{(D)}$ as the same coefficient in the Einstein–Hilbert action.
- Under KK dimensional reduction, we obtain $G_N^{(D)} = G_N^{(4)} V^{(n)}$, and we define $1/(8\pi G_N^{(D)}) = [M_{\rm Pl}^{(D)}]^{D-2}$.
- For $r \ll R$, we have for the gravitational acceleration $|\vec{g}(r)| \sim G_N^{(D)} M/r^{D-2}$, whereas for $r \gg R$, we have $|\vec{g}(r)| \sim G_N^{(D)}/[(2\pi R)^{D-4} r^2]$.
- For a KK dimensional reduction, the truncation can be inconsistent, due to terms in the action linear in the modes to be put to zero.
- The truncation becomes consistent if the integral of the Y_n harmonic with the other Y_0 harmonics gives zero, which happens in the torus case, as well as in the case when we keep *all* the singlets under the symmetry group G in the truncation.
- Sometimes nonlinearly redefining the fields, or equivalently, for a nonlinear KK reduction ansatz, can make the truncation consistent.
- To obtain the Einstein–Hilbert action from the same in the higher dimension, we need to nonlinearly redefine $g_{\mu\nu}^{(D)} = g_{\mu\nu}^{(4)}[\det g_{\mu\nu}/\det g_{\mu\nu}^{(0)}]^{-1/(d-2)}$.

Further reading: For consistent truncations of supergravity, see [14].

Exercises

(1) Calculate the gravitational potential of a thin shell of radius R and width $\Delta R \ll R$ and density ρ in 4 dimensions, in all relevant radial regions.
(2) Consider an insulating sphere in 4 dimensions (shell only), of surface charge density σ and radius R, situated between 2 parallel conducting plates, each at a distance $2R$ from the center of the sphere. Calculate the potential at the position of one of the conductors.
(3) Consider the Einstein–Hilbert action coupled nonminimally with the scalar ϕ, with action

$$S = \frac{M_{\rm Pl}^2}{2}\int d^4x\sqrt{-g}\left[R(1+\alpha\phi) - \beta\phi^{1+\frac{1}{n}}\right]. \tag{13.53}$$

Knowing the formula for the general Weyl rescaling in d dimensions

$$R[g_{\mu\nu}] = \Omega^{-2} R[g^E_{\mu\nu}] - 2(d-1) g_E^{\mu\nu} \nabla^E_\mu \nabla^\nu_E \ln \Omega - (d-2)(d-1) g_E^{\mu\nu} (\nabla^E_\mu \ln \Omega)(\nabla^E_\nu \ln \Omega) , \tag{13.54}$$

find the transformation that reduces the action to the Einstein–Hilbert action minimally coupled to a canonical dynamical scalar.

(4) Consider a charge q in between two perfectly conducting infinite conducting planes, situated both at a distance R from it. Calculate the electric potential on one of the conductors, at the minimum distance point from the charge. Write the result in arbitrary number of dimensions, for planes of codimension one.

(5) Prove that for the nonlinear ansatz for the original KK reduction from 5 dimensions, we obtain the Maxwell action for electromagnetism.

Chapter 14
$\mathcal{N} = 1$ Supergravity in 4 Dimensions

In this chapter we will study $\mathcal{N} = 1$ supergravity in 4 dimensions. Supergravity can be defined as a supersymmetric theory of gravity, or otherwise as a theory of local supersymmetry. In practice, we will use a bit of both definitions to find it. Before that however, we need to define what supersymmetry is. After that, we will count degrees of freedom, in order to see what supergravity models we can have. After defining the vielbein-spin connection formulation of gravity, within which supergravity is defined, we will finally define the model of $\mathcal{N} = 1$ supergravity.

14.1 Rigid Supersymmetry

The simplest model to understand supersymmetry in is the Wess–Zumino model in 2 dimensions. Indeed, supersymmetry is defined as a symmetry between bosons and fermions, but the symmetry must relate a degree of freedom with another. In the bosonic case, $\delta\phi^i = \epsilon\phi^i$ implicitly relates a bosonic degree of freedom to another, but it is kind of trivial, so we don't usually think about it that way. But in the case of supersymmetry, it will relate a bosonic degree of freedom with a fermionic one. As such, we can consider either on-shell supersymmetry, or off-shell supersymmetry, and the corresponding numbers of bosonic and fermionic degrees of freedom must match.

Here we will only describe on-shell supersymmetry. A Dirac fermion has $n = 2^{[d/2]}$ complex components. But in supersymmetry, one usually deals with Majorana fermions, which can be thought of as real. Thus an off-shell Majorana fermion has n real components, $n = 2$ for $d = 2$. On-shell, the Dirac equation $(\partial\!\!\!/ + m)\psi = 0$ is a matrix equation that relates half of the components with the other half, so an on-shell Majorana fermion has $n/2$ real components, 1 in the case of $d = 2$.

Therefore in $d = 2$ we have the simplest model, with a Majorana fermion ψ with a real on-shell degree of freedom, that can match the degree of freedom of a real

H. Năstase, *Cosmology and String Theory*, Fundamental Theories of Physics 197, https://doi.org/10.1007/978-3-030-15077-8_14

scalar ϕ. Then the simplest supersymmetric model is the free Wess–Zumino model, with the free action of a real scalar and a Majorana fermion,

$$S = -\frac{1}{2}\int d^2x[(\partial_\mu\phi)^2 + \bar{\psi}\partial\!\!\!/\psi]. \tag{14.1}$$

Since we have the mass dimensions $[d^2x] = -2$ and $[\partial_\mu] = 1$, the mass dimensions of the scalar and fermion are $[\phi] = 0$ and $[\psi] = 1/2$.

We must define the supersymmetry transformation rules. We take it as a definition that the scalar varies into the fermion times a parameter ϵ. For Lorentz invariance, ϵ must be also a spinor, and the variation must be

$$\delta\phi = \bar{\epsilon}\psi \equiv \bar{\epsilon}_\alpha\psi^\alpha \equiv \epsilon^\beta C_{\beta\alpha}\psi^\alpha. \tag{14.2}$$

A Majorana spinor satisfies the condition that the Majorana conjugate χ^C equals the Dirac conjugate $\bar{\chi}$, i.e.,

$$\chi^C \equiv \chi^T C = \bar{\chi} \equiv \chi^\dagger i\gamma_0. \tag{14.3}$$

Here C is the charge conjugation matrix, which can be defined generally in any dimension, but in 2 and 4 dimensions, is defined as an antisymmetric matrix that transforms γ_μ into $-\gamma_\mu^T$, i.e.,

$$C^T = -C; \quad C\gamma^\mu C^{-1} = -(\gamma^\mu)^T. \tag{14.4}$$

In dimensions other than 2 and 4, we can have either plus or minus for either of the two relations.

From the definition of supersymmetry, $\delta\phi = \bar{\epsilon}\psi$, we see that $[\epsilon] = -1/2$. If we want to write $\delta\phi$, it should be proportional to ϕ and ϵ. Now the Lorentz transformations match, but the dimensions don't match, we need to introduce an object of mass dimension 1. It could be ∂_μ, but we need to contract the μ index in order not to upset the Lorentz transformation property, so we multiply by $\partial\!\!\!/ = \gamma_\mu\partial_\mu$ instead, so that

$$\delta\psi = \partial\!\!\!/\phi\epsilon. \tag{14.5}$$

We now prove that the free WZ action is invariant under the supersymmetry (susy) transformation laws that we defined.

Majorana Spinor Identities

Before that however, we must prove some identities for Majorana spinor identities, valid both in 2 and 4 dimensions. There are more similar ones we could write, but those depend on dimension. We have

$$\bar{\epsilon}\chi = \bar{\chi}\epsilon \tag{14.6}$$

$$\bar{\epsilon}\gamma_\mu\chi = -\bar{\chi}\gamma_\mu\epsilon. \tag{14.7}$$

The first relation is proven as follows,

$$\bar{\epsilon}\chi = \epsilon^\alpha C_{\alpha\beta}\chi^\beta = -\chi^\beta C_{\alpha\beta}\epsilon^\alpha = +\chi^\beta C_{\beta\alpha}\epsilon^\alpha \equiv \bar{\chi}\epsilon \,, \tag{14.8}$$

where in the second equality the minus comes from permuting the order of the two fermions, and in the third we used $C^T = -C$. The second relation is proved similarly, by first multiplying the second equation defining the C matrix (14.4) by C, to find

$$C\gamma_\mu = -\gamma_\mu^T C = +\gamma_\mu^T C^T = +(C\gamma_\mu)^T \,, \tag{14.9}$$

thus that $(C\gamma_\mu)$ is symmetric. Then we have

$$\bar{\epsilon}\gamma_\mu\chi \equiv \epsilon^\alpha (C\gamma_\mu)_{\alpha\beta}\chi^\beta = -\chi^\beta (C\gamma_\mu)_{\alpha\beta}\epsilon^\alpha = -\chi^\beta (C\gamma_\mu)_{\beta\alpha}\epsilon^\alpha \equiv -\bar{\chi}\gamma_\mu\epsilon. \tag{14.10}$$

We now return to the invariance of the action. Varying (14.1), we obtain

$$\begin{aligned} \delta S &= -\int d^2x \left[-\delta\phi\Box\phi + \frac{1}{2}\delta\bar{\psi}\partial\!\!\!/\psi + \frac{1}{2}\bar{\psi}\partial\!\!\!/\delta\psi \right] \\ &= -\int d^2x[-\delta\phi\Box\phi + \delta\bar{\psi}\partial\!\!\!/\psi] \,, \end{aligned} \tag{14.11}$$

where we have used the relation (14.6). Replacing the susy rules inside the variations, we find

$$\begin{aligned} \delta S &= -\int d^2x[-\bar{\epsilon}\psi\Box\phi + \bar{\psi}\partial\!\!\!/\partial\!\!\!/\phi\epsilon] \\ &= -\int d^2x[-(\bar{\epsilon}\psi)\Box\phi + (\bar{\psi}\epsilon)\Box\phi] \\ &= 0 \,, \end{aligned} \tag{14.12}$$

where in the second equality we used the relation

$$\partial\!\!\!/\partial\!\!\!/ = \gamma^\mu\gamma^\nu\partial_\mu\partial_\nu = \frac{1}{2}\{\gamma^\mu, \gamma^\nu\}\partial_\mu\partial_\nu = \partial^2 \equiv \Box. \tag{14.13}$$

In conclusion, we find that the action (14.1) is invariant under the susy rules *without using the equations of motion*, i.e., off-shell. That is however strange, since we have an on-shell supersymmetry only. But the point is that in general for a symmetry it is not enough to have an action that is invariant under it, we also need the algebra of the symmetries to close on the fields. In the case of usual, bosonic, symmetries, this is somewhat trivial, since the Lie algebra means that the commutator of two symmetries is another symmetry of the same type, so

$$[T_a, T_b] = f_{ab}{}^c T_c \Rightarrow [\epsilon_1^a T_a, \epsilon_2^b T_b] = (\epsilon_1^a f_{ab}{}^c \epsilon_2^b T_c) \Rightarrow [\delta_{\epsilon_1}, \delta_{\epsilon_2}] = \delta_{\epsilon_1 f \epsilon_2}. \tag{14.14}$$

This is usually satisfied. But in the case of supersymmetry, this is less obvious. The susy algebra that we need to satisfy is

$$\{Q^i_\alpha, Q^j_\beta\} = 2(C\gamma^\mu)_{\alpha\beta} P_\mu + \dots , \quad (14.15)$$

which means that we need to have on fields (multiplying by $\epsilon_1^\alpha \epsilon_2^\beta$, and considering that P_μ is represented by ∂_μ)

$$[\delta_{\epsilon_1}, \delta_{\epsilon_2}] = 2\bar{\epsilon}_2 \gamma^\mu \epsilon_1 \partial_\mu (+\dots) \quad (14.16)$$

It turns out that the algebra is realized on ϕ without the use of the equations of motion (off-shell), which is left as an exercise to prove. That is how we derive the algebra above. On the other hand, on the fermions, we need to use the equations of motion, i.e., it is on-shell, but we will not show it here.

14.2 Supergravity

We are now ready to define supergravity, as a supersymmetric theory of gravity, and a theory of local supersymmetry. In general, if we want to make a rigid symmetry local, we need to introduce a gauge field. For instance, if we have a complex scalar ϕ with action $S = -\frac{1}{2} \int d^d x |\partial_\mu \phi|^2$, invariant under $\phi \to e^{i\alpha}\phi$, we can make the invariance local, $\alpha \to \alpha(x)$, by introducing a gauge field, transforming as $\delta A_\mu = \partial_\mu \alpha$, and replace the derivative with the covariant derivative, $\partial_\mu \to D_\mu = \partial_\mu - i A_\mu$.

In the case of supersymmetry then, if we make it local by $\epsilon \to \epsilon(x)$, we need to introduce a "gauge field of local supersymmetry", "A_μ^α"$\to \psi_\mu^\alpha$, that has a spinor index α since the parameter ϵ does. This field ψ_μ^α is called the gravitino, and it is a spin 3/2 field (more precisely, on-shell we need to remove its gamma-trace, as we will see later). In fact, it turns out that the gravitino is also the superpartner of the graviton, so we can write the simplest $\mathcal{N} = 1$ supergravity just from the graviton and the gravitino.

Counting Degrees of Freedom

Before we construct supergravity, we want to describe the degrees of freedom in it. We have seen the counting of degrees of freedom of a spin 1/2 fermion, and the counting of the spin 1 gauge field is also familiar: off-shell, we have the d degrees of freedom, minus the one degree of freedom corresponding to gauge invariance, $\delta A_\mu = \partial_\mu \alpha$, giving $d - 1$ off-shell degrees of freedom. On shell, in the Lorenz gauge $\partial^\mu A_\mu = 0$, the equation of motion is simply the KG equation, $\Box A_\mu = 0$, which doesn't change the number of degrees of freedom, but the Lorenz condition itself is one condition, so on shell we have $d - 2$ degrees of freedom.

Graviton

1. *Off-shell*

For the graviton, the metric $g_{\mu\nu}$ is a symmetric matrix, so it has $d(d+1)/2$ components, but they are subject to the general coordinate transformation invariance, with parameter $\xi^\mu(x)$, i.e., we need to subtract d degrees of freedom, so we have $d(d-1)/2$ off-shell degrees of freedom.

2. *On-shell*

The linearized action for the graviton $h_{\mu\nu}$ ($g_{\mu\nu} = \eta_{\mu\nu} + 2\kappa_N h_{\mu\nu}$) coming from the Einstein–Hilbert action is the Fierz–Pauli action,

$$\mathcal{L} = \frac{1}{2}h^2_{\mu\nu,\rho} + h^2_\mu - h^\mu h_{,\mu} + \frac{1}{2}h^2_{,\mu} \,, \tag{14.17}$$

where

$$h_\mu \equiv \partial^\nu h_{\nu\mu}; \quad h \equiv h^\mu_\mu \,, \tag{14.18}$$

and, $\mu \equiv \partial_\mu$.

Just like in the case of the Lorenz gauge for electromagnetism, we can now choose a gauge, namely the *de Donder gauge*,

$$\partial^\nu \bar{h}_{\mu\nu} = 0 \,, \tag{14.19}$$

where

$$\bar{h}_{\mu\nu} \equiv h_{\mu\nu} - \frac{h}{2}\eta_{\mu\nu} \,, \tag{14.20}$$

in which the equation of motion becomes simply Klein-Gordon,

$$\Box \bar{h}_{\mu\nu} = 0. \tag{14.21}$$

We leave proving this as a simple exercise. The KG equation doesn't change the number of degrees of freedom, simply imposes $k^2 = 0$ in momentum space, but the de Donder gauge condition amounts to d local conditions, thus eliminating d degrees of freedom.

In conclusion, on-shell for the graviton we have $d(d-1)/2 - d = (d-1)(d-2)/2 - 1$ degrees of freedom, corresponding to a transverse ($d-2$ components only) symmetric traceless matrix.

Gravitino

1.*Off-shell*

Off-shell, $\psi_{\mu\alpha}$ has nd components, where $n = 2^{[d/2]}$. But it is also invariant under a gauge-like transformation, namely local supersymmetry (supergravity), which we will see that it is $\delta\psi_\mu = D_\mu\epsilon$, so it removes n components, leaving $n(d-1)$ components off-shell.

2.*On-shell*

Naively, we would say that, since $\psi_{\mu\alpha}$ is a gauge field, but also a fermion, on-shell we would have $n/2(d-2)$ components, where $n/2$ comes from the fact that the Dirac-like gravitino equation relates again half of the components to the other half.

However, there is a subtlety. As we already mentioned, the indices of $\psi_{\mu\alpha}$ mean that we make a product of the spin 1 and spin 1/2 representations, but that in fact decomposes as two irreducible representations,

$$1 \otimes 1/2 = 3/2 \oplus 1/2. \tag{14.22}$$

The point is that the gamma-trace, $\gamma^\mu \psi_\mu$, is a spin 1/2 fermion that transforms into itself, so lives in a separate irreducible representation, and only the gamma-traceless part represents the spin 3/2 representation. That means that the correct number of on-shell degrees of freedom of the gravitino is actually

$$\frac{n}{2}(d-2) - \frac{n}{2} = \frac{n}{2}(d-3). \tag{14.23}$$

We can now verify that in 4 dimensions, on-shell the graviton has $(d-1)(d-2)/2 - 1 = 3 \cdot 2/2 - 1 = 2$ degrees of freedom, and the gravitino has $2^{[d/2]}(d-3)/2 = 2^2 \cdot 1/2 = 2$ degrees of freedom also. That means that indeed, we can construct a $\mathcal{N} = 1$ supergravity model just from the graviton and the gravitino.

14.3 Vielbein-Spin Connection Formulation of General Relativity

But in order to define the supergravity model, we need to use the vielbein-spin connection formulation of general relativity, since there are fermions in the theory, that couple to gravity using the spin connection only.

The vielbein e_μ^a is a sort of square root of the metric $g_{\mu\nu}$, which is written as

$$g_{\mu\nu}(x) = e_\mu^a(x) e_\nu^b(x) \eta_{ab}. \tag{14.24}$$

Here μ, ν are "curved" indices, in the curved space, whereas a, b are "flat" indices, in a space tangent to the curved space at each point. As such, there is a *local Lorentz* symmetry acting on them, with Minkowski metric η_{ab}.

Therefore the local Lorentz (l.L.) and general coordinate (g.c.) transformations of the vielbein are

$$\begin{aligned} \delta_{l.L.} e_\mu^a(x) &= \lambda^a{}_b(x) e_\mu^b(x) \\ \delta_{g.c.} e_\mu^a(x) &= (\xi^\rho \partial_\rho) e_\mu^a + (\partial_\mu \xi^\rho) e_\rho^a. \end{aligned} \tag{14.25}$$

Here note that the local Lorentz transformation acts in the usual way on the index a, and the general coordinate transformation acts as on a vector on e_μ.

A note on the nomenclature. The term "vielbein" was originally defined in 4 dimensions only, where it was "vierbein" from the German "vier" = four and "bein" = leg. Then it was extended to ein-, zwei-, drei-bein (one, two, three legs in one, two, three dimensions), until finally it was generalized to "vielbein" for "viel" = many in an arbitrary dimension.

The *spin connection* is a kind of gauge field ("connection" in mathematical language) for curved space, specifically for the covariant derivative acting on spinors. The generator of Lorentz transformations in the spinor representation is $\frac{1}{4}\Gamma_{ab}$, so the covariant derivative on the spinors is defined as

$$D_\mu\psi = \partial_\mu\psi + \frac{1}{4}\omega_\mu^{ab}\Gamma_{ab}\psi. \tag{14.26}$$

But in order to obtain the usual general relativity, we cannot have both the vielbein e_μ^a and the spin connection ω_μ^{ab} be independent, we must write ω_μ^{ab} in terms of e_μ^a, otherwise we have too many degrees of freedom.

In fact, in terms of e_μ^a, we can check that we already have the correct number of degrees of freedom, since we have d^2 components, minus the d components of the general coordinate transformations defined by ξ^μ, minus the $d(d-1)/2$ components of the local Lorentz transformations, leading to the same $(d-2)(d-1)/2 - 1$ degrees of freedom.

The condition that we need to impose is known as the no-torsion constraint, or the "vielbein postulate". The torsion is defined as the antisymmetric part of the covariant derivative of the vielbein, $T_{\mu\nu}^a \equiv D_{[\mu}e_{\nu]}^a$, and in general the covariant derivative is

$$D_\mu e_\nu^a = \partial_\mu e_\nu^a - \Gamma_{\mu\nu}^\lambda e_\lambda^a + \omega_\mu^{ab} e_\nu^b\,, \tag{14.27}$$

but the Chistoffel symbol is symmetric in $(\mu\nu)$, so by taking the antisymmetric part we obtain

$$T_{\mu\nu}^a = D_{[\mu}e_{\nu]}^a = \partial_{[\mu}e_{\nu]}^a + \omega_{[\mu}^{ab}e_{\nu]}^b = 0. \tag{14.28}$$

The solution of this equation is an $\omega_\mu^{ab} = \omega_\mu^{ab}(e)$, which is left as an exercise to prove.

We can define the field strength of the "gauge field" ω_μ^{ab} (for the $SO(1,3)$ local Lorentz group) is

$$R_{\mu\nu}^{ab}(\omega) = \partial_\mu\omega_\nu^{ab} - \partial_\nu\omega_\mu^{ab} + \omega_\mu^{ac}\omega_\nu^{cb} - \omega_\nu^{ac}\omega_\mu^{cb}. \tag{14.29}$$

With this definition, this field strength, in the case the vielbein is $\omega = \omega(e)$, is just the Riemann tensor (in terms of the Christoffel tensor) with two indices flattened by the vielbein, so

$$R_{\mu\nu}^{ab}(\omega(e)) = R^\mu{}_{\nu\rho\sigma}(\Gamma(g(e)))e_\mu^a e^{-1\nu b}. \tag{14.30}$$

Since moreover, as a matrix we have

$$g = e \cdot \eta \cdot e \Rightarrow \det g = -(\det e)^2 , \quad (14.31)$$

in the Einstein–Hilbert action we can replace $\sqrt{-g}$ with $\det e$, and the Ricci scalar with the contraction of the field strength of ω, obtaining

$$S_{EH} = \frac{1}{16\pi G_N} \int d^4x (\det e) R_{\mu\nu}^{ab}(\omega(e))(e^{-1})_a^\mu (e^{-1})_b^\nu . \quad (14.32)$$

But since, as we can check,

$$\det e (e^{-1})_a^\mu (e^{-1})_b^\nu = \epsilon_{abcd} \epsilon^{\mu\nu\rho\sigma} e_\rho^c e_\sigma^d , \quad (14.33)$$

the Einstein–Hilbert action can be rewritten as

$$\begin{aligned} S_{EH} &= \frac{1}{2\kappa_N^2} \int d^4x \epsilon_{abcd} \epsilon^{\mu\nu\rho\sigma} R_{\mu\nu}^{ab}(\omega) e_\rho^c e_\sigma^d \\ &\equiv \frac{1}{2\kappa_N^2} \int \epsilon_{abcd} R^{ab}(\omega) \wedge e^c \wedge e^d , \end{aligned} \quad (14.34)$$

where on the last line we have written things in terms of forms, like $\omega^{ab} = \omega_\mu^{ab} dx^\mu$. Now we notice that in fact, we can consider $\omega = \omega(e)$, as it was implicit until now, or not, the result is the same. Indeed, considering now ω as an independent variable, we can vary with respect to it, and the resulting equation of motion is

$$\epsilon_{abcd} De^c \wedge e^d = 0 \Rightarrow De^c = 0 : \quad D_{[\mu} e_{\nu]}^a = T_{\mu\nu}^a = 0 , \quad (14.35)$$

which is the same as the vielbein postulate defining $\omega(e)$!

14.4 $\mathcal{N} = 1$ Supergravity in 4 Dimensions

To construct the action of $\mathcal{N} = 1$ supergravity, we need to add to the Einstein–Hilbert action for the graviton an action for the gravitino.

In flat space, Rarita and Schwinger wrote the action for a spin 3/2 field, in a general dimension

$$S_{RS} = -\frac{1}{2} \int d^d x \bar{\psi}_\mu \gamma^{\mu\nu\rho} \partial_\nu \psi_\rho . \quad (14.36)$$

In 4 dimensions, the action can be rewritten using the relation

$$i \epsilon^{\mu\nu\rho\sigma} \gamma_5 \gamma_\nu = \gamma^{\mu\rho\sigma} , \quad (14.37)$$

where

$$\gamma_5 = i\gamma_0\gamma_1\gamma_2\gamma_3 , \quad (14.38)$$

as

$$S_{RS,4d} = -\frac{i}{2}\int d^4x \epsilon^{\mu\nu\rho\sigma}\bar{\psi}_\mu\gamma_5\gamma_\nu\partial_\rho\psi_\sigma. \quad (14.39)$$

We can easily write the action in curved space, thus coupling it to the graviton, by putting det e in the volume and replacing $\partial_\mu \to D_\mu$, so

$$S_{RS} = -\frac{1}{2}\int d^dx (\det e)\bar{\psi}_\mu\gamma^{\mu\nu\rho}D_\nu\psi_\rho , \quad (14.40)$$

and in 4 dimensions also

$$S_{RS,4d} = -\frac{i}{2}\int d^4x \epsilon^{\mu\nu\rho\sigma}\bar{\psi}_\mu\gamma_5\gamma_\nu D_\rho\psi_\sigma. \quad (14.41)$$

The supergravity action is then simply the sum of the two terms,

$$S_{\mathcal{N}=1\ sugra} = S_{EH}(e,\omega) + S_{RS}(\psi_\mu,\omega). \quad (14.42)$$

The supersymmetry transformation laws are written remembering the fact that supergravity is a supersymmetric theory of gravity, and a theory of local supersymmetry. Being a supersymmetric theory of gravity, the vielbein, standing in for gravity, must vary into the gravitino, like we had $\delta\phi = \bar{\epsilon}\psi$ for the 2 dimensional WZ model. Now the index a on e_μ^a doesn't correspond to anything on the right hand side, so we must introduce the constant matrix γ^a to compensate for it, giving finally

$$\delta e_\mu^a = \frac{\kappa_N}{2}\bar{\epsilon}\gamma^a\psi_\mu. \quad (14.43)$$

Note that the factor of $\kappa_N/2$ is conventional, it could have been absorbed in the definition of ϵ. Since ψ_μ is considered as a gauge field of local supersymmetry, and the local Lorentz group is nonabelian, we expect a nonabelian gauge field transformation of the type $\delta\psi_\mu = D_\mu\epsilon$. Indeed, we can write

$$\delta\psi_\mu = \frac{1}{\kappa_N}D_\mu\epsilon , \quad (14.44)$$

where as before, we have

$$D_\mu\epsilon = \partial_\mu\epsilon + \frac{1}{4}\omega_\mu^{ab}\Gamma_{ab}\epsilon. \quad (14.45)$$

Now that we have defined the supergravity action (14.42) and the transformation rules, one can check whether the action is invariant under them, and find that it is indeed. However, there is a subtlety. If we consider the spin connection ω as a dependent quantity, we cannot however consider $\omega = \omega(e)$ as in the absence of fermions. This would be inconsistent, i.e., in particular would not lead to a supersymmetric theory.

Indeed, now, if we consider ω as independent in (14.42), we see that the equation of motion is modified due to the ω in D_μ action on fermions, to $D_{[\mu}e^a_{\nu]} = \bar{\psi}\psi$ terms, so its solution is

$$\omega = \omega(e, \psi) = \omega(e) + \bar{\psi}\psi \text{ terms.} \tag{14.46}$$

This would be a *first order* formulation of supergravity. If we want to consider instead a *second order* formulation of supergravity, with a dependent ω, by consistency we need to consider $\omega = \omega(e, \psi)$ instead of $\omega = \omega(e)$.

Complete Transformation Laws of Supergravity

To finish, we summarize the complete set of transformation laws of $\mathcal{N} = 1$ supergravity in 4 dimensions.

The supersymmetry transformation laws are

$$\begin{aligned} \delta_S e^a_\mu &= \frac{\kappa_N}{2}\bar{\epsilon}\gamma^a\psi_\mu \\ \delta_S \psi_\mu &= \frac{1}{\kappa_N} D_\mu \epsilon. \end{aligned} \tag{14.47}$$

In a first order formulation, we would need also the transformation laws for ω_μ^{ab}, but we will not write them here.

The Einstein (general coordinate) transformation laws are

$$\begin{aligned} \delta_E e^a_\mu &= \xi^\nu \partial_\nu e^a_\mu + (\partial_\mu \xi^\nu) e^a_\nu \\ \delta_E \omega_\mu^{ab} &= \xi^\nu \partial_\nu \omega_\mu^{ab} + (\partial_\mu \xi^\nu)\omega_\nu^{ab} \\ \delta_E \psi_\mu &= (\xi^\nu \partial_\nu)\psi_\mu + (\partial_\mu \xi^\nu)\psi_\nu. \end{aligned} \tag{14.48}$$

Finally, the local Lorentz transformation laws are

$$\begin{aligned} \delta_{I.L.} e^a_\mu &= \lambda^a{}_b e^b_\mu \\ \delta_{I.L.} \omega_\mu^{ab} &= D_\mu \lambda^{ab} = \partial_\mu \lambda^{ab} + \omega_\mu^{ac}\lambda^{cb} - \omega_\mu^{bc}\lambda^{ca} \\ \delta_{I.L.} \psi_\mu &= -\frac{1}{4}\lambda^{ab}\gamma_{ab}\psi_\mu. \end{aligned} \tag{14.49}$$

Important Concepts to Remember

- Supergravity is a theory of local supersymmetry, or a supersymmetric theory of gravity.
- The gauge field of the local susy, and the superpartner of the graviton, is the gravitino ψ_μ.
- Supersymmetry is a symmetry between bosons and fermions, that matches the number of degrees of freedom.

- The susy transformation rules are of the type $\delta\phi = \bar{\epsilon}\psi$ and $\delta\psi = \partial\!\!\!/\phi\epsilon$.
- On-shell supersymmetry uses the equations of motion and matches degrees of freedom on-shell, and off-shell supersymmetry matches degrees of freedom off-shell.
- In the off-shell case, the algebra of transformations needs to be realized on the fields, and in the on-shell case this only happens up to equations of motion.
- The gravitino ψ_μ contains also a spin 1/2 on-shell representation, the gamma trace. The spin 3/2 part is the gamma-traceless part.
- One can formulate general relativity in terms of vielbein, a type of square root of the metric, with a flat index, in the tangent space at a local point, acted upon by local Lorentz transformations; and the spin connection, the "gauge field" appearing in the covariant derivative that acts on spinors.
- The Einstein–Hilbert action is written in terms of the field strength of ω and the vielbein e, and the equation of motion for ω gives the "vielbein constraint" that is solved by $\omega = \omega(e)$ and reduces it to the EH action.
- The $\mathcal{N} = 1$ supergravity action is the sum of the EH action for gravity and the Rarita-Schwinger action for the gravitino.
- The supersymmetry rules are $\delta e_\mu^a = \kappa_N/2\bar{\epsilon}\gamma^a\psi_\mu$ and $\delta\psi_\mu = D_\mu\epsilon/\kappa_N$.
- In the second order formulation of supergravity, $\omega = \omega(e, \psi)$ is the solution of the equations of motion of the action with independent ω.

Further reading: See [15, 16] for more on supersymmetry and supergravity. For a recent complete work on supergravity see [17]. For a very good review of supergravity, see [18].

Exercises

(1) Check that we can represent the susy algebra on the scalar ϕ of the 2 dimensional WZ model without the use of the equations of motion.
(2) Check that the spin connection as a function of the vielbein,

$$\omega_\mu^{ab}(e) = \frac{1}{2}(\partial_\mu e_\nu^a - \partial_\nu e_\mu^a) - \frac{1}{2}e^{b\nu}(\partial_\mu e_\nu^a - \partial_\nu e_\mu^a) - \frac{1}{2}e^{a\rho}e^{b\sigma}(\partial_\rho e_{c\sigma} - \partial_\sigma e_{c\rho})e_\mu^c\,, \tag{14.50}$$

satisfies the no-torsion (vielbein) constraint $T_{\mu\nu}^a \equiv 2D_{[\mu}e_{\nu]}^a = 0$.
(3) Prove that by imposing the de Donder gauge, the Fierz–Pauli action gives the KG equation for $\bar{h}_{\mu\nu}$.
(4) Find the spectrum (field content) and check the number of on-shell degrees of freedom for $\mathcal{N} = 2$ supergravity in 4 dimensions, as a sum of the $\mathcal{N} = 1$ supergravity multiplet and another $\mathcal{N} = 1$ multiplet that starts with the gravitino (gravitino multiplet). Repeat the same exercise in 3 dimensions. Can you guess off-shell fields that need to be added to have also matching of degrees of freedom off-shell (off-shell susy)?
(5) Show that at the linearized level, the Einstein–Hilbert action reduces to the Fierz–Pauli action.

Chapter 15
KK Compactification of Supergravity Models

In this chapter, we will study the KK compactification of supergravity models in higher dimensions.

Supergravity models in higher dimensions have:

- a graviton, described by the metric $g_{\mu\nu}$ (which can be written as $e^a_\mu \eta_{ab} e^b_\nu$), except for the coupling to fermions, which is written in terms of $\omega_\mu^{ab}(e)$.
- gravitino(s) $\psi^i_{\mu\alpha}$, one for each supersymmetry. Indeed, each supersymmetry ϵ^i takes us from the graviton to a different gravitino, $\delta e^a_\mu = \bar{\epsilon}^i \gamma^a \psi^i_\mu$.
- other fields: scalars ϕ^I, vectors A^I_μ, spinors λ^I_α, and also antisymmetric tensors $A_{\mu_1 \dots \mu_r}$.

The antisymmetric tensors are generalizations of the gauge fields (Maxwell fields) A_μ, that have also a field strength,

$$F_{\mu_1 \dots \mu_{r+1}} = (n+1)\partial_{[\mu_1} A_{\mu_2 \dots \mu_{r+1}]} , \tag{15.1}$$

and so satisfy a gauge invariance

$$\delta A_{\mu_1 \dots \mu_r} = \partial_{[\mu_1} \Lambda_{\mu_2 \dots \mu_r]} . \tag{15.2}$$

The gauge invariant action for the antisymmetric tensor field is

$$S = -\frac{1}{2(r+1)!} \int d^d x \sqrt{-\det g} F^2_{\mu_1 \dots \mu_{r+1}} . \tag{15.3}$$

H. Năstase, *Cosmology and String Theory*, Fundamental Theories of Physics 197, https://doi.org/10.1007/978-3-030-15077-8_15

15.1 KK Reduction

We have already seen how to do KK reduction of scalars, metrics and gauge fields, but let us start by reviewing the latter. The gauge field $A_M(\vec{x}, t; \vec{y})$ splits as $(A_\mu(\vec{x}, t; \vec{y}), A_m(\vec{x}, t; \vec{y}))$, where $M = (\mu m)$ and $\mu = 0, 1, 2, 3$, $m = 4, \ldots, 4+n$. The A_μ components transform as a vector under $\Lambda_\mu{}^\nu \in SO(1, 3)$ and A_m as scalars, i.e.,

$$\begin{aligned} A'_\mu(\vec{x}', t'; \vec{y}) &= \Lambda_\mu{}^\nu A_\nu(\vec{x}, t; \vec{y}) \\ A'_m(\vec{x}, t; \vec{y}) &= A_m(\vec{x}, t; \vec{y}). \end{aligned} \tag{15.4}$$

The KK expansion (which is simply a parametrization of the general function on the compact space) is

$$\begin{aligned} A_\mu(\vec{x}, t; \vec{y}) &= \sum_{q, I_q} A_\mu^{q, I_q}(\vec{x}, t) Y_q^{I_q}(\vec{y}) \\ A_m(\vec{x}, t; \vec{y}) &= \sum_{q, I_q} A^{q, I_q}(\vec{x}, t) Y_m^{q, I_q}(\vec{y}). \end{aligned} \tag{15.5}$$

Antisymmetric Tensors

Now we can easily generalize to the antisymmetric tensors, where $A_{M_1 \ldots M_p}(\vec{x}, t; \vec{y})$ splits into a set of components

$$\{A_{\mu_1 \ldots \mu_r m_{r+1} \ldots m_p}(\vec{x}, t; \vec{y})\}\,, \tag{15.6}$$

where $p - r \neq n \equiv dim(K_n)$ and $r \leq 4$, since we cannot have more antisymmetrized indices than the dimension of the space.

The KK expansion in spherical harmonics is

$$A_{\mu_1 \ldots \mu_r m_{r+1} \ldots m_p}(\vec{x}, t; \vec{y}) = \sum_{q, I_q} a^{q, I_q}_{\mu_1 \ldots \mu_r}(\vec{x}, t) Y^{q, I_q}_{m_{r+1} \ldots m_p}(\vec{y}). \tag{15.7}$$

Fermions

For fermions, if we would dimensionally reduce, i.e., keep just the zero mode, on a torus, we would write

$$\lambda_A(\vec{x}, t; \vec{y}) = \lambda^i_\tau(\vec{x}, t)\,, \tag{15.8}$$

where the index A in the spinor representation of $SO(1, 3+n)$ would split into (τ, i), with τ an index in the spinor representation of $SO(1, 3)$, and i just an index filling up A.

In general, on a $M_4 \times K_n$ space, for the dimensional reduction we would write

$$\lambda_A(\vec{x}, t; \vec{y}) = \lambda^I_\tau(\vec{x}, t)\eta^I_i(\vec{y})\,, \tag{15.9}$$

where I is an index in the symmetry group G of the compact space K_n, and again A splits as $A = (\tau i)$. But moreover, now $\eta_i^I(\vec{y})$ on spaces with symmetries (symmetry group G) are *Killing spinors*. These are a kind of "square root" of a Killing vector.

A Killing vector (named after Wilhelm Killing), associated with an isometry of a space K_n, is a vector V_μ^{AB} that satisfies the Killing vector equation

$$D_{(\mu}V_{\nu)}^{AB} = 0. \tag{15.10}$$

A Killing spinor on a sphere satisfies the condition

$$D_\mu \eta_i^I = c(\gamma_\mu \eta^I)_i \ , \tag{15.11}$$

where c is usually taken to be $i/2$. It is a square root of a Killing vector in the sense that

$$V_\mu^{AB} = c' \bar{\eta}^I \gamma_\mu \eta^J (\gamma^{AB})_{IJ}. \tag{15.12}$$

Usually, c' is taken to be −i/8.

On a more general space, a Killing spinor can be defined in supersymmetric theories as the condition that the variation of the gravitino is zero, leading to the covariant derivative of the spinor being proportional to some gamma matrices times some constant fields.

Note that we want to have spinor fields in both $4+n$ dimensions and in 4 dimensions. But then the spin-statistics theorem says that these fields must both be anticommuting. However, then the reduction relation (15.9) means that the Killing spinors must be *commuting* spinors (that is fine, since they are thought of as constants, not fields).

The Killing vector V_μ^{AB} is associated with the isometries of the space, which as we said turn into gauge symmetries of the reduced theory. Specifically, *at the linearized level*, the KK reduction ansatz for the off-diagonal part of the metric is

$$g_{\mu m}(\vec{x}, t; \vec{y}) = B_\mu^{AB}(\vec{x}, t) V_m^{AB}(\vec{y}). \tag{15.13}$$

At the nonlinear level, we must multiply by a factor involving the compact space metric.

15.2 $\mathcal{N} = 1$ Supergravity Coupled to Matter

We consider models of $\mathcal{N} = 1$ supergravity in 4 dimensions, coupled to matter. We are interested only in $\mathcal{N} = 1$ models for phenomenology reasons. We eventually want our models to contain the Standard Model of particle physics, but then we cannot have $\mathcal{N} = 2$ supersymmetry, since in it, chiral fermions come in opposite pairs, whereas we know that low energy physics is chiral (contains chiral fermions).

So in 4 dimensions, we can only have $\mathcal{N} = 1$ unbroken supersymmetry if we are to obtain the Standard Model after supersymmetry breaking.

Wess–Zumino Multiplets

One type of matter multiplets coupled to supergravity are the Wess–Zumino or chiral multiplets, composed of a complex scalar field ϕ and a (Majorana) fermion λ. We have seen in the last chapter that in 2 dimensions, the WZ model contains a Majorana fermion and a real scalar. But a fermion in 4 dimensions has twice as many degrees of freedom as one in 2 dimensions (since $n = 2^{[d/2]}$ is the number of components of a Dirac fermion), whereas a scalar has the same ones. That means that in 4 dimensions we need two real scalars, combining into a complex scalar, in order to form a multiplet, but otherwise the multiplet still has spin content $(1/2, 0)$. Moreover, in the last chapter, we have shown how to write the *free* WZ multiplet, but we can write also an interacting one, that includes self-interactions.

The interactions of the chiral multiplet are encoded into a function called the *superpotential* $W(\Phi)$. It comes from the superspace formulation of supersymmetry, which will not be described here in detail, other than to say that a chiral superfield $\Phi(x, \theta)$ is expanded in the fermionic coordinate θ as

$$\Phi(x,\theta) = \phi(x) + \sqrt{2}\theta\lambda(x) + \theta\theta F(x) + \cdots \ , \tag{15.14}$$

where the dots represent terms dependent on the ones written, and $F(x)$ is an *auxiliary field* (with quadratic action with no derivatives). The superpotential W is a holomorphic function of the superfield Φ (so is independent on $\bar{\Phi}$). Then, the most general renormalizable model has the cubic superpotential

$$W = \lambda\Phi + \frac{m}{2}\Phi^2 + \frac{g}{3}\Phi^3. \tag{15.15}$$

However, if we consider a quantum effective theory, then W can be anything, leading to a non-renormalizable theory.

The potential for the scalars comes from solving the equation of motion for the auxiliary field F and replacing in the action F^2, giving, in rigid supersymmetry and in the case of canonical scalars (with canonical kinetic terms)

$$V = F^2 = \sum_i \left|\frac{\partial W}{\partial\phi^i}\right|^2 . \tag{15.16}$$

Because of its origin, this is generally called an "F term".

In this case, the scalar kinetic Lagrangian is the usual one,

$$\mathcal{L}_{\text{kin,sc.}} = -(D_\mu\phi^i)(D^\mu\phi^i)^\dagger \ , \tag{15.17}$$

and the fermionic terms in the Lagrangian are

$$\mathcal{L}_{\text{fermi}} = -i\bar{\psi}\gamma^\mu\partial_\mu\psi - \frac{1}{2}\frac{\partial^2 W}{\partial\phi^i\partial\phi^j}\psi_i\psi_j - \frac{1}{2}\frac{\partial^2 \bar{W}}{\partial\phi^{\dagger i}\partial\phi^{\dagger j}}\bar{\psi}_i\bar{\psi}_j. \tag{15.18}$$

But the chiral multiplets can also have nontrivial (yet still second derivative) kinetic terms. This is described by another function, known as the *Kähler potential* $K(\Phi, \bar{\Phi})$. Unlike the superpotential, this is not a holomorphic function. In its presence, the scalar and fermion kinetic terms are modified to

$$\mathcal{L}_{\text{kinetic}} = g_{i\bar{j}}[(D_\mu\phi^i)(D^\mu\phi^j)^* + \psi^i \not{D}\bar{\psi}^{\bar{j}}] , \tag{15.19}$$

where we have defined the metric in scalar field space (Kähler metric)

$$g_{i\bar{j}} = \frac{\partial^2 K}{\partial\Phi^i\partial\bar{\Phi}^{\bar{j}}}. \tag{15.20}$$

Gauge Multiplets

The next possibility to couple to supergravity is the gauge multiplet, composed of a gauge field A_μ^a (in the adjoint of some gauge group G) and a fermion λ_α^a, in the same representation, so having spin content $(1, 1/2)$. The multiplet is defined by a function $\mathcal{F}(\{\Phi\})$ depending on all the chiral scalar multiplets Φ. Defining in the case of dependence on adjoint scalars Φ^a the quantity

$$\mathcal{F}_{ab} = \frac{\partial^2\mathcal{F}}{\partial\Phi^a\partial\Phi^b} , \tag{15.21}$$

the action for the gauge multiplets becomes

$$-\frac{1}{4}\text{Re}[\mathcal{F}_{ab}(\phi)F^a_{\mu\nu}F^{b\mu\nu}] - \frac{1}{2}\text{Re}[\mathcal{F}_{ab}(\phi)\bar{\lambda}^a \not{D}\lambda^b] + \frac{\partial W}{\partial\phi^i}\left(\frac{\partial\mathcal{F}_{ab}}{\partial\phi^j}\right)^*\bar{\lambda}^a\lambda^b. \tag{15.22}$$

Here we have assumed that $\mathcal{F}_{ab}(\Phi)$ depends on all the scalars, ϕ^i, not just those in the adjoint representation.

Moreover, the presence of the gauge fields, coupled to all the scalars, means that we actually add another term to the scalar potential, coming from the auxiliary field in the vector multiplet. We have, in a certain gauge, $V^a = \cdots + \theta^2\bar{\theta}^2 D^a$, so the last component of the vector superfield V^a is another auxiliary field called D^a. This will again lead to a new term in the scalar potential, known as "D term",

$$\frac{1}{2}|D_a|^2 = \frac{1}{2}|\Phi^{\dagger i}(T_a)_{ij}\Phi^j|^2. \tag{15.23}$$

There could be in principle also the gravitino multiplet, with spin content $(3/2, 1)$, but those generally are associated with extra supersymmetry for the graviton, which as we said we want to avoid, so that is it for fields with spins ≤ 2.

Coupling Supergravity to Matter

To couple to supergravity, we must consider a finite M_{Pl} (decoupled gravity corresponds to $M_{\text{Pl}} \to \infty$). We first define a kind of covariant derivative,

$$D_i = \frac{\partial}{\partial \phi^i} + \frac{1}{M_{\text{Pl}}^2} \frac{\partial K}{\partial \phi^i}\,, \tag{15.24}$$

in which case the potential for the scalars coupled to $\mathcal{N} = 1$ supergravity is

$$\begin{aligned} V = |F|^2 + |D^a|^2 = e^{\frac{K}{M_{\text{Pl}}^2}} \left[\sum_{i\bar{j}} (g^{-1})^{i\bar{j}} D_i W (D_j W)^* - \frac{3}{M_{\text{Pl}}^2} |W|^2 \right] \\ + \frac{1}{2} (\mathcal{F}^{-1})^{ab} \left(\frac{\partial K}{\partial \phi^i} (T_a)_{ij} \phi^j \right) \left(\frac{\partial K}{\partial \phi^k} (T_b)_{kl} \phi^l \right). \end{aligned} \tag{15.25}$$

The gauge action stays unchanged,

$$\mathcal{L}_{\text{gauge}} = -\frac{1}{4} \text{Re}[\mathcal{F}_{ab}(\phi) F^a_{\mu\nu} F^{b\mu\nu}]\,, \tag{15.26}$$

whereas the gaugino (superpartner of the gauge field) action gets modified as well, to

$$\mathcal{L}_{\text{gaugino}} = -\frac{1}{2} \text{Re}[F_{ab}(\phi) \bar{\lambda}^a \not{D} \lambda^b] + \frac{e^{\frac{K}{2M_{\text{Pl}}^2}}}{2} \text{Re} \left[\sum_{i\bar{j}} (g^{-1})^{i\bar{j}} D_W \left(\frac{\partial \mathcal{F}_{ab}}{\partial \phi^j} \right)^* \bar{\lambda}^a \lambda^b \right]. \tag{15.27}$$

We have now described the most general action for $\mathcal{N} = 1$ supergravity coupled to matter.

15.3 Supergravity Compactifications

Supergravity Compactification on Tori

If we compactify supergravity theories on tori, we obtain abelian theories in 4 dimensions. Indeed, the torus $K_n = T^n$ has only abelian symmetries (of course, it has the nonabelian Lorentz symmetry $SO(n)$, but we are referring to isometries here, which are associated to gauge fields under dimensional reduction; $SO(n)$ gives only the usual global symmetry by dimensional reduction). Indeed, we saw that there wasn't a Killing spinor for the torus, or rather, we could think of the trivial (abelian) Killing spinors $\eta_i^I = \delta_i^I$.

Then in principle we have fields of the same spin coming from various different fields in the higher dimension. For instance, a vector field can come from $g_{\mu m}$, as well as from components of a gauge field A_M in the higher dimension, or of antisymmetric tensors, components $A_{\mu m_1 \dots m_r}$. But all these different fields will combine to form a multiplet of some *global* symmetry group in the KK reduced theory. As far as I know, there is no systematics to find the global symmetry group G, it is trial and error (case by case): we see what fields we get, and we try to fit them into representations of some symmetry group. For instance, the maximal $\mathcal{N} = 8$ supergravity in 4 dimensions has a global $SO(8)$ symmetry that transforms the gauge fields, but also a local composite symmetry $SU(8)$ and a global symmetry E_7 transforming the scalars. So the abelian gauge fields in general transform in some global symmetry group.

Nontrivial Compactifications and Gauged Supergravity

If we dimensionally reduce on a nontrivial space instead, like for instance a sphere $K_n = S^n$, the abelian Killing spinors of the torus turn into the nonabelian Killing spinors of K_n, and correspondingly the abelian gauge fields transforming in a *global* symmetry group G now turn into nonabelian (non-commuting) gauge fields of parts of, or all of the group G. This corresponds to making part of, or all the symmetry G local, or *gauging* it, with a nontrivial coupling constant g. The result is a *gauged supergravity*, and the ansatz needed to obtain it via dimensional reduction will be *nonlinear reduction ansatz*.

In the gauged supergravity, the susy transformation law for the gravitinos is modified by the addition of coupling constant (g) terms, to

$$\delta\psi^i_\mu = D_\mu(\omega(e,\psi))\epsilon^i + g\gamma_\mu\epsilon^i + gA_\mu\epsilon^i. \tag{15.28}$$

We see that there is a constant term in the transformation law, which is a nonlinear term. But a usual transformation law is linear, a field varying into another (times ϵ), and a nonlinearity like a constant term usually signifies the breaking of the symmetry, perhaps spontaneously. Indeed, in the Higgs effect for instance, $\delta\phi^i = M^i{}_j\phi^j\epsilon$ acquires a constant term under the split into VEV plus fluctuation, $\delta\varphi^i = M^i{}_j(v+\varphi)^j\epsilon$, that signals the breaking of the symmetry. However what happens now is that the action is still supersymmetric, but we need to add a *constant* term in the gravity part, that depends on g, i.e., a *cosmological constant* term $\int \det e\Lambda$. But it turns out that the sign of Λ is fixed to be negative, $\Lambda < 0$, in order to have supersymmetry.

Therefore gauged supergravity is AdS supergravity. And we can only get a de Sitter background by breaking supersymmetry (and a Minkowski background either by breaking susy, or by having an ungauged model).

15.4 11 Dimensional Supergravity

It turns out that all supergravities can be obtained from a single, unique, $\mathcal{N} = 1$ supergravity in 11 dimensions, by KK dimensional reduction and the use of symmetries. There doesn't seem to be a theorem about this, just an observation verified in all examples. The unique supergravity in 11 dimensions has fields $g_{\mu\nu}$, $\psi_{\mu\alpha}$ (graviton and gravitino), but now, unlike in 4 dimensions, we need to add the antisymmetric 3-form field $A_{\mu\nu\rho}$. We can verify (by generalizing the counting of degrees of freedom to include antisymmetric tensors) that we have the same number of bosonic and fermionic degrees of freedom. The action is

$$\mathcal{L} = -\frac{e}{2}R(e,\omega) - \frac{e}{2}\bar{\psi}_\mu \Gamma^{\mu\nu\rho} D_\nu \psi_\rho - \frac{e}{48}F^2_{\mu\nu\rho} - \frac{1}{6}\epsilon^{\mu_1\dots\mu_{11}} A_{\mu_1\mu_2\mu_3} F_{\mu_4\dots\mu_7} F_{\mu_8\mu_{11}} , \tag{15.29}$$

and the field strength of the antisymmetric tensor is

$$F_{\mu\nu\rho\sigma} = 24\partial_{[\mu} A_{\nu\rho\sigma]}. \tag{15.30}$$

The first three terms are kinetic (Einstein–Hilbert, Rarita–Schwinger and Maxwell) and the last is a Chern–Simons term for $A_{\mu\nu\rho}$.

The reason that this supergravity is unique is the following. In 4 dimensions, the maximum amount of supersymmetry that doesn't involve spins greater than 2 is 8, since a supersymmetry changes the helicity by 1/2, and so by starting with helicity -2 (the minimum possible with spins up to 2) and acting 8 times, we reach $+2$. Otherwise, we would get helicities (and hence spins) higher than 2. We want to have spins ≤ 2, since there is no known interacting theory of a finite number of spins higher than 2 in flat space. Of course, string theory corresponds to an infinite tower of interacting higher spins, and there is the so-called Vasiliev theory of higher spins living in AdS background, but for a finite number of fields in flat space, there is no known interacting theory.

But then, the $\mathcal{N} = 8$ theory, with 8 gravitinos, can lift up only until 11 dimensions, since there the 8 minimal spinors of 4 dimensions (the gravitinos) combine to form a minimal spinor of 11 dimensions. In higher dimensions, we would not get a complete spinor, so we would break Lorentz invariance. That gives the 11 dimensional theory its uniqueness. On the other hand, not all supergravities are obtained by compactification alone. The simplest example is compactification on a circle, which gives the massless type IIA supergravity. But there is also the type IIB supergravity, obtained by a symmetry called T-duality.

The 11 dimensional supergravity has as its backgrounds preserving the maximal amount of supersymmetry 11 dimensional Minkowski space Min_{11}, the spaces $AdS_4 \times S^7$, $AdS_7 \times S^4$, as well as the pp waves obtained as limits ("Penrose limits") of these. The fact that there is no other maximally supersymmetric background is actually a theorem. Note that the $AdS_4 \times S^7$ solution, called the Freund–Rubin ansatz (with $F_{\mu\nu\rho\sigma} \propto \epsilon_{\mu\nu\rho\sigma}$, for μ, ν, ρ, σ in 4 dimensions) or spontaneous compactification, raised the hope that the theory itself will choose a compactified space with only 4 dimensions non-compact. In this solution however, the cosmological constant is related to the radius of the compact space, so that is clearly ruled out experimentally, and it was then proven that we cannot deform this situation, and always the cosmological constant will be related to the parameters of the compact space, in this "spontaneous compactification" type ansatz.

Important Concepts to Remember

- Supergravity models have gravitons, gravitinos (as many as supersymmetries), scalars, vectors, spinors and antisymmetric tensors.
- Fermions KK reduce to lower dimensional fermions times commuting Killing spinors.
- Killing spinors are square roots of Killing vectors, associated with the isometries of the space.
- The most general coupling of supergravity with matter in the $\mathcal{N}=1$ context is defined by the superpotential and Kähler potential of chiral multiplets, and the $\mathcal{F}$ coupling function for vector multiplets.
- KK reduction of supergravity models on tori gives ungauged supergravity models, with only nonabelian global symmetries.
- KK reduction of supergravity models on nontrivial spaces (with nonabelian symmetries) gives gauged supergravity, where part or all of the global symmetry is gauged.
- Gauged supergravity is AdS supergravity, and we can only get de Sitter by breaking susy, and Minkowski by breaking susy or by ungauged models.
- All models can be obtained from the unique 11 dimensional supergravity, with fields $g_{\mu\nu}$, $\psi_{\mu\alpha}$ and $A_{\mu\nu\rho}$.

Further reading: See [13] for the KK reduction of supergravity, as well as [18]. For supergravity coupled to matter in a modern formulation, see [17, 19].

Exercises

(1) Consider a chiral multiplet, with the general renormalizable superpotential

$$W=\lambda\Phi+\frac{m}{2}\Phi^2+\frac{g}{3}\Phi^3\,, \tag{15.31}$$

coupled to supergravity, with Kähler potential

$$K=-3\ln(\Phi+\bar{\Phi}). \tag{15.32}$$

Calculate the scalar potential V.

(2) Consider the KK reduction of 11 dimensional supergravity on a circle S^1, with metric ansatz

$$ds^2=G^{(1)}_{MN}dx^M dx^N=e^{-\frac{2\Phi}{3}}G^{S,(10)}_{\mu\nu}dx^\mu dx^\nu+e^{\frac{4\Phi}{3}}(dx^{10}+A_\mu dx^\mu)^2. \tag{15.33}$$

Calculate the field content in 10 dimensions, and the kinetic terms in the case Φ is constant.

(3) If

$$Y^A=\frac{1}{4}\bar{\eta}^I\gamma_5\eta^J(\gamma^A)_{IJ} \tag{15.34}$$

for a 4-sphere S^4, show that

$$V_\mu^{AB} = Y^{[A} D_\mu Y^{B]} \tag{15.35}$$

is the Killing vector.

(4) Calculate the field content of the dimensional reduction of $\mathcal{N} = 1$ supergravity on a T^4 torus, down to 7 dimensions. How many supersymmetries are there in 7 dimensions?

Chapter 16
The Relativistic Point Particle

In this chapter, we will study the relativistic point particle, from the point of view of its "worldline", i.e., its path as it moves in spacetime. The purpose it to put it in a way that we can then generalize, to define string theory. Indeed, if we only think clearly enough about some mechanism for the particle, we can then easily do apply it to string theory.

The "worldline formalism" is a *first quantized* way to describe the relativistic quantum mechanics. The formalism of quantum field theory, sometimes called "second quantization", since we can think of it as a quantization of the wave function, is the standard way to describe relativistic quantum mechanics, but anything we can do with it, we can in principle do in the first quantized worldline formalism.

We start with writing the action for the particle, on its worldline.

The action for a nonrelativistic, free, point particle is

$$S = \int L dt = \int dt \, \frac{m\vec{v}^2}{2} = \int dt \, \frac{m\dot{\vec{x}}^2}{2} \,, \tag{16.1}$$

since $L = T - U$, and $U = 0$. We see that in the final form, it is written as an action on the worldline, defined by $x = x(t)$.

16.1 Relativistic Second Order Action

For a relativistic point particle, we need to find an action that reduces to the above in the nonrelativistic limit. As we know, the action is obtained by replacing dt with the relativistic proper time $d\tau$, with

$$d\tau^2 = -\eta_{\mu\nu} dx^\mu dx^\nu \,, \tag{16.2}$$

and the kinetic energy with minus the rest energy mc^2, i.e.,

H. Năstase, *Cosmology and String Theory*, Fundamental Theories of Physics 197, https://doi.org/10.1007/978-3-030-15077-8_16

$$S = -m(c^2)\int d\tau. \tag{16.3}$$

But we need to rewrite it in terms of the trajectory of the free particle, $X^\mu = X^\mu(\tau)$. From the definition of τ, by dividing both sides with $d\tau^2$, we see that

$$\frac{dX^\mu}{d\tau}\frac{dX^\nu}{d\tau}\eta_{\mu\nu} = -1. \tag{16.4}$$

This allows us to rewrite the action of the free particle as (writing $\dot{X}^\mu = dX^\mu/d\tau$)

$$S_1 = -m(c^2)\int d\tau\sqrt{-\dot{X}^\mu\dot{X}^\nu\eta_{\mu\nu}}. \tag{16.5}$$

In the nonrelativistic limit, *on the path* $X^i = X^i(t)$ *of the free particle*, we have

$$d\tau^2 = dt^2 - \frac{d\vec{x}^2}{c^2} = dt^2\left(1 - \frac{v^2}{c^2}\right), \tag{16.6}$$

so the action becomes

$$S_1 \simeq -mc^2\int dt\sqrt{1 - \frac{v^2}{c^2}} \simeq \int dt\left[-mc^2 + \frac{m\vec{v}^2}{2}\right], \tag{16.7}$$

which is the nonrelativistic action plus a constant term.

Symmetries

The action S_1 is invariant under *reparametrizations*, that is, we can change the parametrization of the worldline, from the proper time τ to any other function $\tau' = \tau'(\tau)$. Under it, the coordinates X^μ are invariant (transform as scalars on the worldline), i.e., $X'^\mu(\tau'(\tau)) = X^\mu(\tau)$, since the reparametrization of the line doesn't modify the path itself. We write

$$\frac{dX^\mu}{d\tau} = \frac{dX^\mu}{d\tau'}\frac{d\tau'}{d\tau} = \frac{dX'^\mu}{d\tau'}\frac{d\tau'}{d\tau}, \tag{16.8}$$

so by plugging it into S_1, we obtain

$$\begin{aligned} S_1 &= -m\int d\tau\sqrt{-\eta_{\mu\nu}\frac{dX^\mu}{d\tau}\frac{dX^\nu}{d\tau}} \\ &= -m\int d\tau\frac{d\tau'}{d\tau}\sqrt{-\eta_{\mu\nu}\frac{dX'^\mu}{d\tau'}\frac{dX'^\nu}{d\tau'}} \\ &= -m\int d\tau'\sqrt{-\eta_{\mu\nu}\frac{dX'^\mu}{d\tau'}\frac{dX'^\nu}{d\tau'}}. \end{aligned} \tag{16.9}$$

Equations of Motion

The variation of the action S_1 gives

$$\begin{aligned}\delta S_1 &= -m\int d\tau\delta\left(\sqrt{-\dot{X}^\mu\dot{X}^\nu\eta_{\mu\nu}}\right)\\ &= -m\int d\tau\left[-\frac{\eta_{\mu\nu}\dot{X}^\mu\delta\dot{X}^\nu}{\sqrt{-\dot{X}^\rho\dot{X}_\rho}}\right]\\ &= -m\int d\tau\frac{d}{d\tau}\left[\frac{\eta_{\mu\nu}\dot{X}^\nu}{\sqrt{-\dot{X}^\mu\dot{X}^\nu\eta_{\mu\nu}}}\right]\delta X^\mu + \delta X^\mu m\frac{dX_\mu}{d\tau}\bigg|_{\tau_i}^{\tau_f}\,,\end{aligned} \tag{16.10}$$

where in the third equality we have used partial integration. Considering that $-\dot{X}^2 = 1$ and defining the momentum of the particle

$$p^\mu = mu^\mu = m\frac{dX^\mu}{d\tau}\,, \tag{16.11}$$

the equation of motion, coming from putting to zero the bulk of the variation δS_1, is

$$\frac{dp_\mu}{d\tau} = 0\,, \tag{16.12}$$

whereas the boundary condition, coming from the vanishing of the boundary term, is

$$\delta X^\mu p_\mu(\tau_f) - \delta X^\mu p_\mu(\tau_i) = 0. \tag{16.13}$$

The equation of motion is the equation of the free particle, and in this case it implies also the vanishing of the boundary term. The solution looks somewhat trivial, in the sense that it is the straight line of a free particle. But if we consider the motion in a *curved* spacetime, with $\eta_{\mu\nu}$ replaced with a general metric $g_{\mu\nu}$, we obtain instead the motion on a geodesic in the curved spacetime.

This is now nontrivial, and it corresponds to the interaction of a particle with the gravitational field. It is part of a more general situation, where background fields (like the metric here) appearing in the particle action correspond to interaction effects of the particle with the fields.

As another example, in order to couple to a background gauge field (gauge field), we add to the action the coupling to a current, $\int A_\mu j^\mu$. But normally, in the quantum field theory formalism (second quantization), this is a function of the spacetime, $\int d^4x A_\mu(X) j^\mu(X)$, and the current is a spatial delta function centered on the path of the particle,

$$j^\mu(X^\rho) = q\frac{dX^\mu}{d\tau}\delta^3(X^\rho - X^\rho(\tau)). \tag{16.14}$$

That means that the coupling becomes

$$\int d^4x A_\mu(X^\rho) j^\mu(X^\rho) = \int d^4x A_\mu(X^\rho) q \frac{dX^\mu}{d\tau} \delta^3(X^\rho - X^\rho(\tau))$$
$$= \int d\tau A_\mu(X^\rho(\tau)) \left(q \frac{dX^\mu(\tau)}{d\tau} \right). \tag{16.15}$$

16.2 Quantum Field Theory in the Worldline Formalism

Having defined a particle action and its coupling to the background fields, we can define the formalism of first quantization, i.e., the worldline formalism. This is somewhat unusual and is seldom taught in quantum field theory, since it is a bit cumbersome, and it works mostly in simple theories, like scalar theories or at most QED. Instead of starting from a field theory action, and deriving Feynman rules, and from them the Feynman diagrams, we must define the Feynman rules by hand.

- The propagator (from x to y) is defined from the particle action, as we will see shortly.
- The vertex factor at x and y, instead of being derived from the field theory action, must now be postulated. We have many ways of defining the vertex factor, corresponding to many possible (consistent) quantum field theories. In string theory however, as we will see, the difference is that the vertex is not defined ad hoc, but is completely constrained by the quantum consistency of the theory (in particular, by conformal invariance) and as a such, there is a unique result, corresponding to the fact that there is a unique consistent string theory.
- Finally, we must define rules about how to integrate the points. Here, it is rather obvious, namely $\int d^4x \int d^4y$, but in string theory it is less so. In fact, there are many subtleties for integration, related to possible overcounting and singularities, that make the general case very difficult.

Propagator from Action

To find the *spacetime* propagator between x and y, we need the *worldline* action.

We first note that we can write, for an operator as well as for a constant (number),

$$\langle y|\Box^{-1}|x\rangle = \int_0^\infty d\tau \langle y|e^{-\tau\Box}|x\rangle. \tag{16.16}$$

We then note that the Hamiltonian for a nonrelativistic particle of mass $m = 1/2$ is

$$H = \frac{\vec{p}^2}{2m} = \frac{\Box}{2m} = \Box. \tag{16.17}$$

That means that the transition amplitude for (Euclidean) time separation τ obtained with this Hamiltonian can be written as a path integral,

$$\langle y|e^{-\tau H}|x\rangle = \int_x^y \mathcal{D}X(t)e^{-\frac{1}{4}\int_0^\tau dt\dot{X}^2}\ , \tag{16.18}$$

so that the particle propagator,

$$\left\langle y|\frac{1}{\Box}|x\right\rangle = \int_0^\infty d\tau\langle y|e^{\tau\Box}|x\rangle = \int_0^\infty d\tau\int_x^y \mathcal{D}X(t)e^{-\frac{1}{2}S_P}\ , \tag{16.19}$$

where the (massless) particle action is, as we will see shortly,

$$S_P = -\frac{1}{2}\int_0^\tau dt\ \dot{X}^2. \tag{16.20}$$

In conclusion, once we know the action, we can calculate the propagator (and then we need to define the vertices and rules for integration).

16.3 First Order Particle Action

In order to have the propagator (and in general, in order to quantize), we need an action. But the action S_1 is highly nonlinear in $X^\mu(\tau)$, so it is hard to quantize, and it gives a complicated propagator.

Instead of it, we must define a first order formulation of the action, which is quadratic in the fields (and perhaps in derivatives as well). To general procedure on how to achieve that is to introduce some auxiliary fields. Here, the auxiliary field must be an independent (intrinsic) "worldline metric" in one dimension, $\gamma_{\tau\tau}(\tau)$. Better yet, consider the "vielbein" (the name is however inappropriate now, since we are in one dimension, so we should rather use "einbein")

$$e(\tau) = \sqrt{-\gamma_{\tau\tau}(\tau)} > 0. \tag{16.21}$$

Then we note that in one dimension we have $\sqrt{-\det\gamma} = e(\tau)$, and

$$\sqrt{-\det\gamma}\gamma^{\tau\tau} = -e^{-1}(\tau). \tag{16.22}$$

We could continue by trying to find the action that gives the second order form S_1, but it is easier instead to write the action as a general quadratic action for the scalar fields $X^\mu(\tau)$. One invariant term is the scalar kinetic term $-1/2\int d\tau\sqrt{-\det\gamma}\gamma^{\tau\tau}\frac{dX^\mu}{d\tau}\frac{dX^\nu}{d\tau}\eta_{\mu\nu}$, and another is the cosmological constant term $-\int d\tau\sqrt{-\det\gamma}m^2$. The action then becomes

$$S_P = \frac{1}{2}\int d\tau \left[e^{-1}(\tau)\frac{dX^\mu}{d\tau}\frac{dX^\nu}{d\tau}\eta_{\mu\nu} - em^2\right], \tag{16.23}$$

and is invariant under reparametrizations. Indeed, reparametrization invariance means

$$e'(\tau')d\tau' = e(\tau)d\tau \Rightarrow \frac{e'^{-1}(\tau')}{d\tau'} = \frac{e^{-1}(\tau)}{d\tau}. \tag{16.24}$$

The action we have obtained indeed is equivalent to S_1, since the equation of motion for $e(\tau)$ is

$$-\frac{1}{e^2}\dot{X}^2 - m^2 = 0, \tag{16.25}$$

with the solution

$$e = \frac{1}{m}\sqrt{-\eta_{\mu\nu}\dot{X}^\mu\dot{X}^\nu}. \tag{16.26}$$

Replacing this solution in S_P, we obtain

$$S_P = -m\int d\tau\sqrt{-\eta_{\mu\nu}\dot{X}^\mu\dot{X}^\nu} = S_1. \tag{16.27}$$

Therefore S_P is the first order form of S_1; they are classically equivalent, though quantum mechanically they can in principle be different.

But S_P is much simpler, and in particular is quadratic in fields, so we can base our quantization on it. Moreover, we are really interested in the massless limit $m \to 0$, and the nonlinear action S_1 doesn't have a good $m \to 0$ limit (it vanishes under it), whereas S_P does: it is

$$S_{P,m\to 0} = \frac{1}{2}\int d\tau \left[e^{-1}(\tau)\dot{X}^\mu\dot{X}^\nu\eta_{\mu\nu}\right]. \tag{16.28}$$

The action still has reparametrization invariance ("gauge invariance"), under the transformation $e'(\tau')d\tau' = e(\tau)d\tau$. That means that we can fix a gauge by performing such a transformation, and in this gauge $e(\tau)$ is anything that we want. The simplest choice is $e(\tau) = 1$, for which we have the simplest quadratic action

$$S_P = \frac{1}{2}\int d\tau\ \dot{X}^\mu\dot{X}^\nu\eta_{\mu\nu}. \tag{16.29}$$

Equations of Motion and Constraints

The equations of motion of S_P in the gauge $e(\tau) = 1$ is

$$\frac{d}{d\tau}\left(\frac{dX^\mu}{d\tau}\right) = 0. \tag{16.30}$$

But since now we have fixed a gauge, we need to impose the equation of motion of the field that was fixed, $e(\tau)$, as a constraint. This is similar to what we do in electromagnetism, in the gauge $A_0 = 0$, where we need to impose the Gauss constraint, i.e., the would-be equation of motion of A_0.

Now the constraint is

$$-e^2\frac{\delta S}{\delta e} = \frac{\delta S}{\delta e^{-1}} = 0 \Rightarrow \frac{dX^\mu}{d\tau}\frac{dX^\nu}{d\tau}\eta_{\mu\nu} = 0. \tag{16.31}$$

But this gives the energy-momentum tensor

$$T_{\gamma\gamma} \equiv \frac{2}{\sqrt{-\det\gamma}}\frac{\delta S}{\delta\gamma^{\tau\tau}} = -\frac{2}{e}\frac{\delta S}{\delta e^{-2}} = e^2\frac{\delta S}{\delta e}, \tag{16.32}$$

so the constraint is the vanishing of the energy-momentum tensor,

$$T_{\gamma\gamma} \equiv T = 0. \tag{16.33}$$

16.4 Quantization of the Particle in the Light-Cone Gauge

We could continue with the quantization of the particle in the gauge $e = 1$, but we will leave that as an exercise. Instead, we will quantize it in the light-cone gauge.

Considering light-cone coordinates

$$X^\pm = \frac{t \pm X^1}{\sqrt{2}}, \tag{16.34}$$

we fix a gauge for reparametrization invariance by imposing a light-cone gauge condition

$$X^+ = \frac{1}{m^2}p^+\tau. \tag{16.35}$$

This fixes the reparametrization for τ, but it is not the usual proper time, nor is it the same as in the $e = 1$ gauge. Indeed, now we obtain $\dot{X}^+ = p^+/m^2$, but on the other hand we have in general (from S_1, or from S_P by eliminating e)

$$p_\mu = \frac{\partial L}{\partial \dot{X}^\mu} = \frac{m\dot{X}_\mu}{\sqrt{-\dot{X}^2}}, \tag{16.36}$$

so that $p^2 + m^2 = 0$. By comparing the two forms for p^μ, we obtain

$$p^+ = \frac{m\dot{X}^+}{\sqrt{-\dot{X}^2}} = m^2\dot{X}^+ \Rightarrow \dot{X}^2 = -\frac{1}{m^2}. \tag{16.37}$$

Replacing back in the definition of p_μ, we find

$$p^\mu = m^2 \dot{X}^\mu. \tag{16.38}$$

But as we saw, $e = \sqrt{-\dot{X}^2}/m$, so that now we find

$$e = \frac{1}{m}\sqrt{-\dot{X}^2} = \frac{1}{m^2}. \tag{16.39}$$

On the other hand, the mass-shell condition $p^2 + m^2 = 0$ becomes in light-cone coordinates

$$-2p^+p^- + p^I p^I + m^2 = 0\,, \tag{16.40}$$

where p^I is the transverse momentum, solved by

$$p^- = \frac{1}{2p^+}(p^I p^I + m^2). \tag{16.41}$$

Then p^-, as well as its canonically conjugate x^+, is a dependent quantity. The independent variables, from $p^\mu = m^2 \dot{X}^\mu$, are found to be

$$\begin{aligned}\frac{dX^-}{d\tau} &= \frac{p^-}{m} \Rightarrow X^-(\tau) = x_0^- + \frac{p^-}{m^2}\tau\\ \frac{dX^I}{d\tau} &= \frac{p^I}{m^2} \Rightarrow X^I(\tau) = x_0^I + \frac{p^I}{m^2}\tau.\end{aligned} \tag{16.42}$$

Schrödinger Picture

The quantum phase space variables are (X^I, x_0^-, p^I, p^+) (or x_0^I instead of X^I), and we can postulate the canonical commutation relations for them, as usual replacing the Poisson brackets with commutators

$$\begin{aligned}&[X^I, p^J] = i\eta^{IJ} = i\delta^{IJ}\\ &[x_0^-, p^+] = i\eta^{-+} = -i.\end{aligned} \tag{16.43}$$

The other commutators are zero.

Heisenberg Picture

In the Heisenberg picture, the quantum phase space variables turn into time-dependent operators $(X^I(\tau), X^-(\tau), p^I(\tau), p^+(\tau))$, and the canonical commutation relations are

$$\begin{aligned}&[X^I(\tau), p^J(\tau)] = i\eta^{IJ} = i\delta^{IJ}\\ &[X^-(\tau), p^+(\tau)] = i\eta^{-+} = -i\,,\end{aligned} \tag{16.44}$$

and the rest are zero.

Besides these variables, we can define also the dependent variables as operators,

$$\begin{aligned} X^+(\tau) &\equiv \frac{p^+}{m^2}\tau \\ p^- &= \frac{1}{2p^+}(p^I p^I + m^2). \end{aligned} \tag{16.45}$$

In fact, in light cone coordinates, one customarily defines x^+ as light-cone "time", and therefore the generator associated with translations in it,

$$p^- \leftrightarrow \frac{\partial}{\partial X^+}\,, \tag{16.46}$$

as light-cone energy. On the other hand, the Hamiltonian is the generator of τ translations, hence it is

$$\frac{\partial}{\partial \tau} = \frac{p^+}{m^2}\frac{\partial}{\partial X^+} \leftrightarrow \frac{p^+}{m^2}p^-. \tag{16.47}$$

Then we must define the Heisenberg picture Hamiltonian as

$$H(\tau) = \frac{p^+(\tau)}{m^2}p^-(\tau) = \frac{1}{2m^2}(p^I(\tau)p^I(\tau) + m^2) \leftrightarrow \frac{\Box + m^2}{2m^2}. \tag{16.48}$$

We still need to define the state space on which the Heisenberg operators act. A basis of states is the momentum basis $|p^+, \vec{p}_T\rangle \equiv |p^+, p^I\rangle$. On it, the Hamiltonian acts as

$$H|p^+, \vec{p}_T\rangle = \frac{1}{2m^2}(p^I p^I + m^2)|p^+, \vec{p}_T\rangle. \tag{16.49}$$

A general state then is defined by a wavefunction

$$|\psi, \tau\rangle = \int dp^+ d\vec{p}_T \psi(\tau, p^+, \vec{p}_T)|p^+, \vec{p}_T\rangle\,, \tag{16.50}$$

so that the Schrödinger equation is

$$i\frac{\partial}{\partial \tau}\psi(\tau, p^+, \vec{p}_T) = \frac{1}{2m^2}(p^I p^I + m^2)\psi(\tau, p^+, \vec{p}_T). \tag{16.51}$$

Important Concepts to Remember

- The action of a relativistic particle is $S_1 = -m\int d\tau = -m\int d\tau\sqrt{-\dot{X}^2}$.
- The action is invariant under reparametrizations and its equation of motion is the free motion.
- Coupling to background fields results in the motion of the particle interacting with these fields.

- In quantum field theory in the worldline formalism, we define Feynman diagrams ad hoc: the propagator is given from the particle action, the vertex is defined separately, as is the rule for integrating over vertices.
- In field theory, consistency doesn't restrict the form of vertices, hence we have many interacting quantum field theories. In string theory, it is completely constrained, so we will find a unique string theory.
- We can write a first order particle action S_P that is quadratic in fields and derivatives, and has a smooth $m \to 0$ limit.
- The action has reparametrization invariance, which can be fixed by the gauge choice $e = 1$, where $S_P = 1/2 \int d\tau \dot{X}^2$. The Gauss constraint is the vanishing of the energy-momentum tensor $T_{\gamma\gamma} = 0$.
- In light-cone gauge $X^+ = p^+\tau/m^2$, quantization is in terms of the independent (physical) variables (X^I, X^-, p^I, p^+), and p^- acts as light-cone energy.

Further reading: See [20] (at an advanced undergraduate/beginning graduate student level), [21] (at advanced graduate level) and [22] (at an advanced graduate level; very good for basics, however it doesn't have modern topics) for more on string theory.

Exercises

(1) Calculate the geodesic equation for motion of a free particle in a gravitational field, coming out of the particle action S_1, with $\eta_{\mu\nu} \to g_{\mu\nu}$.
(2) Write down a particle action, second order in derivatives, for motion on a circle.
(3) Quantize covariantly the particle in the gauge $e = 1$.
(4) Show that the equation of motion for the particle action plus the coupling $\int d^4x j^\mu A_\mu$ is the relativistic Lorentz force law. Repeat the derivation in the non-relativistic limit.

Chapter 17
Relativistic Strings

In this chapter we will finally start describing string theory. String theory is a theory of relativistic strings. That means, it is not like violin strings, or cosmic strings, which are essentially non-relativistic and as a result can have compression modes, along the string. Also, unlike them, they represent an idealization, and have zero thickness, i.e., they are truly one-dimensional, and moreover their endpoints will move at the speed of light. But like them, will be characterized by a certain string tension, or energy per unit length, T.

17.1 String Worldsheet and Tension

Note that the idealization is one step better than the idealization of having point particles, which is a very counter-intuitive one when we think about it, yet only a little bit so: in the direction transverse to the string, it is still point-like. Because of this, the string moving in time will span a two-dimensional surface called a *worldsheet*. The surface will be characterized by *intrinsic* coordinates on the worldsheet, like always in curved spaces, σ for worldsheet length (along the string) and τ for worldsheet time, see Fig. 17.1. On the other hand, in spacetime we will have a path depending on (σ, τ), $X^\mu = X^\mu(\sigma, \tau)$, generalizing the case of $X^\mu(\tau)$ for the particle.

Just like the action for the particle was a mass parameter $-m$ times the length of its worldline, $\int d\tau$, the action of a string is expected to be a the mass parameter $-T$ times the area of its worldsheet. In this way, the equation of motion will minimize the area of the worldsheet, just like the equation of motion for the particle minimizes the length of the path, giving the geodesic.

For historical reasons, the string tension is written as $T = 1/(2\pi\alpha')$. Indeed, string theory did not start the way we describe it here, as a theory of strings, but rather with an amplitude (the Veneziano amplitude) that was found to describe hadronic physics somewhat. It was then found that it reproduces a linear relation between spin and mass squared of hadrons, $J = \alpha' M^2 + J_0$, that can be obtained from rotating strings. Then one finds that $T = 1/(2\pi\alpha')$ is the tension of the strings. Therefore the

H. Năstase, *Cosmology and String Theory*, Fundamental Theories
of Physics 197, https://doi.org/10.1007/978-3-030-15077-8_17

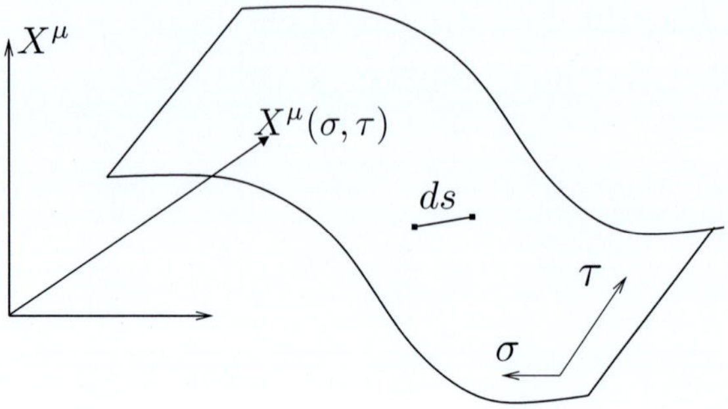

Fig. 17.1 String worldsheet moving in spacetime

action is

$$S = -T \int dA = -\frac{1}{2\pi\alpha'} \int dA. \quad (17.1)$$

We see that α' has dimension of length squared. But as we know from general relativity, the invariant area of the worldsheet is $d^2\xi\sqrt{-\det(\gamma_{ab})}$, where $\{\xi\} = (\sigma, \tau)$ for $a, b = 0, 1$. The action would then be

$$S = -\frac{1}{2\pi\alpha'} \int d\sigma d\tau \sqrt{-\det(\gamma_{ab})} = -\frac{1}{2\pi\alpha'} \int_{\tau_i}^{\tau_f} d\tau \int_0^l d\sigma \sqrt{-\gamma_{00}\gamma_{11} + \gamma_{12}^2} \,, \quad (17.2)$$

but γ_{ab} would be the intrinsic metric on the worldsheet.

17.2 Induced Metric and the Nambu–Goto Action

However, we want that the action defines the embedding of the string worldsheet in spacetime, $X^\mu X^\mu(\xi)$, i.e., the string trajectory.

For simplicity, to understand the embedding, we start with a spatial 2-dimensional surface embedded into an Euclidean space, with metric

$$ds^2 = d\vec{X} \cdot d\vec{X}. \quad (17.3)$$

Consider a surface spanned by $\{\xi_i\}$, $i = 1, 2$, with embedding $\vec{X} = \vec{X}(\xi)$. Then on the embedding, we have

$$d\vec{X} = \frac{\partial \vec{X}}{\partial \xi^i} d\xi^i \,, \quad (17.4)$$

and replacing it inside the Euclidean metric, we obtain the induced line element,

$$ds^2_{\text{induced}} = \left(\frac{\partial \vec{X}}{\partial \xi^i} \cdot \frac{\partial \vec{X}}{\partial \xi^j} \right) d\xi^i d\xi^j \equiv g_{ij}(\xi) d\xi^i d\xi^j \,, \quad (17.5)$$

where the induced metric is

$$g_{ij}(\xi) = \frac{\partial \vec{X}}{\partial \xi^i} \cdot \frac{\partial \vec{X}}{\partial \xi^j}. \tag{17.6}$$

We can now repeat this for a 1+1 dimensional (Minkowski signature) surface, embedded into an arbitrary space of Minkowski signature, with metric $g_{\mu\nu}(X)$, and

$$ds^2 = g_{\mu\nu} dX^\mu dX^\nu. \tag{17.7}$$

As before, the embedding $X^\mu(\xi^a)$ implies that on the embedding we have

$$dX^\mu = \frac{\partial X^\mu}{\partial \xi^a} d\xi^a\,, \tag{17.8}$$

so the line element is

$$ds^2_{\text{induced}} = g_{\mu\nu}(X) \frac{\partial X^\mu}{\partial \xi^a} \frac{\partial X^\nu}{\partial \xi^b} d\xi^a d\xi^b. \tag{17.9}$$

Finally then, the induced metric on the worldsheet, or "pullback" of the spacetime metric is

$$h_{ab}(\xi) = \partial_a X^\mu \partial_b X^\nu g_{\mu\nu}(X). \tag{17.10}$$

We are now ready to write the action for the string in the form found by Nambu and Goto. Note that Yoichiro Nambu got the Nobel prize mostly for the Nambu–Goldstone boson, but the string action was also cited by the Nobel committee, so to date this is the only Nobel prize related to string theory. The Nambu–Goto action is written with the above induced metric instead of γ_{ab}, so

$$S_{NG} = -\frac{1}{4\pi\alpha'} \int d\sigma\, d\tau \sqrt{-\det(h_{ab}(\xi))}. \tag{17.11}$$

Symmetries

The Nambu–Goto action is

- reparametrization (general coordinate, or diffeomorphism) invariant, like the particle action was. The action in terms of the intrinsic metric γ_{ab} is explicitly reparametrization invariant (as we know from general relativity), but so is the action in terms of the induced metric $h_{ab}(\xi)$, since for a transformation $\xi'^a = \xi'^a(\xi^b)$, it transforms as

$$h_{ab}(\xi) = g_{\mu\nu} \frac{\partial X^\mu}{\partial \xi'^c} \frac{\partial X^\nu}{\partial \xi'^d} \frac{\partial \xi'^c}{\partial \xi^a} \frac{\partial \xi'^d}{\partial \xi^b} = h'_{cd}(\xi') \frac{\partial \xi'^c}{\partial \xi^a} \frac{\partial \xi'^d}{\partial \xi^b}. \tag{17.12}$$

That means that (by taking the determinant on both sides)

$$\sqrt{-\det(h_{ab})} = \sqrt{-\det(h'_{ab})}\left|\frac{\partial\xi'}{\partial\xi}\right| \Rightarrow d^2\xi\sqrt{-\det(h_{ab})} = d^2\xi'\sqrt{-\det(h'_{ab})}\,, \tag{17.13}$$

so indeed the action is invariant.

- spacetime Poincaré invariant, in the case that we will study further, of flat spacetime, $g_{\mu\nu}(X) = \eta_{\mu\nu}$.

17.3 The First Order Polyakov Action

Like in the case of the particle, we note that the Nambu–Goto action is highly nonlinear, so it is not ideal for quantization. For that, we must write a first order form that is quadratic in the fields. In fact, the coordinates $X^\mu(\xi)$ act as scalars in the quantum field theory on the two dimensional worldsheet. In order to write a first order action, we must introduce auxiliary fields, and like in the case of the particle, we know that we can use an intrinsic worldsheet metric γ_{ab} for that (though in the particle case, we have written the action instead in terms of the einbein, since it was somewhat easier).

We can easily write the first order action simply by thinking of what we want to obtain. As we said, we will want to write the action in flat spacetime ($g_{\mu\nu} = \eta_{\mu\nu}$). We need a quadratic action for two dimensional massless scalars, with a prefactor $1/(2\pi\alpha')$. The unique result is the *Polyakov action*, (which was first written by Brink, Di Vecchia, Howe, Deser and Zumino...)

$$S_P[X,\gamma] = -\frac{1}{4\pi\alpha'}\int_M d\sigma\, d\sigma\sqrt{-\gamma}\gamma^{ab}\partial_a X^\mu\partial_b X^\nu\eta_{\mu\nu}. \tag{17.14}$$

The action is equivalent to the Nambu–Goto action upon solving for the intrinsic metric γ_{ab}.

Varying S_P with respect to γ^{ab}, we obtain (using $\delta\det\gamma_{ab}/\det\gamma_{ab} = \gamma^{ab}\delta\gamma_{ab} = -\gamma_{ab}\delta\gamma^{ab}$)

$$\delta S_P = -\frac{1}{4\pi\alpha'}\int_M d\sigma\, d\tau\sqrt{-\gamma}\gamma^{ab}\left[\partial_a X^\mu\partial_b X^\nu\eta_{\mu\nu} - \frac{1}{2}\gamma_{ab}(\gamma^{cd}\partial_a X^\mu\partial_b X^\nu\eta_{\mu\nu})\right]. \tag{17.15}$$

The equation of motion of γ^{ab}, $\delta S_P/\delta\gamma^{ab}$=0 is, identifying h_{ab}=$\partial_a X^\mu\partial_b X^\nu\eta_{\mu\nu}$,

$$h_{ab} - \frac{1}{2}\gamma_{ab}(\gamma^{cd}h_{cd}) = 0. \tag{17.16}$$

Multiplying by h^{ab}, we obtain $(\gamma^{ab}h_{ab})^2 = 4$, which is solved by

$$\gamma_{ab} = h_{ab} = \partial_a X^\mu\partial_b X^\nu\eta_{\mu\nu}. \tag{17.17}$$

Then indeed, solving the equation of motion for γ_{ab} and it putting back in S_P, we obtain $S_P = S_{NG}$, hence the Polyakov action is a first order form for Nambu–Goto.

Symmetries

The Polyakov action has the following symmetries:

- spacetime Poincaré invariance, as before.
- worldsheet reparametrization (diffeomorphism) invariance, defined by 2 transformations $\sigma'(\sigma,\tau)$ and $\tau'(\sigma,\tau)$. Under it, the coordinates X^μ are scalars, as we already said, so $X'^\mu(\sigma',\tau') = X^\mu(\sigma,\tau)$, whereas the intrinsic metric transforms in the usual way,

$$\gamma_{ab}(\sigma,\tau) = \gamma'_{cd}(\sigma',\tau')\frac{\partial\xi'^c}{\partial\xi^a}\frac{\partial\xi'^d}{\partial\xi^b}. \tag{17.18}$$

- Weyl invariance. This is an extra symmetry with respect to the Nambu–Goto action, and is also a local invariance, that acts on the metric, but not on the two dimensional coordinates ξ^a, or the scalars X^μ, i.e.,

$$\begin{aligned} X'^\mu(\sigma,\tau) &= X^\mu(\sigma,\tau) \\ \gamma'_{ab}(\sigma,\tau) &= e^{2\omega(\sigma,\tau)}\gamma_{ab}(\sigma,\tau); \quad \forall\omega(\sigma,\tau). \end{aligned} \tag{17.19}$$

Indeed, in two dimensions, we easily see that $\sqrt{-\det\gamma_{ab}} \to e^{2\omega}\sqrt{-\det\gamma_{ab}}$, but $\gamma^{ab} \to e^{-2\omega}\gamma^{ab}$.

This extra Weyl symmetry will be very important in the following, showing that the Polyakov form of the string action is more powerful.

17.4 Equations of Motion, Boundary Conditions and Constraints

We now vary the action, in order to obtain the equations of motion (from the bulk term), boundary conditions (from the boundary term) and constraints (from gauge fixing).

Before considering the complete boundary conditions, we can make an observation about them. The strings have spatial extension, so we need boundary conditions for the endpoints, which then can be *open*, i.e., the they are different points in spacetime, or *closed*, which means that they are at the same points in spacetime.

Before continuing, we note that the energy-momentum tensor (or more precisely, the Belinfante tensor) is obtained by varying the action with respect to the metric. In string theory, one usually considers a different normalization that normal, with an extra factor of 2π, i.e.,

$$T_{ab}(\sigma,\tau) = -\frac{4\pi}{\sqrt{-\gamma}}\frac{\delta S_P}{\delta\gamma^{ab}}. \tag{17.20}$$

In string theory, it is conventional to multiply by this 2π, in order for the result to be simpler, namely

$$T_{ab}(\sigma,\tau) = \frac{1}{\alpha'}\left[\partial_a X^\mu \partial_b X_\mu - \frac{1}{2}\gamma_{ab}\partial_c X^\mu \partial^c X_\mu\right]. \tag{17.21}$$

Note that normally, this would not be possible, since the energy-momentum tensor couples to gravity through Einstein's equation. But in two dimensions, we cannot add a kinetic term for gravity, i.e., an Einstein–Hilbert term on the worldsheet, since this term,

$$\frac{1}{4\pi}\int_M d^2\xi\sqrt{-\gamma}\mathcal{R}^{(2)} = \chi = 2(1-g)\ , \tag{17.22}$$

is a topological invariant, the Euler number, an integer described by $g = 0, 1, 2, \ldots$, the *genus* of the surface M. That means that the variation with respect to the metric gives zero, not the Einstein tensor $G_{ab} = R_{ab} - \frac{1}{2}g_{ab}\mathcal{R}^{(2)}$ as in higher dimensions. In other dimensions, the Einstein equation $\delta S/\delta\gamma^{ab} = 0$ is $G_{ab} = 8\pi G_N T_{ab}$, so rescaling T_{ab} matters, but in two dimensions, the equation of motion is

$$T_{ab} = 0\ , \tag{17.23}$$

for which a rescaling is irrelevant. As usual, the energy-momentum tensor is conserved,

$$\nabla_a T^{ab} = 0. \tag{17.24}$$

Since the action S_P is Weyl invariant, and Weyl invariance corresponds to multiplying all metric components by the same value, Weyl invariance is equivalent to the statement that

$$\gamma^{ab}\frac{\delta S_P}{\delta\gamma^{ab}} = 0 \Rightarrow T^a{}_a = 0\ , \tag{17.25}$$

i.e., that the energy-momentum tensor is *traceless off-shell*. But note that it was essential to have Weyl invariance, otherwise the energy-momentum tensor need not be traceless off-shell.

We now vary the action with respect to $X^\mu(\sigma,\tau)$, for which we need to partially integrate, obtaining

$$\delta_X S_P = \frac{1}{2\pi\alpha'}\int d\tau\int_0^l d\sigma\sqrt{-\gamma}\delta X^\mu \nabla^2 X_\mu - \frac{1}{2\pi\alpha'}\int d\tau\sqrt{-\gamma}\delta X^\mu\ \partial_\sigma X_\mu\Big|_{\sigma=0}^{\sigma=l}. \tag{17.26}$$

We have considered here only a boundary in σ, not in τ, for the string worldsheet, leading to the above boundary term.

Boundary Conditions

The vanishing of the boundary term leads to the boundary conditions,

$$\delta X^\mu \, \partial_\sigma X_\mu \Big|_{\sigma=0}^{\sigma=l} = 0, \tag{17.27}$$

which can be split as follows:

–*closed strings*: The boundary conditions are periodic,

$$X^\mu(l,\tau) = X^\mu(0,\tau); \quad \gamma_{ab}(l,\tau) = \gamma_{ab}(0,\tau)\ , \tag{17.28}$$

or more generally,

$$X^\mu(\sigma+l,\tau) = X^\mu(\sigma,\tau); \quad \gamma_{ab}(\sigma+l,\tau) = \gamma_{ab}(\sigma,\tau). \tag{17.29}$$

–*open strings*. For them, we can either consider

- Neumann boundary conditions (vanishing of the derivative of the field at the boundary),

$$\partial_\sigma X^\mu(l,\tau) = \partial_\sigma X^\mu(0,\tau) = 0\ , \tag{17.30}$$

where the endpoints of the string are free, and as we will shortly see, they move at the speed of light.

- Dirichlet boundary conditions (the field at the boundary is constant),

$$\delta X^\mu(l,\tau) = \delta X^\mu(0,\tau) = 0. \tag{17.31}$$

In this case, the endpoints of the string are fixed, and we will see in the chapter after next that this corresponds to the strings ending on objects called D-branes. We will therefore not study them further in this chapter.

Equations of Motion

The vanishing of the bulk variation of the action gives the equation of motion, which is

$$\nabla^2 X^\mu = 0\ , \tag{17.32}$$

the wave equation in 1+1 dimensions.

Gauge Fixing and Constraints

We have seen that the Polyakov action has 3 local worldsheet invariances, i.e., arbitrary functions of (σ,τ):

- two diffeomorphisms, $\sigma'(\sigma,\tau)$ and $\tau'(\sigma,\tau)$
- one Weyl, $\omega(\sigma,\tau)$.

On the other hand, the metric is a symmetric matrix, so has also 3 independent components, γ_{00}, γ_{01} and γ_{11}. That means that we can fix the metric to anything that we want. Of course, one should check that we can indeed reach the domain of values for the components in which the desired value resides, but that is true in the cases of interest.

The simplest choice is usually called "conformal gauge", and is

$$\gamma_{ab} = \eta_{ab}. \tag{17.33}$$

Note that sometimes this is called (more rigorously) "unit gauge", but we will use the term conformal gauge in the following.

In the conformal gauge, the Polyakov action becomes

$$S_P = -\frac{1}{4\pi\alpha'}\int d^2\sigma\eta_{\mu\nu}\eta^{ab}\partial_a X^\mu \partial_b X^\nu \ , \tag{17.34}$$

and correspondingly, the equations of motion become

$$\Box X^\mu = \left(-\frac{\partial^2}{\partial\tau^2}+\frac{\partial^2}{\partial\sigma^2}\right)X^\mu = 0. \tag{17.35}$$

But when we fix a gauge condition, we must impose the equation of motion of the gauge fixed object as a constraint. Since we have fixed γ_{ab}, we must impose its equation of motion,

$$T_{ab} = 0 \ , \tag{17.36}$$

as a constraint.

Static Gauge and the String Tension

Another useful gauge is the static gauge, in which the string looks static, and we fix time X^0 to be equal to the worldsheet time τ,

$$X^0 = \tau; \quad \vec{X} = \vec{X}(\sigma). \tag{17.37}$$

We could have chosen also $X^1 = \sigma$, but we want to keep things a bit more general. Then for a static string $\vec{X} = \vec{X}(\sigma)$, we find

$$S_{NG} = -\frac{1}{2\pi\alpha'}\int d\sigma d\tau\sqrt{(\dot{X}\cdot X')^2 - \dot{X}^2 X'^2} = -\frac{1}{2\pi\alpha'}\int_0^l d\sigma\int d\tau\sqrt{|\vec{X}'|} \ , \tag{17.38}$$

where dot means time ($\tau = X^0$) derivative and prime means σ derivative. Then the action becomes

$$S = -\Delta t T\int d\sigma\frac{|d\vec{X}|}{d\sigma} = -\Delta t|\Delta\vec{X}|T \ , \tag{17.39}$$

and since on a static configuration, the action equals minus the energy times Δt, we find

$$E = |\Delta\vec{X}|T \ , \tag{17.40}$$

i.e., T is energy per unit length, deserving thus the name of string tension.

Open String Endpoints

We finally prove that the open string endpoints, in the case of Neumann (free) boundary conditions, move at the speed of light.

In conformal gauge, $\gamma_{ab} = \eta_{ab}$, the energy-momentum tensor is

$$T_{ab} = \frac{1}{\alpha'}\left[\partial_a X^\mu \partial_b X_\mu - \frac{1}{2}\eta_{ab}\partial_c X^\mu \partial^c X_\mu\right]. \tag{17.41}$$

Using the notation above, we find

$$\alpha' T_{01} = \alpha' T_{10} = \dot{X}\cdot X'; \quad \alpha' T_{00} = \alpha' T_{11} = \frac{1}{2}(\dot{X}^2 + X'^2). \tag{17.42}$$

Since at endpoints, the Neumann boundary condition is $\partial_\sigma X^\mu(\sigma,\tau) \equiv X' = 0$, the conditions become simply

$$0 = \dot{X}^2 = \frac{\partial X^\mu}{\partial\tau}\frac{\partial X^\nu}{\partial\tau}\eta_{\mu\nu}\,, \tag{17.43}$$

which amounts to the fact that the endpoints move on $ds^2 = 0$, i.e., at the speed of light.

17.5 Coupling of String to Background Fields

We have seen how to couple the fields to spacetime gravity in the Polyakov action. We simply need to replace $\eta_{\mu\nu}$ with the general metrix $g_{\mu\nu}(X)$. But we need to maintain the local symmetries of the original action. In fact, in string theory, we start with an action in a particular background, here flat space, and quantize the theory (as we will do in the next chapter) in this background. But then we will see that the massless states of the string are fields $A_{\mu\nu}$, whose symmetric traceless part is identified with the graviton $h_{\mu\nu}$, the antisymmetric part with a tensor $B_{\mu\nu}$, and the trace with a field ϕ called the dilaton. These fields will condense and modify the background, but this means that the modified background should still satisfy the initial symmetries, if there are no anomalies (if the symmetries are preserved at the quantum level, as they should be, for a *local* symmetry). We have then consistency conditions on the possible backgrounds.

With this caveat, we can turn to writing the general Polyakov action in a consistent background of $g_{\mu\nu}$, $B_{\mu\nu}$ and ϕ. Since $B_{\mu\nu}$ is the antisymmetric part of the field whose symmetric traceless part is $g_{\mu\nu}$, it will couple in the action in the same way, via the *pull-back* from spacetime to the worldsheet of $B_{\mu\nu}$,

$$B_{\mu\nu}\partial_a X^\mu \partial_b X^\nu\,, \tag{17.44}$$

and since moreover this object is antisymmetric in (ab), we can make an invariant out of it by multiplying with ϵ^{ab} instead of $\sqrt{-\gamma}\gamma^{ab}$, for an action

$$S_P = -\frac{1}{4\pi\alpha'}\int d^2\sigma\partial_a X^\mu \partial_b X^\nu \left[\sqrt{-\gamma}\gamma^{ab} g_{\mu\nu}(X) + \alpha'\epsilon^{ab} B_{\mu\nu}(X)\right]. \quad (17.45)$$

The α' in the second term is conventional, to make $B_{\mu\nu}$ dimensionful.

Finally, to add a coupling to the dilaton, we must consider a term that would vanish for *constant* ϕ, since ϕ is not necessarily zero. It cannot be a normal scalar, since that would have the interpretation of an extra coordinate X. The only possibility, once we think about it, is to multiply the $\mathcal{R}^{(2)}$, whose integral gives the topological invariant (Euler number), i.e., to add

$$\frac{1}{4\pi}\int d^2\sigma\phi(X)\mathcal{R}^{(2)} \quad (17.46)$$

to S_P.

Important Concepts to Remember

- String theory is a theory of relativistic idealized strings, defined on a worldsheet (σ, τ), with the coordinates appearing as worldsheet scalars $X^\mu(\sigma, \tau)$.
- The Nambu–Goto action is the minus the string tension times the invariant area of the worldsheet, calculated with the induced metric on the worldsheet, or the "pullback" of the spacetime metric.
- The first order form of the Nambu–Goto action is the Polyakov action, with an independent (intrinsic) metric γ_{ab}, and with the form of a kinetic action for the scalars $X^\mu(\sigma, \tau)$.
- The Polyakov action is spacetime Poincaré invariant, diffeomorphism invariant and Weyl invariant.
- The boundary conditions for the string are: closed, open Neumann and open Dirichlet.
- The equations of motion for X^μ are simply the two dimensional wave equation.
- We can fix a gauge for diffeomorphisms and Weyl invariance, fixing the metric γ_{ab} to anything, in particular to the conformal gauge $\gamma_{ab} = \eta_{ab}$.
- In a gauge like the conformal gauge, the metric γ_{ab} equation of motion, $T_{ab} = 0$, is imposed as a constraint.
- The open strings move at the speed of light, and in a static gauge, the string energy per unit length is the string tension.
- We can write the Polyakov action in a background for $g_{\mu\nu}$, $B_{\mu\nu}$, ϕ.

Further reading: See [20–22].

Exercises

(1) Calculate the Hamiltonian and the equations of motion for the Nambu–Goto string in static gauge.

(2) Consider the Polyakov action

$$S_P = -\frac{1}{4\pi\alpha'}\int d\sigma d\tau[\sqrt{-\gamma}\gamma^{ab}\partial_a X^\mu \partial_b X^\nu g_{\mu\nu}(X) + \epsilon^{ab}\partial_a X^\mu \partial_b X^\nu B_{\mu\nu}(X)] , \tag{17.47}$$

where $g_{\mu\nu}(X)$ is a metric and $B_{\mu\nu}(X)$ is antisymmetric. Derive the Nambu–Goto action corresponding to it.
(3) Expand the Polyakov action in background fields $g_{\mu\nu}$, $B_{\mu\nu}$, ϕ in terms of gauge covariant objects around a spacetime point, using the tangent space at the point.
(4) Can one add a "cosmological constant" term on the 1+1 dimensional worldsheet of the Nambu–Goto string? Why? Compare with the case of a 2+1 dimensional "worldvolume".

Chapter 18
Light-Cone Gauge Strings and Quantization

In the previous chapter, we have described the classical theory of strings, from the point of view of their worldsheet. Here we will describe the way to quantize them, using a particular gauge called light-cone gauge.

18.1 Light-Cone Gauge

We have seen that the Polyakov action, with diffeomorphism and Weyl invariance, can be gauge fixed to put the metric in the Minkowski form, in the conformal gauge, $\gamma_{ab} = \eta_{ab}$. The action becomes

$$S_P = -\frac{1}{4\pi\alpha'}\int_{-\infty}^{+\infty} d\tau \int_0^l d\sigma \eta^{ab}\partial_a X^\mu \partial_b X^\nu \eta_{ab} , \tag{18.1}$$

and the equation of motion is the flat space wave equation,

$$\left(-\frac{\partial^2}{\partial\tau^2} + \frac{\partial^2}{\partial\sigma^2}\right) X^\mu(\sigma, \tau) = 0. \tag{18.2}$$

We have used thus 3 arbitrary functions to fix 3 components, so we might think that the gauge invariance is completely fixed. But we know from the case of electromagnetism that this is not so. For instance, in Lorenz gauge or the $A_0 = 0$ gauge, even though it uses one function to fix one component of the gauge field, we can still have *residual* gauge invariance, where the parameter λ of the transformation is not anymore an arbitrary function, but has a special dependence on coordinates (this allows us to fix both $A_0 = 0$ and $\partial^\mu A_\mu = 0$, for the radiation gauge condition).

Similarly now, we have a residual symmetry, a combination of reparametrization (diffeomorphism, or general coordinate) invariance and Weyl invariance, with a

H. Năstase, *Cosmology and String Theory*, Fundamental Theories of Physics 197, https://doi.org/10.1007/978-3-030-15077-8_18

specific coordinate dependence, that leaves the gauge fixed action invariant. Defining

$$\sigma^+ \equiv \tau + \sigma; \quad \sigma^- \equiv \tau - \sigma\,, \tag{18.3}$$

a subset of general coordinate transformations that we consider is

$$\begin{aligned} \sigma^+ &\to \tilde{\sigma}^+ = f(\sigma^+) \\ \sigma^- &\to \tilde{\sigma}^- = g(\sigma^-). \end{aligned} \tag{18.4}$$

Invariance *of flat space* under these transformations is called "conformal invariance", and it leads to a modification of the flat space metric by multiplication by a factor,

$$ds^2 = -d\sigma^+ d\sigma^- = -\frac{d\tilde{\sigma}^+}{f'}\frac{d\tilde{\sigma}^-}{g'} = -(f'g')^{-1}(\tilde{\sigma}^+, \tilde{\sigma}^-)d\tilde{\sigma}^+ d\tilde{\sigma}^-. \tag{18.5}$$

In general, conformal invariance is invariance under a subset of general coordinate transformations *of flat space* that multiplies the metric with a conformal factor,

$$ds^2 = d\vec{x}^2 = e^{2\omega(\vec{x}')} d\vec{x}'^2\,, \tag{18.6}$$

such that the action remains the same action *with the original flat metric*.

The string action in conformal gauge is conformal invariant, since after the conformal transformation we can make a compensating Weyl transformation $\gamma_{ab} \to e^{2\omega}\gamma_{ab}$, under which $\sqrt{-\gamma}\gamma^{ab}$ is invariant in 2 dimensions, that gets rid of the conformal factor $(f'g')^{-1} \equiv e^{-2\omega}$, so we are back at the flat space metric.

After the conformal transformation, we have (defining also $\tilde{\sigma}^\pm \equiv \tilde{\tau} \pm \tilde{\sigma}$),

$$\begin{aligned} \tilde{\tau} &= \frac{1}{2}\left(\tilde{\sigma}^+(\sigma^+) + \tilde{\sigma}^-(\sigma^-)\right) \\ \tilde{\sigma} &= \frac{1}{2}\left(\tilde{\sigma}^+(\sigma^+) - \tilde{\sigma}^-(\sigma^-)\right). \end{aligned} \tag{18.7}$$

and, since now $\partial_\pm = \frac{1}{2}(\partial_\tau \pm \partial_\sigma)$, we obtain that $\tilde{\tau}$ satisfies also the free wave equation,

$$\left(-\frac{\partial^2}{\partial\tau^2} - \frac{\partial^2}{\partial\sigma^2}\right)\tilde{\tau} = -\left(\frac{\partial}{\partial\tau} - \frac{\partial}{\partial\sigma}\right)\left(\frac{\partial}{\partial\tau} + \frac{\partial}{\partial\sigma}\right)\tilde{\tau} = -4\frac{\partial}{\partial\sigma^+}\frac{\partial}{\partial\sigma^-}\tilde{\tau}. \tag{18.8}$$

Defining light-cone coordinates in spacetime by

$$X^\pm = \frac{X^0 \pm X^{D-1}}{\sqrt{2}}, \tag{18.9}$$

we see that $\tilde{\tau}$ satisfies the same equation as X^μ, in particular as X^+ (since linear combinations of X^μ's satisfy the same equation). That means that we can choose a linear relation between them, specifically

$$\tilde{\tau} = \frac{X^+}{p^+} + \text{const.}\,, \tag{18.10}$$

or

$$X^+(\sigma, \tau) = x^+ + p^+\tau. \tag{18.11}$$

This is called the *light-cone gauge* condition, and is a gauge fixing for conformal invariance, besides the conformal gauge that fixed diff and Weyl invariances.

Then the general solution of the wave equation is obtained like the form for $\tilde{\tau}$ was written also, namely as a sum of an arbitrary function of σ^+ and an arbitrary function of σ^-,

$$X^\mu = X_R^\mu(\sigma^-) + X_L^\mu(\sigma^+). \tag{18.12}$$

Here X_R^μ are called "right-moving modes" and X_L^μ are "left-moving modes", referring to the fact that these are wave modes moving one way or the other along the string.

18.2 String Mode Expansions

Closed String Mode Expansion

For a closed string, we have periodicity, $X^\mu(\tau, l) = X^\mu(\tau, 0)$, and we can choose a scale (by constant reparametrizations on the worldsheet) such that $l = 2\pi$. The reason is that we can now parametrize the closed string by (σ being) an angle, and correspondingly we can use the Fourier theorem to decompose $X_L(\tau + \sigma)$ and $X_R(\tau - \sigma)$ in Fourier modes.

The general solution of the equations of motion is then

$$\begin{aligned} X_R^\mu(\tau - \sigma) &= \frac{1}{2}x^\mu + \frac{\alpha'}{2}p^\mu(\tau - \sigma) + \frac{i}{2}\sqrt{2\alpha'}\sum_{n\neq 0}\frac{1}{n}\alpha_n^\mu e^{-in(\tau-\sigma)} \\ X_L^\mu(\tau + \sigma) &= \frac{1}{2}x^\mu + \frac{\alpha'}{2}p^\mu(\tau + \sigma) + \frac{i}{2}\sqrt{2\alpha'}\sum_{n\neq 0}\frac{1}{n}\tilde{\alpha}_n^\mu e^{-in(\tau+\sigma)}. \end{aligned} \tag{18.13}$$

Note that we could have written different constants for the two fields, we chose to consider half of the constant for $X^\mu = X_L^\mu + X_R^\mu$. Also note that the sum over n is for $n \neq 0$, but we can put in the linear term as the $n = 0$ term, defining

$$\alpha_0^\mu = \sqrt{\frac{\alpha'}{2}}p^\mu = \tilde{\alpha}_0^\mu. \tag{18.14}$$

Then the total X^μ is

$$X^\mu(\sigma,\tau) = x^\mu + \alpha' p^\mu \tau + \frac{i}{2}\sqrt{2\alpha'}\sum_{n\neq 0}\frac{1}{n}[\alpha_n^\mu e^{-in(\tau-\sigma)} + \tilde{\alpha}_n^\mu e^{-in(\tau+\sigma)}]. \quad (18.15)$$

Since X_L^μ, X_R^μ must be real, it follows that x^μ, $p^\mu \in \mathbb{R}$, and moreover

$$\alpha_{-n}^\mu = (\alpha_n^\mu)^\dagger; \quad \tilde{\alpha}_{-n}^\mu = (\tilde{\alpha}_n^\mu)^\dagger. \quad (18.16)$$

Open String Mode Expansion

We next consider the open string with Neumann boundary conditions (Dirichlet boundary conditions will be considered in the next chapter),

$$X'^\mu\big|_{\sigma=0,\pi} = 0. \quad (18.17)$$

Note that we have fixed the length of the open string to be $l = \pi$, again conventionally. Then the most general solution amounts to taking $\alpha_n^\mu = \tilde{\alpha}_\mu^n$ in (18.15), such that there are only $\sin n\sigma$ modes in X', which vanish at $\sigma = 0, \pi$,

$$X^\mu(\sigma,\tau) = x^\mu + 2\alpha' p^\mu \tau + i\sqrt{2\alpha'}\sum_{n\neq 0}\frac{1}{n}\alpha_n^\mu e^{-in\tau}\cos n\sigma. \quad (18.18)$$

Again we can absorb the linear term as the $n = 0$ term in the sum, by defining

$$\alpha_0^\mu = \sqrt{2\alpha'}p^\mu. \quad (18.19)$$

18.3 Constraints and Hamiltonian

We saw that the constraints coming from fixing the conformal gauge are $T_{ab} = 0$, or $\alpha' T_{01} = \alpha' T_{10} = \dot{X}\cdot X' = 0$ and $\alpha' T_{00} = \alpha' T_{11} = \frac{1}{2}(\dot{X}^2 + X'^2) = 0$. In the $\sigma^\pm$ coordinates, they are

$$\begin{aligned}\alpha' T_{++} &= \frac{\alpha'}{2}(T_{00} + T_{01}) = \partial_+ X\cdot\partial_+ X = \dot{X}_L^2\\ \alpha' T_{--} &= \frac{\alpha'}{2}(T_{00} - T_{01}) = \partial_- X\cdot\partial_- X = \dot{X}_R^2\,,\end{aligned} \quad (18.20)$$

where the last form is valid only on-shell.

The action in conformal gauge is

$$S_P = \frac{1}{4\pi\alpha'}\int_{-\infty}^{+\infty} d\sigma(\dot{X}^2 - X'^2)\,, \quad (18.21)$$

leading to the worldsheet momentum

$$P_\tau^\mu = \frac{\dot{X}^\mu}{2\pi\alpha'} , \tag{18.22}$$

and Hamiltonian

$$H = \int_0^l d\sigma(\dot{X}\cdot P_\tau - \mathcal{L}) = \frac{1}{4\pi\alpha'}\int_0^l d\sigma(\dot{X}^2 + X'^2) = \frac{1}{2\pi}\int_0^l d\sigma T_{00}. \tag{18.23}$$

–For the *on-shell open string*, $l = \pi$, and substituting the expansion of X^μ and the orthonormality relations

$$\int_0^\pi d\sigma \cos n\sigma \cos m\sigma = \delta_{n+m,0}$$
$$\int_0^\pi d\sigma \sin n\sigma \sin m\sigma = \delta_{n+m,0} , \tag{18.24}$$

we find the Hamiltonian

$$H = \frac{1}{2}\sum_{n=-\infty}^{+\infty} \alpha_{-n}^\mu \alpha_n^\mu. \tag{18.25}$$

–For the *on-shell closed string*, $l = 2\pi$, and substituting the expansion of X^μ and the orthonormality relations for the exponentials, we find the Hamiltonian

$$H = \frac{1}{2}\sum_{n=-\infty}^{+\infty} [\alpha_{-n}^\mu \alpha_n^\mu + \tilde{\alpha}_{-n}^\mu \tilde{\alpha}_n^\mu]. \tag{18.26}$$

Constraint Modes

The *Virasoro constraints* $T_{++} = 0$ and $T_{--} = 0$ are still functions on the circle (for the closed string) or on the interval (for the open string), and as such can be expanded into Fourier modes.

–*closed string*. In the closed string case, the Fourier modes are defined as

$$L_m = \frac{1}{2\pi}\int_0^{2\pi} d\sigma e^{-im\sigma} T_{--} = \frac{1}{2\pi\alpha'}\int_0^{2\pi} d\sigma e^{-im\sigma}\dot{X}_R^2 = \frac{1}{2}\sum_{n=-\infty}^{+\infty}\alpha_{m-n}^\mu \alpha_n^\mu$$
$$\tilde{L}_m = \frac{1}{2\pi}\int_0^{2\pi} d\sigma e^{-im\sigma} T_{++} = \frac{1}{2\pi\alpha'}\int_0^{2\pi} d\sigma e^{-im\sigma}\dot{X}_L^2 = \frac{1}{2}\sum_{n=-\infty}^{+\infty}\tilde{\alpha}_{m-n}^\mu \tilde{\alpha}_n^\mu , \tag{18.27}$$

where after the definition, we have used the on-shell form. Note that the factor of $1/(2\pi)$ is due to the conventional factor in the energy-momentum tensor that was added in front.

–open string. In the open string case, the Fourier modes are defined as

$$
\begin{aligned}
L_m &= \frac{1}{2\pi}\int_0^{2\pi} d\sigma (e^{im\sigma}T_{++} + e^{-im\sigma}T_{--}) = \frac{1}{2\pi}\int_{-\pi}^{+\pi} d\sigma e^{im\sigma}T_{++} \\
&= \frac{1}{8\pi\alpha'}\int_{-\pi}^{+\pi} d\sigma e^{im\sigma}(\dot{X}+X')^2 = \frac{1}{2}\sum_{n=-\infty}^{+\infty} \alpha^\mu_{m-n}\alpha^\mu_m .
\end{aligned} \tag{18.28}
$$

We note then that the Hamiltonian can be written as

$$
\begin{aligned}
H &= L_0 \text{ open} \\
&= L_0 + \tilde{L}_0 \text{ closed}
\end{aligned} \tag{18.29}
$$

Then for the open string, the $H = L_0 = 0$ constraint translates into (writing the sum for $n \geq 1$ and for $n \leq -1$ as twice the sum for $n \geq 1$, and separating the $n = 0$ term)

$$
M^2 \equiv -p^\mu p_\mu = -\frac{\alpha_0^2}{2\alpha'} = \frac{1}{\alpha'}\sum_{n\geq 1}\alpha^\mu_{-n}\alpha^\mu_n , \tag{18.30}
$$

and similarly in the closed string case, the $H = L_0 + \tilde{L}_0 = 0$ constraint becomes

$$
M^2 \equiv -p^\mu p_\mu = -\frac{\alpha_0^2 + \tilde{\alpha}_0^2}{\alpha'^2} = \frac{2}{\alpha'}\sum_{n\geq 1}(\alpha^\mu_{-n}\alpha^\mu_n + \tilde{\alpha}^\mu_{-n}\tilde{\alpha}^\mu_n). \tag{18.31}
$$

Light-Cone Gauge

Until now, we had considered only the conformal gauge, but now we also impose the fact that we are in the light-cone gauge

$$
X^+(\sigma,\tau) = x^+ + p^+\tau. \tag{18.32}
$$

By comparing with the general expansion (18.15) and the equivalent one for the open string, we see that we must have

$$
\alpha_n^+ = 0, \quad \text{for} \quad n \neq 0. \tag{18.33}
$$

We also obtain then $\dot{X}^+ \pm X'^+ = p^+$.

Then the Virasoro constraints $(\dot{X} \pm X')^2 = 0$ become, in spacetime light-cone coordinates (where $v \cdot w = -v^-w^+ - v^+w^- + v^iw^i$) and in the light-cone gauge become

$$(\dot{X}^- \pm X'^-) = \frac{(\dot{X}^i \pm X'^i)^2}{2(\dot{X}^+ \pm X'^+)} = \frac{(\dot{X}^i \pm X'^i)^2}{2p^+}. \tag{18.34}$$

It is easy to see that in components we obtain

$$\alpha_n^- = \frac{1}{2p^+} \sum_{m \in \mathbb{Z}} \alpha_{n-m}^i \alpha_n^i. \tag{18.35}$$

That means that the light-cone gauge puts $\alpha_n^+ = 0$ and fixes (through the Virasoro constraints) α_n^- from α_n^i. The only independent oscillators are now α_n^i. Indeed, then also the mass formula becomes (for the open string)

$$M^2 = 2p^+p^- - p^i p^i = \frac{1}{\alpha'} \sum_{n \geq 1} \alpha_{-n}^i \alpha_n^i \,, \tag{18.36}$$

and a similar sum only over the transverse indices i for the closed string.

18.4 Quantization of String Modes

We next move to describing quantization. We proceed first with general issues, before focusing on the light-cone gauge quantization. Before the light-cone gauge, the Polyakov action is (18.21), with the canonical momentum $P_\tau^\mu = \dot{X}^\mu/(2\pi\alpha')$.

We can calculate the equal time Poisson brackets, and find

$$\begin{aligned} &[P^\mu(\sigma,\tau), X^\nu(\sigma',\tau)]_{P.B.} = -\delta(\sigma-\sigma')\eta^{\mu\nu} \\ &[X^\mu(\sigma,\tau), X^\nu(\sigma',\tau)]_{P.B.} = [P^\mu(\sigma,\tau), P^\nu(\sigma',\tau)]_{P.B.} = 0. \end{aligned} \tag{18.37}$$

From them, we can deduce the Poisson brackets for the modes of the string,

$$\begin{aligned} &[\alpha_m^\mu, \alpha_n^\nu]_{P.B.} = [\tilde{\alpha}_m^\mu, \tilde{\alpha}_n^\nu]_{P.B.} = -im\delta_{m+n,0}\eta^{\mu\nu} \\ &[p^\mu, x^\nu]_{P.B.} = \eta^{\mu\nu} \\ &[\alpha_m^\mu, \tilde{\alpha}_n^\nu]_{P.B.} = 0. \end{aligned} \tag{18.38}$$

These however need to be supplemented with the Virasoro constraints $(\dot{X} \pm X')^2 = 0$ ($T_{++} = T_{--} = 0$), or their modes $L_n = 0$ and $\tilde{L}_n = 0$.

Then quantization proceeds as usual, by replacing the Poisson brackets $[,]_{P.B.}$ with the commutator $-i[,]$.

The canonical (equal time) commutation relations are

$$\begin{aligned} &[P^\mu(\sigma,\tau), X^\nu(\sigma',\tau)] = -i\delta(\sigma-\sigma')\eta^{\mu\nu} \\ &[X^\mu(\sigma,\tau), X^\nu(\sigma',\tau)] = [P^\mu(\sigma,\tau), P^\nu(\sigma',\tau)] = 0. \end{aligned} \tag{18.39}$$

From them, one finds the canonical commutation relations for the modes α^μ_m, $\tilde{\alpha}^\nu_m$, x^μ, p^μ,

$$\begin{aligned}
&[\alpha^\mu_m, \alpha^\nu_n] = [\tilde{\alpha}^\mu_m, \tilde{\alpha}^\nu_n] = m\delta_{m+n,0}\eta^{\mu\nu} \\
&[p^\mu, x^\nu] = \eta^{\mu\nu} \\
&[\alpha^\mu_m, \tilde{\alpha}^\nu_n] = 0.
\end{aligned} \tag{18.40}$$

We see then that the string modes are usual oscillators with a rescaling,

$$\alpha^\mu_m = \sqrt{m} a^\mu_m; \quad \alpha^\mu_{-m} = \sqrt{m} a^{\dagger\mu}_m; \quad m > 0. \tag{18.41}$$

But we note that as it stands, we have too many modes, as seen from the fact that the timelike modes have negative norm (their commutator has the wrong term). As in electromagnetism, the point is that we still need to impose the constraints, now $L_m = 0$ and $\tilde{L}_m = 0$. In a covariant type of quantization, like the Gupta–Bleuler of electromagnetism, we need to impose the constraints as operators acting on physical states (for the covariant quantization of electromagnetism, Gupta–Bleuler type, one imposes $\partial^\mu A^{(+)}_\mu|\psi\rangle = 0$, the positive frequency part of the operatorial Lorenz condition on physical states), so one would need to impose $L_m|\psi\rangle = \tilde{L}_m|\psi\rangle = 0$.

18.5 Light-Cone Gauge Quantization

But instead, I will continue with the quantization in the light-cone gauge. We saw that there, at the classical level, the Virasoro constraints were solved by writing α^-_n in terms of α^i_n, and we also have $\alpha^+_n = 0$. At the quantum level we must care about the ordering of the oscillators α^i_n, since they don't commute. The correct solution at the quantum level is written in terms of normal ordering (defined as usual for a and $a^\dagger$'s, by putting the $a^\dagger$'s to the left and a's to the right) as

$$\alpha^-_n = \frac{1}{p^+}\left[\frac{1}{2}\sum_{i=1}^{D-2}\sum_{m\in\mathbb{Z}} :\alpha^i_{n-m}\alpha^i_m: - a\delta_{n,0}\right], \tag{18.42}$$

since only the $n = 0$ terms don't commute. As before, we have $\alpha^+_m = 0$.

In this way, X^+ and X^- are eliminated completely, and there are no constraints left. That means that we can quantize in a physical gauge, meaning only the physical, independent oscillators α^i_n obey canonical commutation relations.

Open String Spectrum

For $n = 0$, the Virasoro constraint mode $L_n = 0$ gives now, considering carefully the order of oscillators,

$$M^2 \equiv 2p^+p^- - p^ip^i = \frac{1}{\alpha'}(N-a)\,, \tag{18.43}$$

where N is a kind of number operator,

$$N = \sum_{n\geq 1} \alpha^i_{-n}\alpha^i_n = \sum_{n\geq 1} n a^{\dagger i}_n a^i_n\,, \tag{18.44}$$

and n is called the *level*, and a is an ordering constant, calculated as follows. In fact, in L_0, we have the terms (rewriting the $n \geq 1$ and $n \leq -1$ through flipping n)

$$\sum_i \sum_{n\geq 1} n \frac{a^{\dagger i}_n a^i_n + a^i_n a^{\dagger i}_n}{2} = N + \sum_{i=1}^{D-2}\sum_{n\geq 1}\frac{n}{2} = N + \frac{D-2}{2}\sum_{n\geq 1} 1. \tag{18.45}$$

But the infinite sum over n is regularized using the usual zeta function regularization of quantum field theory. The Riemann zeta function,

$$\zeta(s) = \sum_{n\geq 1}\frac{1}{n^s}\,, \tag{18.46}$$

is actually a well-defined function on the complex plane, minus the negative integers, but can be analytically continued to a function defined over the whole complex plane, and then one finds that at $s = -1$ one has

$$\zeta(-1) = \sum_{n\geq 1} n = -\frac{1}{12}. \tag{18.47}$$

Replacing above, we find the normal ordering constant

$$a = \frac{D-2}{24}. \tag{18.48}$$

We see from the mass-shell relation $M^2 = (N-a)/\alpha'$ that string theory is a theory of an infinite number of different kinds of particles, labelled by the level n, somewhat similar to the case of a solid, with phonons with Hamiltonian $H = \sum_n \omega_n N_n$. Just that in the solid case, the different kinds of phonons, with different ω_n's, are not considered really as different kinds of particles, but here it is so, since we are in flat space.

But then we notice that for $N = 1$, more precisely for states $\alpha^i_{-1}|0\rangle = a^{\dagger i}_1|0\rangle$, which have eigenvalue 1 for N, the mass is $M^2 = 1 - a$, yet there are $i = 1, \ldots, D-2$ of them, like for a massless vector (which has only transverse components, unlike a massive vector). That means that we must have $M^2 = 0$ in this case, or

$$a = 1. \tag{18.49}$$

That then implies

$$D = 26\,, \tag{18.50}$$

so our (bosonic) string must live in 26 dimensions. We will see later that a supersymmetric version of the string must instead live in 10 dimensions, by a similar argument.

We have also seen in the particle case that in light-cone coordinates, x^+ plays the role of (light-cone) time, and therefore its canonical conjugate, p^-, is the Hamiltonian. The light-cone Hamiltonian for the string is then

$$H_{l.c.} = p^- = \frac{p^i p^i}{2p^+} + \frac{1}{2\alpha' p^+}(N-1). \tag{18.51}$$

The vacuum of the open string is the one annihilated by all the oscillators (like for the phonon case), but since we have also p^μ among the quantum operators, the vacuum must also come with an eigenvalue for it, i.e., with momentum.

Therefore the vacuum state is $|0, \vec{k}\rangle$, and it satisfies

$$\begin{aligned} p^+|0,\vec{k}\rangle &= k^+|0,\vec{k}\rangle \\ p^i|0,\vec{k}\rangle &= k^i|0,\vec{k}\rangle \\ \alpha^i_m|0,\vec{k}\rangle &= 0, \end{aligned} \tag{18.52}$$

A general state is then

$$|N,\vec{k}\rangle = \left[\prod_{i=1}^{D-2}\prod_{n\geq 1}\frac{(a_n^{\dagger i})^{N_{in}}}{\sqrt{N_{i_n}!}}\right]|0,\vec{k}\rangle = \left[\prod_{i=1}^{D-2}\prod_{n\geq 1}\frac{(\alpha_n^{\dagger i})^{N_{in}}}{\sqrt{n^{N_{in}}N_{i_n}!}}\right]|0,\vec{k}\rangle. \tag{18.53}$$

We note that the vacuum is a spacetime scalar $|0,\vec{k}\rangle$, with $M^2 = -1/\alpha'$, i.e., it is a *tachyon*. This simply says that the potential for this scalar ϕ has negative curvature at zero, $V''(\phi = 0) < 0$, i.e. the vacuum is *unstable*. The bosonic string is therefore actually unstable around the usual vacuum. But there could be some nonperturbative vacuum at some $\phi_0 \neq 0$, which is a true minimum, and around which bosonic string theory is stable. This is not known until now, despite various tentatives at understanding it.

As we argued before, the next state, $a_1^{\dagger i}|0\rangle$, is a vector state corresponding to some gauge field A_μ, which is massless.

Closed String Spectrum

For the closed string, there are two constraints $L_0 - a = \tilde{L}_0 - a = 0$, and it turns out that the constant a is the same $a = 1$ as before (by a similar argument). Then we can write the M^2 formula in two ways,

$$M^2 = \frac{4}{\alpha'} \sum_{n\geq 1} (\alpha^i_{-n}\alpha^i_n - 1) = \frac{4}{\alpha'} \sum_{n\geq 1} \tilde{\alpha}^i_{-n}\tilde{\alpha}^i_n - 1) \ , \tag{18.54}$$

but we must supplement it with the *level-matching condition* (matching the levels on the left and right sectors) $L_0 = \tilde{L}_0$, which means

$$\sum_{n\geq 1} \alpha^i_{-n}\alpha^i_n = \sum_{n\geq 1} \tilde{\alpha}^i_{-n}\tilde{\alpha}^i_n. \tag{18.55}$$

That means that we can write the mass-shell condition also, symmetrically, as

$$M^2 = \frac{2}{\alpha'} \sum_{n\geq 1} (\alpha^i_{-n}\alpha^i_n + \tilde{\alpha}^i_{-n}\tilde{\alpha}^i_n) = \frac{2}{\alpha'}(N + \tilde{N} - 2) \ , \tag{18.56}$$

but supplemented with the level matching condition

$$N = \tilde{N}. \tag{18.57}$$

In this case the vacuum is still a tachyon $|0, \vec{k}\rangle$, but now with $M^2 = -4/\alpha'$. On the other hand, because of the level-matching condition, the first excited states is the one with $N = \tilde{N} = 1$, which is therefore again massless ($M^2 = 0$). This is a state

$$a_1^{\dagger i}\tilde{a}_1^{\dagger j}|0, \vec{k}\rangle = \alpha^i_{-1}\tilde{\alpha}^j_{-1}|0, \vec{k}\rangle \equiv |ij\rangle. \tag{18.58}$$

So this is a state with two transverse indices $|ij\rangle$, $i, j = 1, \ldots, D-2$. In terms of irreducible representations of the Lorentz ($SO(D)$) group, this splits into the symmetric traceless state $((ij))$, the antisymmetric state $[ij]$ and the trace ii.

String theory started as an attempt to describe hadronic (strong interaction) physics via the Veneziano amplitude. Only later it was realized that this amplitude comes from strings, and then it was found that closed strings have this massless symmetric traceless (spin 2) state in the spectrum. It was however known that there are no hadronic states that have spin 2 and are massless, so for a while the interest in string theory waned (also since QCD came along as a very good theory of strong interactions), until it was realized (by John Schwarz) that this simply means that *string theory is a theory of quantum gravity* (since it includes a graviton in its spectrum), and is not a theory of strong interactions.

Now we can therefore say that the massless states of the closed bosonic string are the graviton $g_{\mu\nu}$, an antisymmetric tensor $B_{\mu\nu}$ and a scalar ϕ called the dilaton.

We can then self-consistently consider that these massless modes of the string condense and create a background for the string, like we did in the previous chapter. But we need the consistency condition that when we do so, the symmetries of the action are preserved at the quantum level by the introduction of this background. And we saw in particular that the Polyakov action had conformal invariance. Imposing conformal invariance at the quantum level for a general background gives some consistency conditions that imply equations of motion for the spacetime fields.

Thus, even though we are doing worldsheet string physics, we are in fact able to obtain spacetime information, just by imposing quantum consistency of the theory in a general background for the string. We will however not show here how to do that, since it is a bit complicated.

Important Concepts to Remember

- The Polyakov action in light-cone gauge has conformal invariance, which is invariance under a coordinate transformation that acts on the flat space metric by multiplication by a common factor.
- The light-cone gauge is $X^+ = x^+ + p^+\tau$ and it fixes the longitudinal modes, leaving only transverse modes.
- The constraints $T_{++} = T_{--} = 0$ give the infinite number of Virasoro modes $L_n = \sum_m \alpha^\mu_{n-m}\alpha^\mu_m = 0$ and $\tilde{L}_n = \sum_m \tilde{\alpha}^\mu_{n-m}\tilde{\alpha}^\mu_{n-m} = 0$.
- For the on-shell closed string in conformal gauge, the Hamiltonian is $H = \frac{1}{2}\sum_n(\alpha^\mu_{-n}\alpha^\mu_n + \tilde{a}^\mu_{-n}\tilde{a}^\mu_n)$, and for the on-shell open string we have only the modes without tilde.
- Under quantization, the open string mass-shell condition is $M^2 = 2p^+p^- - \vec{p}^2 = \frac{1}{\alpha'}(N - a)$ and the closed string one is $M^2 = 2p^+p^- - \vec{p}^2 = \frac{2}{\alpha'}(N + \tilde{N} - 2a)$, supplemented with the level-matching condition $N = \tilde{N}$.
- The light-cone gauge Hamiltonian is $H = p^+$ defined by the mass-shell condition.
- From zeta function regularization of the sum of zero point energies, we obtain $a = (D - 2)/24$, and from the consistency of the massless spectrum we obtain $D = 26$.
- The strings have a tachyonic (unstable) ground state, and the closed massless modes are a graviton, antisymmetric tensor and dilaton, meaning that string theory is a theory of quantum gravity unified with other interactions.

Further reading: See [20, 21] for more details.

Exercises

(1) Write down the most general solution for the open string $X^\mu(\sigma, \tau)$ with one Neumann (free) boundary condition and one Dirichlet (fixed) boundary condition.
(2) Write down the states of the first 2 *massive* levels of the bosonic open string and closed string.
(3) Calculate the algebra satisfied classically by the L_m, $\tilde{L}_m$, via Poisson brackets.
(4) Construct a coherent state of the graviton (symmetric traceless massless mode of the closed string), and show that it can be thought of as a gravitational background.
(5) Write the (naive) generator of the Lorentz symmetry $J_{\mu\nu}$ in terms of the oscillators of the string α^μ_n.

Chapter 19
D-Branes and Gauge Fields

In this chapter we will explore the consequences of having Dirichlet boundary conditions for open strings. It will imply the existence of objects called D-branes, that have gauge fields on their "worldvolume".

The idea of having Dirichlet boundary conditions arose early on, through the argument we have, of vanishing the boundary term in the variation of the string action. However, initially it was thought it was a bad idea, since a Dirichlet boundary condition meant that the endpoints of strings are fixed at one point in spacetime, which would break translational invariance, one of the most obvious symmetries of quantum field theory. For a while, the idea of having Dirichlet boundary conditions in *all* space and time directions was considered, what is now called a D-instanton, since this was clearly a type of soliton, and very like a usual instanton.

But then it came to be that other objects were considered. If we have Dirichlet boundary conditions in $D - p - 1$ space dimensions and Neumann boundary conditions in $p + 1$ spacetime dimensions, the open string ends on a $p + 1$ dimensional "wall", a p-brane (a term coined from a generalization of the word membrane), or more precisely a Dp-brane (the D standing in for Dirichlet). More precisely, each endpoint of the string can be on the same, or on different Dp-branes. The wall will be called a "worldvolume", generalizing the worldline of the particle and the worldsheet of the string. Note that really, we are assuming that both endpoints of the string have the same type of boundary conditions for the same coordinates, i.e., that we have NN or DD boundary conditions, though in principle it is possible to have ND coordinates as well (and then, the Dp-branes on which the endpoints of the string lie are of different types).

We will use light-cone coordinates, with worldvolume coordinates $\{X^+, X^-, X^a\}$ for the $p + 1$ NN coordinates parallel to the D-brane, and $\{X^i\}$ for the DD coordinates, transverse to the D-brane, therefore we split the lightcone transverse coordinates as $X^I = (X^a, X^i)$. From the point of view of the D-brane, the transverse coordinates act as scalar fields on the worldvolume, in the same way we saw this happening for the particle or the string.

The reason that Dp-branes were finally considered as interesting, is that in a seminal paper in 1989, Dai, Leigh and Polchinski proved that the Dirichlet boundary

H. Năstase, *Cosmology and String Theory*, Fundamental Theories of Physics 197, https://doi.org/10.1007/978-3-030-15077-8_19

conditions are not simply inert, but rather that this "wall" is a dynamical object, somewhat like a soliton, the D-brane. The paper proved (through a calculation in string theory) that spacetime closed string modes can collide with this "wall" and excite open string modes on it, including the transverse coordinates, thus making it vibrate.

19.1 The D-Brane Action

If the transverse coordinates are also scalars on the worldvolume, the action for the D-brane must start with an action minimizing the volume of the D-brane, just like the action for the string minimizes the surface of the worldsheet, so with

$$S = -T_p \int d^{p+1}\xi \sqrt{-\det(h_{ab})}\ , \tag{19.1}$$

where as there, h_{ab} is the induced metric on the worldvolume, or "pull-back" of the spacetime metric,

$$h_{ab} = \frac{\partial X^\mu}{\partial \xi^a}\frac{\partial X^\nu}{\partial \xi^b} g_{\mu\nu}(X). \tag{19.2}$$

Here T_p is a D-brane tension, generalizing the string tension T, as energy per unit volume of the worldvolume.

Moreover, also like in the case of the string, we have seen that in string theory $g_{\mu\nu}$ and $B_{\mu\nu}$ are the symmetric and antisymmetric parts of the same field $\alpha^i_{-1}\tilde{\alpha}^j_{-1}|0, \vec{k}\rangle$, so actually the action must have them in the combination $g_{\mu\nu}(X) + \alpha' B_{\mu\nu}(X)$. That means that the action for coupling of X^μ to $g_{\mu\nu}(X)$ and $B_{\mu\nu}(X)$ is

$$S = -T_p \int d^{p+1}\xi \sqrt{-\det\left(\frac{\partial X^\mu}{\partial \xi^a}\frac{\partial X^\nu}{\partial \xi^b}[g_{\mu\nu}(X) + \alpha' B_{\mu\nu}(X)]\right)}. \tag{19.3}$$

Like the Nambu–Goto action for the string, the action has diffeomorphism (general coordinate, or reparametrization) invariance, which can be fixed using a gauge. In the case of the string, we have suggested the use of a *static gauge*, which in this more general case of D-branes can be made more precise. The static gauge means identifying the worldvolume coordinates with $p + 1$ of the spacetime coordinates, so

$$X^a = \xi^a;\quad a = 0, 1, \ldots, p\ , \tag{19.4}$$

whereas the transverse coordinates are identified with scalar fields $\phi^a(\xi)$ on the worldvolume through a constant rescaling,

$$X^i(\xi^a) = \frac{\phi^i(\xi^a)}{\sqrt{T_p}}. \tag{19.5}$$

We also expand the spacetime metric in a graviton perturbation,

$$g_{\mu\nu}(X) = \eta_{\mu\nu} + 2\kappa_N h_{\mu\nu}(X). \tag{19.6}$$

Firstly, we consider the case of only the scalars being nontrivial, and $h_{\mu\nu} = B_{\mu\nu} = 0$. Then the action is

$$S = -T_p \int d^{p+1}\xi \sqrt{-\det\left(\eta_{ab} + \frac{\partial_a \phi^i \partial_b \phi^i}{T_p}\right)}, \tag{19.7}$$

and expanding the determinant we obtain

$$\begin{aligned}
&-\det\left(\eta_{ab} + \frac{\partial_a \phi^i \partial_b \phi^i}{T_p}\right)\\
&= -\frac{1}{(p+1)!}\epsilon^{a_1 \dots a_{p+1}}\epsilon^{b_1 \dots b_{p+1}}(\eta_{a_1 b_1} + \partial_{a_1}\phi^i \partial_{b_1}\phi^i)\dots(\eta_{a_{p+1}b_{p+1}} + \partial_{a_{p+1}}\phi^i \partial_{b_{p+1}}\phi^i)\\
&= -\frac{1}{(p+1)!}\epsilon^{a_1 \dots a_{p+1}}\epsilon_{a_1 \dots a_{p+1}} - \frac{(p+1)}{(p+1)!}\epsilon^{a_1 \dots a_p a_{p+1}}\epsilon_{a_1 \dots a_p b_{p+1}}\frac{\partial_{a_{p+1}}\phi^i \partial^{b_{p+1}}\phi^i}{T_p} + \cdots\\
&= 1 + \frac{\partial_a \phi^i \partial^a \phi^i}{T_p} + \cdots
\end{aligned} \tag{19.8}$$

Here in the first equality we wrote the definition of the determinant, in the second equality we dropped the terms containing more than one factor of $\partial_a \phi^i \partial_b \phi^i$ (if there is a single ϕ, since $\partial_{[a}\phi \partial_{b]}\phi = 0$, there are no such higher order terms), and in the third we used the equality (easily proved in components)

$$\epsilon^{a_1 \dots a_k a_{k+1} \dots a_n}\epsilon_{a_1 \dots a_k b_{k+1} \dots b_n} = -k!(n-k)!\delta^{[a_{k+1} \dots a_n]}_{[b_{k+1} \dots b_n]}. \tag{19.9}$$

Therefore finally, expanding the action in the scalar fields we get

$$\begin{aligned}
S &= -T_p \int d^{p+1}\xi \sqrt{1 + \frac{\partial_a \vec{\phi} \cdot \partial^a \vec{\phi}}{T_p} + \cdots}\\
&\simeq -\int d^{p+1}\xi \left[T_p + \frac{\partial_a \vec{\phi} \cdot \partial^a \vec{\phi}}{2} + \cdots\right].
\end{aligned} \tag{19.10}$$

Next, considering the action with a graviton $h_{\mu\nu}$ in it, and expanding the determinant to first order in the graviton and first order in the scalars ϕ^i,

$$
\begin{aligned}
&-\det\left((\eta_{\mu\nu}+2\kappa_N h_{\mu\nu})(\delta^\mu_a+\partial_a\phi^\mu)(\delta^\nu_b+\partial_b\phi^\nu)\right)\\
&=-\det\left(\eta_{ab}+2\kappa_N h_{ab}+4\kappa_N h_{ai}\frac{\partial_b\phi^i}{\sqrt{T_p}}+\cdots\right)\\
&=1+h^a{}_a+4\kappa_N h_{ai}\frac{\partial^a\phi^i}{\sqrt{T_p}}+\cdots\,,
\end{aligned}
\tag{19.11}
$$

and ignoring the graviton trace $h^a{}_a$ (supposing it is zero), we find for the action (and keeping the leading kinetic term as well)

$$
S\simeq-\int d^{p+1}\xi\left[T_p+\frac{\partial_a\vec{\phi}\cdot\partial^a\vec{\phi}}{2}+2\kappa_N\sqrt{T_p}h_{ai}\partial^a\phi^i+\cdots\right].
\tag{19.12}
$$

From it, we find the vertex for ϕ^j with h^a_i,

$$
V^{aj}_i=-2\kappa_N\sqrt{T_p}ik^a\delta^j_i\,.
\tag{19.13}
$$

This vertex represents a coupling of a graviton h_{ai}, which is a closed string mode (living in the bulk of the spacetime) to ϕ^j, which is an open string mode, living on the worldvolume of the D-brane (since the endpoints of the open string are restricted to the D-brane, the open string mode lives on it). Thus this vertex represents exactly what we discussed before, the fact that a closed string can collide with the D-brane and excite an open string mode, thus making it vibrate (indeed, ϕ^i corresponds to a fluctuation in the transverse position of the D-brane), as in Fig. 19.1.

By matching with a string theory calculation, coming from a worldsheet with the topology of a disk, that ends on the D-brane, and contains a very thin tube going to infinity (as the closed string mode coming from infinity to collide with the D-brane), we find the same vertex, and we can calculate the tension as a function of string theory objects,

$$
T_p=\frac{1}{(2\pi\alpha')^2g^2_{p+1}}\,,
\tag{19.14}
$$

where g_{p+1}, the coupling of the $p+1$ dimensional theory on the Dp-brane, is

$$
g^2_{p+1}=(2\pi)^{p-2}g_s\alpha'^{\frac{p-3}{2}}\,.
\tag{19.15}
$$

We still need to consider the coupling of the D-brane with the last massless mode of the closed bosonic string, the dilaton ϕ, which appears as the trace of the same mode $\alpha^i_{-1}\tilde{\alpha}^i_{-1}|0,\vec{k}\rangle$ as for the graviton and B-field. But the dilaton gives the string coupling g_s through its VEV, by $e^{\langle\phi\rangle}=g_s$. Indeed, string theory has no dimensionless parameter (and only one dimensional parameter, α'), so the only dimensionless parameters come from the vacuum. An action for closed string modes then would be

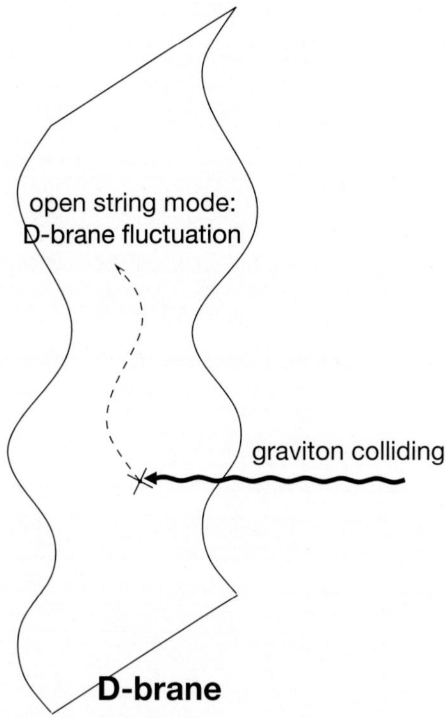

Fig. 19.1 Graviton collides with D-brane and excites the open string mode living on it, making it vibrate

expected to have $1/g_s^2$ in front, and an action for open string modes, like the D-brane action is, to have $1/g_o^2$, g_o being the open string coupling constant.

We can see pictorially that a closed string interaction is the square of an open string interaction, (closed string) = (open string)^2. Indeed, if we cut a basic closed string interaction, a "pair of pants", for a closed string splitting into two, we obtain two identical open string interactions, as in Fig. 19.2, which means that

$$g_s = g_o^2. \tag{19.16}$$

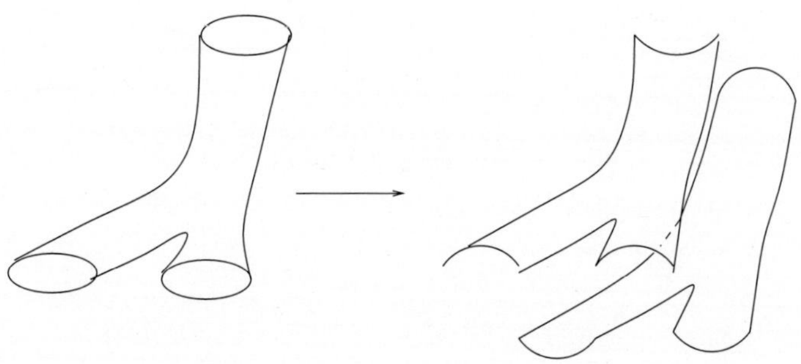

Fig. 19.2 Cutting a basic closed string interaction, a "pair of pants", into two pieces, we obtain two basic open string interactions

That means that we expect to find in front of the D-brane action a factor $\frac{1}{g_o^2} = \frac{1}{g_s} = e^{-\langle\phi\rangle}$. But it follows logically that the full dependence of the D-brane action on ϕ must be via a factor of $e^{-\phi}$, as

$$S_p = -T_p \int d^{p+1}\xi e^{-\phi} \sqrt{-\det\left(\frac{\partial X^\mu}{\partial \xi^a}\frac{\partial X^\nu}{\partial \xi^b}[g_{\mu\nu}(X) + \alpha' B_{\mu\nu}(X)]\right)}. \tag{19.17}$$

In this way we have shown the coupling of the D-brane with all the massless modes of the closed bosonic string. But we have only shown how it couples to ϕ^i, the massless scalars on the worldvolume, whereas there are other massless open string fields. In particular, as we saw in general for Neumann boundary conditions in the last chapter, there are massless vectors coming from $\alpha^i_{-1}|0, \vec{k}\rangle$, and on the D-brane we have the same boundary conditions. The mode A_a should couple gauge invariantly to the D-brane, i.e., through F_{ab}, and for small fields, we expect to have the usual Maxwell action $-\frac{1}{4}F_{ab}F^{ab}$. Moreover, it should fit inside the determinant.

Luckily, there is such a simple action, the Born–Infeld action for electromagnetism found in the 1930s, as an action without the singularities of the Maxwell action, that has a maximum electric field (and never reaches infinite values). It amounts to adding $\alpha' F_{\mu\nu}$ to the metric inside the determinant, so the full Dirac–Born–Infeld (DBI) action is

$$S_{DBI,Dp} = -T_p \int d^{p+1}\xi e^{-\phi} \sqrt{-\det\left[\frac{\partial X^\mu}{\partial \xi^a}\frac{\partial X^\nu}{\partial \xi^b}(g_{\mu\nu}(X) + \alpha' B_{\mu\nu}(X)) + \alpha' F_{ab}\right]}. \tag{19.18}$$

The action at $B_{ab} = X^i = \phi = 0$ is the Born–Infeld action, and the action for the scalars X^i's written before was defined (and studied) also by Dirac.

To the DBI action one in general needs to add another term, the "Wess–Zumino" (WZ) term that describes the coupling to antisymmetric tensor fields, called Ramond–Ramond (RR) fields, that appear for the supersymmetric string. There, the product of two spinorial (Ramond, or R) modes is decomposed in spacetime antisymmetric tensors $A_{\mu_1\ldots\mu_{p+1}}$.

The coupling to them is of the type current times gauge field, like in electromagnetism. There, we have the $\int d^d x j^\mu(\vec{x})A_\mu(\vec{x})$ term, for an electron source $j^0(\vec{x}) = e\delta^{(3)}(\vec{x} - \vec{x}(\tau))$, giving $\int d\tau A_0(\vec{x}(\tau))$. Now similarly, the "gauge field" $A_{\mu_1\ldots\mu_{p+1}}(x)$ is sourced by a $p+1$ dimensional brane, which gives a source

$$j^{01\ldots p}(\vec{x}) = \mu_p \delta^{(D-p-1)}(\vec{x} - \vec{x}(\xi)), \tag{19.19}$$

so the simplest WZ term is

$$\int d^D x j^{\mu_1\ldots\mu_{p+1}}(\vec{x})A_{\mu_1\ldots\mu_{p+1}}(\vec{x}) = \mu_p \int d^{p+1}\xi A_{01\ldots p}(\vec{x}(\xi)). \tag{19.20}$$

Here μ_p is charge density (charge per unit p-dimensional volume) of the D-brane (in fact $\mu_p = T_p e^{\phi-\phi_0}$). This term is the leading part of the WZ term. In the supersymmetric theory, on a supersymmetric background, this term gives a $+T_p$ term that cancels the $-T_p$ constant in the expansion of the DBI term. In fact, through the presence of the $-T_p$ term in the DBI term we can see the need for another term in the action for the supersymmetric case, since supersymmetry requires zero energy for the ground state.

19.2 Chan–Paton Factors and Several D-Branes

We have seen how to write the full action for a D-brane, but we can in fact have more than one D-brane. We introduce them through so-called Chan–Paton factors. Indeed, it was noticed that we can add labels $|i\rangle$ to each of the open string endpoints, having thus $|i\rangle|j\rangle$, with $i, j = 1, \dots, N$. But then one endpoint can be thought of being as in the **N** representation of $U(N)$, and the other in the $\bar{\mathbf{N}}$ representation, which means that the string itself can be thought to be in the $\mathbf{N} \times \bar{\mathbf{N}} = \text{adj}$, the adjoint representation of $U(N)$. Using the basis of adjoint matrices λ^a_{ij}, the wavefunctions of the states are

$$|\vec{k}, a\rangle = \sum_{i,j=1}^{N} |\vec{k}, ij\rangle\lambda^a_{ij}. \tag{19.21}$$

For the massless vector states, the fact that the Chan–Paton factors give a state in the adjoint representation means that we have nonabelian gauge fields.

In the case of Neumann and Dirichlet boundary conditions, giving D-branes, the index labels D-branes, so there are N D-branes, on which lives a $U(N)$ gauge theory. At the quadratic level, we have the action (already assuming the cancellation of the constant term through a supersymmetric WZ term)

$$S = \int d^{p+1}\xi(-2T_p)\,\text{Tr}\left[-\frac{1}{2}D_a\vec{\phi}\cdot D^a\vec{\phi} - \frac{1}{4}F_{ab}F^{ab}\right]. \tag{19.22}$$

19.3 Quantizing Open Strings on Dp-Branes

We consider open string stretching between 2 parallel Dp-branes, situated at transverse positions, x_1^i and x_2^i, i.e., with Dirichlet boundary conditions for the DD directions

$$X^i(\tau, \sigma = 0) = x_i^i; \quad X^i(\tau, \sigma = \pi) = x_2^i. \tag{19.23}$$

We first review what we did in the usual case, for NN directions. The Virasoro constraints were solved in light-cone coordinates by

$$\dot{X}^- \pm X'^- = \frac{1}{2\alpha'}\frac{1}{2p^+}(\dot{X}^I \pm X'^I)^2 = \frac{1}{2\alpha'}\frac{1}{2p^+}[(\dot{X}^i \pm X'^i)^2 + (\dot{X}^a \pm X'^a)^2]. \tag{19.24}$$

Here we have split the lightcone transverse coordinates as $X^I = (X^a, X^i)$ as before. But from solving the wave equation for the NN coordinates, expanding the solution in modes, we found

$$\dot{X}^a \pm X'^a = \sqrt{2\alpha'}\sum_{n\in\mathbb{Z}} \alpha_n^a e^{-in(\tau\pm\sigma)}. \tag{19.25}$$

Then substituting in the Virasoro constraint above, and expanding in modes, we found in particular the mass-shell condition from the $n = 0$ mode of the equation. The quantization of the modes was

$$[\alpha_n^a, \alpha_m^b] = m\eta^{ab}\delta_{m+n,0}. \tag{19.26}$$

We now turn to the DD coordinates and try to find a similar formula. Again, the wave equation for the X^i coordinates is

$$\partial_+\partial_- X^i = 0. \tag{19.27}$$

As before, the most general solution is a sum of a general function of $\tau + \sigma$ and a general function of $\tau - \sigma$,

$$X^i = f^i(\tau+\sigma) + g^i(\tau-\sigma). \tag{19.28}$$

Imposing the boundary condition at $\sigma = 0$, we find that $g^i(x) = x_1^i - f^i(x)$, so

$$X^i(\tau,\sigma) = x_1^i + f^i(\tau-\sigma) - f^i(\tau-\sigma). \tag{19.29}$$

Imposing also the boundary condition at $\sigma = \pi$, we find

$$f^i(\tau+\pi) - f^i(\tau-\pi) = x_2^i - x_1^i \Rightarrow f^i(x+2\pi) = f^i(x) + x_2^i - x_1^i. \tag{19.30}$$

This periodicity condition means that we can expand the functions in sines and cosines, plus a linear term,

$$f^i(x) = \frac{x_2^i - x_1^i}{\pi} + \sum_{n\geq 1}(f_n^i(x)\sin n\sigma + \tilde{f}_n^i \cos n\sigma). \tag{19.31}$$

In turn, we find then the coordinates

$$X^i(\tau,\sigma) = x_1^i + \frac{x_2^i - x_1^i}{\pi}\sigma + \sum_{n\geq 1}(f_n^i \cos n\tau - \tilde{f}_n^i \sin n\tau)\sin n\sigma. \tag{19.32}$$

Redefining the coefficients in order to find $e^{+in\tau}$ and $e^{-in\tau}$ modes, we find

$$X^i(\tau,\sigma) = x_1^i + \frac{x_2^i - x_1^i}{\pi}\sigma + \sqrt{2\alpha'}\sum_{n\geq 0}\frac{1}{n}\alpha_n^i e^{-in\tau}\sin n\sigma. \tag{19.33}$$

We can include the $n = 0$ modes in the sum by defining

$$\sqrt{2\alpha'}\alpha_0^i = \frac{x_2^i - x_1^i}{\pi}. \tag{19.34}$$

Note that this is different than the NN case, in which case α_0^a were related to the momenta p^a. In the DD case, since the endpoints are fixed, there is no center of mass momentum for the string, so α_0^i is written instead in terms of the D-brane positions. With this definition, we find

$$X'^i \pm \dot{X}^i = \sqrt{2\alpha'}\sum_{n\in\mathbb{Z}}\alpha_n^i e^{-in(\tau\pm\sigma)}\ , \tag{19.35}$$

which up to an irrelevant overall sign is the same as we found in the NN case. That means that replacing in the Virasoro constraint we find the same relation as a function of α_n^i.

Then we can quantize in the same way as for the NN directions, with

$$[\alpha_n^i, \alpha_m^j] = m\delta^{ij}\delta_{m+n,0}. \tag{19.36}$$

Finally, that means that the mass-shell condition, coming from the modes of the Virasoro constraint, is the same when written in terms of α_n^i and α_n^a, the only difference being that α_0^i is now not related to $p^i = 0$. We thus obtain

$$2p^+p^- = \frac{1}{\alpha'}\left[\alpha' p^a p^a + \frac{1}{2}\alpha_0^i\alpha_0^i + \sum_{n\geq 1}(\alpha_{-n}^a\alpha_n^a + \alpha_{-n}^i\alpha_n^i) - 1\right]. \tag{19.37}$$

This can be rewritten as an equation for M^2,

$$\begin{aligned} M^2 \equiv 2p^+p^- - p^a p^a &= \frac{1}{2\alpha'}\alpha_0^i\alpha_0^i + \frac{1}{\alpha'}\left[\sum_{n\geq 1}(\alpha_{-n}^a\alpha_n^a + \alpha_{-n}^i\alpha_n^i) - 1\right] \\ &= \left(\frac{x_2^i - x_1^i}{2\pi\alpha'}\right)^2 + \frac{1}{\alpha'}(N^\perp - 1)\ , \end{aligned} \tag{19.38}$$

where the transverse number operator is

$$N^\perp \equiv \sum_{n\geq 1}\sum_a n a_n^{a\dagger}a_n^a + \sum_{m\geq 1}\sum_i m a_m^{i\dagger}a_m^i. \tag{19.39}$$

The extra term in M^2 is understood as coming from the string extended between the 2 D-branes. Since the energy of the string is $E = TL$, and $T = 1/(2\pi\alpha')$ is the string tension and L is the length of the string, we obtain exactly the extra term.

The ground state of the string is obtained for $N^\perp$ having eigenvalue 0, with momentum and Chan–Paton factors $[\tilde{i}\tilde{j}]$, $\tilde{i}, \tilde{j} = 1, 2$ for the 2 D-branes,

$$|p^+, p^a; [\tilde{i}\tilde{j}]\rangle \,, \tag{19.40}$$

and it has the mass squared of

$$M^2 = -\frac{1}{\alpha'} + \left(\frac{x_2^1 - x_1^i}{2\pi\alpha'}\right)^2 . \tag{19.41}$$

The general state is obtained by acting with creation operators onto it, so has the form

$$|\psi\rangle = \left[\prod_n \prod_i \frac{(a_n^{a\dagger})^{N_{an}}}{\sqrt{N_{a_n}!}}\right]\left[\prod_m \prod_a \frac{(a_m^{i\dagger})^{N_{im}}}{\sqrt{N_{i_m}!}}\right] |p^+, p^a; [\tilde{i}\tilde{j}]\rangle. \tag{19.42}$$

We see that the scalar ground state is not necessarily tachyonic anymore, it can be massless for

$$|x_2^i - x_1^i| = 2\pi\sqrt{\alpha'}. \tag{19.43}$$

On the other hand, the next excited states are now

$$\phi^{i[12]} \equiv a_1^{i\dagger}|p^+, p^a; [12]\rangle \,, \qquad A_a^{[12]} \equiv a_1^{a\dagger}|p^+, p^a; [12]\rangle \,, \tag{19.44}$$

and correspond to the gauge field A_a and the transverse scalars (positions of the D-brane) ϕ^i, with Chan–Paton labels 1, 2 and with

$$M^2 = \left(\frac{x_2^1 - x_1^i}{2\pi\alpha'}\right)^2 . \tag{19.45}$$

If we consider coincident branes, these excited states are massless, and the vectors

$$a_1^{a\dagger}|p^+, p^a; [\tilde{i}\tilde{j}]\rangle \tag{19.46}$$

are Yang–Mills fields for a $U(N)$ group, while the scalars are charged adjoint scalars

$$a_1^{i\dagger}|p^+, p^a; [\tilde{i}\tilde{j}]\rangle. \tag{19.47}$$

Note that the scalars being in the adjoint, there are N^2 of them. But if we separate the branes, only the diagonal ones remain massless (corresponding to the unbroken $U(1)^N$ gauge group on the N D-branes), and off-diagonal ones get a mass, "eating" the corresponding scalars via the Higgs mechanism. In this way, only the N diagonal scalars, corresponding to the N D-branes positions, remain massless, while the other ones become massive via the Higgs mechanism.

But when the branes are coincident, we have N^2 "coordinates" for the N D-branes, which means that classical geometry loses its meaning, and we have some sort of quantum geometry, where the space becomes matriceal.

Important Concepts to Remember

- D-branes are endpoints of open strings with Dirichlet boundary conditions in $D-p-1$ directions and Neumann boundary conditions in $p+1$ directions.
- They are dynamical objects in string theory, with well defined tension and charge. Their worldvolume modes can be excited by a mode coming in from the bulk of spacetime.
- Their worldvolume action is the sum of a DBI (Dirac–Born–Infeld) and a WZ (Wess–Zumino) or CS term.
- The DBI term is $S_{DBI} = -T_p \int e^{-\phi}\sqrt{-\det(g_{ab} + \alpha' B_{ab} + \alpha' F_{ab})}$, where g_{ab} and B_{ab} are the pullback onto the worldvolume of the spacetime fields, and F_{ab} is the field strength of the YM fields on the worldvolume.
- In a static gauge, the transverse coordinates become worldvolume scalars ϕ^i, and one can find the couplings of the worldvolume fields A_a, ϕ^i with the spacetime fields $g_{\mu\nu}$, $B_{\mu\nu}$, ϕ, etc.
- The WZ term has the coupling of the spacetime RR fields with the worldvolume of the D-brane. At leading order, it is $\mu_p \int A_{01\ldots p}$.
- When we have several D-branes on top of each other, this corresponds to Chan–Paton factors on the strings, and Yang–Mills massless fields.
- The mass of the open strings stretched between two D-branes has a term involving the distance between the D-branes, $M^2 = \frac{(\Delta x)^2}{2\pi\alpha'} + \frac{1}{\alpha'}(N^\perp - 1)$.

Further reading: See [20, 21] for more details.

Exercises

(1) Consider a Dp-brane moving in a general space (of a doubly-Wick rotated nonextremal Dp-brane), with fields

$$\begin{aligned} ds^2 &= H_p^{-1/2}(r)(-dt^2 + d\vec{x}_{p-1}^2 + f(r)dx_p^2) + H_p^{1/2}\left(\frac{dr^2}{f(r)} + r^2 d\Omega_{8-p}^2\right) \\ e^{2\phi} &= g_s^2 H_p^{-\frac{p-3}{2}} \\ C_p &= \frac{1}{g_s} H_p^{-1} dt \wedge dx^1 \wedge \cdots \wedge dx^p. \end{aligned} \tag{19.48}$$

Write a general formula for the potential $V_p(r)$ felt by the Dp-brane.

(2) Find the coupling between 2 gauge fields A_a, A_b and a dilaton ϕ in the D-brane action.

(3) Find the first 2 excited states for a string stretching between 2 D-branes situated at a distance $L = 4\pi\sqrt{\alpha'}$.

(4) Assuming that the nonabelian version of the DBI action, for X^I (transverse coordinates) and A_a that are $N \times N$ matrices, is obtained as a symmetrized trace of the usual DBI action, but for the nonabelian fields, write down explicitly this action up to quartic level in the fields (expanding the square root of the determinant). [Note: up to quartic order in the fields, this action is the correct D-brane action, but not up to arbitrary order.]

Chapter 20
Electromagnetic Fields on D-Branes and $\mathcal{N} = 4$ SYM. T-Duality of Closed Strings

In this chapter we continue the description of the field theory on D-branes, and we describe an important symmetry of string theory called T-duality, in the closed string case.

We have seen that the theory on the D-branes involves gauge fields A_μ^a of $U(N)$ and adjoint scalars ϕ^{ia}, with the standard quadratic action

$$S_{\text{lin.}} = -\frac{1}{g_{p+1}^2}\int d^{p+1}x(-2)\,\text{Tr}\left[\frac{1}{2}D_\mu\phi^i D^\mu\phi^i + \frac{1}{4}F_{\mu\nu}F^{\mu\nu}\right] \tag{20.1}$$

Here we have defined the trace with the normalization $\text{Tr}[T_a T_b] = -(1/2)\delta_{ab}$, and $D_\mu = \partial_\mu + g[A_\mu, .]$.

20.1 Supersymmetric String Theory

Until now, we have considered only the bosonic string, but as we have seen, the bosonic string has tachyons, so at least perturbatively it is unstable. Because of this, it is useful to consider a supersymmetric string theory.

There are several ways to obtain it. We are interested in a theory that has supersymmetry in spacetime, but it is easier to work with a theory with supersymmetry on the worldsheet. This is the Neveu–Schwarz–Ramond (NSR) formulation, or "spinning string". The spectrum turns out to be a truncation ("GSO projection") of the bosonic case, plus some superpartners; this truncation removes the tachyon, but keeps the massless states, and obtains a spacetime supersymmetric spectrum. There is also a formulation with spacetime supersymmetry, called the Green–Schwarz formulation, or the superstring; after gauge fixing a certain symmetry ("kappa symmetry"), it has also a worldsheet supersymmetry. It turns out to have the same spectrum as the NSR formulation, so is actually the same theory. There is also a formulation involving "pure spinors" (spinors involving some constraints on them), for the Berkovits

H. Năstase, *Cosmology and String Theory*, Fundamental Theories of Physics 197, https://doi.org/10.1007/978-3-030-15077-8_20

superstring, introduced in order to quantize covariantly the superstring (string with spacetime supersymmetry).

In all these formulations, one finds, in an analogous way to what we found in the bosonic case, that we need to have $D = 10$ spacetime dimensions in order to have a sensible quantum theory.

We have seen that the simplest example of *on-shell* supersymmetry (there is also *off-shell* supersymmetry that one can define by introducing some auxiliary fields) lives in 1+1 dimensions, of one real scalar ϕ and one Majorana fermion ψ, also with one on-shell degree of freedom. The *free* action was

$$S = -\frac{1}{2}\int d^2x[(\partial_\mu\phi)^2 + \bar{\psi}\partial\!\!\!/\psi] , \tag{20.2}$$

and it had supersymmetry

$$\begin{aligned}\delta\phi &= \bar{\epsilon}\psi\\ \delta\psi &= \partial\!\!\!/\phi\epsilon.\end{aligned} \tag{20.3}$$

Here the dimensions were $[\phi] = 0$, $[\psi] = 1/2$, $[\epsilon] = -1/2$. The NSR string has 10 scalars X^μ in 1+1 dimensions; the physical modes are the transverse X^I, $I = 1, \dots, 8$ only, so one needs 8 on-shell fermionic degrees of freedom to have on-shell supersymmetry, but the action is constructed by taking multiples of the above one, reorganizing the fermions, and adding interactions.

20.2 D3-Brane Action and $\mathcal{N} = 4$ SYM

In superstring theory, the D-branes are supersymmetric (this could be proven by analyzing the boundary conditions on the open superstring, giving the D-brane theory). We will focus on D3-branes, since their worldvolume is 3+1 dimensional, so they are most relevant for us. The D3-branes, like the whole superstring theory, will live in 10 dimensions.

But as we saw, the gauge theory on them has the gauge fields A_μ^a, with 2 on-shell degrees of freedom, and the 6 scalars ϕ^i, $i = 1, \dots, 6$, corresponding to the 6 coordinates transverse to the D3-brane, for a total of 8 on-shell bosonic degrees of freedom. Since in 4 dimensions, a minimal spinor has 2 on-shell degrees of freedom (the number of components of a Dirac fermion is $2^{[D/2]}$, which on-shell gives for the *minimal spinor* one real component in 2 dimensions, and two real components in 4 dimensions), we need to introduce 4 spinors, ψ^I, $I = 1, \dots, 4$. The spinors will transform under $SU(4)$ acting on the index I. Since $SU(4) = SO(6)$, we can write the fundamental representation of $SO(6)$, to which ϕ^i belongs, as the antisymmetric representation $[IJ]$ of $SU(4)$, thus replacing $\phi^i \to \phi^{[IJ]}$. Since a supersymmetry transforms the gauge field into a spinor, we can guess that the theory will have $\mathcal{N} = 4$ supersymmetries, each (with a parameter ϵ^I) transforming A_μ^a into a different ψ^{aI}.

It turns out that for theories with spins ≤ 1 in 4 dimensions, there is a maximum number of supersymmetries equal to 4. The reason is that for $s \leq 1$, we have maximum helicity $+1$ and minimum -1, and one supersymmetry changes the helicity by 1/2. The same argument for spins $s \leq 2$, i.e. for supergravity theories, with helicities $\in [-2, +2]$, gives a maximum number of 8 supersymmetries. The gauge theory with $\mathcal{N} = 4$ supersymmetries, and with a maximum of two derivatives (usual kinetic terms), is actually unique, so the "linearized" (or more precisely, quadratic in derivatives) action on the D3-branes is this $\mathcal{N} = 4$ **SYM theory**.

The supersymmetry transformation rules for the bosons are easy to write down, based on the basic transformation $\delta\phi = \bar{\epsilon}\psi$, generalized first to $\delta\psi = \bar{\epsilon}_I\psi^I$ for 4 susies, and then to our case, when we just need to match indices (and as a result, to introduce a gamma function, i.e., a constant matrix with a μ index, for the gauge field), as

$$
\begin{aligned}
\delta A_\mu^a &= \bar{\epsilon}_I \gamma_\mu \psi^{Ia} \\
\delta\phi^{[IJ]} &= \frac{i}{2}\epsilon^{[I}\psi^{J]}.
\end{aligned}
\tag{20.4}
$$

The fermion transformation law is more complicated, and will not be written here. The action for $\mathcal{N} = 4$ SYM is then

$$
\begin{aligned}
S = \int d^4x(-2)\,\mathrm{Tr}\Bigg[&-\frac{1}{2}D_\mu\phi_{[IJ]}D^\mu\phi^{[IJ]} - \frac{1}{4}F_{\mu\nu}F^{\mu\nu} - \frac{1}{2}\bar{\psi}_I \not{D}\psi^J \\
&-g\bar{\psi}^I[\phi_{[IJ]}, \psi^J] - \frac{g^2}{4}[\phi_{[IJ]}, \phi_{[KL]}][\phi^{[IJ]}, \phi^{[KL]}]\Bigg].
\end{aligned}
\tag{20.5}
$$

The terms on the first line are just the kinetic terms (for the "quadratic" action), and the terms on the second line are necessary in order to maintain supersymmetry once we have introduced the minimal coupling of the gauge fields to the scalars and fermions. In fact, this action can be simply obtained from the action for $\mathcal{N} = 1$ SYM in 10 dimensions, which is just the minimally coupled action of a fermion coupled to a gauge fields,

$$
S = \int d^{10}x(-2)\,\mathrm{Tr}\left[-\frac{1}{4}F_{MN}F^{MN} - \frac{1}{2}\bar{\psi}\not{D}\psi\right],
\tag{20.6}
$$

under dimensional reduction. Indeed, then the gauge field splits as $A_M = (A_\mu, \phi_i)$, and the field strength as $F_{MN} = (F_{\mu\nu}, F_{\mu i} = D_\mu\phi_i, F_{ij} = g^2[\phi_i, \phi_j])$, with $\phi_i \leftrightarrow \phi_{[IJ]}$, and the spinor splits as $\psi_A = \psi_{\alpha I}$.

We have only shown the action quadratic in derivative on the N D3-branes, namely $SU(N)\ \mathcal{N} = 4$ SYM. But while for the $U(1)$ case the nonlinear DBI action is known, for the non-Abelian case, the full action is not known. For a while there was a conjecture that the result was the symmetric trace of the nonabelian version of the DBI action, but it turned out not to be the case beyond a few orders.

20.3 Nonlinear Born–Infeld Action

We therefore move on to some considerations about the nonlinear $U(1)$ action. In this case the full nonlinear supersymmetric action is known (it was found by M. Aganagic, C. Popescu and J. Schwarz), though it is quite complicated, and will not be reproduced here.

We consider the Born–Infeld action, the action originally found by Born and Infeld, in terms of only the Maxwell field (electromagnetism), i.e., the DBI action for $e^{\phi}=e^{\langle\phi\rangle}=g_s$, $\phi^i=0$, $B_{\mu\nu}=0$,

$$S_{BI}=-T_3\int d^4x\left[\sqrt{-\det\left(\eta_{ab}+2\pi\alpha' F_{ab}\right)}-1\right], \tag{20.7}$$

where $1/g_s$ was absorbed in T_3 and we have considered already the -1 coming from the WZ term. Then we write explicitly in terms of $E_i=F_{0i}$ and $B_i=\frac{1}{2}\epsilon_{ijk}F_{jk}$ the matrix

$$M_{ab}=\eta_{ab}+2\pi\alpha' F_{ab}=\begin{pmatrix}-1 & -E_1 & -E_2 & -E_3\\ E_1 & 1 & B_3 & -B_2\\ E_2 & -B_3 & 1 & B_1\\ E_3 & B_2 & -B_1 & 1\end{pmatrix}, \tag{20.8}$$

where in the last equality we have put $2\pi\alpha'=1$ for simplicity. Taking the determinant, we find

$$-\det M_{ab}=1-(\vec{E}^2-\vec{B}^2)-(\vec{E}\cdot\vec{B})^2\,, \tag{20.9}$$

and on the other hand we have, defining the dual electromagnetic field $\tilde{F}^{\mu\nu}=\frac{1}{2}\epsilon^{\mu\nu\rho\sigma}F_{\rho\sigma}$,

$$\begin{aligned}\frac{1}{2}(\vec{E}^2-\vec{B}^2)&=-\frac{1}{4}F_{\mu\nu}F^{\mu\nu}\\ \vec{E}\cdot\vec{B}&=-\frac{1}{8}\epsilon^{\mu\nu\rho\sigma}F_{\mu\nu}F_{\rho\sigma}\equiv-\frac{1}{4}\tilde{F}_{\mu\nu}F^{\mu\nu}.\end{aligned} \tag{20.10}$$

The D3-brane action then becomes (reinstating $2\pi\alpha'$)

$$S_{BI,D3}=-T_3\int d^4x\left[\sqrt{1+\frac{1}{2}(2\pi\alpha')^2F_{\mu\nu}F^{\mu\nu}-(2\pi\alpha')^4\left(\frac{1}{4}\tilde{F}_{\mu\nu}F^{\mu\nu}\right)^2}-1\right]. \tag{20.11}$$

We note that by expanding to first nontrivial order, we get an overall factor of $-\frac{T_3(2\pi\alpha')^2}{4}$, to be identified with $-\frac{1}{4g_{3+1}^2}$ as in the linearized action (20.1). Therefore we obtain the D3-brane tension in terms of the SYM coupling as

$$T_3 = \frac{1}{(2\pi\alpha')^2 g_{3+1}^2}. \tag{20.12}$$

The Born–Infeld action for the D3-brane, in terms of electric and magnetic fields is then

$$S_{BI} = -\frac{1}{(2\pi\alpha')^2 g_{3+1}^2} \int d^4x \left[\sqrt{1-(2\pi\alpha')^2(\vec{E}^2-\vec{B}^2)-(2\pi\alpha')^4(\vec{E}\cdot\vec{B})^2}-1\right]. \tag{20.13}$$

At zero magnetic field, the purely electric action is

$$S_{BI} = -\frac{1}{(2\pi\alpha')^2 g_{3+1}^2} \int d^4x \left[\sqrt{1-(2\pi\alpha')^2\vec{E}^2}-1\right], \tag{20.14}$$

and we see that it admits a maximum electric field,

$$|\vec{E}| \leq \frac{1}{2\pi\alpha'} \equiv E_{\max}. \tag{20.15}$$

This was part of the motivation of Born and Infeld, who wanted a nonlinear action for electromagnetism that reduces to the usual Maxwell one for small field, but doesn't have singularities. An electron solution in Maxwell electromagnetism has infinite electric field at the electron source, but in BI electromagnetism we have only $E_{\max}$.

20.4 Closed Strings on Compact Spaces

With the purpose of manifesting an important symmetry of string theory, that will be of use in cosmology, called T-duality, we examine closed strings on compact spaces.

The Circle S^1

The simplest compact space is a circle, so consider that X^{25} is compactified on a circle, i.e., that we have the identification

$$X^{25} \sim X^{25} + 2\pi R. \tag{20.16}$$

On the other hand, closed strings are periodic in σ, so they obey

$$X^\mu(\tau, \sigma + 2\pi) = X^\mu(\tau, \sigma). \tag{20.17}$$

But because of the nontrivial identification, in the direction X^{25} we can have periodicity only up to an identification, i.e.,

$$X^{25}(\tau, \sigma + 2\pi) = X^{25}(\tau, \sigma) + 2\pi m R. \tag{20.18}$$

The interpretation of this is that the closed string winds m times around the compact circle direction. We will define the *winding*

$$w = \frac{mR}{\alpha'}. \tag{20.19}$$

This winding means that the mode expansion of X^{25} will get modified. As before, X^{25} satisfies a wave equation

$$\partial_+\partial_- X^{25} = 0\,, \tag{20.20}$$

solved by a sum of an arbitrary function of left movers and an arbitrary function of right movers,

$$X^{25}(\tau, \sigma) = X_L^{25}(\tau + \sigma) + X_R^{25}(\tau - \sigma). \tag{20.21}$$

We define the variables $u \equiv \tau + \sigma$ and $v = \tau - \sigma$. Then the winding-modified periodicity condition on X^{25} becomes

$$\begin{aligned} X_L^{25}(u + 2\pi) + X_R^{25}(v - 2\pi) &= X_L^{25}(u) + X_R^{25}(v) + 2\pi\alpha' w \Rightarrow \\ X_L^{25}(u + 2\pi) - X_L^{25}(u) &= X_R^{25}(v) - X_R^{25}(v - 2\pi) + 2\pi\alpha' w. \end{aligned} \tag{20.22}$$

But the left hand side is a function of u, and the right hand side is a function of v, so the solution is periodicity for both $X_L^{25}(u)$ and $X_R^{25}(v)$, which therefore can be expanded as usual in exponentials, but only up to linear term, i.e.,

$$\begin{aligned} X_L^{25}(u) &= x_{0,L}^{25} + \sqrt{\frac{\alpha'}{2}}\tilde{\alpha}_0^{25} u + i\sqrt{\frac{\alpha'}{2}}\sum_{n\neq 0}\frac{\tilde{\alpha}_n^{25}}{n}e^{-inu} \\ X_R^{25}(v) &= x_{0,R}^{25} + \sqrt{\frac{\alpha'}{2}}\alpha_0^{25} v + i\sqrt{\frac{\alpha'}{2}}\sum_{n\neq 0}\frac{\alpha_n^{25}}{n}e^{-inv}\,, \end{aligned} \tag{20.23}$$

with the constraint

$$\tilde{\alpha}_0^{25} - \alpha_0^{25} = \sqrt{2\alpha'}w. \tag{20.24}$$

We have seen that the momentum density was $\mathcal{P} = \dot{X}/(2\pi\alpha')$, so the momentum in the 25 direction is

$$p \equiv p^{25} = \frac{1}{2\pi\alpha'}\int_0^{2\pi} d\sigma(\dot{X}_L^{25} + \dot{X}_R^{25}). \tag{20.25}$$

Since $\dot{u} = \dot{v} = 1$, and the nonzero modes integrate to zero, from the explicit expressions for X_L^{25} and X_R^{25}, we find

$$p = \frac{\alpha_0^{25} + \tilde{\alpha}_0^{25}}{\sqrt{2\alpha'}}. \tag{20.26}$$

We invert these equations for w and p, to find

$$\alpha_0^{25} = \sqrt{\frac{\alpha'}{2}}(p - w); \quad \tilde{\alpha}_0^{25} = \sqrt{\frac{\alpha'}{2}}(p + w). \tag{20.27}$$

As we saw, $w = mR/\alpha'$ is written in terms of an integer m, but so is p. Indeed, on a compact space, e^{ipx} must be single valued around the circle, so $e^{2\pi i Rp} = 1$, or

$$p = \frac{n}{R}. \tag{20.28}$$

Finally, substituting α_0^{25}, $\tilde{\alpha}_0^{25}$ in X^{25}, we find

$$X^{25}(\tau, \sigma) = X_L^{25}(\tau + \sigma) + X_R^{25}(\tau - \sigma) = x_0^{25} + \alpha' p\tau + \alpha' w\sigma + \sum_{n\neq 0} \frac{e^{-in\tau}}{n}(\alpha_n^{25} e^{in\sigma} + \tilde{\alpha}_n^{25} e^{-in\sigma}), \tag{20.29}$$

where

$$x_0^{25} = x_{0,L}^{25} + x_{0,R}^{25}. \tag{20.30}$$

Virasoro Constraints and Mass Spectrum on Circle

Now we are in a position to calculate the mass spectrum of closed strings on the circle, from the Virasoro constraints. We remember that the modes of the Virasoro constraints are L_n and $\tilde{L}_n$, and in the $n = 0$ case we have $L_0 - 1 = \tilde{L}_0 - 1 = 0$, and from these latter ones we obtain the mass spectrum. The only difference with respect to the calculation in previous chapters is in the form of α_0^{25} and $\tilde{\alpha}_0^{25}$, but otherwise we have the same result. In lightcone coordinates $(I, 25) = (+, -, i, 25)$, we have as before

$$\begin{aligned} L_0 &= \frac{1}{2}(\alpha_0^I \alpha_0^I + \alpha_0^{25}\alpha_0^{25}) + N^\perp = \frac{\alpha'}{4}(-2p^+p^- + p^i p^i) + \frac{\alpha'}{4}(p - w)^2 + N^\perp \\ \tilde{L}_0 &= \frac{1}{2}(\tilde{\alpha}_0^I \tilde{\alpha}_0^I + \tilde{\alpha}_0^{25}\tilde{\alpha}_0^{25}) + \tilde{N}^\perp = \frac{\alpha'}{4}(-2p^+p^- + p^i p^i) + \frac{\alpha'}{4}(p + w)^2 + \tilde{N}^\perp. \end{aligned} \tag{20.31}$$

Then, as usual, the Hamiltonian constraint $H = L_0 + \tilde{L}_0 - 2 = 0$ gives the mass spectrum, and the (worldsheet) momentum constraint $P = L_0 - \tilde{L}_0$ gives the level matching condition.

The worldsheet momentum constraint is now

$$L_0 - \tilde{L}_0 = -\alpha' pw + N^\perp - \tilde{N}^\perp - 2 = 0 \Rightarrow N^\perp - \tilde{N}^\perp = \alpha' pw = nm, \tag{20.32}$$

so the new level matching is only up to the integer nm.

The mass spectrum is given by the condition $L_0 + \tilde{L}_0 - 2 = 0$, or

$$\alpha'(-2p^+p^- + p^i p^i) + \frac{\alpha'}{2}(p^2 + w^2) + N^\perp + \tilde{N}^\perp - 2 = 0, \tag{20.33}$$

which implies that the mass squared in the compactified space is

$$M^2_{\text{compactified}} \equiv 2p^+ p^- - p^i p^i = p^2 + w^2 + \frac{2}{\alpha'}(N^\perp + \tilde{N}^\perp - 2). \qquad (20.34)$$

The extra term in the spectrum compared to the uncompactified case is the $w^2 = (mR/\alpha')^2$ term, which comes from the length of the wound string. Indeed, we have seen in the previous chapter that an open string stretched between two D-branes had a mass contribution coming from the tension times the length of the string. A similar thing happens here, since the contribution to the mass is

$$\Delta M = T \Delta L = \frac{1}{2\pi\alpha'} 2\pi m R = \frac{mR}{\alpha'} = w. \qquad (20.35)$$

20.5 T-Duality of Closed Strings

We can write the spectrum in terms of integers as

$$M^2_{\text{compactified}} = \left(\frac{n}{R}\right)^2 + \left(\frac{mR}{\alpha'}\right)^2 + \frac{2}{\alpha'}(N^\perp + \tilde{N}^\perp - 2). \qquad (20.36)$$

If we do so, we notice a symmetry of the spectrum, called *T-duality*, which exchanges n with m, and R with

$$\tilde{R} = \frac{\alpha'}{R}, \qquad (20.37)$$

so also p with w, i.e., momentum with winding. More precisely, we find that

$$M^2_{\text{compactified}}(R; n, m) = M^2_{\text{compactified}}(\tilde{R}; m, n). \qquad (20.38)$$

This is a symmetry that exists only in string theory, as it exchanges the momentum modes with winding modes, which only exist for the string. It says that a certain *vacuum*, characterized by a value of the radius R, is equivalent with another vacuum, characterized by another radius $\tilde{R}$. The radius is assumed to be a *modulus* in this case, i.e., to have no potential (perturbatively, non-perturbatively there is a potential), so it costs no energy to change it. Then, effectively, theories with $R < \sqrt{\alpha'} \equiv l_s$ (string length) are equivalent with theories with $R > l_s$, since when the radius becomes small, while the momenta have increasing gaps between them, the windings become continuous, just like the momenta are in the decompactified limit.

Exchanging n with m and R with $\tilde{R}$, thus p with w has the effect to act on the zero mode oscillators as

$$\tilde{\alpha}_0^{25} \leftrightarrow \tilde{\alpha}_0^{25}; \quad \alpha_0^{25} \leftrightarrow -\alpha_0^{25}. \qquad (20.39)$$

But then we note that in fact, the T-duality symmetry of the spectrum is more general, since we can also exchange the nonzero mode in the same way as the zero modes, i.e., by

$$\tilde{\alpha}_n^{25} \leftrightarrow \tilde{\alpha}_n^{25}; \quad \alpha_n^{25} \leftrightarrow -\alpha_n^{25}. \tag{20.40}$$

Moreover, we can also exchange the constant pieces as

$$x_{0,L}^{25} \leftrightarrow x_{0,L}^{25}; \quad x_{0,R}^{25} \leftrightarrow -x_{0,R}^{25}, \tag{20.41}$$

so that finally the whole X_L^{25} and X_R^{25} are changed under the symmetry as

$$X_L^{25}(\tau+\sigma) \leftrightarrow X_L^{25}(\tau+\sigma); \quad X_R^{25}(\tau-\sigma) \leftrightarrow -X_R^{25}(\tau-\sigma), \tag{20.42}$$

so $X^{25}(\tau,\sigma)$ is exchanged with

$$\begin{aligned} X'^{25}(\tau,\sigma) &= X_L^{25}(\tau+\sigma) - X_R^{25}(\tau-\sigma) \\ &= q_0^{25} + \alpha' w\tau + \alpha' p\sigma + i\sqrt{\frac{\alpha'}{2}}\sum_{n\neq 0}\frac{e^{-in\tau}}{n}(\tilde{\alpha}_n^{25}e^{-in\sigma} - \alpha_n^{25}e^{in\sigma}), \end{aligned} \tag{20.43}$$

where

$$q_0^{25} = x_{0,L}^{25} - x_{0,R}^{25}. \tag{20.44}$$

We have defined T-duality only as a symmetry of the mass spectrum, i.e., of the free theory, but in fact is a symmetry of the interactions as well, so that in fact is a symmetry of the full string theory.

As such, it can also be considered as a symmetry of the theory in background created by the condensation of the string modes, i.e., in supergravity backgrounds. But then the string theory in some background will be related to a string theory in another background, related by the *Buscher rules*. The original Buscher rules, written for the background of the massless modes of the closed bosonic string, $(g_{\mu\nu}, B_{\mu\nu}, \phi)$, are given by (here 0 refers to the T-duality direction, not to time, and i refers to the other directions)

$$\begin{aligned} \hat{g}_{00} &= \frac{1}{g_{00}} \\ \hat{g}_{0i} &= \frac{B_{0i}}{g_{00}} \\ \hat{g}_{ij} &= g_{ij} - \frac{g_{0i}g_{0j} - B_{0i}B_{0j}}{g_{00}} \\ \hat{B}_{0i} &= \frac{g_{0i}}{g_{00}} \\ \hat{B}_{ij} &= B_{ij} + \frac{g_{0i}B_{0j} - g_{0j}B_{0i}}{g_{00}} \end{aligned}$$

$$\hat{\phi} = \phi - \frac{1}{2} \log g_{00}. \tag{20.45}$$

Note that the first relation is the usual $\tilde{R} = \alpha'/R$ for $\alpha' = 1$, since we can use a 2π periodicity, but put R^2 in the metric as g_{00}, so $\hat{g}_{00} = 1/g_{00}$ corresponds to $\hat{R}^2 = 1/R^2$.

T-duality can be generalized also for a general compact space K_n. In particular, for a torus $T^n = (S^1)^n$, we have T-duality on each circle.

Important Concepts to Remember

- Superstring theory has several formulations: Green–Schwarz has supersymmetry in spacetime and κ symmetry on worldsheet, which when gauge-fixed leads to worldsheet supersymmetry. Neveu–Schwarz–Ramond has supersymmetry on the worldvolume ("spinning string"). And Berkovits's "pure spinor" formulation is a way to covariantly quantize the string with spacetime supersymmetry.
- Supersymmetry can be defined on-shell or off-shell, and relates bosons with fermions, and degree of freedom with degree of freedom.
- The superstring lives in 10 dimensions, though a modification of the argument from the previous chapter about 26 dimensions for the bosonic string.
- D3-branes have a 3+1 dimensional worldvolume, on which the maximum number of supersymmetries (actually realized) for spins ≤ 1 is 4. The unique $\mathcal{N} = 4$ SYM theory is the linearized theory on the worldvolume of the D3-branes.
- The $\mathcal{N} = 4$ SYM has 6 scalars transforming in $SO(6) = SU(4)$, a gauge field and 4 fermions in the fundamental of $SU(4)$.
- There is a maximum electric field on the D3-brane, $E_{\max} = 1/(2\pi/\alpha')$.
- Closed strings on compact spaces have not just momentum $p = n/R$ but also winding $w = mR/\alpha'$.
- The mass spectrum of closed strings on a circle has *T-duality*, $M^2_{\text{compactified}}(R; n, m) = M^2_{\text{compactified}}(\tilde{R}; m, n)$, where $\tilde{R} = \alpha'/R$.
- The T-duality relation exchanges momentum with winding, n with m, and changes the sign of X_R, while leaving X_R invariant.
- T-duality has an action on the massless modes of the string, which implies a relation on the supergravity background fields, the Buscher rules.

Further reading: See [20–22] on string theory, as well as the more specialized [23] on D-branes.

Exercises

(1) Is there a maximum *magnetic* field on the brane? Perform Lorentz transformations and show consistency with a maximum electric field.

(2) Consider the closed string spectrum for compactification on the torus T^6 of type II string theory. Write down the mass formula for the string, and specialize it to $R_i = \sqrt{\alpha'}$.

(3) Calculate the solution T-dual (on X^1) to

$$\begin{aligned} ds^2 &= dx_1^2 + dx_2^2 + dx_3^2 \\ e^{\phi} &= e^{\phi_0} \\ B_{12} &= m x_3 \end{aligned} \tag{20.46}$$

using the Buscher rules.

(4) Show explicitly that the action of $\mathcal{N} = 4$ SYM in 4 dimensions is the dimensional reduction of the action for $\mathcal{N} = 1$ SYM in 10 dimensions.

Chapter 21
T-Duality of Open Strings. M-Theory and the Duality Web

In the last chapter, we considered T-duality for closed strings. Now we start off by considering T-duality for open strings, and see that it will lead to D-branes. Then, we define the low energy supergravity limits of string theory, the strong coupling limit of string theory known as M-theory, and finally show that all the various incarnations of string theory are related into a single duality web, as facets of the same unique theory.

21.1 T-Duality of Open Strings

We consider open strings with free endpoints, i.e., with Neumann boundary conditions on both endpoints (NN strings) for the compact direction X^{25}. Then the mode expansion for X^{25} is the one we already found for NN strings,

$$X^{25}(\tau,\sigma) = x_0^{25} + \sqrt{2\alpha'}\alpha_0^{25} + i\sqrt{2\alpha'}\sum_{n\neq 0}\frac{\alpha_n^{25}}{n}e^{-in\tau}\cos n\sigma\ , \tag{21.1}$$

just that the zero mode α_0^{25}, still expressible in terms of the momentum in the 25th direction, is now quantized,

$$\alpha_0^{25} = \sqrt{2\alpha'}p = \sqrt{2\alpha'}\frac{n}{R}. \tag{21.2}$$

Note that the strings are open and free, so there is no winding mode.

As before, the above formula can be actually found by splitting X^{25} into left- and right-movers, and then seeing that for the boundary condition we need to have $\alpha_n^{25} = \tilde{\alpha}_n^{25}$. In any case, we can split X^{25} as

$$X^{25}(\tau,\sigma) = X_L^{25}(\tau+\sigma) + X_R^{25}(\tau-\sigma)\ , \tag{21.3}$$

H. Năstase, *Cosmology and String Theory*, Fundamental Theories of Physics 197, https://doi.org/10.1007/978-3-030-15077-8_21

where each of the modes is given by the free closed string expansion,

$$X_L^{25} = \frac{x_0^{25} + q_0^{25}}{2} + \sqrt{\frac{\alpha'}{2}}\alpha_0^{25}(\tau + \sigma) + i\sqrt{\frac{\alpha'}{2}}\sum_{n\neq 0}\frac{\alpha_n^{25}}{n}e^{-in\tau}e^{-in\sigma}$$
$$X_R^{25} = \frac{x_0^{25} - q_0^{25}}{2} + \sqrt{\frac{\alpha'}{2}}\alpha_0^{25}(\tau - \sigma) + i\sqrt{\frac{\alpha'}{2}}\sum_{n\neq 0}\frac{\alpha_n^{25}}{n}e^{-in\tau}e^{+in\sigma}\ , \quad (21.4)$$

where we have introduced a q_0^{25}, since the zero modes of X_L and X_R are in principle independent.

Now we can do the same T-duality transformation that we did in the case of closed strings, i.e., replace $X^{25}(\tau, \sigma)$ with

$$X'^{25}(\tau, \sigma) = X_L^{25}(\tau + \sigma) - X_R^{25}(\tau - \sigma) = q_0^{25} + \sqrt{2\alpha'}\alpha_0^{25}\sigma + \sqrt{2\alpha'}\sum_{n\neq 0}\frac{\alpha_n^{25}}{n}e^{-in\tau}\sin n\sigma. \quad (21.5)$$

But now we notice that this is the same as the expansion for an open string stretching between 2 D-branes, just with

$$\alpha_0^{25} = \sqrt{2\alpha'}\frac{n}{R} = \frac{1}{\sqrt{2\alpha'}}\frac{x_2^{25} - x_1^{25}}{\pi}. \quad (21.6)$$

This means that the distance between the possible D-branes is

$$X'^{25}(\tau, \sigma = \pi) - X'^{25}(\tau, \sigma = 0) = 2\alpha'\frac{n}{R} = 2\pi\tilde{R}n\ , \quad (21.7)$$

where in the last equality we have used the radius of the T-dual space, $\tilde{R} = \alpha'/R$. And we can also say that

$$X'^{25}(\tau, \sigma = 0) = q_0^{25}\ , \quad (21.8)$$

which can be therefore identified with the position of the D-brane on the T-dual circle. The difference of $2\pi\tilde{R}n$ means that the T-dual string winds n times around the T-dual circle, and then comes back to the same D-brane at q_0^{25}.

We see then that under T-duality, a D25-brane (free Neumann boundary conditions for the whole space) turned into a D24-brane.

Moreover, the boundary conditions are interchanged (Neumann with Dirichlet), since

$$\partial_\sigma X^{25}(\tau, \sigma) = \frac{dX_L^{25}(u = \tau + \sigma)}{du} - \frac{dX_R^{25}}{du}(u = \tau - \sigma) = \partial_\tau X'^{25}(\tau, \sigma)$$
$$\partial_\tau X^{25}(\tau, \sigma) = \frac{dX_L^{25}(u = \tau + \sigma)}{du} + \frac{dX_R^{25}}{du}(u = \tau - \sigma) = \partial_\sigma X'^{25}(\tau, \sigma)\ , \quad (21.9)$$

which means that Neumann conditions become Dirichlet, and Dirichlet become Neumann,

$$\partial_\sigma X^{25}(\tau,\sigma)=0\rightarrow\partial_\tau X^{25}(\tau,\sigma)=0$$
$$\partial_\tau X^{25}(\tau,\sigma)=0\rightarrow\partial_\sigma X^{25}(\tau,\sigma)=0. \tag{21.10}$$

Then we can say more generally, that if we take a Dp-brane and make a T-duality along a parallel direction to the brane, we turn the Dp into D$(p-1)$-brane, and if we make a T-duality along a transverse direction to the brane, we turn it into a D$(p+1)$-brane. Either way, the dimensionality of the brane changes.

21.2 T-Duality with Chan–Paton Factors: Several D-Branes

We have seen what happens when we perform T-duality on a empty space with a single open string: we obtain a D-brane. But we want to ask what happens when we perform T-duality on an open string with Chan–Paton factors for a group $U(N)$.

We need to introduce the concept of Wilson line, or loop (for a closed one) associated with a path in spacetime. For a closed path C and an abelian ($U(1)$) group, it is

$$W\equiv e^{iq\oint_C dxA_x}\equiv e^{iw}. \tag{21.11}$$

Here we have concentrated on a single coordinate x, which will be moreover chosen to be compact, and the contour C to wrap around it. In general, we would have $\int_C dx^\mu A_\mu$ in the exponent.

The Wilson loop is gauge invariant. Indeed, consider gauge transformations

$$A_x\rightarrow A_x+\partial_x\lambda\;, \tag{21.12}$$

and a periodic gauge parameter, modulo $2\pi m/q$,

$$q\lambda(x+2\pi R)=q\lambda(x)+2\pi m. \tag{21.13}$$

Then the gauge transformation $U=e^{iq\lambda}=1$, and the variation in the exponent of the Wilson loop is

$$\delta q\oint_C dxA_x=\oint dx\partial_x\lambda=\lambda(2\pi R)-\lambda(0)=2\pi m\;, \tag{21.14}$$

which means that the Wilson loop is invariant,

$$W\rightarrow W\cdot e^{2\pi im}=W. \tag{21.15}$$

Moreover, because of the periodicity,

$$w \equiv q \oint dx A_x \equiv \theta \in [0, 2\pi] \tag{21.16}$$

is an angle.

Given a value for the the Wilson loop, it can be realized in the simplest way by a constant A_x, namely

$$qA_x = \frac{\theta}{2\pi R}. \tag{21.17}$$

But when adding a gauge coupling qA_μ to a particle, the momentum is changed to the gauge invariant form,

$$p_\mu \to p_\mu - qA_\mu , \tag{21.18}$$

so in our case

$$p_{25} \to p_{25} - qA_{25} \Rightarrow \frac{n}{R} \to \frac{n}{R} - \frac{\theta}{2\pi R}. \tag{21.19}$$

Consider now an open string with $U(N)$ Chan–Paton factors, and a general Wilson loop. It can be realized by a constant $U(N)$ matrix A_{25}, which can therefore be diagonalized by gauge transformations, to a form

$$A_{25} = \frac{1}{2\pi R}\text{diag}(\theta_1, \ldots, \theta_N). \tag{21.20}$$

By its presence, we break the $U(N)$ invariance to only $U(1)^N$. But then the Chan–Paton state $|ij\rangle$ has charge $+1$ under $U(1)_i$ and charge -1 under $U(1)_j$ (since we can think of one end other string as belonging to the fundamental representation, and the other to the antifundamental; when considering a $U(1)$ subgroup, one corresponds to $+1$ charge, the other -1 charge). That means that the gauge invariant form of the momentum is now

$$\frac{2\pi n + \theta_i - \theta_j}{2\pi R} = \frac{1}{2\pi\alpha'}\tilde{R}(2\pi n + \theta_i - \theta_j). \tag{21.21}$$

Substituting this into the usual open string spectrum, we now find

$$M^2 = \left(\frac{2\pi n - (\theta_j - \theta_i)}{2\pi R}\right)^2 + \frac{1}{\alpha'}(N^\perp - 1). \tag{21.22}$$

We see then that now $\tilde{R}\theta_i$ plays the role of position of an ith D-brane in X'^{25} (which has radius $\tilde{R}$), and the T-dual picture is one where we have N D24-branes, at the locations θ_i, as in Fig. 21.1. Moreover, we can now consider

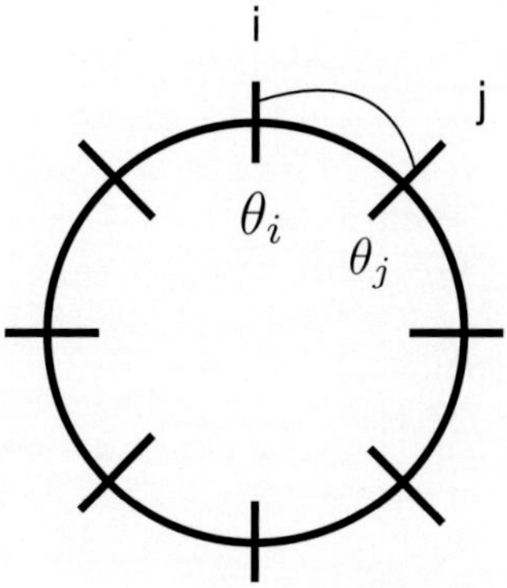

Fig. 21.1 The T-dual of Wilson lines is a set of D-branes at positions $R\theta_i$, $R\theta_j$ in the compact dimension, in between which we have open strings

$$X'^{25}(\tau,\sigma=\pi) - X'^{25}(\tau,\sigma=0) = \int_0^\pi d\sigma\partial_\sigma X'^{25} = \int_0^\pi d\sigma\partial_\tau X^{25} = 2\pi\alpha' p$$
$$= 2\pi\alpha'\left(\frac{n}{R} + \frac{\Delta\theta}{2\pi R}\right) = \tilde{R}(2\pi n + \Delta\theta)\ , \quad (21.23)$$

so indeed the string stretches between two D-branes at θ_i and θ_j, after winding n times around the circle.

21.3 Supergravity Actions

We want to study supersymmetric string theory, i.e., superstring theory, which lives in $D = 10$ dimensions for consistency, as we saw. Moreover, it has $\mathcal{N} = 2$ supersymmetry, which is the maximal possible in order to have spins $s \leq 2$ in 4 dimensions, since as we argued that implied $\mathcal{N} \leq 8$ in 4 dimensions, and one 10 dimensional minimal spinor has the same number of degrees of freedom as 8 spinors in 4 dimensions, i.e., 16. But if there are two supersymmetries, and each one relating the bosons to a different fermion, there are two fermions as well, which means that there are two types of theories, depending on the chirality (eigenvalue of $\Gamma_{D+1} = \Gamma_1 \dots \Gamma_D$) of the spinors,

$$\Gamma_{11}\psi_i = \pm\psi_i. \quad (21.24)$$

- If the spinors have the same chirality,

$$\Gamma_{11}\psi_i = +\psi_i\ , i = 1, 2\ , \quad (21.25)$$

  we have the **type IIB** theory, and
- If the spinors have opposite chirality,

$$\Gamma_{11}\psi_i = (-1)^i\psi_i\ , i = 1, 2\ , \quad (21.26)$$

  we have **type IIA** theory.

At low energies, i.e., for $E \ll 1/\sqrt{\alpha'}$ (since the unique energy scale of string theory is made from α' in this way), the string theory turns into a supersymmetric theory that involves $g_{\mu\nu}$ (since as we saw the closed string massless modes contain $g_{\mu\nu}, B_{\mu\nu}, \phi$), i.e., *supergravity*. Correspondingly, at low energy we have type IIA supergravity or type IIB supergravity.

In terms of the superstring modes, which as we said were composed of the products (for left- and right-movers) of either the bosonic Neveu–Schwarz (NS) sector or the fermionic Ramond (R) sector, we have the fields

- The NS-NS modes are bosons, and the massless modes are the same as the modes of the closed bosonic string, $(g_{\mu\nu}, B_{\mu\nu}, \phi)$. They are common to the IIA and IIB theories.
- The NS-R and the R-NS sectors are fermions, but we will not study their massless modes.
- The R-R modes are also bosons, but their massless modes differ between the IIA and IIB theories. Due to the fact that they are constructed, they are associated to objects of the type $\bar{\psi}\Gamma_{\mu_1\ldots\mu_{p+1}}\chi$, where ψ, χ are fermions, so they are antisymmetric tensor fields $A_{(p+1)}$.
  type IIA. In type IIA theory, we have odd form potentials: $A_{(1)}$ or A_μ, a usual gauge field, with field strength $F_{(2)}$ or $F_{\mu\nu} = \partial_\mu A_\nu - \partial_\nu A_\mu$, which couples electrically to sources = D0-branes ("D-particles"). In general as we saw, Dp-branes have a WZ term that includes the electric-type coupling $\mu_p \int A_{(p+1)}$, so D-particles couple to gauge fields in the usual way. We also have a $A_{(3)}$, or $A_{\mu\nu\rho}$, with field strength $F_{\mu_1\ldots\mu_4} = 3\partial_{[\mu_1} A_{\mu_2\mu_3\mu_4]}$, and coupling electrically to D2-branes. We also have D4-branes, coupling magnetically (i.e., through the 5-form dual to $A_{(3)}$, $A_{(5)} = *A_{(3)}$) to $A_{(3)}$, and D6-branes, coupling magnetically to $A_{(1)}$.
  type IIB. In type IIB theory, we have even form potentials: the (pseudo)scalar (axion) a with field strength $F_{(1)}$ or $F_\mu = \partial_\mu a$, coupling electrically to a D(-1)-brane, or "D-instanton". This is an object with Dirichlet boundary conditions in all space *and time* directions. We also have a 2-form $A_{(2)}$ or $A_{\mu\nu\rho}$ with field strength $F_{(3)}$ or $F_{\mu\nu\rho}$ and coupling electrically to D1-branes ("D-strings"), and a self-dual 4-form $A^+_{(4)}$ or $A^+_{\mu\nu\rho\sigma}$, with field strength $F^+_{\mu_1\ldots\mu_5}$ ($F^+_{(5)}$) that is self dual,

$$F^+_{\mu_1\ldots\mu_5} = \frac{1}{5!}\epsilon_{\mu_1\ldots\mu_5}{}^{\mu_6\ldots\mu_{10}} F_{\mu_6\ldots\mu_{10}} , \tag{21.27}$$

  and that couples to a D3-brane. Note that we impose the self-duality condition in order to obtain an irreducible representation (if not, we can reduce the representation into self-dual and anti-self-dual parts). We also have D5-branes, coupling magnetically to $A_{(2)}$, and D7-branes, coupling magnetically to a. There is no magnetic D3-brane, since the D3-brane is self-dual (so both electric and magnetic).

We now write supergravity actions, in the so-called "string frame", which means that we consider instead of the "Einstein metric" $G^{(E)}_{\mu\nu}$, for which the gravity action is the Einstein–Hilbert term $\frac{1}{2\kappa_N^2}\int d^Dx\sqrt{-G}R$, the action $\frac{1}{2\kappa_N^2}\int d^Dx\sqrt{-G^{(S)}}e^{-2\phi}R$. This is what one naturally obtains in string theory, since the $e^{-2\phi}$ factor has as a VEV $e^{-2\langle\phi\rangle} \equiv 1/g_s^2$, as expected from a (closed) string action.

In the string frame, the *bosonic type IIA supergravity action* is

$$S_{IIA}^{\text{bose}} = \frac{1}{2\kappa_{10}^2} \int d^{10}x \left\{ \sqrt{-G} \left[e^{-2\Phi} \left(R + 4\partial_\mu \Phi \partial^\mu \Phi - \frac{1}{2}|H_3|^2 \right) \right. \right.$$
$$\left. - \frac{1}{2}|F_2|^2 - \frac{1}{2}|\tilde{F}_4|^2 \right]$$
$$\left. - \frac{1}{2} B_2 \wedge F_4 \wedge F_4 \right\} , \tag{21.28}$$

where

$$\tilde{F}_4 \equiv dA_3 - A_1 \wedge F_3$$
$$\int d^D x \sqrt{-G} |F_p|^2 \equiv \int d^D x \sqrt{-G} \frac{1}{p!} G^{M_1 N_1} \dots G^{M_p N_p} F_{M_1 \dots M_p} F_{N_1 \dots N_p} , \tag{21.29}$$

and the terms on the first line in the action, with $e^{-2\Phi}$, are the NS-NS terms (common to IIA and IIB), the terms on the second line are the R-R terms (just for IIA) and the term on the third line is a CS term.

The *bosonic type IIB supergravity action* in string frame has a caveat. Since the field $A^+_{(4)}$ is self-dual, there is no covariant 2-derivative action. One has either to write an action without the self-duality and make the self-duality condition a constraint to be added to the equations of motion, or one breaks explicit Lorentz covariance, or one introduces auxiliary fields. It is customary to present the first form, which is what we will do here. The action is then

$$S_{IIB}^{\text{bose}} = \frac{1}{2\kappa_{10}^2} \int d^{10}x \left\{ \sqrt{-G} \left[e^{-2\Phi} \left(R + 4\partial_\mu \Phi \partial^\mu \Phi - \frac{1}{2}|H_3|^2 \right) \right. \right.$$
$$\left. - \frac{1}{2}|F_1|^2 - \frac{1}{2}|\tilde{F}_3|^2 - \frac{1}{4}|\tilde{F}_5|^2 \right]$$
$$\left. - \frac{1}{2} A_4 \wedge H_3 \wedge F_3 \right\} , \tag{21.30}$$

where

$$\tilde{F}_3 \equiv dA_2 - A_0 \wedge H_3$$
$$\tilde{F}_5 \equiv F_5 - \frac{1}{2} A_2 \wedge H_3 + \frac{1}{2} B_2 \wedge F_3 . \tag{21.31}$$

The self-duality of $\tilde{F}_5$ is imposed as a constraint, in addition to the equations of motion,

$$\tilde{F}_5 = *\tilde{F}_5. \tag{21.32}$$

Note that the extra factor of 1/2 in front of the kinetic action for $\tilde{F}_5$ is due to the fact that in the action we have not imposed the self-duality constraint, so we have twice as many degrees of freedom as we should.

S-Duality of Type IIB Theory

We now describe an important duality symmetry of type IIB theory. We start with the IIB supergravity action, which is invariant under $Sl(2,\mathbb{R})$. To write the action in an invariant way, we first define the combination of the axion (pseudo)scalar C_0 and the scalar dilaton

$$\tau \equiv C_0 + ie^{-\Phi}\,, \tag{21.33}$$

and then out of it, the matrix

$$\mathcal{M}_{ij} \equiv \frac{1}{\text{Im}\tau}\begin{pmatrix} |\tau|^2 & -\text{Re}\tau \\ -\text{Re}\tau & 1 \end{pmatrix}. \tag{21.34}$$

We also define the column vector of the NS-NS and R-R field strengths

$$F_3^i = \begin{pmatrix} H_3 \\ F_3 \end{pmatrix}. \tag{21.35}$$

We then write the action *in the Einstein frame*, i.e. in terms of the Einstein metric $G_{\mu\nu}^{(E)}$, as

$$\begin{aligned} S_{IIB}^{\text{bose}} &= \frac{1}{2\kappa_{10}^2}\int d^{10}x\sqrt{-G_E}\left[R_E - \frac{\partial_\mu\bar{\tau}\partial^\mu\tau}{2(\text{Im}\tau)^2} - \frac{\mathcal{M}_{ij}}{2}F_3^iF_3^j - \frac{1}{4}|\tilde{F}_5|^2\right] \\ &\quad - \frac{\epsilon_{ij}}{8\kappa_{10}^2}\int d^{10}x A_4 \wedge F_3^i \wedge F_3^j. \end{aligned} \tag{21.36}$$

In this way, the action is manifestly invariant under the $Sl(2,\mathbb{R})$ group transformation

$$\begin{pmatrix} a & b \\ c & d \end{pmatrix}, \tag{21.37}$$

(with determinant one, $ad - bc = 1$) acting on τ by

$$\tau \to \tau' = \frac{a\tau + b}{c\tau + d}\,, \tag{21.38}$$

and on F_3^i by the rearranged matrix

$$\Lambda^i{}_j \equiv \begin{pmatrix} d & c \\ b & a \end{pmatrix}, \tag{21.39}$$

as

$$F_3^{i} = \Lambda^{i}{}_{j} F_3^{j}. \tag{21.40}$$

Under this transformation, the matrix $\mathcal{M}_{ij}$ transforms as

$$\mathcal{M} \to \mathcal{M}' = (\Lambda^{-1})^T \mathcal{M} \Lambda^{-1} , \tag{21.41}$$

making the F_3^i kinetic term invariant. Also F_5 and $G_{\mu\nu}^{(E)}$ are invariant.

In this way, the type IIB supergravity action is invariant, or rather the equations of motion are, since we need to impose the self-duality constraint on them. But in the full quantum string theory, only a $Sl(2, \mathbb{Z})$ subgroup (with integer coefficients) survives as full quantum duality symmetry of string theory.

Included in this $Sl(2, \mathbb{Z})$ transformations is the important transformation with matrix

$$\begin{pmatrix} a & b \\ c & d \end{pmatrix} = \begin{pmatrix} 0 & -1 \\ 1 & 0 \end{pmatrix} \in Sl(2, \mathbb{Z}) , \tag{21.42}$$

which gives the transformation

$$\tau' = -\frac{1}{\tau}. \tag{21.43}$$

If $C_0 = 0$, this transformation becomes

$$\Phi' = -\Phi \Rightarrow g_s' = \frac{1}{g_s} , \tag{21.44}$$

which means that this is a nonperturbative duality that exchanges weak and strong coupling. Since now $\Lambda = \begin{pmatrix} 0 & 1 \\ -1 & 0 \end{pmatrix}$, the transformation exchanges NS-NS and R-R fields,

$$H_3' = F_3, \quad F_3' = -H_3. \tag{21.45}$$

The 5-form field is invariant $F_5' = F_5$, and also the Einstein metric is invariant.

21.4 M-Theory

It turns out that type IIA string theory at strong coupling, $g_s = e^{\langle \phi \rangle} \to \infty$, becomes 11 dimensional, and is called M-theory. This was proven by Ed Witten in a 1995 paper ("String theory in various dimensions"), in which it was also proven that various string theories are related by dualities, and thus in fact there is a single unified theory, as we will see at the end of the chapter.

We don't know too much about M-theory, in particular we don't have a quantum formulation, not even a perturbative one like in the case of string theory. The one thing that we know is about the low energy theory, i.e., in the limit of energies much smaller than the 11 dimensional Planck scale, the theory can only be a supersymmetric field theory including gravity, so it must be the unique 11 dimensional supergravity. The theory has bosonic fields the metric G_{MN} and a 3-form antisymmetric tensor $A_{MNP} = A_{(3)}$, with field strength

$$F_{M_1 \ldots M_4} = 24\partial_{[M_1} A_{M_2 M_3 M_4]} : \quad F_{(4)}. \tag{21.46}$$

The fermionic field is the gravitino $\psi_{\mu\alpha}$, but we will not need it in the following. We consider only the bosonic action, which is

$$S_{11}^{\text{bose}} = \frac{1}{2\kappa_{11}^2} \int d^{11}x \sqrt{-G} \left(R - \frac{1}{2} |F_4|^2 \right) - \frac{1}{6\kappa_{11}^2} \int d^{11}x A_{(3)} \wedge F_{(4)} \wedge F_{(4)}. \tag{21.47}$$

Under dimensional reduction, the theory must reduce to type II supergravity (the low energy limit of superstring theory). Specifically, since the 11 dimensional spinor splits into two 10 dimensional spinors of opposite chirality, it must be type IIA supergravity. The dimensional reduction ansatz for the metric is

$$\begin{aligned} ds_{11}^2 &= G_{MN}^{(11)} dx^M dx^N \\ &= e^{-\frac{2\Phi}{3}} G_{\mu\nu}^{s,(10)} dx^\mu dx^\nu + e^{\frac{4\Phi}{3}} (dx^{10} + A_\mu dx^\mu)^2 , \end{aligned} \tag{21.48}$$

and the reduction ansatz for the 3-form $A_{MNP} = (A_{\mu\nu\rho}, A_{\mu\nu 10})$ is that $A_{\mu\nu 10} = B_{\mu\nu}^{IIA}$ is the NS-NS B-field, whereas $A_{\mu\nu\rho} = A_{\mu\nu\rho}^{IIA}$ is the R-R 3-form, completing the bosonic fields of type IIA supergravity.

Relation Among Parameters

In the dimensional reduction metric, the coefficient of $(dx^{10})^2$ must be the radius squared in Planck units, $(R/l_P)^2$, but it is actually $e^{4\Phi/3}$. which means that

$$\left(\frac{R}{l_P} \right)^2 = e^{\frac{4\langle\Phi\rangle}{3}} = g_s^{4/3} \Rightarrow R = l_P g_s^{4/3}. \tag{21.49}$$

Here l_P is the 11 dimensional Planck length, so $2\kappa_{11}^2 = (2\pi)^8 l_P^9$. Dimensionally reducing the 11 dimensional Einstein–Hilbert term, we get

$$\frac{2\pi R}{2\kappa_{11}^2} \int d^{10}x e^{-2\Phi} R = \frac{1}{2\kappa_{10}^2} \int d^{10}x e^{-2(\Phi - \langle\Phi\rangle)} R \, , \tag{21.50}$$

where $2\kappa_{10}^2 = g_s^2 (2\pi)^7 (\sqrt{\alpha'})^8$. Identifying the two sides, we get

$$(2\pi)^7 l_P^9 = R g_s^2 (\sqrt{\alpha'})^8 (2\pi)^7 \Rightarrow l_P = g_s^{1/3} \sqrt{\alpha'}. \tag{21.51}$$

Combining with the previous relation, $R = l_P g_s^{2/3}$. we get

$$R = g_s\sqrt{\alpha'}\ , \tag{21.52}$$

i.e., the radius is the string coupling in string units, as we had advertised.

21.5 The String Duality Web

We have seen that under T-duality, Dp-branes change to D$(p+1)$-branes or D$(p-1)$-branes, and that type IIA string theory has even Dp-branes, whereas type IIB has odd Dp-branes. We can prove moreover that in fact the full type IIA string theory maps to type IIB string theory under T-duality. We also saw that type IIB string theory is invariant under S-duality.

Besides type II string theories, we also have type I and heterotic string theories, both of which reduce at low energies to $\mathcal{N}=1$ supergravity, coupled to $\mathcal{N}=1$ gauge theories for gauge groups $SO(32)$ (for both type I and heterotic) and $E_8 \times E_8$ (only for heterotic).

In fact, as we mentioned, it was proven by Witten that there is only one theory, and the various string theories and M-theory are connected to it, making a duality web. We try to describe here these dualities.

- S duality applied to the type I $SO(32)$ theory gives the $SO(32)$ heterotic theory, giving

$$\Phi_I = -\Phi_h \Rightarrow g_s^I = \frac{1}{g_s^{\text{heterotic}}}. \tag{21.53}$$

  The Einstein metric is invariant, $G_{\mu\nu}^{E,I} = G_{\mu\nu}^{E,h}$, the gauge fields also, $A_{(1)}^I = A_{(1)}^h$, and the 3-forms are identified, $\tilde{F}_{(3)}^I = \tilde{H}_{(3)}^h$.
- T-duality applied to the $E_8 \times E_8$ heterotic string (on S^1) gives the $SO(32)$ heterotic string.
- T-duality of type I $SO(32)$ string theory on S^1 gives type IIA string theory on the interval $S^1/\mathbb{Z}_2$ (obtained by identifying the circle under $x \leftrightarrow -x$), which is M-theory on $S^1 \times (S^1/\mathbb{Z}_2)$.
- The strong coupling limit of the $E_8 \times E_8$ heterotic string is M-theory on the interval $S^1/\mathbb{Z}_2$, with "M9-branes" (10 dimensional walls in 11 dimensions, similar to the D-branes in string theory, having gauge fields on the worldvolume) at the endpoints. Each M9-brane carries a group E_8 on its worldvolume, for a total of $E_8 \times E_8$.

The relations above then connect all possible string theories, together with M-theory, in a "duality web", so that all are in fact corners in moduli space of the same (unknown) theory. Indeed, under S-duality, two theories are equivalent, but when one is perturbative, the other is nonperturbative, so only one is a good (calculable) description at any point. Similarly, under T-duality, among the equivalent theories only one (with $R > \sqrt{\alpha'}$) is a good perturbative description.

Important Concepts to Remember

- T-duality for open strings changes $X = X_L + X_R$ into $X' = X_L - X_R$, and the open string now stretches between two D-branes.
- Thus T-duality exchanges Neumann with Dirichlet boundary conditions, and turns a Dp-brane into a D$(p+1)$-brane (for T-duality on a transverse direction) or D$(p-1)$-brane (for T-duality on a parallel direction)
- T-duality with Chan–Paton factors, i.e., with nonabelian gauge fields, in the presence of Wilson lines, leads to D-branes in the nonabelian gauge group.
- For a Wilson loop breaking $U(N) \to U(1)^N$ by $A = \text{diag}(\theta_1, \dots, \theta_N)$, the D-branes are at different positions $\theta_1, \dots, \theta_N$.
- We thus have a quantum geometry, with "positions" for N D-branes that are now $N \times N$ matrices.
- Low energy superstring theory in 10 dimensions gives type II supergravity, specifically IIA if the spinors have opposite chiralities, and IIB if they have different chiralities.
- The field content of supergravity is the NS-NS sector, $(g_{\mu\nu}, B_{\mu\nu}, \phi)$, common with the bosonic string, the fermionic NS-R and R-NS sectors, and the antisymmetric tensors in the R-R sector: odd p-form potentials in type IIA, even in type IIB.
- The type IIB supergravity action can only be written covariantly (and without auxiliary fields) by imposing the self-duality constraint for the 5-form field strength, $F_5 = *F_5$, by hand.
- The IIB action has S-duality, $Sl(2, \mathbb{R})$ invariance of the equations of motion and $Sl(2, \mathbb{Z})$ restriction at the quantum level, acting on $\tau = C_0 + ie^{-\phi}$. In it, we have the transformation $\tau' = -1/\tau$ that inverts the string coupling and changes the RR with the NS-NS form fields.
- Type IIA string theory at strong coupling becomes 11 dimensional M-theory, with the 11th dimension being the string coupling, $R = g_s\sqrt{\alpha'}$, or $R = g_s^{2/3} l_P$.
- The 11 dimensional supergravity, the low energy limit of M-theory, KK reduces on a circle to the 10 dimensional type IIA supergravity.
- All string theories, type IIA, IIB, type I, heterotic $E_8 \times E_8$ and $SO(32)$, as well as M-theory, are related in a string duality web by T-duality, S-duality and dimensional reduction.

Further reading: See [21] for more details.

Exercises

(1) *(Buscher T-duality)*

Consider the string Polyakov action for $B_{\mu\nu} = 0$, $\phi = \phi_0$,

$$S_P = -\frac{1}{4\pi\alpha'} \int d^2\sigma \sqrt{-\gamma}\gamma^{ab} g_{\mu\nu} \partial_a X^\mu \partial_b X^\nu \,, \tag{21.54}$$

with $g_{\mu\nu}$ split into (g_{ij}, g_{i0}, g_{00}), and 0 is a T-duality direction.

Find a first order form for X^0 by replacing it with a $V_a = \partial_a X^0$ variable and adding a Lagrange multiplier. By solving for X^0, find the dual action in terms of the Lagrange multipliers $\hat{X}^0$.
(2) Check that under KK dimensional reduction of 11 dimensional supergravity, the terms involving $A_{(3)}$ reduce to the type IIA supergravity terms involving $B_{(2)}$ and $A_{(3)}$.
(3) Prove that "D0-branes" ("D-particles") in type IIA string theory, objects charged under the type IIA A_μ field, with masses $M_n = \frac{|n|}{g_s\sqrt{\alpha'}}$, $n \in \mathbb{Z}$, are momentum modes in the 11 dimensional supergravity on a circle.
(4) Show that both type IIB and type IIA supergravity bosonic actions give the same action by dimensional reduction on a circle to 9 dimensions.
(5) Consider D-branes situated at $\Delta\theta$ distance between them. When do we find massless modes for the strings between them, and when these modes are the ground state ("tachyons")?

Chapter 22
String Theory and Particle Physics

In this next to last chapter of part II of this book, we finally consider how to obtain particle physics from string theory. Indeed, string theory is supposed to describe the real world, and since moreover we want to describe cosmology, we need to check whether what we obtain is consistent with the particle physics that we know, i.e., the Standard Model.

–The standard way to obtain particle physics is via dimensional reduction on the compact space.

–The alternative is to consider a *braneworld* in the string theory context, i.e., the particle physics lives on a brane in some curved space, whereas gravity can live in the bulk of spacetime. In the context of string theory, the brane refers to a D-brane, or more precisely for the intersection of D-branes.

22.1 Dimensional Reduction on Compact Spaces

We start the analysis with the standard case, with KK dimensional reduction. In 10 dimensional superstring theory, we consider the standard set-up, $M_{10} = M_4 \times K_6$. Moreover, we want to obtain $\mathcal{N} = 1$ supersymmetry in 4 dimensions after compactification.

The reason is that in 4 dimensions it is very hard to break $\mathcal{N} > 1$ to $\mathcal{N} = 1$ only, and we need only $\mathcal{N} = 1$ for phenomenological reasons: $\mathcal{N} = 2$ and higher supermultiplets come in chiral pairs (non-complex representation), whereas in the real world we have chiral fermions. The only other possibility is no supersymmetry at all ($\mathcal{N} = 0$), but then, what is the point of string theory, if we don't even use supersymmetry to solve the standard problems of particle physics?

$\mathcal{N} = 1$ unbroken susy means that the vacuum $|\psi\rangle$ is invariant under the susy, i.e.,

$$Q|\psi\rangle = 0 , \qquad (22.1)$$

which on the fields means that

H. Năstase, *Cosmology and String Theory*, Fundamental Theories of Physics 197, https://doi.org/10.1007/978-3-030-15077-8_22

$$\delta_Q \text{fields} = 0. \tag{22.2}$$

But since δ bosons = fermions, and fermions have zero VEV, in vacuum the variation of the bosons is automatically satisfied, and we only need to impose as a constraint

$$\delta_Q \text{fermions} = 0. \tag{22.3}$$

The most popular string theory for compactifications leading to Standard Model-like theories, and also the first to be well established, is the $\mathcal{N}=1$ *heterotic superstring*, which in the supergravity limit has the fermion variations

$$\begin{aligned}
\delta\psi_M &= \frac{1}{\kappa_N} D_M\eta + \frac{\kappa_N}{2g^2\phi}\left(\Gamma_M{}^{NPQ} - 8\delta_M^N\Gamma^{PQ}\right)\eta H_{NPQ} + (\text{fermi})^2\eta \\
\delta\chi^a &= -\frac{1}{4g\sqrt{\phi}}\Gamma^{MN}F^a_{MN}\eta + (\text{fermi})^2\eta \\
\delta\lambda &= \frac{1}{\sqrt{\phi}}(\Gamma\cdot\partial\phi)\eta + \frac{\kappa_N}{6\sqrt{2}g^2\phi}\Gamma^{NPQ}\eta H_{NPQ} + (\text{fermi})^2.
\end{aligned} \tag{22.4}$$

Here ψ_M is the gravitino (superpartner of the graviton), χ^a is the gluino (superpartner of the gauge fields) and λ is the dilatino, and $H = dB$ is the field strength of the NS-NS B-field. For supersymmetry, we need to put to zero all three variations (of the gravitino, gluino and dilatino). We have not written explicitly the fermion squared terms, since in the vacuum (as VEV) they must vanish.

The standard way to solve the vanishing of all 3 variations (but by no means the only one; recent models obtaining the Standard Model spectrum violate this simple case) is via $H = d\phi = 0$. In this case, we can easily see that the solution of the equations is

$$\begin{aligned}
&D_i\eta = 0 \\
&\Gamma^{ij}F_{ij}\eta = 0\,,
\end{aligned} \tag{22.5}$$

where i, j are indices in K_6, the compact space (on M_4, the equations are trivial in the vacuum). Since we want a *single* supersymmetry, we need to have a *single covariantly constant spinor* in K_6.

Holonomy and Calabi–Yau Spaces

We can define *parallel transport* of a spinor along a curve in curved spacetime by the condition of vanishing covariant derivative (covariantly constant spinor),

$$D_i\eta = 0 \Rightarrow \eta(x+dx) = (1+\omega\cdot dx)\eta(x)\,, \tag{22.6}$$

which leads to the finite form

$$\begin{aligned}
\eta^\alpha &\to U^\alpha{}_\beta\eta^\beta \\
U(\gamma) &= P\exp\int_\gamma \omega\cdot dx.
\end{aligned} \tag{22.7}$$

Here $\omega \cdot dx = \omega_\mu dx^\mu = \omega_\mu^{ab} \frac{1}{4} \Gamma^{ab} dx^\mu$ is given by the spin connection.

On a closed path γ, we call U the *holonomy* for the spinor. Since $\omega_\mu^{ab} \in SO(n)$ (is in the Local Lorentz group), then *generically* $U \in SO(n)$. That means that the *holonomy group*, the group formed by all possible holonomies of the space, is $\subseteq SO(n)$.

We can define a *complex manifold* by the statement that there is a matrix $J^i{}_j$ such that $J^2 = -\mathbb{1}$ (this technically means an almost complex manifold), plus an extra condition (involving the so-called Nijenhuis tensor). The statement is that locally, around a point, we can define the notion of i, and thus write complex coordinates.

We can further define a *Kähler manifold*, as the case that the holonomy group is $\subseteq U(N)$, where $N = n/2$ is the complex dimension of the manifold. In that case, $J^i{}_j$ is covariantly constant. An equivalent definition of the Kähler manifold is as a complex manifold where we can locally define a function K, called the Kähler potential, such that

$$g_{i\bar{j}} = \partial_i \partial_{\bar{j}} K. \tag{22.8}$$

In that case, after dimensional reduction on the manifold, we will obtain $\mathcal{N} = 1$ supersymmetry, where we have the same Kähler potential for the scalars corresponding to the compact coordinates, so the name is not a coincidence.

In the case of a Kähler manifold, the spin connection is in $U(N) \simeq SU(N) \times U(1)$ (modulo topological issues). But if the holonomy group is actually $SU(N)$, it means that the $U(1)$ part is topologically trivial (pure gauge, with zero field strength), or mathematically we say that the *first Chern class of the manifold K is zero*, $c_1(K) = 0$.

But Calabi and Yau proved a theorem saying that having a Kähler manifold with vanishing first Chern class $c_1(K) = 0$ is $\Leftrightarrow$ having a unique Kähler metric of $SU(N)$ holonomy. Such a space is then called a *Calabi–Yau manifold*.

On a Calabi–Yau manifold, there is a unique covariantly constant spinor η, $D_i \eta = 0$. But then,

$$[D_i, D_{\bar{j}}]\eta = 0 \Rightarrow R_{i\bar{j}} = 0\,, \tag{22.9}$$

so the manifold is Ricci-flat (the commutator of two covariant derivatives is proportional to the field strength, and for ω the field strength $R_{ij}(\omega)$ is proportional to the Ricci tensor).

So if we take a Calabi–Yau space when compactifying string theory, we not only get a unique covariantly constant spinor, guaranteeing $\mathcal{N} = 1$ supersymmetry in 4 dimensions, but we also satisfy the Einstein equation in the compact space, since $R_{i\bar{j}} = 0$.

If η is the unique covariantly constant spinor, then

$$k_{ij} = \bar{\eta} \Gamma_{ij} \eta \tag{22.10}$$

is the Kähler form, such that when contracted with the metric it gives the complex structure,

$$J^i{}_j = g^{ik} k_{kj}. \quad (22.11)$$

If moreover we consider a 3-dimensional Calabi–Yau CY_3, we also have a *holomorphic 3-form*

$$\Omega_{ijk} = \bar{\eta}\Gamma_{ijk}\eta. \quad (22.12)$$

On CY_n, we will have a holomorphic n-form.

Moduli Space of CY_3

A modulus is a scalar that has no perturbative potential, so it costs no energy to change it. In the case of KK compactifications, parameters that don't have a potential become moduli. For CY_3, we ask also that they don't change the CY_3 structure; then, we have a number of such deformations corresponding to moduli, that fall within two categories:

–*complex structure moduli* are moduli than deform the shape of the space. The simplest example of a Calabi–Yau space is an even torus, and the simplest example of that is T^2. In this case, there is one complex structure modulus, the ratio of radii, R_2/R_1. More precisely, since the modulus needs to be a complex, the "ratio" includes an angle for the periodicities of the identified directions on the plane that give the torus, and is the modular parameter τ of the torus.

–*Kähler structure moduli* are moduli that deform the size of the space. In the case of T^2, the only size modulus is the volume, $R_1 R_2$, or more precisely a complexified version that includes a B-field on the torus.

The moduli (deformations of the CY, which in particular is Kähler) will form a scalar space with no (perturbative) potential, but with nontrivial kinetic term. In fact, they will also form a Kähler space, which is why their kinetic term is written in terms of a Kähler potential, $g_{i\bar{j}}\partial_\mu\phi^i\partial^\mu\phi^{\bar{j}}$, with $g_{i\bar{j}} = \partial_i\partial_{\bar{j}}K$.

Moreover, it is an important result, due to Strominger, that while the Calabi–Yau compactification preservers $\mathcal{N} = 1$ supersymmetry, the moduli preserve $\mathcal{N} = 2$ supersymmetry, and they fall into $\mathcal{N} = 2$ multiplets, in particular $\mathcal{N} = 2$ vector multiplets (made up of $\mathcal{N} = 1$ scalar (WZ) and vector multiplets) and $\mathcal{N} = 2$ hypermultiplets (made up of two $\mathcal{N} = 1$ scalar, or WZ, multiplets).

22.2 Calabi–Yau Compactification of String Theories

Type IIB String Theory on CY_3

In this case, the low energy theory has the following $\mathcal{N} = 2$ supermultiplets: the $\mathcal{N} = 2$ supergravity multiplet, $n_v = b_3$ (here b_3 is the "third Betti number", a topological quantity that characterizes the manifold) vector multiplets containing the complex structure moduli, and $n_h = b_2 + 1$ hypermultiplets (b_2 is the second Betti number, also a topological number) containing Kähler structure moduli. There is also an extra complex structure modulus, the complex scalar τ made up of the axion a and the dilaton ϕ by

$$\tau = a + ie^{-\phi}. \tag{22.13}$$

Type IIB string theory by itself is not phenomenologically interesting, and the interest in it only arose once people started to introduce *G-fluxes*, which mean the nonzero integral over a cycle in the geometry of an antisymmetric tensor G defined as

$$G = F^{\text{RR}} - \tau H^{\text{NS-NS}}. \tag{22.14}$$

Here H is the field strength of the NS-NS B-field, and F is the field strength of the RR 2-form. The case of theories with G-flux was defined by Giddings, Kachru and Polchinski, and then Gukov, Vafa and Witten calculated the resulting superpotential in 4 dimensions, and found it to be

$$W = \int_{K_6} \Omega \wedge G\,, \tag{22.15}$$

where Ω is the holomorphic 3-form.

Always there is at least one Kähler modulus, the complexified volume ρ. We also have at least a complex structure τ, as we saw, and the rest are called τ_α. In this case, the Kähler potential is

$$\begin{aligned} K(\rho, \tau, \tau^\alpha) &= K(\rho) + K(\tau, \tau^\alpha) \\ K(\rho) &= -3 \ln[-i(\rho - \bar{\rho})] \\ K(\tau, \tau^\alpha) &= -\ln[-i(\tau - \bar{\tau})] - \ln\left[-i \int_{K_6} \Omega \wedge \bar{\Omega}\right]. \end{aligned} \tag{22.16}$$

Heterotic $E_8 \times E_8$ String Theory on CY_3

The first compactification model to become popular was the heterotic $E_8 \times E_8$ model, specifically on CY_3. We have seen the general conditions for $\mathcal{N} = 1$ susy in 4 dimensions. But the NS-NS field strength has an anomalous Bianchi identity. Instead of $dH = 0$ (since $H = dB$), in the heterotic $E_8 \times E_8$ model we have in fact

$$dH = \text{Tr}(F \wedge F) - \text{Tr}(R \wedge R)\,, \tag{22.17}$$

where in the first term, the trace refers to the $E_8 \times E_8$ YM group, for F_{ij}^{AB}, and in the second term, the trace refers to the local Lorentz group, for R_{ij}^{ab}. We have seen that the simplest model we considered had $H = 0$, which means from the above that we must have $\text{Tr}(F \wedge F) = \text{Tr}(R \wedge R)$. The simplest way to satisfy this is using "$F = R$". Specifically, this can be done by "embedding the spin connection into the gauge group," i.e., identifying the 4 dimensional fields coming from the gauge field A_i and the metric g_{ij} (or more precisely the spin connection ω_i^{ab}), by (here A, B are $SO(16) \subset E_8$ indices)

$$A_i^{AB} = \begin{pmatrix} 0 & 0 \\ 0 & \omega_i^{ab} \end{pmatrix}. \tag{22.18}$$

This is a common thing in KK compactifications, as we said: the various fields with the same index structure (same spin) mix up, and thus here the two fields in A_i and g_{ij} have the same lower dimensional field in their ansatz. For this embedding, the gauge group is broken to the subgroup of E_8 commuting with the ansatz (i.e., with the $SO(6)$ of the spin connection), namely $SO(10)$.

One can also consider nontrivial Wilson loops in the geometry, i.e.,

$$U_\gamma \equiv P \exp \oint_\gamma A \cdot dx \neq 0 \tag{22.19}$$

for some γ. As we said also in the previous chapter, this breaks the gauge group further (for a general Wilson line, this would break $U(N)$ to $U(1)^N$, but in specific cases, we can break it less), towards the Standard Model group, $SU(3) \times SU(2) \times U(1)$.

This concludes our analysis of KK compactification models. There is also the possibility of string theory near some special singularities, but we will not study that. Then there is only the idea of braneworld.

22.3 Intersecting D-Branes: Braneworld

In this construction, within string theory, the Standard Model of particle physics is supposed to live on the intersection of D-branes, whereas gravity exists everywhere. This seems to have a chance of working, since the Standard Model gauge group is $SU(3) \times SU(2) \times U(1)$, and on N coincident branes we have an $U(N) \simeq SU(N) \times U(1)$ group, where the $U(1)$ multiplet contains the scalar parametrizing the center of mass motion of the N D-branes.

Since we want to obtain the Standard Model in 3+1 dimensions on the intersection of branes, we need to have Dp-branes with $p > 3$. In fact, it turns out that the most flexible case is $p = 6$, since then we have more possibilities. Thus the intersecting D6-branes models are the most popular.

As a simple exercise (not even remotely phenomenologically correct), we consider D6-brane on the torus T^6, and even more specifically, we consider two D6-branes, D6 and D6'. In reality, we will have some discrete identification of T^6 directions (using maybe $\mathbb{Z}_n$ identifications), and usually more D-branes, for a more realistic case. Splitting the T^6 into 3 tori for coordinates (x_4, x_5), (x_6, x_7) and (x_8, x_9), we consider the directions filled by D6 and D6' to be denoted with a cross,

	0	1	2	3	4	5	6	7	8	9
$D6$	×	×	×	×	×	0	×	0	×	0
$D6'$	×	×	×	×	0	×	0	×	0	×

(22.20)

and the D6 and D6' intersect in each T^2 in the middle of the square torus.

More generally, we can consider intersections of D6-branes at angles, defined by some integers. For instance, the D-brane with $l = (3, 1)$ is identified over 3 fundamental domains in the x direction and one domain in the y direction.

When D-branes intersect, there are extra massless open string states that appear at the intersection. Indeed, when we stretch open strings between D-branes i and j we have a mass $M_{ij} = \frac{1}{2\pi\alpha'} r_{ij}$, so if the distance $r_{ij} = 0$, also $M_{ij} = 0$. Thus indeed, at the intersection (and only there), where $r_{ij} = 0$ between the intersecting branes, we have massless states. However, for D-branes at angles different than $\pi/2$, the bifundamental scalars coming from r_{ij} are generically massive, whereas the bifundamental fermions are always massless. If $i \in SU(m)$ (for m coincident branes on the left) and $j \in SU(q)$ (for q coincident branes on the right), the state (ij) is bifundamental in $SU(m) \times SU(q)$. If for instance, we manage to "freeze" the $SU(q)$ gauge fields, so that $SU(q)$ becomes a global symmetry and $SU(p)$ remains local, then this matter is really like the quark matter in the Standard Model, with one local and one global index.

22.4 Moduli Stabilization

We have seen that in string theory we have many moduli, i.e., scalars that are massless or nearly massless perturbatively. But experimentally, we know there aren't any, at least on the scales of the Earth, since otherwise we would detect a fifth force. There could be such scalars on cosmological scales, hidden from Earth experiments, in the so called chameleon theories, that will be studied later on, but certainly there aren't on Earth scales.

That means that we need to "stabilize" these moduli, by introducing some nonperturbative potentials. But in string theory, that is obviously difficult, as we will discuss in part III of the book, since we only have a perturbative definition of string theory, and nonperturbative quantities we can compute are few, like ones found through dualities, or nonperturbative contributions to the superpotential coming from wrapped branes (since the D-brane action is proportional to $1/g_s$, by having terms of the e^{iS} type, we get nonperturbative contributions).

The moduli we need to stabilize are:

- shape and size moduli, i.e., complex and Kähler structure moduli.
- the dilaton is a special case, since it defines the string coupling constant, and is usually considered separately.
- if we have D-branes, they come with their own moduli, for instance for their positions.
- if we have fluxes, the fluxes usually have integers values, but sometimes they also lead to moduli.

Generically, it is very hard to stabilize *all* the moduli. Usually, the last modulus to stabilize turns out to be difficult to do so.

Important Concepts to Remember

- Particle physics is obtained from string theory either by KK dimensional reduction, or in a braneworld scenario.
- For phenomenological reasons we need $\mathcal{N}=1$ supersymmetry in 4 dimensions, which means that we need to put to zero the supersymmetry variation of the fermions in the vacuum, $\delta\psi=0$.
- For heterotic supergravity, the standard way is to use $H=d\phi=0$, giving $D_i\eta=0$ and $F_{ij}\Gamma^{ij}\eta=0$.
- A Kähler manifold is a complex manifold that has holonomy inside $U(N)$.
- If the holonomy is actually $SU(N)$, so $c_1(K)=0$ (first Chern class of K is zero), then there is a unique Kähler metric of $SU(N)$ holonomy, and the space is called Calabi–Yau manifold.
- On a Calabi–Yau manifold, there is a unique covariantly constant spinor, $D_i\eta=0$, so we obtain $\mathcal{N}=1$ supersymmetry in 4 dimensions, and $R_{i\bar{j}}=0$, so the manifold is Ricci-flat.
- Calabi–Yau spaces have complex structure moduli, that deform the shape, and Kähler structure, that deform the size.
- For type IIB on CY_3 we can introduce G-flux, for $G=F^{\text{R-R}}-\tau H^{\text{NS-NS}}$, with superpotential $W=\int\Omega\wedge G$, where Ω is the unique holomorphic 3-form on CY_3.
- For heterotic $E_8\times E_8$ string theory on CY_3, we can embed the spin connection in the gauge group, and put Wilson lines, to break the gauge group to $SU(3)\times SU(2)\times U(1)$ or a GUT group.
- A braneworld model can be obtained for instance by intersecting D6-branes in the compact space.
- The moduli must be stabilized for phenomenological reasons.

Further reading: See [20–22] for more details.

Exercises

(1) If $H_{MNP}=R\epsilon_{MNP}$, and we are on a sphere, write down the condition for the spinor η that preserves susy.
(2) Assuming that (in type IIB on CY_3)

$$\int_{K_6}\Omega\wedge\bar{\Omega}=0\,,\quad\int_{K_6}\Omega\wedge G=\rho e^{a\rho}\,,\tag{22.21}$$

calculate the scalar potential (the above is not a very realistic case).
(3) In the model with D6, D6'-brane on T^6, calculate the open string spectrum on the common intersection.
(4) Show that by "embedding the spin connection in the gauge group", which as we said leads to the group $SO(10)$, we can neatly fit the Standard Model particles, plus a right-handed neutrino, into a single representation of the unbroken gauge group.
(5) Is a sphere a Calabi–Yau space? Why?

Chapter 23
Holography and the AdS/CFT Correspondence

In this chapter, we will describe the AdS/CFT correspondence, which relates a gravity theory, usually in an Anti-de Sitter (AdS) $d+1$ dimensional background to a field theory with conformal invariance (CFT) in d dimensions, living at a boundary of the space. Therefore the correspondence is an example of a holographic relation, the gravitational physics in the bulk being described holographically ("projected" without losing information) onto the non-gravitational boundary of one dimension less. The AdS/CFT correspondence will be used later on, in part III, for constructing cosmological models. In order to understand that, we will describe the AdS/CFT map for fields and observables.

23.1 AdS Space and Its Holography

In cosmology, the relevant gravitational spacetimes, besides flat space and the generic FLRW metric, are de Sitter space, approximated during the inflationary period, and its deformations, perhaps relevant for today's accelerated expansion, in the case of a cosmological constant. The cosmological constant, both for inflation, and for today's accelerated expansion, is positive.

However, one could consider also a negative cosmological constant, $\Lambda < 0$, leading to an Anti-de Sitter space. This would not be relevant as a cosmological model, but it turns out to be relevant in string theory, and to lead to an application of string theory not as a fundamental theory of Nature, but as a mathematical tool to solve difficult, i.e., strong coupling, field theory problems.

Anti-de Sitter space in d dimensions (AdS_d) is a maximally symmetric, homogenous and isotropic, gravitational space of constant negative curvature. In Euclidean signature, the maximally symmetric spaces, besides flat space, are the sphere S^d, which is a space of constant positive curvature $R > 0$, and the Lobachevsky space, which is a space of constant negative curvature $R < 0$.

H. Năstase, *Cosmology and String Theory*, Fundamental Theories of Physics 197, https://doi.org/10.1007/978-3-030-15077-8_23

The sphere is defined via embedding in $d+1$ dimensional Euclidean space

$$ds^2 = dX_1^2 + \cdots + dX_d^2 + dX_{d+1}^2 \tag{23.1}$$

by the constraint with the same $SO(d+1)$ symmetry,

$$(X_1)^2 + \cdots (X_d)^2 + (X_{d+1})^2 = R^2. \tag{23.2}$$

Lobachevsky space is defined similarly, but by embedding in $d+1$ dimensional Minkowski space (despite having Euclidean signature)

$$ds^2 = dX_1^2 + \cdots + dX_d^2 - dX_{d+1}^2 \tag{23.3}$$

by the constraint with the same $SO(d, 1)$ symmetry,

$$(X_1)^2 + \cdots (X_d)^2 - (X_{d+1})^2 = -R^2. \tag{23.4}$$

Moving on to Minkowski signature spaces, the sphere becomes de Sitter space, as the space of constant positive curvature (so constant cosmological constant), defined by embedding in $d+1$ dimensional Minkowski space

$$ds^2 = -dX_0^2 + dX_1^2 + \cdots dX_{d-1}^2 + dX_{d+1}^2 \tag{23.5}$$

by the constraint with the same $SO(d, 1)$ symmetry,

$$-(X_0)^2 + (X_1)^2 + \cdots (X_{d-1})^2 + (X_{d+1})^2 = R^2. \tag{23.6}$$

Thus the isometry of de Sitter is $SO(d, 1)$, and "Euclidean de Sitter" is the sphere, $EdS_d = S^d$.

Then Lobachevsky space becomes Anti-de Sitter, as the space of constant negative curvature (so negative cosmological constant), defined by embedding in a $d+1$ dimensional space with two "times",

$$ds^2 = -dX_0^2 + dX_1^2 + \cdots dX_{d-1}^2 - dX_{d+1}^2 \tag{23.7}$$

by the constraint with the same $SO(d-1, 2)$ symmetry,

$$-(X_0)^2 + (X_1)^2 + \cdots (X_{d-1})^2 - (X_{d+1})^2 = -R^2. \tag{23.8}$$

That means that AdS space has $SO(d-1, 2)$ isometry, and "Euclidean Anti-de Sitter" $EAdS_d$ is Lobachevsky space.

Besides the implicit metric obtained by solving, say for X_{d+1} in terms of the others, there are other forms of the AdS metric. One relevant one is the Poincaré metric,

$$ds^2 = \frac{R^2}{x_0^2}\left(-dt^2 + \sum_{i=1}^{d-2} dx_i^2 + dx_0^2\right). \tag{23.9}$$

Here $t, x_i \in \mathbb{R}$ and $0 < x_0 < +\infty$, yet this metric only covers part of AdS space, the "Poincaré patch". Another form of the Poincaré metric is obtained by $x_0/R = e^{-y}$, giving

$$ds^2 = e^{2y}\left(-dt^2 + \sum_{i=1}^{d-2} dx_i^2\right) + R^2 dy^2. \tag{23.10}$$

The boundary of the space (or rather, of the Poincaré patch) is at $y \to \infty$, i.e., at $x_0 = 0$. But it takes a finite time for light (moving on $ds^2 = 0$) to reach it from a finite point, since then

$$t = R\int_{y_0}^{\infty} e^{-y} dy = Re^{-y_0} < \infty. \tag{23.11}$$

That means that the boundary of the Poincaré patch *is a finite time away*.

Coordinates that cover the whole of AdS space are the "global coordinates", with metric

$$ds^2 = R^2(-\cosh^2\rho\, d\tau^2 + d\rho^2 + \sinh^2\rho d\Omega_{d-2}^2). \tag{23.12}$$

The further coordinate transformation $\tan\theta = \sinh\rho$ leads to another form for the global coordinate metric,

$$ds^2 = \frac{R^2}{\cos^2\theta}(-d\tau^2 + d\theta^2 + \sin^2\theta d\Omega_{d-2}^2)\,, \tag{23.13}$$

and the range of the coordinates is $0 \leq \theta \leq \pi/2$ and arbitrary τ.

To understand this space, we construct the Penrose diagram, which encodes causal and topological information. For it, we can drop conformal factors, obtaining Einstein's static Universe

$$ds^2 = -d\tau^2 + d\theta^2 + \sin^2\theta d\Omega_{d-2}^2. \tag{23.14}$$

Thus the Penrose diagram is an infinite strip, $\mathbb{R}_\tau \times [0, \pi/2]$, rotated around the sphere directions, giving a cylinder. The Poincaré patch is just a triangle region of the infinite strip, with the base at $\theta = 0$, see Fig. 23.1.

In this form, it is obvious that the boundary of space is the surface of the cylinder, at $\theta = \pi/2$,

$$ds^2 = -d\tau^2 + d\Omega_{d-2}^2. \tag{23.15}$$

The time it takes light to go from a finite point at $\theta = \theta_0$ to the boundary of space $\theta = \pi/2$ in global coordinates is now again finite,

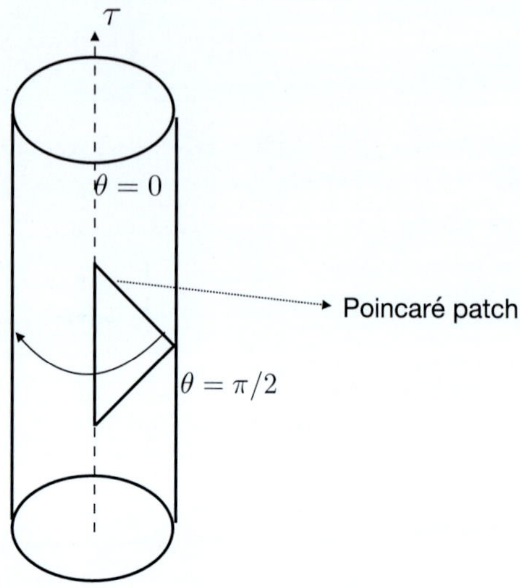

Fig. 23.1 Penrose diagram for global AdS, with the triangle corresponding to the Poincaré patch on it. The diagram is a cylinder of revolution

$$\tau = \frac{\pi}{2} - \theta_0. \tag{23.16}$$

Since the boundary of AdS_d space is a finite time away, light can go the boundary and back in finite time, which means that logically there is the possibility that we can project holographically the physics in the bulk of the space onto the boundary, which is $\mathbb{R}_\tau \times S^{d-2}$ in global coordinates and $\mathbb{R}^4$ in Poincaré coordinates. These two spaces are related by a conformal factor in Euclidean signature, since

$$ds^2 = \sum_{i=1}^{d} (dx_i)^2 = d\tilde{\rho}^2 + \tilde{\rho}^2 d\Omega_{d-1}^2 = \tilde{\rho}^2 \left(\frac{d\tilde{\rho}^2}{\tilde{\rho}^2} + \Omega_{d-1}^2 \right) = e^{-2y}(dy^2 + d\Omega_{d-1}^2) , \tag{23.17}$$

where in the last step we wrote $\tilde{\rho} = e^{-y}$.

Moreover, we saw that the isometry group of AdS_d space is $SO(d-1, 2)$, and it will turn out that this is the conformal symmetry group in $d-1$ dimensions, suggesting that the holographic theory on the boundary of AdS space is a conformal field theory.

23.2 Conformal Field Theories

To understand the relation, we need to understand conformal field theories. Conformal transformations are generalizations of scale transformations of flat space, $x'^\mu = \lambda x^\mu$, which relate metrics as

$$ds^2 = (dx'^\mu)^2 = \lambda^2 (dx^\mu)^2. \tag{23.18}$$

For conformal transformations, the constant conformal factor λ^2 above is replaced by a local conformal factor. Thus conformal transformations are coordinate transformations $x^\mu \to x'^\mu(x^\nu)$ of flat space, such that the flat space metric gets a local conformal factor

$$ds^2 = (dx'^\mu)^2 = [\Omega(x)]^2 (dx^\mu)^2. \tag{23.19}$$

It turns out that most, if not all, theories with scale invariance at the quantum level have also conformal invariance. In fact, a characteristic of conformal invariant theories is that there are no parameters with mass scales in them (they are massless theories). The other observation is that conformal invariance is an invariance of flat space. So, even though a conformal transformation is part of general coordinate transformations, conformal invariance means that the flat space action, after the conformal transformation, is still a flat space action (it is independent of the conformal factor, so it is still written with the flat space metric).

Conformal field theories in $d = 2$ spacetime dimensions are special, since in 2 dimensions, any holomorphic transformation, $z \to z' = f(z)$, where $z = x_1 + ix_2$, is conformal:

$$ds^2 = dz' d\bar{z}' = |f'(z)|^2 dz d\bar{z}. \tag{23.20}$$

That amounts to an infinitely dimensional conformal group of transformations, which highly restricts possible theories.

But we will be interested in dimensions other than 2, $d > 2$, in which case the conformal group (group of conformal transformations) is finite dimensional. For an infinitesimal transformation $x'^\mu = x^\mu + v^\mu(x)$, with an infinitesimal $\Omega(x) \simeq 1 - \sigma_v(x)$, we obtain

$$\partial_\mu v_\nu + \partial_\nu v_\mu = 2\sigma_v \delta_{\mu\nu} \Rightarrow \sigma_v(x) = \partial^\mu v_\mu / d. \tag{23.21}$$

In $d > 2$, the most general solution of this equation is

$$v_\mu(x) = a_\mu + \omega_{\mu\nu} x_\nu + \lambda x_\mu + b_\mu x^2 - 2x_\mu b \cdot x\ , \tag{23.22}$$

giving $\sigma_v(x) = \lambda - 2b \cdot x$. Here λ is a scale transformation, corresponding to a dilatation generator D, a_μ is a translation, corresponding to the generator P_μ, the antisymmetric $\omega_{\mu\nu}$ corresponding to $J_{\mu\nu}$, and b_μ corresponds to a new type of transformation, a "special conformal transformation", for a generator K_μ. Together, the parameters fit into an antisymmetric matrix Ω_{MN}, with $M = (\mu, d+1, d+2)$, corresponding to generators of $SO(d, 2)$

$$\bar{J}_{MN} = \begin{pmatrix} J_{\mu\nu} & \bar{J}_{\mu,d+1} & \bar{J}_{\mu,d+2} \\ -\bar{J}_{\mu,d+1} & 0 & D \\ -\bar{J}_{\mu,d+2} & -D & 0 \end{pmatrix}. \tag{23.23}$$

Here $D = \bar{J}_{d+1,d+2}$ and

$$\bar{J}_{\mu,d+1} = \frac{K_\mu - P_\mu}{2}, \quad \bar{J}_{\mu,d+2} = \frac{K_\mu + P_\mu}{2}. \tag{23.24}$$

Thus indeed, as we have already reported, the isometry of AdS_{d+1} is the same as the conformal group in d flat spacetime dimensions, $SO(d, 2)$.

23.3 AdS/CFT Motivation (Heuristic Derivation)

The AdS/CFT correspondence was "motivated" (heuristically derived) by Maldacena in 1997 based on equating two descriptions of a large number of D-branes, in a certain decoupling limit that we will define.

We consider a near-extremal (i.e., only slightly non-extremal, mass only slightly larger than the charge) system of (a large number) N D-branes. As it was proven by comparing thermodynamic quantities, the temperature and entropy, given a mass and charge, first by Strominger and Vafa, then by Callan and Maldacena (both in early 1996), this system of D-branes describe the gravitational system of a near-extremal black hole.

But then in late 1996 Maldacena and Strominger go further, and prove explicitly that the Hawking radiation spectrum obtained from the near-extremal black hole matches the emission spectrum of the near-extremal large N system of D-branes, which means that at least in this toy model case, the Hawking radiation is unitary, specifically coming from a unitary string process happening on the D-branes. We can therefore understand this process of Hawking radiation from string theory as follows: two open strings, ending on a D-brane, collide, form a closed string, which can then move off the D-brane, as in Fig. 23.2. The closed string mode is a particle radiated by the black hole.

Then in his famous 1997 paper, Maldacena realized that by thinking of the two descriptions of this Hawking radiation, one gravitational, the other one in terms of D-branes, and taking a decoupling limit, we find the equivalence of a field theory living on the D-brane with a gravitational theory living near it. We present here the two points of view, and the equivalence resulting from it.

Point of View nr. 1: D-Branes, i.e., Open-String Endpoints

Thinking of Hawking radiation of the near-extremal D-branes as an open string process, the action describing the system is

$$S = S_{\text{bulk}} + S_{\text{brane}} + S_{\text{interaction}}. \tag{23.25}$$

On the brane we have open strings, in the bulk we have closed strings, and the two are coupled by an interaction that gives the Hawking radiation decay process. This interaction (for a closed string to go into two open strings) is of order g_s, the closed

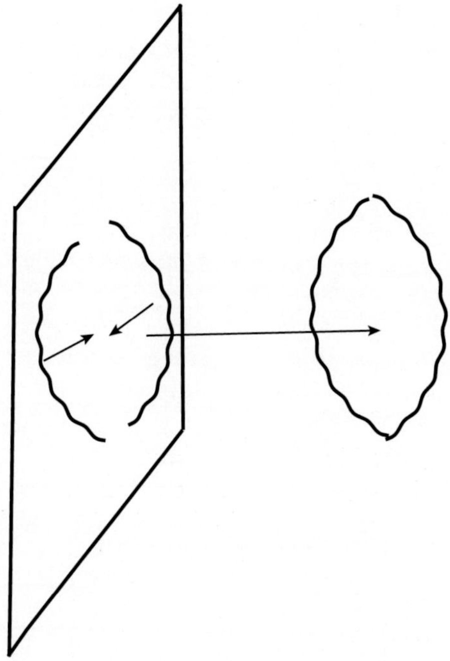

Fig. 23.2 Hawking radiation as a string process on a D-brane: two open strings on the D-brane join and move off the D-brane as a closed string, giving Hawking radiation

string coupling constant. But this is a gravitational interaction, proportional to the 10 dimensional Newton constant κ_N, which is proportional to g_s and to α'^2 (since in 10 dimensions $2\kappa_N^2$ has dimension -8, and α' has dimension -2). But that means that the interaction vanishes, in the "decoupling limit" $\alpha' \to 0$ (which means $\alpha' E^2 \to 0$ for relevant energies E), when moreover $\kappa_N \to 0$, so we obtain free (super)gravity, and also the D-brane action becomes just $\mathcal{N} = 4$ SYM.

Thus in this point of view, in the decoupling limit $\alpha' E^2 \to 0$ (where E is some relevant energy in the field theory) we have:

- free 10 dimensional gravity in the bulk, far from the branes (at $\delta r \to \infty$).
- $\mathcal{N} = 4$ SYM in 4 dimensions on the boundary.

Point of View nr. 2: (Near-)Extremal p-Brane Solutions, i.e., Black Holes

Now, replacing the N D-branes with the gravitational space curved by them, namely the type IIB supergravity solution

$$
\begin{aligned}
ds^2 &= H^{-1/2}(r) d\vec{x}_{||}^2 + H^{1/2}(r)(dr^2 + r^2 d\Omega_5^2) \\
F_5 &= (1 + *)(dt \wedge dx_1 \wedge dx_2 \wedge dx_3 \wedge (dH^{-1})) \\
H(r) &= 1 + \frac{R^4}{r^4} \\
R &= 4\pi g_s N \alpha'^2 \,, \quad Q = g_s N \,,
\end{aligned}
\tag{23.26}
$$

we have only a gravitational system, but it contains several regions. We consider the decoupling limit $\alpha' E^2 \to 0$, or $E\sqrt{\alpha'} \to 0$ but, since we are in gravitational theory, we need to keep the energy at a point p in the bulk, of radius r, fixed. Considering E as the energy measured at $r \to \infty$, the two are related as

$$E_p \sim \frac{d}{d\tau} = \frac{1}{\sqrt{-g_{00}}}\frac{d}{dt} \sim \frac{E}{\sqrt{-g_{00}}}\,, \tag{23.27}$$

which, since at $r \to 0$, $H(r) \sim R^4/r^4 \propto \alpha'^2/r^4$, means

$$\sqrt{\alpha'}E \sim (\sqrt{\alpha'}H^{-1/4})E_p \sim rE_p \to 0 \tag{23.28}$$

happens for $r \to 0$ if E_p is fixed.

So the decoupling limit is a limit of low energy at infinity, and then at infinity (for $\delta r \to \infty$) we get free gravity (since $\kappa_N \propto \alpha'^2 \to 0$, as before), but now also at $r \to 0$ we get $E \to 0$. That means that now we also have two decoupled systems:

- free gravity in the bulk at $\delta r \to \infty$
- low energy gravity excitations for $r \to 0$.

AdS/CFT Correspondence in Decoupling Limit

Identifying the two free systems in the two points of view, we see that one is the same (free gravity at $\delta r \to \infty$), which means the others must be identified:

$\mathcal{N} = 4$ **SYM with gauge group** $SU(N)$ = **gravitational theory in the D3-brane background at** $r \to 0$ and $\alpha' \to 0$.

We now interpret (23.28) at $r \to 0$ as

$$E \propto \frac{(E_p\sqrt{\alpha'})r}{\alpha'} \tag{23.29}$$

and, now identifying E with the energy of the SYM theory instead of the energy at infinity in the gravity description, we consider that, for the purposes of the comparison of SYM with gravity, we must keep fixed both the SYM energy E and the energy in the gravity (string) description in string units, so $\sqrt{\alpha'}E_p$, leading to

$$U = \frac{r}{\alpha'} \tag{23.30}$$

fixed in the $r \to 0$, $\alpha' \to 0$ limit.

In this limit, $H \sim R^4/r^4$, so we obtain

$$ds^2 \simeq \frac{r^2}{R^2}(-dt^2 + d\vec{x}_3^2) + \frac{R^2}{r^2}dr^2 + R^2 d\Omega_5^2\,, \tag{23.31}$$

or, with the change of variables $r/R = R/x_0$,

$$ds^2 = R^2 \frac{-dt^2 + d\vec{x}_3^2 + dx_0^2}{x_0^2} + R^2 d\Omega_5^2 , \tag{23.32}$$

corresponding to the space $AdS_5 \times S^5$ of radius R. Writing it in terms of the variable $U = r/\alpha' \propto E/(E_p\sqrt{\alpha'})$, thus identified with the energy scale in SYM given a fixed energy in gravity, we obtain

$$ds^2 = \alpha' \left[\frac{U^2}{\sqrt{4\pi g_s N}} (-dt^2 + d\vec{x}_3^2) + \sqrt{4\pi g_s N} \left(\frac{dU^2}{U^2} + d\Omega_5^2 \right) \right] . \tag{23.33}$$

In this form, there is a vanishing overall factor α', but the rest is finite. One can show that the string coupling g_s is related to the SYM coupling g_{YM} by

$$4\pi g_s = g_{YM}^2 . \tag{23.34}$$

The proportionality comes becomes two basic open string interactions (proportional to g_{open}, i.e., g_{YM}) can be glued into a basic closed string interaction, or "pair of pants" (proportional to g_{closed}, i.e., g_s), and the proportionality constant can be calculated.

In this way, the $AdS_5 \times S^5$ metric is written in terms of SYM variables N, g_{YM} and U. The SYM field theory in some sense lives at the boundary of $AdS_5 \times S^5$, which is $\mathbb{R}^4$ (the S^5 shrinks to zero when going to the boundary of AdS_5), and happens as we saw for $x_0 \to 0$, or $r \to \infty$. Thus the AdS/CFT correspondence is *holographic*.

23.4 AdS/CFT Definition and Limits

We have obtained the AdS/CFT correspondence as an equality between $\mathcal{N} = 4$ SYM and string theory in the $AdS_5 \times S^5$ background given in (23.33), where N is either the rank of $SU(N)$, or the brane charge, and $4\pi g_s = g_{YM}^2$.

One could say that is all, and that is in fact the most general understanding of the AdS/CFT correspondence, currently believed to be correct. But as things stand, there are some approximations that put the heuristic derivation above on a stronger footing. Namely, we should be able to use the gravitational limit of string theory, which means that α' corrections and g_s corrections should be small.

Having small α', or string worldsheet, corrections has the invariant meaning of having large curvatures with respect to the string length $\sqrt{\alpha'}$, i.e., $\alpha'/R^2 \ll 1$. For the $AdS_5 \times S^5$ space, that becomes

$$\frac{R^2}{\alpha'} = \sqrt{4\pi g_s N} = \sqrt{g_{YM}^2 N} \gg 1 , \tag{23.35}$$

which therefore translates in the SYM theory into a very large 't Hooft coupling,

$$\lambda \equiv g_{YM}^2 N \gg 1. \tag{23.36}$$

Indeed, 't Hooft showed that for an $SU(N)$ gauge theory the effective loop parameter is λ, so perturbation theory corresponds to $\lambda \ll 1$. That means that the perturbative gravity limit is an ultra-nonperturbative limit in SYM. Then the AdS/CFT correspondence is an example of *duality*, that relates strong coupling to weak coupling and vice versa.

The next condition is that quantum string corrections are small, which amounts to g_s small, $g_s \ll 1$.

Putting together the two conditions, we obtain $N \to \infty$, which means that the limit is really the one considered by 't Hooft, $N \to \infty$, $g_{YM} \to 0$, but with $\lambda = g_{YM}^2 N$ fixed but very large ($\gg$1). This 't Hooft limit corresponds in the string side of the correspondence to the gravity limit of string theory.

In terms of the SYM variables, small string corrections correspond to

$$\frac{\alpha'}{R^2} = \frac{1}{\sqrt{\lambda}} \ll 1\,, \quad 4\pi g_s = \frac{\lambda}{N} \ll 1. \tag{23.37}$$

In principle, we have 3 possibilities for the AdS/CFT correspondence:

- (weakest) Matching between the two sides only in the $\alpha'/R^2 \to 0$, $g_s \to 0$ limit.
- (stronger) α'/R^2 corrections match, but g_s corrections don't.
- (strongest) both α'/R^2 and g_s corrections match, so matching of the two sides is at the full string theory level, or equivalently for any N and g_{YM}.

As we said, the strongest form of the correspondence is actually believed to be correct, and the matching is at the full string level. Many α' and g_s corrections have been calculated and found to match.

23.5 Operator/State Map and GKPW Construction

Having been described what the equivalence is, we now proceed to describe the mapping of objects between the two sides.

The first observation is that the symmetries of the gravitational $AdS_5 \times S^5$ background and of the $\mathcal{N}=4$ SYM CFT match between the two sides: the $SO(4,2)$ isometry group of AdS_5 becomes the conformal group in (3, 1) flat Minkowski dimensions, and the $SO(6)$ isometry group of S^5 becomes the $SO(6)$ global "R-symmetry" group of the $\mathcal{N}=4$ SYM theory.

Then the fields in AdS_5, in the KK expansion on the compact space S^5, thus in representations of the isometry group $SO(6)$, should correspond to *gauge invariant operators* $\mathcal{O}$ in the SYM, in the same representation of the global symmetry group $SO(6)$. More precisely, KK expanding *scalar* fields as

$$\phi(x,y) = \sum_n \sum_{I_n} \phi^{I_n}_{(n)}(x) Y^{I_n}_{(n)}(y)\ , \tag{23.38}$$

where $Y^{I_n}_{(n)}(y)$ are spherical harmonics on S^5 in the representation I_n of $SO(6)$, with level n, measuring the eigenvalue of $\Box$ on S^5, the fields $\phi^{I_n}_{(n)}(x)$ in AdS_5, of mass m, correspond to the operators $\mathcal{O}^{I_n}_n$, with conformal dimension

$$\Delta = \frac{d}{2} + \sqrt{\frac{d^2}{4} + m^2 R^2} \tag{23.39}$$

and d is the dimension of the CFT, for us $d = 4$. For p-form fields in AdS_{d+1}, we have the relation

$$(\Delta - p)(\Delta + p - d) = m^2 R^2\ , \tag{23.40}$$

solved by

$$\Delta = \frac{d}{2} + \sqrt{\frac{d^2}{4} + m^2 R^2 + p(p-d)}. \tag{23.41}$$

For $m = 0$, we obtain

$$\Delta = \frac{d}{2} + \frac{d-2p}{2} = d - p. \tag{23.42}$$

One important observation is that in AdS space, m^2 need not be just positive as in flat space, which is a condition from normalizability at infinity of localized excitations: $e^{ik\cdot x} = e^{imt+\cdots} = e^{+t\sqrt{-m^2}+\cdots}$ gives $m^2 \geq 0$. In AdS space, normalizability of localized excitations gives the so-called *Breitenlohner–Freedman bound*, which for scalars is

$$m^2 R^2 \geq -\frac{d^2}{4}. \tag{23.43}$$

Among the possible field-operator pairs, two will be of importance to us:

- A *Noether current* J_μ for a global symmetry group G corresponds to a gauge field A_μ for the same symmetry group in the dual gravitational background, just that now it is *local*. From the above general analysis for $p = 1$ (one-form field), we see that a massless gauge field corresponds to a current J_μ of dimension $d - 1$, as it should be. Gauging a global symmetry involves adding a coupling term $\int d^d x J^\mu(x) A_\mu(x)$, so this is a consistent choice.
- An important particular case of the above is the case of the *energy-momentum tensor* $T_{\mu\nu}$, the Noether current for translations, with generator P_μ. The gauge field for them, making locally translation invariant fields, i.e., general coordinate transformation invariant ones, is known to be the metric $g_{\mu\nu}$. Indeed, making a translationally invariant theory into a general coordinate invariant one involves adding a term $-\frac{1}{2}\int d^d x \sqrt{-g} T_{\mu\nu}(x) \delta g^{\mu\nu}(x)$, justifying the mapping.

Having described the map from SYM operators to gravitational states (fields), we should describe how to relate observables on the two sides. This was done by Gubser, Klebanov, Polyakov, and separately by Witten, thus we call it the GKPW construction. Or, since the map described here is the one of Witten, simply the Witten map.

A massless field in AdS space, $\phi(\vec{x}, x_0)$, can be shown to go to a constant at the boundary $x_0 \to 0$, $\phi_0(\vec{x})$, which means that the natural interpretation of $\phi_0(\vec{x})$ is as a source for the boundary operator $\mathcal{O}$ that couples to it. Indeed, we saw above that on the boundary, gauging a symmetry meant adding a term $\int d^d x J^\mu(x) A_\mu(x)$, where A_μ is the source for the current J^μ, and also the boundary value of the field in the dual gravitational background. For massive fields, and other types of fields (general p-forms, fermions, etc.), the analysis is a bit more complicated, but it is similarly done.

In order to define observables in the $\mathcal{N} = 4$ SYM, we are led to define the partition function for the operator $\mathcal{O}$, with a source ϕ_0 for it, which in Euclidean signature is

$$Z_{\mathcal{O}}[\phi_0] = \int \mathcal{D}[\text{SYM fields}] e^{-S_{\text{SYM}} + \int d^d x \mathcal{O}(x)\phi_0(x)}. \tag{23.44}$$

The correlation functions of $\mathcal{O}$ are found from it by differentiation,

$$\langle \mathcal{O}(x_1) \dots \mathcal{O}(x_n) \rangle = \frac{\delta^n}{\delta\phi_0(x_1) \dots \delta\phi_0(x_n)} Z_{\mathcal{O}}[\phi_0] \Big|_{\phi_0 = 0}. \tag{23.45}$$

Since AdS/CFT relates the CFT theory with the AdS one, it is natural to propose that the partition function above is the same as the partition function in AdS space, with boundary value $\phi_0(\vec{x})$ for the fields $\phi(\vec{x}, x_0)$, i.e.,

$$Z_{\mathcal{O}}^{\text{CFT}}[\phi_0] = Z_{\text{string}}^{\text{AdS}}[\phi[\phi_0]]. \tag{23.46}$$

This should be valid generally in the string theory case, but in particular in the supergravity limit of string theory (for $\alpha' \to 0$, $g_s \to 0$), the string partition function should become just the exponent of the supergravity action, on-shell for the fields ϕ as a function of their boundary values ϕ_0, thus for $\phi = \phi[\phi_0]$, that is,

$$Z_{\text{string}}^{\text{AdS}}[\phi[\phi_0]] \to e^{-S_{\text{supergravity}}[\phi[\phi_0]]}. \tag{23.47}$$

Finally then, the Witten map in the supergravity limit is

$$Z_{\mathcal{O}}^{\text{CFT}}[\phi_0] = Z_{\text{string}}^{\text{AdS}}[\phi[\phi_0]] = e^{-S_{\text{supergravity}}[\phi[\phi_0]]}. \tag{23.48}$$

23.6 Generalization to Non-conformal Theories: Gauge/Gravity Duality

We will be interested in a somewhat more general set-up than the one of $AdS_5 \times S^5$ versus $\mathcal{N} = 4$ SYM, which have conformal symmetry and maximal supersymmetry, so we must generalize for other cases.

First, we note that there are other cases, for other dimensions, with conformal symmetry and (near) maximal supersymmetry: the "ABJM model", a 2+1 dimensional model with $\mathcal{N} = 6$ supersymmetry and an $SU(N) \times SU(N)$ gauge group versus string theory in $AdS_4 \times \mathbb{CP}^3$; the "(0, 2) theory" (named after the amounts of supersymmetry) in 6 dimensions versus M-theory in $AdS_4 \times S^7$ background. But then it was realized that we can consider more general cases, with either less supersymmetry, or with no conformal symmetry, and still a version of the correspondence is valid, under the new name of the "gauge/gravity duality", the gravitational background being called the "gravity dual".

Some relevant properties of these gravity duals that will be of use are:

(1) The gauge theory lives at the boundary of the gravity dual, usually reduced on a compact space X_d, giving a holographic duality, between a $d + 1$ dimensional gravity theory and a d dimensional gauge theory.

(2) The map is a duality, relating a strongly coupled theory on one side of the duality with a weakly coupled one on the other. In particular, we must have a large N (rank of the gauge group) gauge theory, corresponding to a $g_s \to 0$ theory, and a large $\lambda = g_{YM}^2 N$ theory, corresponding to an $\alpha' \to 0$ limit.

(3) The symmetries must match, so global symmetries of the field theory match the local symmetries in the gravity dual, which usually come, via KK expansion, from the global symmetries of the compact space.

(4) The energy scale in the field theory is $U = r/\alpha'$, where r is the holographic coordinate in the gravity dual. Therefore one geometrizes motion in energy (RG flow) as motion in an extra coordinate.

(5) Supergravity fields in the gravity dual correspond to gauge invariant field theory operators made up from adjoint fields, i.e., "glueball" type operators.

(6) The mass spectrum of a tower of glueballs associated with an operator $\mathcal{O}$ correspond to the mass spectrum of the wave equation for the supergravity field ϕ coupling to $\mathcal{O}$ in the gravity dual.

Important Concepts to Remember

- Anti-de Sitter (AdS) space is a maximally symmetric, homogenous and isotropic space of constant negative curvature, with negative cosmological constant.
- Lobachevsky space is the Euclidean signature space equivalent to the sphere, but embedded in Minkowski space. de Sitter space is the Minkowskian signature equivalent of the sphere, and Anti-de Sitter space is the Minkowskian signature equivalent of Lobachevsky space.
- AdS space has a Poincaré metric, with boundary Poincaré symmetry, that only covers a Poincaré patch, and a global metric, that covers the whole space.

- The boundary of AdS space, in both Poincaré and global metrics, is a finite time away (by a light signal).
- The boundary of Poincaré AdS is flat $\mathbb{R}^4$, and of global AdS is $S^3 \times \mathbb{R}_t$, which are related by a conformal factor in Euclidean signature.
- Conformal transformations are transformations of flat space that change the metric by a position-dependent scale factor, $d\vec{x}'^2 = \Omega^2(x)d\vec{x}^2$. In 2 dimensions, they form an infinite group, but in $d > 2$ Minkowski space, they form the group $SO(d, 2)$.
- The first observation leading to AdS/CFT is the holographic property of AdS space, and the fact that the symmetry group of AdS_{d+1}, $SO(d, 2)$, is the same as the conformal group in d dimensions.
- We derive AdS/CFT heuristically, by considering two points of view for D-branes: as the endpoints of open strings, and as gravitational objects that curve space.
- In the decoupling limit $\alpha' \to 0$ and $r \to 0$, we obtain the equivalence of $\mathcal{N} = 4$ SYM in 3+1 dimensions, with the gauge group $SU(N)$, to string theory in $AdS_5 \times S^5$ background, with N the charge in the background, and $4\pi g_s = g_{YM}^2$.
- The general proposal, found until now to be true, is that the equality is valid at all N and g_{YM}, but initially it was thought that it is true only for small g_s and α', that is, in the classical supergravity limit, corresponding to $N \to \infty$, $g_s \to 0$, with $\lambda = g_{YM}^2 N \gg 1$ and fixed. The correspondence is $\alpha'/R^2 = 1/\sqrt{\lambda}$ and $4\pi g_s = \lambda/N$.
- Gauge invariant operators in $\mathcal{N} = 4$ SYM correspond to fields in AdS (KK dimensionally reduced on S^5), with the same symmetry, and dimension Δ related to the mass squared $m^2 R^2$ of the dual field.
- A Noether current corresponds to a gauge field for the same symmetry group, and the energy-momentum tensor to the gravity field (metric).
- The CFT path integral for operators $\mathcal{O}$ with sources ϕ_0 corresponds to the string path integral with boundary value ϕ_0 for the field ϕ dual to $\mathcal{O}$, $Z_{\mathcal{O}}[\phi_0] = Z_{\text{string}}^{\text{AdS}}[\phi_0]$. In the supergravity limit, we obtain $e^{-S_{\text{sugra}}[\phi[\phi_0]]}$. Correlators are obtained by differentiation with respect to ϕ_0.
- AdS/CFT is generalized to gauge/gravity duality, for dual pairs of a field theory and a gravitational background.

Further reading: The AdS/CFT correspondence was defined by Maldacena in [24]. The prescription for calculating correlation functions was defined by Gubser, Klebanov and Polyakov in [25] and by Witten in [26]. For a review of the AdS/CFT correspondence, see the books [27, 28].

Exercises

(1) Show that the coordinate transformation

$$\begin{aligned} X_0 &= R\cosh\rho\cos\theta \\ X_i &= R\sinh\rho\Omega_i \\ X_{d+1} &= R\cosh\rho\sin\tau \end{aligned} \tag{23.49}$$

is an embedding of the AdS space in global coordinates (23.12) inside the flat space embedding in signature $(d-1, 2)$.
(2) Show that the coordinate transformation $x'_\mu = x_\mu + v_\mu$ of flat space, with v_μ in (23.22), is a conformal transformation with $\Omega = 1 - \sigma_v(x)$, where $\sigma_v = \partial_\mu v^\mu / d$.
(3) Solve the KG equation $(\Box - m^2)\phi = 0$ in Poincaré coordinates of AdS_{d+1}, near its boundary at $x_0 = 0$, and show that the two independent solutions are $\sim x_0^{2h_\pm}$, with

$$2h_\pm = \frac{d}{2} \pm \sqrt{\frac{d^2}{4} + m^2 R^2}. \tag{23.50}$$

(4) Consider the supergravity solution for N extremal M5-branes,

$$\begin{aligned} ds^2 &= f_5^{-1/3}(r)[-dt^2 + d\vec{x}_5^2] + f_5^{2/3}(r)[dr^2 + r^2 d\Omega_4^2] \\ f_5(r) &= 1 + \frac{\pi N l_P^3}{r^3} \\ F_4 &= *(dt \wedge dx^1 \wedge \cdots \wedge dx^5 \wedge df_5^{-1}). \end{aligned} \tag{23.51}$$

Show that the decoupling limit $r \to 0, l_P \to 0$, with $U^2 \equiv r/l_P^3$ fixed gives an AdS/CFT duality, with an $AdS_7 \times S^4$ metric and constant F_4. Calculate R_{S^4} and R_{AdS_7}.
(5) Consider the classical solution for a scalar ϕ_i in AdS_{d+1} space in Poincaré coordinates, written as a function of the boundary value ϕ_{0i} as

$$\phi_i(z_0, \vec{z}) = \int d^d z K_{B,\Delta_i}(z_0, \vec{z}; \vec{x}) \phi_{0i}(\vec{x}). \tag{23.52}$$

Use the Witten (GKPW) prescription in the classical supergravity limit, in order to show that, if we have 3 scalars ϕ_1, ϕ_2, ϕ_3 with interaction term $\mathcal{L}_{\text{int}} = \lambda \phi_1 \phi_2 \phi_3$, the 3-point correlator for the operators dual to it is

$$\langle \mathcal{O}_1(x_1) \mathcal{O}(x_2) \mathcal{O}(x_3) \rangle = -\lambda \int \frac{d^d z dz_0}{z_0^{d+1}} K_{B,\Delta_1}(z_0, \vec{z}; \vec{x}_1) K_{B,\Delta_2}(z_0, \vec{z}; \vec{x}_2) K_{B,\Delta_3}(z_0, \vec{z}; \vec{x}_3). \tag{23.53}$$

Part III
Introduction to String Cosmology

Chapter 24
Problems of String Inflation

We now move on to the study of string cosmology, i.e., applications of string theory to cosmology. We have to start, of course, with trying to obtain the most popular cosmological model, inflation. But it turns out that strictly speaking (applying the fully top-down mentality of deriving everything from string theory), we have no good model of string inflation yet. That is to say, there are models that when we consider some approximations seem to give inflation, but in the end, corrections not taken into account can spoil the nice features. Therefore in this chapter we will examine the problems that appear when we try to apply string theory to find inflationary models.

The first question to ask is, why do we want to obtain string inflation? Of course, inflation being the most popular cosmological model we need to see whether it needs to be embedded into something else. It turns out that in fact, inflation is quite UV sensitive. We will review in more detail the relevant argument when discussing the Brandenberger-Vafa "string gas cosmology", but the crux of it is that, for a sufficiently large number of e-folds of inflation, the last scales to re-enter the horizon (the CMBR scales) spent a large part of the early history (before going outside the horizon and being frozen-in) being sub-Planckian, hence they must have necessarily been influenced by trans-Planckian physics, i.e., quantum gravity. While this does not happen for models with only 50–60 e-folds of inflation, it points out the UV sensitivity of inflation, hence the need to have it at the very least embedded into a consistent theory of quantum gravity.

Inflation is an effective field theory description, which means one needs to integrate out all modes of the string higher than a cut-off scale. In phenomenological effective field theory, one writes all the operators consistent with symmetries, valid below a cut-off, and puts arbitrary coefficients for them, to be fixed by experiments, but in a top-down approach like the string theory one, one must integrate out higher modes, and obtain the correct coefficients in the effective field theory.

What do we need to have in order to obtain inflation? Inflation was presented as a more general paradigm, for accelerated expansion governed by evolution in a certain field, but the most popular incarnation is in terms of a scalar field that slowly rolls down a flat potential. Therefore, we need to obtain a scalar with a nearly flat potential.

H. Năstase, *Cosmology and String Theory*, Fundamental Theories of Physics 197, https://doi.org/10.1007/978-3-030-15077-8_24

But the problem in string theory is not to obtain such a scalar, but rather that we have too many, coming from the various parameters of compactification. These are *moduli*, i.e., scalars that perturbatively have no potential (it takes no energy to deform the compactification along these directions), but could have a nonperturbative potential.

24.1 Moduli of Type IIB String Compactifications on CY_3

In particular, for a popular concrete example, we will consider type IIB theory compactified on a CY_3 space. We will not describe the most general topological description of the Calabi-Yau spaces. We will just say some things that we need.

The CY_3 space has a number of topologically nontrivial 3-dimensional surfaces, or 3-cycles, the number of independent ones being given by b_3, the "third Betti number", characterizing *homology* (the nontrivial way in which surfaces interesect). A homology basis for 3-cycles is given by (A_I, B^J), where $I, J = 1, \ldots, b_3/2$. We can define the *intersection number of surfaces* as the number of nontrivial "encirclings" of one surface by the other. It is defined as an antisymmetric function of surfaces, and then we have (for the so-called A-cycles and B-cycles)

$$A_I \cap B^J = -B^J \cap A_I = \delta_I^J; \quad A_I \cap A_J = 0 = B^I \cap B^J. \tag{24.1}$$

The basis is unique up to symplectic transformations with integer coefficients acting on the cycles (A_I, B^J), i.e., $Sp(b_3, \mathbb{Z})$ transformations, that leave invariant the intersection numbers.

The homology basis above is dual to s basis of 3-forms on the space, for the *cohomology* group $H^3(X, \mathbb{R})$, given by (α_I, β^J), such that

$$\int_{A_I} \beta^J = \delta_I^J; \quad \text{etc.} \tag{24.2}$$

Complex Structure Moduli

Then the complex structure moduli, or shape moduli (in the case of the square 2-torus T^2, the complex structure modulus is R_2/R_1) are obtained as integrals of the unique holomorphic 3-form Ω on the homology basis (A_I, B^J), i.e.,

$$F_I = \int_{A_I} \Omega; \quad Z^J = \int_{B^J} \Omega. \tag{24.3}$$

However, since the normalization of Ω is irrelevant, we actually have $h^{2,1} = b_3 - 1$ independent complex structure moduli (there are some more nontrivial topological numbers called Hodge numbers $h^{p,q}$ that will not be explained here).

Kähler Structure Moduli

The Kähler structure moduli, or size moduli (in the case of the square 2-torus T^2, the Kähler structure modulus is the volume $R_1 R_2$) are obtained in a similar way, but by integrating a unique 2-form on a basis of 2-cycles.

On a Calabi-Yau space there is a unique covariantly constant spinor η, as we saw in the last chapter, and from it we can construct the Kähler form

$$K_{ij} = \bar{\eta}\Gamma_{ij}\eta, \tag{24.4}$$

and in turn the complex structure

$$J^i{}_j = g^{ik} K_{kj} , \tag{24.5}$$

or sometimes written as a 2-form (since it is antisymmetric)

$$J = g_{i\bar{j}} dz^i \wedge dz^{\bar{j}}. \tag{24.6}$$

But then in string theory we also the NS-NS 2-form field $B_{\mu\nu}$, so we can define the *complexified Kähler class*

$$K = J + iB. \tag{24.7}$$

We can define also the homology of 2-cycles, by finding the basis of topologically nontrivial 2-dimensional surfaces (2-cycles) $(A'_{I'}, B'^{J'})$, $I', J' = 1, \dots, b_2/2$, with orthonormal intersection numbers

$$A'_{I'} \cap B'^{J'} = \delta_{I'}^{J'}, \quad A'_{I'} \cap A'_{J'} = 0 = B'^{I'} \cap B'^{J'}. \tag{24.8}$$

Here there are $b_2 = h^{1,1}$ 2-cycles, where b_2 is the second Betti number (and $h^{1,1}$ is a Hodge number).

Then as before, the Kähler structure moduli are given by the integrals of the 2-form K over the basis of 2-cycles,

$$X_{I'} = \int_{A_{I'}} K, \quad X^{J'} = \int_{B^{J'}} K. \tag{24.9}$$

These 2 types of modulis exist for whatever CY_3 construction, but in more general cases we can have other types of moduli as well, like:

- **D-brane moduli**, coming from the fields on the worldvolume of D-branes. The most obvious one is the position X_0 of the D-brane in the transverse space, but in general there could be others as well. For instance, a D-brane can wrap a variable cycle in the geometry, parametrized by some scalar field.
- **axions**, coming from fluxes of antisymmetric tensor fields on cycles in the geometry.
- there can be also other moduli in more complicated cases.

Example

One interesting example is provided by the type IIB flux compactification, with G-flux.

We define the complex scalar made up from the RR axionic scalar a and the dilaton ϕ,

$$\tau = a + ie^{-\phi}. \tag{24.10}$$

The G-flux refers then to the combination of the NS-NS 3-form field strength H^{NS-NS} and the R-R 3-form field strength F^{R-R},

$$G = F^{R-R} - \tau H^{NS-NS}. \tag{24.11}$$

In the case that the G-flux is nontrivial on cycles, we have the GVW superpotential

$$W = \int_{CY_3} \Omega \wedge G\,, \tag{24.12}$$

where Ω is the holomorphic 3-form.

We see that we have very many scalar moduli in a string compactification, in fact *too many*. In fact, we have not observed a single light scalar in the real world, which means that we need to stabilize the string moduli, by a nonperturbative (super) potential.

Since today we don't see any light scalars, it means that today *all* the moduli need to be stabilized. There is one possibility around that will be explored towards the end of the book, in the form of so-called chameleon scalars. They arose out of the observation that in fact, we only know that there are no light scalars here on Earth (where we can perform fifth force search experiments, and see that there is none), or in the Solar System (the motion of the planets follows from Newtonian theory, without the need for a fifth force). But we could still have a scalar (like the chameleon) that is light on cosmological scales and massive on Earth and Solar System scales.

But at early stages, we still want one of the scalars to be not stabilized, namely the inflaton. In fact, in the usual picture of new inflation, we would have a flat plateau, and after a drop, we have a minimum of large mass, stabilizing it now.

However, note that this presents a problem, because we want to have all the moduli stabilized, and moreover stabilized at zero potential (no cosmological constant), whereas without the stabilization, the extra moduli would have no potential, that means that we should extend the inflaton potential trivially in all extra moduli directions. But that means that generically the quantum corrections must introduce steep drops (followed by stabilization) in transverse directions, invalidating the flatness of the potential needed for inflation, since the evolution in field space is along the steepest path.

To check whether the above happens, or the potential continues to have flat valleys, we would need to have the full nonperturbative moduli potential (since stabilization is nonperturbative), which would in effect mean that we have to solve the full nonperturbative string theory in some sector, which is clearly impossible. That is why we don't have a good model of string inflation yet: to stabilize moduli we need nonperturbative corrections, so g_s is large(ish), but then we must consider *all* possible corrections, since without it we are not sure whether we are in a true valley, or on top of a hill with a flat narrow top.

24.2 Approximations for String Calculations of the Potential

In doing actual calculations for the potential, one usually does a combination of approximations. The most important approximations, well established in the literature over many years, are as follows.

(1) The *α' expansion,* needed in order to have a perturbative string worldsheet. In this context, it is an expansion in

$$\alpha'/\mathcal{R}^2 \ll 1 , \qquad (24.13)$$

where $\mathcal{R}$ is the scale of variation of the various background fields. It is usually hard to satisfy.

(2) The *string loop expansion,* in $g_s \ll 1$, in order not to have the full nonperturbative quantum string theory. However, as said above, we need to at least partially contradict it for moduli stabilization. This presents a problem, since we must invoke some nonperturbative contributions that stabilize the moduli, yet we cannot calculate the full nonperturbative potential.

(3) The *probe approximation* for D-branes. When treating D-branes, usually one considers them as probes, i.e., they do not backreact on the background. They are only fixed at some point, or wrapping some cycles, in the compact manifold. But often times, the approximation is, or must be, violated.

(4) An opposite approximation is also possible, considering a *large charge approximation,* $N \gg 1$, for the number of branes, which means that the branes drastically curve the background, and one usually ends up in a near-horizon geometry, like in the case of large N number of D3-branes, when we obtain the $AdS_5 \times S^5$ geometry.

(5) The *smeared approximation.* One other possibility to deal with branes, in the case that it is difficult to find the dependence on one of the coordinates, x, is to consider the brane charge to be uniformly distributed, or *smeared* in that direction, such that the fields are independent of x.

(6) The *adiabatic approximation* amounts to a slow motion of branes, done such as not to change the background.

(7) More of a subset of number 6 is the *moduli space approximation*, a general procedure in the case of solitonic objects. When a static configuration of solitons (or here, D-branes) is stable (has no potential), it means that the various positions are moduli (this can be generalized to other kinds of moduli). But if we give a small (relative) velocity to the moduli (i.e., make them time dependent), we don't

have anymore a solution. Instead, one must re-solve the equations of motion and introduce first order corrections in the velocities (time dependence), and one can find an effective Lagrangian on the moduli space, of the type

$$L = g_{ij}\dot{X}^i\dot{X}^j - V(X^k) , \tag{24.14}$$

where X^i are the moduli and g_{ij} is the moduli space metric.

(8) *Mode truncation* (for instance, truncation of KK modes).

What usually happens is that at least one of the above approximations contradicts the needed regime of parameters, so we cannot trust the result.

24.3 Eta Problem in String Theory

The problem is that the stabilization of moduli doesn't decouple from inflation. One considers a model of "string inflation" that works, in the absence of moduli stabilization, but after stabilizing the moduli, we create terms that spoil the flatness of the potential. The "eta problem" is the fact that the corrections introduced this way, or perhaps by quantum corrections (see later) spoil the smallness of the inflationary parameter η (which needs to be $\ll 1$ for slow roll).

Example

Consider open strings stretched between two sets of branes, D-brane sector A and sector B. As we already said, to obtain the effective field theory of inflation, we must integrate out the string modes above a cut-off. Integrating out the open string, of mass M_{AB}, we obtain terms of the type

$$\frac{\mathcal{O}_A^{(\Delta_1)}\mathcal{O}_B^{(\Delta_2)}}{M_{AB}^{\Delta_1+\Delta_2-4}} , \tag{24.15}$$

such as to give a dimension 4 operator. Here $\mathcal{O}_A$ is an operator for fields in the D-brane sector A, of dimension Δ_1 and $\mathcal{O}_B$ in D-brane sector B, of dimension Δ_2.

For instance, consider $\mathcal{O}_A$ to be a constant (independent of the inflaton, rather) V_0, and $\mathcal{O}_B$ to be a mass term for the inflaton, ϕ^2. Since $M_{AB} = \Delta X/(2\pi\alpha')$, and if $\Delta X < \sqrt{\alpha'}$ we need string corrections, we must instead consider $\Delta x > \sqrt{\alpha'}$, which means that $M_{AB} \sim M_{\rm Pl}$ is natural, leading to a term in the effective field theory of type

$$\sim \mathcal{O}_4 \frac{\phi^2}{M_{\rm Pl}^2} . \tag{24.16}$$

Then then η parameter of inflation is

$$\eta \equiv M_{\rm Pl}^2 \frac{V''}{V} \sim \frac{M_{\rm Pl}^2}{\phi^2} . \tag{24.17}$$

This is generically > 1, but moreover, usually one has also independently $\mathcal{O}_4$, so that in total

$$\mathcal{O}_4\left(1+\frac{\phi^2}{M_{\rm Pl}^2}\right), \tag{24.18}$$

which gives $\eta = 1$, or more generally $\eta \sim \mathcal{O}(1)$.

This is the eta problem, at the classical level (from the point of view of the effective field theory). However, there are also (perhaps even more importantly) quantum corrections, that will be analyzed later.

24.4 Energy Scales for Most Controlled Approximations

In practice, we need to have some sort of control over the approximations made in order to calculate (so that we can trust them). In that case, several energy scales appear, in a particular order, that I will explain in the following.

- In general, we have the Planck scale larger than the string scale, $M_{\rm Pl} > M_s = 1/\sqrt{\alpha'}$. This is generic for *perturbative* string theory, i.e., with $g_s \ll 1$, which as we said is needed for the absence of large quantum string corrections. The 4 dimensional Planck scale appears from the coefficient of the Einstein action in 4 dimensions, $M_{\rm Pl}^2/2$. But that arises from KK dimensional reduction from 10 dimensions of the same term, with coefficient $1/(g_s^2 2\kappa_{N,10}^2)$. The $1/g_s^2$ is there since this is a closed string action, and so must have this overall coefficient (g_s is the closed string coupling), and the $1/2\kappa_{N,10}^2$ is the standard gravity coupling, which by dimensional analysis must equal $M_s^8/2$. KK dimensional reduction means that we get the volume of the 6 dimensional compact space $V_{\rm compact}^{(6)} = M_{\rm KK}^{-6}$ as a prefactor. Here we have defined a KK scale (scale of masses of KK modes) by analogy with the one-dimensional (circle) case, where $R = 1/M_{\rm KK}$. Finally, that gives

$$M_{\rm Pl}^2 = \frac{1}{\kappa_{N,10}^2}\frac{1}{g_s^2}V_{\rm compact}^{(6)} \Rightarrow M_{\rm Pl} = \frac{M_s}{g_s}\left(\frac{M_s}{M_{KK}}\right)^3. \tag{24.19}$$

  We see that $g_s \ll 1$ and $M_s \sim M_{\rm KK}$ indeed implies $M_s \ll M_{\rm Pl}$. In a natural compactification, indeed $M_{\rm KK} \sim M_s$ (since the compact space would become compact due to string effects). But if one tries to obtain large extra dimensions, we would have $R \gg \sqrt{\alpha'}$, or $M_s \gg M_{\rm KK}$, which would mean that M_s is even smaller than $M_{\rm Pl}$. The only issue is whether we can have $M_{\rm KK} > M_s$, but that is both unlikely in a general quantum gravity scenario, and also in particular in string theory, we have T-duality equating large and small volume physics, so in that case we can take it as excluding an equivalent scenario.
- Therefore we have also $M_s > M_{\rm KK}$, or $V_{\rm compact}^{(6)} \gg M_s^{-6} = \alpha'^3$. Note that this large volume in string units will also imply small α' corrections, so it is actually a needed condition in order to avoid string worldsheet corrections.

- The energy scale of inflation is the Hubble scale H_i during inflation. For the validity of inflation as an effective field theory, we must have the cut-off scale be larger. In this case, the cut-off refers to $M_{\rm KK}$, beyond which we must consider the whole infinite tower of KK modes, or equivalently the full 10 dimensional theory. Therefore we must have $H_i < M_{\rm KK}$.
- Finally, we usually need to use supersymmetry during inflation, if we are to trust quantum corrections calculated (without susy, there is not much control over quantum corrections). But that means that we must have the scale of susy breaking in 4 dimensions (of the breaking of the $\mathcal{N} = 1$ susy to MSSM) be less than the scale of inflation, $M_{\rm susy\ br.} < H_i$.

All in all, we have the needed hierarchy

$$M_{\rm susy\ br.} < H_i < M_{\rm KK} < M_s < M_{\rm Pl}. \tag{24.20}$$

Quantum Eta Problem Redux

We now can explain the *quantum eta problem* better, as something akin to the hierarchy problem in the Standard Model, namely that the mass of some scalar (in the case of the Standard Model, the Higgs, in the case of inflation, the inflaton) is naively taken by quantum corrections in a generic effective field theory to the scale of the UV cut-off, $\Lambda_{\rm cut-off}$. For that *not* to happen, we need something drastic like supersymmetry that prevents various quantum corrections.

But the effective field theory of inflation is usually only valid in a very small window, since $H_i \in (M_{\rm susy\ br.}, M_{\rm KK})$, and as we saw, at $M_{\rm KK}$ we break most of the susy, being left with the $\mathcal{N} = 1$ needed for phenomenology. Yet nevertheless, one obtains usually susy breaking of the remaining $\mathcal{N} = 1$ at a scale smaller, but not much smaller than $M_{\rm KK}$. That means that we have

$$\Lambda_{\rm cut-off} \gtrsim H_i \ , \tag{24.21}$$

which in turn means that the quantum corrections for the inflaton mass (loops of other fields inserted into the inflaton propagator) give a mass at this cut-off scale, i.e.

$$\Delta m_\phi \sim H. \tag{24.22}$$

Since moreover the constant part of the potential $\mathcal{O}_4 = V_0$ defines the Hubble scale during inflation by $V_0 = 3M_{\rm Pl}^2 H_i^2$, we have the inflaton potential

$$V \sim \mathcal{O}_4 + \Delta m_\phi^2 \frac{\phi^2}{2} = 3M_{\rm Pl}^2 H_i^2 + \Delta m_\phi^2 \frac{\phi^2}{2} \ , \tag{24.23}$$

leading to

$$\Delta\eta = \frac{\Delta m_\phi^2}{3H^2} \sim 1. \tag{24.24}$$

Again, to see whether or not this happens at the quantum level (in the effective field theory of inflation, itself obtain by integrating out massive modes beyond $\Lambda_{\text{cut-off}}$), we would need to have the *full* quantum effective potential. But getting that would be equivalent to solve the full nonperturbative string theory, an impossible task. We would have to be very lucky to have a model where we could calculate everything in a controlled approximation.

24.5 Large Field Inflation

Another issue that can appear in string theory models for inflation is a generic one to most approaches, related to the Lyth bound, found in Chap. 11 (of part I). We saw then that, for a tensor to scalar ratio of perturbations r and a number of e-folds of inflation N_e, we have a minimum inflaton field excursion

$$\Delta\phi_{\Delta N_e} \geq M_{\text{Pl}}\sqrt{r}\frac{\Delta N_e}{\sqrt{8}}. \tag{24.25}$$

This would become a problem if we see a large(ish) r (as of this moment, we have not observed tensor modes, only scalar modes), since it would imply that we need to have a $\Delta\phi$ that is trans-Planckian.

The first problem with that is that it might not be kinematically allowed. In fact, a typical string compactification (with string scale extra dimensions) usually doesn't allow it.

In something like large extra dimensions this would be allowed. In fact, large extra dimensions means that the scalar representing the volume in string units *is* (by definition) trans-Planckian. But some models are extremely constrained (see for instance my papers with Khoury and Hinterbichler, and Khoury, Hinterbichler and Rosenfeld), to the point that it is not clear that there is a good model compatible with experiments.

But moreover, even if one has a kinematically allowed trans-Planckian $\Delta\phi$, that would mean that we need to control dynamically the flatness of the inflaton over these large excursions. But in a generic effective field theory, we would expect corrections to the operator of dimension four (the potential) $\mathcal{O}_4$, of the order of

$$\mathcal{O}_4\left(\frac{\phi}{M_{\text{Pl}}}\right)^n, \tag{24.26}$$

which would naively blow up. So unless they are suppressed by unknown quantum gravity (string theory) mechanisms, these would make any calculation of the potential unreliable.

One set of models that has become popular recently (to be considered later on in the book) is called *axion monodromy* inflation, and is the only known consistent string model with trans-Planckian field excursion where it seems like the quantum corrections are under control.

Important Concepts to Remember

- For string inflation one of the scalar moduli of string theory is the inflaton.
- In string theory, we have complex structure moduli, Kähler moduli, D-brane moduli, axions, etc. Any of them could be the inflaton, with the other ones stabilized.
- But that means that we need an almost flat direction, that cannot have a steep drop in any transverse direction, which is what would generically happen when we stabilize the transverse directions.
- Among approximations that one must use in string theory in order to be able to calculate the potential, we have the α' and g_s string expansions, either the probe or large charge approximation for branes, the smeared approximation, the adiabatic approximation, the moduli space approximation, and KK mode truncations.
- The eta problem in string theory is that quantum corrections to η are of order one, since generic quantum corrections to the potential are of order one in Planck units. They also appear when we try to stabilize the moduli (which is, necessarily, nonperturbative).
- In order to control the appoximations, we need the hierarchy of scales $M_{\text{susybr.}} < H_i < M_{\text{KK}} < M_s < M_{\text{Pl}}$.
- The cut-off scale for quantum corrections must obey $\Lambda_{\text{cut-off}} \gtrsim H_i$, leading to $\Delta m_\phi \sim H$, so $\delta\eta \sim 1$.
- The Lyth bound says that if the tensor to scalar ratio of fluctuations r is found to be large, and ΔN_e is also large, we will have a large field excursion $\Delta\phi$.
- That is hard to obtain in string models, and also can generate large quantum corrections, leading to the η problem, so we would need to explain why that doesn't happen.

Further reading: See the string inflation reviews [29, 30], as well as Sects. 4.1, 4.2 in [31] (arXiv review containing part of the book [32]).

Exercises

(1) Consider an $\mathcal{N} = 1$ susy model with chiral superfield Φ, canonical Kähler potential $K = \Phi^\dagger\Phi$ and superpotential $W = A\Phi^3$ coupled to supergravity, with $|A| \ll 1$. Does it have a *classical* η problem?
(2) Assuming that we see an r (tensor to scalar ratio of perturbations) at the current limit, $r \simeq 0.10$, and we have 60 e-folds of inflation, what is the reduction needed from *generic* quantum corrections to operators $\mathcal{O}_4$?
(3) Consider compactification on the CY_3 space $(T_2)^{\otimes 3}$ (a product of 3 tori), and no D-branes or fluxes. How many moduli there are?
(4) Consider an axion, which is a scalar field with a potential of the type

$$V(\phi) = V_0[1 - \cos(\phi/f)]. \tag{24.27}$$

Does it have a classical eta problem? If we see a tensor to scalar perturbation ratio r, what is the bound on the constant f such that we don't have transplanckian (large field) quantum corrections?

Chapter 25
Problems of the Supergravity Approximation to String Inflation

Because string theory itself is not very calculable, there are few truly stringy models of cosmology. Therefore, one mostly considers the low energy limit of string theory, supergravity, as an effective field theory, and in this chapter we want to note the problems we encounter when we try to obtain inflation in this effective field theory. As usual, for phenomenological reasons we only consider $\mathcal{N} = 1$ supersymmetry, considering that the rest has been lost at a high scale. In $\mathcal{N} = 1$ supersymmetry, besides the supergravity multiplet, we have the chiral multiplet Φ and the vector multiplet V^a.

25.1 Supergravity Plus Chiral Multiplet

We start by the most standard case, of $\mathcal{N} = 1$ supergravity coupled to a scalar (chiral) multiplet Φ. As we saw in part II, the potential for supergravity coupled to chiral superfields Φ^i is

$$V = e^{K/M_{\rm Pl}^2}\left[g^{i\bar{j}} D_i W \overline{D_j W} - \frac{3}{M_{\rm Pl}^2}|W|^2\right]. \tag{25.1}$$

Here $K(\Phi^i, \bar{\Phi}^i)$ is the Kähler potential, giving the kinetic term, and $W(\Phi^i)$ is the superpotential, which is a holomorphic function of its arguments, and the covariant derivative is

$$D_i W = \partial_i W + \frac{\partial_i K}{M_{\rm Pl}^2} W. \tag{25.2}$$

In the following, we will only consider a single chiral superfield Φ.

The simplest assumption is to consider the Kähler potential to be Taylor expandable around zero, and write

$$K(\Phi, \bar{\Phi}) = K_0 + \partial_\Phi \partial_{\bar{\Phi}} K|_0 \Phi\bar{\Phi} + \cdots \tag{25.3}$$

H. Năstase, *Cosmology and String Theory*, Fundamental Theories of Physics 197, https://doi.org/10.1007/978-3-030-15077-8_25

Note that this expansion is not the most general one. We have ignored the possible linear term $\Phi + \bar{\Phi}$ in the expansion, on the grounds that it gives zero metric on field space, but in principle it can be there, and we will see at the end that an important special case has such a nonzero term.

With the above Kähler potential, we obtain the scalar kinetic term

$$-(\partial_\phi \partial_{\bar{\phi}} K)\partial_\mu \phi \partial^\mu \bar{\phi}\,, \tag{25.4}$$

and the scalar potential has the terms

$$V = V_0 \left(1 + g_{\phi\bar{\phi}} \frac{\phi\bar{\phi}}{M_{\rm Pl}^2} + \cdots \right) + \cdots\,, \tag{25.5}$$

coming from the exponential prefactor $e^{K/M_{\rm Pl}^2}$ independently of the superpotential W. But these terms by themselves will give a $\Delta\eta = g_{\phi\bar{\phi}} \sim \mathcal{O}(1)$, i.e., a *superpotential independent* order one contribution to η, as we can easily see. Here

$$V_0 = e^{K_0} \left[(\partial_\phi \partial_{\bar{\phi}} K)^{-1} D_\phi W \overline{D_\phi W} - \frac{3}{M_{\rm Pl}^2} |W|^2 \right] \Bigg|_{\phi=0}\,. \tag{25.6}$$

But moreover, in the low energy limit of string theory (supergravity limit), it is natural to assume that we get a superpotential that, if it can be Taylor expanded, gives

$$W = M_{\rm Pl}^3 \sum_p c_p \left(\frac{\phi}{M_{\rm Pl}} \right)^p\,, \tag{25.7}$$

with $c_p \sim \mathcal{O}(1)$. That in turn, will generically give also contributions of order one to η, at least if we have, together with the first nonzero coefficient, c_p, also the coefficient c_{p+2} to be nonzero (if c_{p+2} is zero, we get a field-dependent η, which can be made to be small; if not, we get a constant that is what it is). Of course, these will interfere with each other in the potential V and with the above generic contribution coming from K, so it could be that they cancel the total contribution to η, but this would then involve some *fine tuning*.

This problem could perhaps be circumvented if there is no simple Taylor expansion for K. For instance, for the volume modulus ρ of the CY_3, a complex Kähler modulus combining the volume and another scalar, the Kähler potential is

$$K(\rho, \bar{\rho}) = -3M_{\rm Pl}^2 \ln[\rho + \bar{\rho}]. \tag{25.8}$$

Then indeed, there is no Taylor expansion around zero. Moreover, if we introduce D3-branes in the CY_3, at positions z_α, $\alpha = 1, 2, 3$ in CY_3, we obtain a modification inside the log,

$$K = -3 \ln[\rho + \bar{\rho} - \Delta k(z_\alpha, \bar{z}_\alpha)]\,, \tag{25.9}$$

and we will see later that in a popular model (the KKLMMT model), the position of a D3-brane in the compact space acts as the inflation.

However, as we have said previous chapter, we actually need to take care to stabilize the remaining moduli, so ρ must in particular be stabilized, by the inclusion of a nonperturbative superpotential. That means that ρ gains a large mass around its minimum. But then from K, this will imply a mass for z_α as well.

This is an example of the eta problem in string theory: we have a good inflationary model, but when we remember to stabilize the non-inflaton moduli, we generate too large a mass for the inflaton, destroying inflation.

One could argue that we need some softly broken symmetry to guarantee the near-flatness of the potential. The obvious one is a shift symmetry,

$$\phi \to \phi + \text{const.} \,, \tag{25.10}$$

softly broken by small power law terms in the potential, i.e., with Lagrangian

$$\mathcal{L} = -\frac{1}{2}(\partial_\mu \phi)^2 - \lambda_p \frac{\phi^p}{M_{\rm Pl}^{p-4}} \,, \tag{25.11}$$

with $\lambda_p \ll 1$. Then at $\lambda_p = 0$ we have an exact shift symmetry, but the potential breaks it softly. However, note that this is really parametrizing what we want to obtain, not really "explaining" anything, so one needs a deeper reason behind the smallness of the coefficients.

25.2 D-Term Inflation

We now move on to the other possibility, of having also an abelian vector superfield V^a, with auxiliary field D^a. Here a is an index in the adjoint representation of the gauge group. The scalar potential in this case is given by solving for the auxiliary field, and replacing it in the "D-term" potential, which for a standard YM kinetic term $-(1/4g^2)F^a_{\mu\nu}F^{a\mu\nu}$ is

$$V_D = \frac{g^2 D^a D^a}{2} \,, \tag{25.12}$$

that is to be added to the F-term considered previously (in rigid susy, the potential for the chiral superfield is given by $F^i F^i/2$, with F^i the auxiliary field in the chiral multiplet).

If we have a coupling of the vector and chiral multiplets Φ^i, with a canonical kinetic term for Φ^i, by solving the equation of motion of the auxiliary field D^a we get

$$D^a = \Phi^\dagger T^a \Phi \equiv \Phi^{\dagger i}(T^a)_i{}^j \Phi_j . \tag{25.13}$$

More generally, for a general gauge kinetic term $-(1/4g_a^2)f_a(\phi)F^a_{\mu\nu}F^{a\mu\nu}$, we obtain the D-term in the potential

$$V_D = \sum_a [\text{Re} f_a(\phi)]^{-1} \frac{g_a^2 D_a^2}{2} , \tag{25.14}$$

and the auxiliary field solution is

$$D^a = \Phi_i (T_a)^i{}_j \frac{\partial K}{\partial \Phi_j}. \tag{25.15}$$

We can also add to the Lagrangian the *Fayet-Iliopoulos (FI) term* defined by a FI parameter ξ^a,

$$\int d^2\theta d^2\bar{\theta} \xi^a V^a , \tag{25.16}$$

which effectively replaces

$$D^a \to D^a + \xi^a \tag{25.17}$$

in the D-term. The presence of the FI term means that we can have a spontaneous breaking of susy, by having a nonzero ground state energy (in a susy theory the energy of the -supersymmetric- vacuum must be zero, since the susy algebra contains $\{Q, Q\} \sim H + \cdots$, and so $Q|\psi\rangle = 0$ implies $H|\psi\rangle = 0$), or at least a nonzero local extremum of the potential.

Example: Fayet-Iliopoulos Model

The original FI model provides a simple example. Consider an Abelian vector multiplet, coupled with two chiral multiplets of $U(1)$ charges ± 1, and with superpotential

$$W = m\Phi_+\Phi_-. \tag{25.18}$$

For standard kinetic terms, we obtain the potential

$$V = m^2|\phi_+|^2 + m^2|\phi_-|^2 + (\xi + e^2|\phi_+|^2 - e^2|\phi_-|^2)^2. \tag{25.19}$$

From its equations of motion, we find that the only extremum is at $\phi_+ = \phi_- = 0$, which gives $V_0 = \xi^2 \neq 0$. That means that we have spontaneous susy breaking, but since the VEVs of the scalars vanish, we don't have a Higgs mechanism, and the $U(1)$ is unbroken.

One can in fact generalize this mechanism to inflation, and consider the $\mathcal{N} = 1$ supergravity multiplet coupled to the vector multiplet and 3 chiral multiplets, a neutral one S, and the two charges ones $\Phi_\pm$, with superpotential

$$W = S\Phi_+\Phi_- , \tag{25.20}$$

and Kähler potential

$$K = |S|^2 + |\Phi_+|^2 + |\Phi_-|^2. \tag{25.21}$$

However one finds that one cannot fix all the constraints on the model. It is left as an exercise to find the potential and its equations of motion.

25.3 Field Redefinitions

We saw that we have only a restricted set of possibilities that give inflation. We willl restrict that further, but here we note that in fact we also have some equivalences. In fact, there is a transformation of the Kähler potential, called Kähler transformations, that preserves the metric on the field space $g_{\phi\bar{\phi}} = \partial_\phi \partial_{\bar{\phi}} K$,

$$K \to K + f_1(\phi) + f_2(\bar{\phi}) , \tag{25.22}$$

where f_1 and f_2 are arbitrary functions. If we want to maintain K real, we need to have $f_1 = f_2 \equiv f$, but it is not really needed, we only need observables to be real. In rigid susy, this transformation of K is a symmetry of the scalar potential, however in rigid symmetry we have the $e^{K/M_{\rm Pl}^2}$ prefactor. We can in fact transform also the superpotential, in the real case $f_1 = f_2 = f$, by

$$W \to e^{-f/M_{\rm Pl}^2} W , \tag{25.23}$$

such that the scalar potential is still invariant. These redefinitions, being actions on the supersymmetric objects K and W, preserve the supersymmetry of the theory.

We can also make redefinitions of just the scalar fields in the multilplets, for instance by going to a canonical kinetic term for the scalars,

$$- g_{i\bar{j}} \partial_\mu \phi^i \partial^\mu \bar{\phi}^{\bar{j}} \equiv -\partial_\mu \tilde{\phi}^i \partial^\mu \bar{\tilde{\phi}}^j , \tag{25.24}$$

but this transformation doesn't extend at the level of K and W, so this is a non-supersymmetric transformation (it is actually not a transformation of the superfield Φ either).

25.4 Example with $\epsilon, \eta \ll 1$

Until now we have shown various things that don't work, but we should also show something that does work, and one can have $\epsilon, \eta \ll 1$. Consider a canonical Kähler potential

$$K = \bar{\Phi}\Phi \tag{25.25}$$

and the simplest nontrivial superpotential, a linear term

$$W = A\Phi. \tag{25.26}$$

Then the scalar field metric is trivial, $g_{\phi\bar{\phi}} = 1$, and we obtain

$$DW = a\left(1 + \frac{\bar{\Phi}\Phi}{M_{\rm Pl}^2}\right). \tag{25.27}$$

The scalar potential is therefore

$$\begin{aligned} V &= e^{\frac{\bar{\phi}\phi}{M_{\rm Pl}^2}}\left[|a|^2\left|1 + \frac{\bar{\phi}\phi}{M_{\rm Pl}^2}\right|^2 - \frac{3|a|^2}{M_{\rm Pl}^2}|\phi|^2\right] \\ &= |a|^2 e^{\frac{\bar{\phi}\phi}{M_{\rm Pl}^2}}\left[1 - \frac{|\phi|^2}{M_{\rm Pl}^2} + \frac{|\phi|^4}{M_{\rm Pl}^2}\right]. \end{aligned} \tag{25.28}$$

Considering small fields, $|\phi| \ll M_{\rm Pl}$, the mass term for ϕ cancels between the bracket and the prefactor coming from K, and we obtain

$$V \simeq |a|^2\left[1 + \frac{1}{2}\frac{|\phi|^4}{M_{\rm Pl}^4} + \mathcal{O}\left(\frac{|\phi|^6}{M_{\rm Pl}^6}\right)\right]. \tag{25.29}$$

Considering the inflaton to be the real part of ϕ, $\sigma \equiv {\rm Re}\phi$, the inflationary parameters are now field dependent,

$$\begin{aligned} \epsilon &\simeq \left(\frac{\sigma}{M_{\rm Pl}}\right)^6 \\ \eta &\simeq \frac{6\sigma^2}{M_{\rm Pl}^2}, \end{aligned} \tag{25.30}$$

so can be put to be $\ll 1$ by having $\sigma \ll M_{\rm Pl}$.

However, note that, despite having small ϵ and η, we don't have new inflation, since we have no *end of inflation*. Indeed, we see that the potential contains a plateau, as needed for new inflation, but it *increases* away from zero, instead of decreasing (so that inflation can end, when the decrease becomes drastic).

Modified Example

We can in fact modify this example to a correct one, by adding a subleading term to the superpotential, to have

$$W = a\Phi\left(1 - b\Phi^n\right), \tag{25.31}$$

where $n \geq 3$, and keeping the canonical K. We obtain now

$$DW = a\left[1-(n+1)b\Phi^n + \frac{\bar{\Phi}\Phi}{M_{\rm Pl}^2}(1-b\Phi^n)\right], \qquad (25.32)$$

leading to a scalar potential

$$V = e^{\frac{\bar{\phi}\phi}{M_{\rm Pl}^2}}|a|^2\left[\left|1-(n+1)b\Phi^n + \frac{\bar{\Phi}\Phi}{M_{\rm Pl}^2}(1-b\Phi^n)\right|^2 - \frac{3|\phi|^2}{M_{\rm Pl}^2}|1-b\phi^n|^2\right]. \qquad (25.33)$$

Again considering only small fields, we now get

$$V \simeq |a|^2 e^{\frac{|\phi|^2}{M_{\rm Pl}^2}}\left[1-\frac{|\phi|^2}{M_{\rm Pl}^2} - (n+1)b(\phi^n+\bar{\phi}^n) + \frac{|\phi|^4}{M_{\rm Pl}^4} + \mathcal{O}\left(\frac{\phi^5}{M_{\rm Pl}^2}\right)\right]. \qquad (25.34)$$

If $n = 3$ and the inflaton is $\sigma =$Re ϕ , we get

$$V \simeq |a|^2\left[1-8b\frac{\sigma^3}{M_{\rm Pl}^3} + \mathcal{O}\left(\frac{\sigma^4}{M_{\rm Pl}^4}\right)\right], \qquad (25.35)$$

and if $n = 4$ we get

$$V \simeq |a|^2\left[1-\left(10b-\frac{1}{2}\right)\frac{\sigma^4}{M_{\rm Pl}^4} + \mathcal{O}\left(\frac{\sigma^5}{M_{\rm Pl}^5}\right)\right]. \qquad (25.36)$$

In both cases, we see that we still have no mass term, so we can get $\epsilon, \eta \ll 1$, but now we also have the leading term in the potential going down, not up. That means that we can arrange to have new inflation, with a plateau followed by a drop to a minimum. This model was introduced by Yzawa and Yanagida.

As we have shown in [33], these are the only possibilities for canonical Kähler potential and small field inflation.

Until now we have assumed the existence of a Taylor or Laurent expansion for K and W, but this is not the only possibility. We can also have a singularity at $\phi = 0$ that is not of the form of a finite order pole, but rather is an essential singularity.

In general, such a case is hard to obtain in string theory, but the exception is the case of a log Kähler potential, discussed before,

$$K = -3\ln[\Phi + \bar{\Phi}], \qquad (25.37)$$

when the inflaton is the real part of ϕ, $\sigma = \text{Re}\,\phi$. This is obtained as follows. In the case of a general KK compactification, the compact space metric must be redefined by a volume field,

$$\bar{g}_{mn}(x, y) = e^{2u(x)} g_{mn}(y), \qquad (25.38)$$

and this field $u(x)$ can be written in the general form

$$\mathrm{Re}\phi = e^{cu(x)}\,, \tag{25.39}$$

with c a number. Then one obtains the kinetic term

$$+\frac{\alpha}{(\phi+\bar{\phi})^2}\partial_\mu\phi\partial^\mu\bar{\phi}\,, \tag{25.40}$$

which is obtained from the Kähler potential

$$K = -\alpha\ln[\Phi+\bar{\Phi}]. \tag{25.41}$$

For the case of a CY_3, $\alpha = 3$.

However, in this case, we find that we cannot obtain inflation, unless we modify K by adding terms inside the log, like in the D3-brane moduli case.

25.5 Special Embedding of Inflationary Potentials in Supergravity

However, it turns out that there is a special embedding of (almost) any inflationary potential inside $\mathcal{N}=1$ supergravity plus a chiral superfield, at the expense of an unusual Kähler potential,

$$K = -3M_{\rm Pl}^2\ln\left(1+\frac{\phi+\bar{\phi}}{\sqrt{3}M_{\rm Pl}}\right). \tag{25.42}$$

Note that by expanding in a Taylor series around zero, we obtain

$$K \simeq -\sqrt{3}M_{\rm Pl}(\Phi+\bar{\Phi}) + \frac{1}{2}(\Phi+\bar{\Phi})^2 + \cdots \simeq \bar{\Phi}\Phi + \left[\frac{1}{2}\Phi^2 - \sqrt{3}M_{\rm Pl}\Phi\right] + \left[\frac{1}{2}\bar{\Phi}^2 - \sqrt{3}M_{\rm Pl}\bar{\Phi}\right] + \cdots\,. \tag{25.43}$$

Then by a Kähler transformation, at this order we can remove the extra terms, and be left with only the canonical potential. But of course, the point is that we do have the linear term (not considered before) *before* the Kähler transformation, and the Taylor series continues ad infinitum, which was essential to the fact that this model was excluded from the previous analysis of a Taylor expansion.

Despite what one might think, the choice of the inflaton is not the real part of ϕ, appearing in the Kähler potential, but rather the imaginary part. The canonically normalized inflaton is

$$\phi_{\rm can} = \sqrt{2}\,\mathrm{Im}\phi. \tag{25.44}$$

Note then that if we would substitute from the beginning in K having just ϕ_{can} (just the imaginary part), we would get $K = 0$. Of course, the correct thing to do is to first take the derivatives, and then put the real part to zero. Thus, the kinetic term for ϕ is

$$-\partial_\phi \partial_{\bar{\phi}} K \partial_\mu \phi \partial^\mu \bar{\phi} = -\frac{\partial_\mu \phi \partial^\mu \bar{\phi}}{\left(1 + \frac{\phi+\bar{\phi}}{\sqrt{3}}\right)^2}$$
$$= -\frac{1}{2} \partial_\mu \phi_{\text{can}} \partial^\mu \phi_{\text{can}} , \quad (25.45)$$

where in the last equality, we have put the real part of ϕ to zero.

Then we find

$$DW = \partial_\phi W + \frac{\sqrt{2}}{1 + \frac{\phi+\bar{\phi}}{\sqrt{3}}} \frac{W}{M_{\text{Pl}}} , \quad (25.46)$$

and if we put the real part of ϕ to zero, we find

$$DW|_{\phi_{\text{can}}} = W'\left(\sqrt{2} i \text{Im}\phi\right) + \frac{\sqrt{3}}{M_{\text{Pl}}} W\left(\sqrt{2} i \text{Im}\phi\right) . \quad (25.47)$$

Note that, *if W is a real function* (with real coefficients), the first term is imaginary, while the second is real, so in the scalar potential the $|W|^2$ terms cancel, giving (for imaginary ϕ, $g_{\phi\bar{\phi}} = 1$ and $K = 0$)

$$V = e^{\frac{K}{M_{\text{Pl}}^2}} \left[|DW|^2 - \frac{3}{M_{\text{Pl}}^2} |W|^2 \right] \Bigg|_{\phi_{\text{can}}} = |W'(\sqrt{2} i \text{Im}\phi)|^2 \equiv \hat{W}'(\phi_{\text{can}})^2 , \quad (25.48)$$

where we have defined the *real function* $\hat{W}$ by

$$W(\Phi) = \frac{1}{\sqrt{2}} \hat{W}(-\sqrt{2} i \phi). \quad (25.49)$$

It follows that we can embed any positive single scalar inflationary potential into the $\mathcal{N} = 1$ supergravity plus chiral superfield scenario. Indeed, if $V \geq 0$, we can take the square root and find $\hat{W}'$, and from it, W. In this way, we can embed new inflation, chaotic inflation (with $V = \lambda \phi^p$), and a lot of other examples.

However, the problem is that the form of K, while looking somewhat simple, is difficult to obtain in string theory, and in this chapter we were not looking at phenomenological supergravity, but rather at the possibility to have a low energy string theory model, via supergravity. From this point of view, the special embedding above is not very good.

Important Concepts to Remember

- Generic Kähler potential and superpotential terms give order one contributions to η.
- The Kähler potential for the volume modulus in KK compactification has the form, for 6 compact dimensions, $K = -3M_{\rm Pl}^2 \ln[\rho + \bar{\rho}]$, which avoids the above constraint.
- Generically, stabilizing the moduli generates quantum corrections that give $\Delta\eta \sim \mathcal{O}(1)$, thus destroying inflation.
- With a vector superfield, we can have D-term inflation, for instance by introducing a FI term to break susy spontaneously.
- One can use field redefinitions and Kähler transformations to restrict the form of possible inflationary models.
- With a canonical Kähler potential $K = \bar{\Phi}\Phi$ and a linear superpotential $W = a\Phi$ we can obtain small ϵ, η, but no end to inflation.
- By adding a subleading term Φ^{n+1}, $n \geq 3$ to W, we get the Izawa-Yanagida model of small field inflation, but here Φ is restricted to remain small.
- We can embed any positive definite potential in supergravity with a chiral superfield using the special embedding with a real function W.

Further reading: See the string inflation reviews [29, 30], and the paper [33].

Exercises

(1) Calculate the inflationary parameters ϵ and η for $\mathcal{N} = 1$ supergravity coupled to a chiral superfield Φ, with Kähler potential

$$K = -3\ln(\Phi + \bar{\Phi}) \tag{25.50}$$

and superpotential

$$W = A\rho. \tag{25.51}$$

(2) Calculate the potential and its equations of motion for the model of $\mathcal{N} = 1$ supergravity coupled to a Super-Maxwell multiplet and 3 chiral superfields $\Phi_\pm$, S, with Fayet-Iliopoulos term $\xi^a V^a$ and superpotential

$$W = S\Phi_+\Phi_- \,, \tag{25.52}$$

and Kähler potential

$$K = |S|^2 + |\Phi_+|^2 + |\Phi_-|^2. \tag{25.53}$$

Here S is Maxwell-neutral and $\Phi_\pm$ have charges ± 1.

(3) Embed chaotic inflation, with $V = \lambda\phi^p$, into $\mathcal{N} = 1$ supergravity coupled to one chiral superfield, via the special embedding.

(4) Consider the $\mathcal{N}=1$ chiral superfield model with canonical Kähler potential and superpotential

$$K = a\Phi(1 - b\Phi^n). \tag{25.54}$$

Calculate the potential exactly, and then expand it in $\sigma = \text{Re}\ \phi$ to find, in the $n=3$ and $n=4$ cases, (25.35) and (25.36), as well as the $n=1$ and $n=2$ cases. Repeat the exercise for the Kähler potential $K = -3\ln(\Phi + \bar{\Phi})$.

Chapter 26
Brane (-Antibrane) Inflation

In this chapter we will examine the possibility that we have inflation due to branes, or perhaps a brane-antibrane system. In string theory, we have branes as we saw, in particular Dp-branes. They move in a higher dimensional space (usually 10 dimensions), and then from the point of view of their worldvolume, the positions in the extra dimensions behave as scalars. If moreover we adopt a point of view that our 3+1 dimensional world lives on the worldvolume of these branes (a scenario known as "braneworld"), then the extra dimensions are 3+1 dimensional scalars. It is therefore natural to ask whether one of them, $X(\xi^a)$, can be the inflaton.

26.1 Brane Probe Inflation

If we consider a single brane in flat space, and a single transverse coordinate, the DBI+WZ action takes the form

$$S_{\rm DBI+WZ} = -T_p \int d^{p+1}\xi \left(\sqrt{1+(\partial_\mu X)^2} - 1\right) , \qquad (26.1)$$

where the -1 is the effect of the WZ term in a supersymmetric configuration. As we saw, if we have more than one scalar, the determinant inside the square root can give more terms, but for a single scalar this is all. Expanding at small fields or energies, we find

$$S_{\rm DBI+WZ} \simeq T_p \int d^{p+1}\xi \left[-\frac{1}{2}(\partial_\mu X)^2 + \frac{1}{8}[(\partial_\mu X)^2]^2 + ...\right] , \qquad (26.2)$$

so the action has no potential, but besides the canonical kinetic term, has terms with higher derivatives. This case then doesn't give the standard type of inflation. But a similar case can give inflation from higher derivative kinetic terms, also known as

H. Năstase, *Cosmology and String Theory*, Fundamental Theories of Physics 197, https://doi.org/10.1007/978-3-030-15077-8_26

k-inflation (after a paper by Armendariz-Picon, Damour and Mukhanov in 1999), as shown in the case of D-branes for instance by Silverstein and Tong in 2003.

However, in a more general background, we can have a potential. In general, the Dp−brane action is

$$S_p = -T_p \int d^{p+1}\xi e^{-\phi}\sqrt{-\det\left(\partial_a X^\mu \partial_b X^\nu [g_{\mu\nu}(X) + \alpha' B_{\mu\nu}(X)]\right)} + \mu_p \int_{V_{p+1}} C_{(p+1)}. \tag{26.3}$$

For the supersymmetric brane probe, like is the case of the Dp-brane, we have $\mu_p = \pm T_p$, with the sign distinguishing D-branes from anti-D-branes (the sign is anyway relative to a background $C_{(p+1)}$). We need some sort of breaking of supersymmetry in order to have a potential, otherwise the brane probe will not feel a potential. This can be obtained for instance by putting an antibrane in a brane background, or a brane in a background that is not made up only of branes. In those cases, the WZ term will not completely compensate the DBI term, and we will get a potential.

Example

One example is a Dp-brane moving in the background of a doubly-Wick rotated nonextremal Dp-brane (or branes, rather). A nonextremal brane background can be thought of as an extremal brane background (made up of N branes), with added D-brane-anti-D-brane pairs. Indeed, a brane has mass an charge, but in a D-brane-anti-D-brane pair the charges sum up to zero, so we are left with only mass. Then the non-extremal brane background can be thought of as adding mass, i.e., the D-brane-anti-D-brane pairs, to the extremal (D-brane) background.

We make an extremal background with harmonic function $H_p(r)$ nonextremal, by adding a blackening factor (function) $f_p(r)$ as $-dt^2 \to -f_p(r)dt^2$ and $+dr^2 \to +dr^2/f_p(r)$. But moreover, double Wick rotation means that we exchange the roles of t and another worldvolume coordinate x_p, so $-f_p(r)dt^2 + dx_p^2$ becomes $-dt^2 + f_p(r)dx_p^2$.

That means that the solution is

$$\begin{aligned}
ds^2 &= H_p^{-1/2}(r)(-dt^2 + d\vec{x}_{p-1}^2 + f_p(r)dx_p^2) + H_p^{+1/2}(r)\left(\frac{dr^2}{f_p(r)} + r^2 d\Omega_{8-p}^2\right)\\
e^{2\phi} &= g_s^2 H_p^{\frac{3-p}{2}}; \quad B_{\mu\nu} = 0\\
C_{(p+1)} &= \frac{1}{g_s} H_p^{-1} dt \wedge dx^1 \wedge \ldots \wedge dx^p\\
f_p(r) &= 1 - \left(\frac{r_H}{r}\right)^{7-p}\\
H_p(r) &= 1 + \alpha_p \left(\frac{r_p}{r}\right)^{7-p}.
\end{aligned} \tag{26.4}$$

Note that $f_p(r)$ and $H_p(r)$ are harmonic function of the transverse space (r, Ω_{8-p}), as it should be.

We moreover consider the standard static gauge,

$$X^0 = \xi^0, \quad X^i = \xi^i, \quad r = r(\xi^i) = r(X^i). \tag{26.5}$$

Then the induced metric is

$$ds^2_{\text{induced}} = H_p^{-1/2}\left[-(d\xi^0)^2 + \left(\delta_{ij} + \frac{H_p(r)}{f_p(r)}\partial_i r \partial_j r\right) d\xi^i d\xi^j + f_p(r) d\xi_p^2\right]. \tag{26.6}$$

Substituting in the action, we find

$$S = -\frac{T_p}{g_s}\int d^{p+1}\xi H_p^{-1}(r)\left[\sqrt{f_p\left(1 + \frac{H_p(r)}{f_p(r)}\partial_i r \partial_i r\right)} - 1\right]. \tag{26.7}$$

Specialize now to the case $p = 4$, and consider the coordinate $x_p = x_4$ as a KK dimensional reduction coordinate (nothing depends on it), and $\int dx_p = 2\pi R$. Then, by expanding the square root, we obtain the kinetic term

$$-\frac{T_4(2\pi R)}{g_s}\int d^{3+1}\xi \frac{1}{2\sqrt{f_4(r)}}\partial_i r \partial_i r \ , \tag{26.8}$$

and the potential

$$V_4(r) = \frac{T_4 2\pi R}{g_s} H_4^{-1}(r)[\sqrt{f_4(r)} - 1] = +\frac{T_4 2\pi R}{g_s}\frac{1}{1 + \alpha_4\left(\frac{r_4}{r}\right)^3}\left[\sqrt{1 - \frac{r_H^3}{r^3}} - 1\right] < 0. \tag{26.9}$$

Here

$$\begin{aligned} r_4^3 &= \pi g_s N \alpha'^{3/2} \\ \alpha_4 &= \sqrt{1 + \left(\frac{r_H^3}{2r_4^3}\right)^2} - \frac{r_H^3}{2r_4^3}. \end{aligned} \tag{26.10}$$

Note that the potential is negative definite.

We will concentrate in the near extremal case, when $r_H \ll r_4$, so $\alpha_4 \simeq 1$.

It is easy to see that the potential increases monotonically (show that $V_p'(r) > 0$), from $r = r_H$, where

$$V(r_H) = -\frac{T_4 2\pi R}{g_s}\frac{1}{1 + \alpha_4\left(\frac{r_4}{r_H}\right)^3} \simeq -\frac{T_4 2\pi R}{g_s}\left(\frac{r_H}{r_4}\right)^3 , \tag{26.11}$$

where in the last equality I used $r = r_H \ll r_4$, and where also

$$V'(r_H) = +\infty\,, \tag{26.12}$$

to $r = \infty$, where $V(r = \infty) = 0$.

For distances $r \gg r_H$, we find

$$V(r) \simeq -\frac{T_4 2\pi R}{2g_s}\left(\frac{r_H}{r}\right)^3 \frac{1+\frac{1}{4}\left(\frac{r_H}{r}\right)^3}{1+\alpha_4\left(\frac{r_4}{r}\right)^3}. \tag{26.13}$$

If moreover $r_4 \gg r \gg r_H$, we find

$$V(r) \simeq -\frac{T_4 2\pi R}{2g_s\alpha_4}\left(\frac{r_H}{r}\right)^3\left[\frac{r_H^3}{4r^3} - \frac{r^3}{\alpha_4 r_4^3}\right]. \tag{26.14}$$

This potential is approximately constant, and it will give inflation in this region. It is left as an exercise to show that we can satisfy the slow roll conditions.

In conclusion, in this model we have obtained inflation out of the motion of a single D-brane in a nonsupersymmetric background. Other cases of brane motion in backgrounds, without supersymmetry can also result in inflation.

The end of inflation is due to the rapid drop in the potential. As we said in part I, inflation can end either with an oscillation around a minimum, leading to the usual reheating, or with a rapid drop in potential, leading to the nonperturbative preheating. This later case is what happens here, since as we saw $V'(r_H) = +\infty$. As we have described it, it is not obvious, but inflation can end with the inflaton disappearing during this preheating phase. It will become more obvious in the case of *brane-antibrane inflation*.

26.2 Brane-Antibrane Inflation: Flat Space Approximation

Instead of a single brane moving in a background, we can consider a brane and an antibrane attracting each other, and finally annihilating after their collision. In this case, the inflaton will be their *relative* position,

$$\Phi \propto \frac{y}{\alpha'};\quad y^m = (x_1 - x_2)^m. \tag{26.15}$$

The Dp-brane tension can be expressed as

$$T_p = \alpha' M_s^{p+1} e^{-\phi}\,, \tag{26.16}$$

where M_s is the string mass $1/\sqrt{\alpha'}$. The DBI term in the D-brane action is expanded as

$$S_{\rm DBI} = -\int d^{p+1}x\sqrt{-\gamma}[T_p + \ldots]\,, \tag{26.17}$$

and when considering the brane and antibrane together, the two WZ terms $\pm\mu_p \int C_{(p+1)}$ will cancel out, and the DBI terms will add up to

$$S = -\int d^{p+1}x\sqrt{-\gamma}T_p\left[2 + \frac{1}{4}g_{mn}\gamma^{ab}\partial_a y^m \partial_b y^n + \dots\right]. \tag{26.18}$$

If the total dimension is written as $D = 4 + n$, the potential energy of the brane-antibrane system per p-dimensional volume is found to be

$$\frac{E}{V_p} = -\beta\left(\frac{e^{2\phi}}{M_s^{2+n}}\right)\frac{T_p^2}{y^{n+1-p}}, \tag{26.19}$$

where the calculation (which can be found in the book by Polchiski or in the book by Clifford Johnson on D-branes) gives

$$\beta = -\pi^{-\frac{n+1-p}{2}}\Gamma\left(\frac{n-p+3}{2}\right). \tag{26.20}$$

Define the volume of space in the *extra dimensions along the branes*

$$V_{||} = r_{||}^{p-3}, \tag{26.21}$$

and the volume of the coordinates transverse to the branes

$$V_\perp = r_\perp^{n-p-3}, \tag{26.22}$$

we find the Planck scale by identifying the coefficient of the Einstein action in 3+1 dimensions, $M_P^2/2$, with the coefficient of the D-dimensional Einstein action in string theory, $e^{-2\phi}M_s^{2+n}/2$ (so that the Einstein action has the right dimension) after integrating out the extra dimensions, so

$$M_P^2 = e^{-2\phi}M_s^{2+n}V_\perp V_{||}. \tag{26.23}$$

In order to trust the calculation, we need to have small α' corrections, which amounts to the volumes $V_{||}$ and $V_\perp$ be large in string units, or

$$M_s r_\perp \gg 1, \quad M_s r_{||} \gg 1, \tag{26.24}$$

which means

$$\frac{1}{r_\perp}, \frac{1}{r_{||}} \ll M_s < M_P. \tag{26.25}$$

Given the kinetic action (26.18), the canonically normalized inflaton is

$$\Phi = y\sqrt{\frac{T_p V_{||}}{2}} = yM_s M_P \sqrt{\frac{\alpha' e^\phi}{2(M_s r_\perp)^{n+3-p}}}. \qquad (26.26)$$

We then find the inflaton potential given by the constant piece in (26.18) and the potential energy (26.19), giving

$$V(y) = A - \frac{B}{y^{n+1-p}}, \qquad (26.27)$$

where

$$\begin{aligned} A &= 2T_p V_{||} = \frac{2\alpha' e^\phi}{(M_s r_\perp)^{n+3-p}} M_s^2 M_P^2 \\ B &= \frac{\beta e^{2\phi}}{M_s^{2+n}} T_p^2 V_{||} = \frac{\alpha^2 \beta e^\phi M_P^2}{M_s^{2(n-p+1)} r_\perp^{n-p+3}}. \end{aligned} \qquad (26.28)$$

On the other hand, the distance y must be much larger than the string scale, so that we don't have string corrections, but it must also be much smaller than the size of the compact directions $r_\perp$, so that the effect of compactification is small (compactification means that we must impose periodicity on the $r_\perp$ scale, which contradicts the assumptions taken here), thus

$$\sqrt{\alpha'} = M_s \ll y \ll r_\perp. \qquad (26.29)$$

For this potential, we can easily find that

$$|\epsilon| \ll |\eta|, \qquad (26.30)$$

and moreover the expression for η is

$$\eta \simeq -\beta(n-p+2)(n-p+1)\left(\frac{r_\perp}{y}\right)^{n-p+3}. \qquad (26.31)$$

Therefore, in order to have small roll inflation, we need $|\eta| \ll 1$, which implies (given that n, p are of order one)

$$y \gg r_\perp. \qquad (26.32)$$

But that contradicts the assumption under which we have calculated the potential, so we cannot trust the calculation.

26.3 Brane-Antibrane Inflation in Compact Space

In order to obtain inflation, we must then consider a situation with several branes and one antibrane, or more precisely we consider the correct compactification situation: imposing periodicity in the compact directions means also considering the image branes under the periodicity. The effect is to consider several branes and one antibrane (when the constant piece of the potential is the same, coming from the tension of the probe antibrane) so

$$V(\vec{r}) = A - \sum_i \frac{B}{|\vec{r} - \vec{r}_i|^{n+1-p}}. \tag{26.33}$$

But we find, not surprisingly, that in order to obtain inflation we must be at a special point in the transverse space, namely near an *extremum.*

Square Torus $(\mathbb{R}/\mathbb{Z})^{n+3-p}$

We consider now the simplest possibility, of a square (equal radii, and 90^0 angles) torus in all extra directions. Then by putting the images of a brane in order to take care of the periodicity and obtain the torus from the hyperplane, we find a hyper-cubic lattice, with the hyper-cubic unit cell. The extremum is the center of the hyper-cubic cell. It turns out that at the center, we can put to zero not only the first derivatives of the potential (since it is an extremum, obviously), but also the second derivatives,

$$\begin{aligned}
\left.\frac{\partial V}{\partial r_a}\right|_{\vec{r}=\vec{r}_0} &= (n+1-p)\sum_i \frac{B(r_0 - r_i)_a}{|\vec{r} - \vec{r}_i|^{n+3-p}} = 0\\
\left.\frac{\partial^2 V}{\partial r_a \partial r_b}\right|_{\vec{r}=\vec{r}_0} &= (n+1-p)\sum_i \frac{B}{|\vec{r}_0 - \vec{r}_i|^{n+3-p}}\left[\delta_{ab} - (n+3-p)\frac{(r_0 - r_i)_a (r_0 - r_i)_b}{|\vec{r}_0 - \vec{r}_i|^2}\right] = 0.
\end{aligned} \tag{26.34}$$

Moreover, just by symmetry, since we are at the center of the cell, we must have all the odd powers of derivatives (giving odd powers of the deviation in the potential) are zero. That means the third derivative terms are also zero, and finally, for $z = r - r_0$,

$$V \simeq A - \frac{1}{4}Cz^4 , \tag{26.35}$$

where

$$C = \gamma M_s^{-(2+n)} e^{2\phi} T_p^2 V_{||} r_\perp^{-(n+5-p)}. \tag{26.36}$$

Here γ is a number of order one, and the parametric dependence can be figured out without a precise calculation.

Then one finds (we leave it as an exercise) the slow-roll parameter

$$\eta \simeq -3\gamma \left(\frac{z}{r_\perp}\right)^2. \tag{26.37}$$

Moreover, defining z_* as the z when the scale in the CMBR exits the horizon (and is frozen in), we find

$$z_* = \frac{r_\perp}{\sqrt{2\gamma N_e}}\,, \tag{26.38}$$

where N_e is the corresponding number of e-folds. Then moreover, one gets the relations

$$n_s - 1 = -\frac{3}{N_e}; \quad \frac{dn}{d\ln k} = -\frac{3}{N_e^2}. \tag{26.39}$$

But the inflaton Φ, considered here to be in the regime where $y \gg \sqrt{\alpha'}$, eventually reaches the string scale, where the preheating happens, and then we have to deal with the fact that the interbrane separation (between D-brane and anti-D-brane) gives also rise to a tachyon T in open string theory. This open string tachyon T has the mass squared

$$M_T^2 = \frac{y^2}{(2\pi\alpha')^2} - \frac{2}{\alpha'}\,, \tag{26.40}$$

so it becomes tachyonic when $y \sim \sqrt{\alpha'}$. That signals an instability, which is the fact that when the D-brane and the anti-D-brane collide, they will annihilate (similar to the annihilation of solitons like monopoles, that can be described in terms of annihilation of fundamental particles, with disappearence of charge, but not of mass). In fact, there is a branch of mathematics called K-theory, relevant for these annihilations, which says that one can in fact obtain a brane of lower dimensionality in this annihilation, namely $p \to p-2$, so that we are consistent with the existence of only odd p in type IIB and even p in type IIA, which is kind of a vortex solution of the worldvolume theory. But the point is that the p-brane charge disappears and mass remains, and can be transferred as energy to various modes.

This generates the reheating (in its preheating form), and it involves many degrees of freedom. In principle, we would need the potential $V(y, T)$, but we know only that at y large T has a large mass, localizing it at $T = 0$ and making it irrelevant, and at small y (in particular $y = 0$), the tachyon dominates instead. But the truncation of the action to the tachyon and specifically the terms with two derivatives, reduced to 4 dimensions, is conjectured to be (supported by many evidences)

$$S_T = -M_s^2 M_P^2 \int d^4x e^{-|T|^2}\left[1 + \kappa_1\left(\frac{2 + \kappa_2|T|^2}{M_s^2}\right)|\partial T|^2\right]. \tag{26.41}$$

When canonically normalized, we obtain a Higgs-type potential, and the false vacuum at $T = 0$ is replaced by the true vacuum, of the energy well equal to the tension of the brane (according to a much tested conjecture by Sen). The branes can either result in vacuum, or a D$(p-2)$-brane, if the tachyon field has a vortex number.

In principle, we can use this action to calculate better (nonperturbatively) the reheating temperature. But a very crude estimate amounts to saying that all the initial energy of the tachyon will be converted into (reheating) temperature of the modes it gives them to, i.e., that

$$T_{RH}^4 \simeq A \Rightarrow T_{RH} = \left[\frac{2\alpha' e^{\phi}}{(M_s r_\perp)^{n+3-p}}\right]^{1/4} \sqrt{M_s M_P}. \tag{26.42}$$

We treated here only the example of the antibrane at the center of a square torus with branes, but other examples can be found for antibranes at special points in the transverse space relative to branes. But always, in order to obtain inflation, we must choose a special point, in order to avoid having a quadratic potential (coming from a nonzero second derivative), which would lead to the usual eta problem of string theory. That amounts to a level of fine-tuning of the initial conditions that is hard to explain reasonably, so is is not a very good model.

Important Concepts to Remember

- We get get inflation on a D-brane probe, either of the k-inflation type (only derivatives in the action), or the usual one with a potential.
- An example is a Dp-brane probe in a doubly-Wick rotated background of a nonextremal Dp-brane solution. Inflation ends in preheating, due to a rapid drop in the potential.
- Brane-antibrane inflation in flat space is impossible, meaning that the potential for the brane and antibrane is too steep (cannot be made to inflate).
- In a compact space, considering the infinite number of mirrors for the branes, we obtain a lattice. Starting at a special, extremum, point on the lattice, we can have inflation: the potential is a quartic deviation from a plateau, $V = A - Cz^4/4$.
- Preheating, for the brane-antibrane annihilation, is described by a p-brane action with vortex solutions, describing the K theory process of the annihilation giving a $(p-2)$-brane.

Further reading: See the original paper on brane-antibrane inflation [34], the example of brane inflation in [35].

Exercises

(1) Repeat the calculation of the potential of a Dp-brane probe in a doubly-Wick rotated nonextremal Dp-brane for a Dp'-brane probe, for $p' < p$ and $p' > p$. How does the result change?

(2) Find the conditions for slow roll inflation coming out of the 4-brane in doubly-Wick rotated nonextremal background.

(3) For the case of multi-brane inflation scenario, prove that

$$\eta \simeq -\gamma \left(\frac{z}{r_+}\right)^2, \tag{26.43}$$

and

$$z_* = \frac{r_\perp}{\sqrt{2\gamma N_e}}. \tag{26.44}$$

(4) For the tachyon action in (26.41), find the canonical potential, and write a tachyon vortex ansatz.

Chapter 27
Braneworld Cosmology and the Israel Junction Conditions

In this chapter we will consider the general scenario of a brane moving in an extra dimension, and the generic cosmology it implies. This is a "Brane Universe" scenario, or more properly, *braneworld cosmology*. It is based on the general idea introduced by ADD of large extra dimensions, or by Randall-Sundrum (versions I and II), where the matter (i.e., particle physics) resides on a brane, whereas gravity is the only field that feels the extra dimension. In this scenario, the cosmology is induced by the motion of the brane in the gravitational spacetime with an extra dimension. This is different than the previous incarnation, where the position of a brane acted as a field on the worldvolume (also containing our real 3+1 dimensional world), identified with the inflaton. Here we are mostly interested in "late time" cosmology, in the era that we have radiation and matter.

There are in principle two ways of describing the cosmology:

- from the point of view of the brane: *the brane formalism*, and
- from the point of view of the bulk: *the bulk formalism*.

They are in fact equivalent, so in this chapter I will concentrate on the brane formalism. There is in fact a debate on whether the cosmology described here makes sense, but I will not enter into this discussion, I will only present the tools necessary to understand the models. Although in principle one can consider more dimensions, the analysis becomes very complicated, so we will restrict to a 5 dimensional embedding space. Also, although we are interested in string theory, I should say that this scenario is very general: we only need a brane moving in some higher dimensional space, with a large direction and a negative cosmological constant. However, the details of the construction are not so easy to rigorously obtain in string theory, as is the case for the Randall-Sundrum constructions (I and II).

H. Năstase, *Cosmology and String Theory*, Fundamental Theories of Physics 197, https://doi.org/10.1007/978-3-030-15077-8_27

27.1 Braneworld Cosmology in the Brane Formalism

We will start with a general 5 dimensional gravity theory with some matter,

$$S = -\frac{1}{2\kappa_{(5)}^2}\int d^5x\sqrt{-G^{(5)}}R^{(5)} + S_{\text{matter}}. \tag{27.1}$$

Here $\kappa_{(5)}^2 = M_{(5)}^{-3}$, where $M_{(5)}$ is the 5-dimensional Planck constant.
We will use a block-diagonal ansatz,

$$ds^2 = G_{AB}dx^A dx^B = g_{\mu\nu}dx^\mu dx^\nu + b^2dy^2 \,, \tag{27.2}$$

or more generally a *induced metric on the brane*

$$g_{\mu\nu}^{(4)} = G_{\mu\nu}^{(5)} - n_\mu n_\nu \,, \tag{27.3}$$

or more precisely,

$$g_{\mu\nu}^{(4)} = G_{AB} - n_A n_B \,, \tag{27.4}$$

since the components with $A = 5$ vanish. Here n_μ is the normal to the brane.

In the brane formalism adopted here, we will find the Gauss–Codazzi equations and the Israel junction conditions, giving the Einstein's equations on the brane. The same result would be obtained in the bulk formalism, by having a brane moving in a bulk. The latter is usually called "mirage cosmology", which comes from the popular interpretation of "mirage" (which is really an image from something far away) as "something that is not really there, despite appearances". Namely, we have an "illusion of an expanding Universe" generated by the motion of a brane in a geometry. However, both formalisms give the same result.

The principal dynamical tool at our disposal is the 5 dimensional Einstein equations (equations of motion for gravity),

$$R_{AB}^{(5)} - \frac{1}{2}g_{AB}^{(5)}R^{(5)} = \kappa_5^2 T_{AB}. \tag{27.5}$$

Besides it, we need to construct the kinematics of the embedding of the brane. The most important object is the *extrinsic curvature*,

$$K_{\mu\nu} \equiv g_\mu^{(4)\alpha} g_\nu^{(4)\beta} \nabla_\alpha n_b \,, \tag{27.6}$$

where ∇_μ is the 5 dimensional covariant derivative, as opposed to D_μ, which is the 4 dimensional one. We can also define the trace of the extrinsic curvature,

$$K = K^\mu{}_\mu. \tag{27.7}$$

Whereas the intrinsic curvature of an embedded manifold needs no embedding to be defined, the extrinsic one needs it, since it is defined in terms of the normal n_μ and its 5-dimensional covariant derivative. The extrinsic curvature appears also in the Gibbons-Hawking boundary term for the gravitational action, but here we consider a surface inside the manifold.

27.2 Embedding Theory

To describe the embedding of a worldvolume $\{\xi^a\}, a = 0, 1, 2, 3$ into a 5 dimensional spacetime X^A, by $X^\mu = X^\mu(\xi^a)$, $\mu = 0, 1, 2, 3$, we will write down the *Gauss–Codazzi equations* for the embedding.

They are composed of the *Gauss equation*, which in the $5 \to 4$ case is

$$ {}^{(4)}R^\alpha{}_{\beta\gamma\delta} = {}^{(5)}R^\mu{}_{\nu\rho\sigma} g_\mu^{(4)\alpha} g_\beta^{(4)\nu} g_\gamma^{(4)\rho} g_\delta^{(4)\sigma} + K^\alpha{}_\gamma K_{\beta\delta} - K^\alpha{}_\delta K_{\beta\gamma} \, , \tag{27.8} $$

and the *Codazzi equation*, which in the $5 \to 4$ is

$$ D_\nu(K_\mu{}^\nu - K\delta_\mu^\nu) = D_\nu K_\mu{}^\nu - D_\mu K = {}^{(5)}R_{\rho\sigma} n^\sigma g_\mu^{(4)\rho} \, , \tag{27.9} $$

and also the equation

$$ R - K^2 + K_{\mu\nu}K^{\mu\nu} = -2\left(R_{\mu\nu}^{(5)} - \frac{1}{2}G_{\mu\nu}^{(5)}R^{(5)}\right) n^\mu n^\nu \, , \tag{27.10} $$

which however we will not use.

For the metric embedding, we can use (27.3), or rather (27.4). The usual induced metric is defined as

$$ g_{ab}^{(4)} = G_{AB}^{(5)} \partial_a X^A \partial_b X^b \to G_{\mu\nu}^{(5)} \partial_a X^\mu \partial_b X^\nu \, , \tag{27.11} $$

where in the second form we assumed that X^5 is fixed ($\partial_a X^5 = 0$). We can also define

$$ V_a^A = \partial_a X^A \to V_a^\mu = \partial_a X^\mu \, , \tag{27.12} $$

and in the case X^5 fixed we can choose the normal vector to be

$$ n_A = (0, 0, 0, 0, 1) \, , \tag{27.13} $$

such that (valid in general)

$$ G_{AB}^{(5)} n^A n^B = 1 \, , \tag{27.14} $$

and

$$G^{(5)}_{AB}n^A V^B_a = 0. \tag{27.15}$$

Then the four dimensional metric with one up and one down index can be written in alternative forms,

$$g^{(4)\alpha}_{\nu} = G^{(5)\mu\alpha}g^{(4)}_{\mu\nu} = \delta^\alpha_\nu - n^\alpha n_\nu = V^\alpha_a V_{a\nu}. \tag{27.16}$$

Substituting this embedding formalism into the 5 dimensional Einstein tensor, we find the relation of the Einstein tensors in 4 and 5 dimensions as

$$\begin{aligned} {}^{(4)}R_{\mu\nu} - \frac{1}{2}g^{(4)}_{\mu\nu}{}^{(4)}R &= \left({}^{(5)}R_{\rho\sigma} - \frac{1}{2}g^{(5)}_{\rho\sigma}{}^{(5)}R\right) g^{(4)\rho}_{\mu} g^{(4)\sigma}_{\nu} + {}^{(5)}R_{\rho\sigma}n^\rho n^\sigma g^{(4)}_{\mu\nu} \\ &+ KK_{\mu\nu} - K_\mu{}^\rho K_{\nu\rho} - \frac{1}{2}g^{(4)}_{\mu\nu}(K^2 - K^{\alpha\beta}K_{\alpha\beta}) - \tilde{E}_{\mu\nu}\,, \end{aligned} \tag{27.17}$$

where

$$\tilde{E}_{\mu\nu} = {}^{(5)}R^\alpha{}_{\beta\rho\sigma}n_\alpha n^\rho g^{(4)\beta}_{\mu} g^{(4)\sigma}_{\nu}. \tag{27.18}$$

This is left as an exercise to prove.

Next we decompose the 5 dimensional Riemann tensor into a Ricci part (tensor and scalar) and Weyl tensor. The Weyl tensor is an object that is invariant under conformal rescalings (rescalings of the metric by a conformal factor). That means that in particular, it is zero in the case of conformally flat metrics, like the AdS case. This will be useful, since as we will see, the bulk will in fact be of the AdS type, so only fluctuations away from it will have nonzero Weyl tensor.

The definition of the Weyl tensor in a general dimension D is given by

$${}^{(D)}C_{\alpha\beta\gamma\delta} = {}^{(D)}R_{\alpha\beta\gamma\delta} - \frac{2}{D-2}\left(g_{\alpha[\gamma}{}^{(D)}R_{\delta]\beta} - g_{\beta[\gamma}{}^{(D)}R_{\delta]\alpha}\right) + \frac{2}{(D-1)(D-2)}{}^{(D)}R g_{\alpha[\gamma}g_{\delta]\beta}. \tag{27.19}$$

Inverting this relation in 5 dimensions, we write

$${}^{(5)}R_{\alpha\beta\gamma\delta} = {}^{(5)}C_{\alpha\beta\gamma\delta} + \frac{2}{3}\left(G_{\alpha[\gamma}{}^{(5)}R_{\delta]\beta} - G_{\beta[\gamma}{}^{(5)}R_{\delta]\alpha}\right) + \frac{1}{6}{}^{(5)}RG_{\alpha[\gamma}G_{\delta]\beta}. \tag{27.20}$$

Substituting this decomposition into the above relation for the 4 dimensional Einstein tensor in terms of the 5 dimensional Einstein tensor, and using the 5 dimensional Einstein's equations (27.5), we find

$$\begin{aligned} {}^{(4)}R_{\mu\nu} - \frac{1}{2}g^{(4)}_{\mu\nu}{}^{(4)}R &= \frac{2\kappa^2_{(5)}}{3}\left[T_{\rho\sigma}g^{(4)\rho}_{\mu} g^{(4)\sigma}_{\nu} + \left(T_{\rho\sigma}n^\rho n^\sigma - \frac{1}{4}T^\rho{}_\rho\right)g^{(4)}_{\mu\nu}\right] \\ &+ KK_{\mu\nu} - K_\mu{}^\sigma K_{\nu\sigma} - \frac{1}{2}g^{(4)}_{\mu\nu}(K^2 - K^{\alpha\beta}K_{\alpha\beta}) - E_{\mu\nu}\,, \end{aligned} \tag{27.21}$$

where now $E_{\mu\nu}$ is a traceless object depending only on the Weyl tensor, so vanishing in pure AdS space

$$E_{\mu\nu} \equiv {}^{(5)}C^{\alpha}{}_{\beta\rho\sigma}n_{\alpha}n^{\rho}g_{\mu}^{(4)\beta}g_{\nu}^{(4)\sigma}; \quad E^{\mu}{}_{\mu} = 0. \tag{27.22}$$

From the Codazzi equation (27.9) and the 5 dimensional Einstein's equations (27.5), we find

$$D_{\nu}K_{\mu}{}^{\nu} - D_{\mu}K = \kappa_{(5)}^2 T_{\rho\sigma}n^{\sigma}g_{\mu}^{(4)\rho}. \tag{27.23}$$

27.3 Braneworld and Cosmology

In order to describe braneworld cosmology, we will assume that the energy-momentum tensor has a component on the brane, and a bulk component that is simply a (negative) bulk cosmological constant, namely

$$T_{\mu\nu} = -\Lambda g_{\mu\nu} + S_{\mu\nu}\frac{\delta(y)}{b}. \tag{27.24}$$

Moreover, we will assume that the brane energy-momentum tensor has also a brane cosmological constant component λ, and a matter component $\tau_{\mu\nu}$, so

$$S_{\mu\nu} \equiv -\lambda g_{\mu\nu}^{(4)} + \tau_{\mu\nu}. \tag{27.25}$$

It is important that we have $\lambda \neq 0$. In fact, in the first paper on the subject, λ was not considered, and then one obtained the non-traditional Friedman equation

$$H^2 = \left(\frac{\rho_t}{6M_{(5)}^3}\right)^2 + \frac{\Lambda_{\text{bulk}}}{6M_{(5)}^3}, \tag{27.26}$$

which does not reduce to the usual cosmology, since there we have $H^2 \propto \rho$ instead. Here ρ_t is the total energy density. It was quickly realized however what the solution is. Consider that the energy density has a cosmological constant component that cancels the bulk one, i.e., that

$$\rho_t = \lambda + \rho, \tag{27.27}$$

and that

$$\Lambda_{\text{bulk}} = -\frac{\lambda^2}{6M_{(5)}^3}. \tag{27.28}$$

Then, one gets the Friedmann equation

$$H^2 = \pm \frac{\rho_\pm}{3M_{\rm Pl,(4)}^2}\left(1 \pm \rho_\pm \frac{M_{\rm Pl(4)}^2}{16M_{(5)}^6}\right), \tag{27.29}$$

which is just the usual one, $H^2 \propto \rho$, just with a ρ^2 correction term.

Note that this is all in the case that the matter component of the brane energy-momentum tensor is the usual diagonal form,

$$\tau_{\mu\nu} = \text{diag}(-\rho, P, P, P) \Rightarrow T_{\mu\nu,\text{brane}} = \frac{\delta(y)}{b}\text{diag}(-\rho, P, P, P). \tag{27.30}$$

For a *static* fifth dimension, the radius of the extra dimenion, b, is constant, $\dot{b} = 0$, so we can choose $b = 1$. Note that b, appearing as $G_{55} = b^2$, is the radius of the extra dimension, or in terms of the 4 dimensional field theory, a *radion*. This condition of $\dot{b} = 0$ means that the *radion is stabilized*. This is in fact needed, since otherwise we would observe a fifth scalar force; this is the same argument we used to say that all the moduli of the string compactification must be stabilized.

But for the radion of a fifth dimension, there is a very simple mechanism for stabilization that was found by Goldberger and Wise. One adds a scalar field Φ, with bulk action

$$S_{\text{bulk}} = \int d^5x\sqrt{-G}\left(-\frac{1}{2}G^{AB}\partial_A\Phi\partial_B\Phi - \frac{m^2}{2}\Phi^2\right), \tag{27.31}$$

as well as a potential term *on the brane*, of the usual Higgs (Mexican hat) type,

$$S_{\text{brane}} = -\int d^4x\sqrt{-g^{(4)}}\lambda(\Phi^2 - v^2)^2. \tag{27.32}$$

This is in fact enough to stabilize the radion (the radius of the fifth -extra-dimension), so $\dot{b} = 0$.

27.4 Israel Junction Conditions

The Israel junction conditions are the equivalent of the Einstein's equations *on the brane*. In fact, even though strictly speaking they are postulated, they can be argued for ("derived") by considering the (second order) Einstein's equations with a delta function in the source (energy-momentum tensor), and integrating them in an infinitesimal region around the brane. We can use then Codazzi's equation and the Einstein equations to obtain the Israel conditions. This is a familiar trick: When having some second order differential equation with a delta function term (for instance, for the Schrödinger equation in one dimension, with a delta function potential), we integrate the equations, and obtain a relation for the "jump" in the first derivative, with respect to the coefficient of the delta function term.

Using the notation

$$[f(x)] \equiv \lim_{\epsilon\to 0}[f(x+\epsilon)-f(x-\epsilon)] \equiv \lim_{x\to 0+} f(x) - \lim_{x\to 0-} f(x) \equiv f^+ - f^- , \tag{27.33}$$

the Israel junction conditions are

$$[g^{(4)}_{\mu\nu}] = 0 , \tag{27.34}$$

i.e., the 4 dimensional metric is continuous across the surface, and

$$[K_\nu{}^\mu - K\delta^\mu_\nu] = -\kappa^2_{(5)} S^\mu_\nu . \tag{27.35}$$

The latter can be rewritten as (taking the trace and replacing)

$$[K_{\mu\nu}] = -\kappa^2_{(5)}\left(S_{\mu\nu} - \frac{1}{3} g^{(4)}_{\mu\nu} S\right) \tag{27.36}$$

Moreover, we can make a standard assumption in braneworld physics, used to simplify the models, namely the assumption of $\mathbb{Z}_2$ symmetry of the physics with respect to the brane, which gives

$$K^+_{\mu\nu} = -K^-_{\mu\nu} = -\frac{\kappa^2_{(5)}}{2}\left(S_{\mu\nu} - \frac{1}{3} g^{(4)}_{\mu\nu} S\right) . \tag{27.37}$$

Then, substituting (27.37) into (27.21), we obtain the gravitational (Einstein's) equations on the 3-brane, in the form

$${}^{(4)}R_{\mu\nu} - \frac{1}{2} g^{(4)}_{\mu\nu} {}^{(4)}R = -\Lambda_4 g^{(4)}_{\mu\nu} + 8\pi G_N \tau_{\mu\nu} + \kappa^2_{(5)} \pi_{\mu\nu} - E_{\mu\nu} , \tag{27.38}$$

where

$$\Lambda_4 = \frac{\kappa^2_{(5)}}{2}\left(\Lambda + \frac{\kappa^2_{(5)}\lambda^2}{6}\right) \tag{27.39}$$

is the residual (4 dimensional) cosmological constant, if the bulk Λ and brane λ don't cancel each other. From the coefficient of the energy-momentum tensor $\tau_{\mu\nu}$ we find the Newton's constant as

$$G_N = \frac{\kappa^4_{(5)}\lambda}{48\pi} , \tag{27.40}$$

and there is an extra term, quadratic in the energy-momentum tensor (giving the ρ^2 term advertised before),

$$\pi_{\mu\nu} = -\frac{1}{4}\tau_{\mu\alpha}\tau_{\nu}{}^{\alpha} + \frac{\tau}{12}\tau_{\mu\nu} + \frac{1}{8}g^{(4)}_{\mu\nu}\tau_{\alpha\beta}\tau^{\alpha\beta} - \frac{1}{24}g^{(4)}_{\mu\nu}\tau^2. \tag{27.41}$$

The energy-momentum tensor $\tau_{\mu\nu}$ satisfies the usual conservation equation. Indeed, from (27.23) and (27.37) we find that

$$D_\nu K_\mu{}^\nu - D_\mu K \propto D_\nu \tau_\mu{}^\nu = 0. \tag{27.42}$$

Here $E_{\mu\nu}$ is the same as before, but cannot be freely specified. Indeed, from the 4 dimensional Bianchi identity

$$D^\mu \left({}^{(4)}R_{\mu\nu} - \frac{1}{2} g^{(4)(4)}_{\mu\nu} R \right) = 0 \,, \tag{27.43}$$

substituted in the equations of motion (27.38) and using the conservation equation for $\tau_{\mu\nu}$, we find

$$\begin{aligned} D^\mu E_{\mu\nu} &= K^{\alpha\beta}(D_\nu K_{\alpha\beta} - D_\beta K_{\nu\alpha}) \\ &= \frac{\kappa^2_{(5)}}{4}\left[\tau^{\alpha\beta}(D_\nu \tau_{\alpha\beta} - D_\beta \tau_{\nu\alpha} + \frac{1}{3}(\tau_{\mu\nu} - g^{(4)}_{\mu\nu}\tau) D^\mu \tau\right]. \end{aligned} \tag{27.44}$$

27.5 FLRW Cosmology

As advertised, within the context of FLRW cosmology, with 4 dimensional metric

$$ds^2 = g_{\mu\nu} dx^\mu dx^\nu = -dt^2 + a^2(t) g^{(3)}_{ij} dx^i dx^j \,, \tag{27.45}$$

and for the usual diagonal energy-momentum tensor

$$\tau^\mu{}_\nu = \text{diag}(-\rho(t), P(t), P(t), P(t)) \,, \tag{27.46}$$

we obtain the modified Friedmann equations

$$\begin{aligned} H^2 + \frac{\kappa}{a^2} &= \frac{\Lambda_4}{3} + \frac{8\pi G_N}{3}\rho\left(1 + \frac{\rho}{2\lambda}\right) + \frac{\mu}{a^4} \\ \dot{H} - \frac{\kappa}{a^2} &= -4\pi G_N(\rho + P)\left(1 + \frac{\rho}{\lambda}\right) - \frac{2\mu}{a^4}. \end{aligned} \tag{27.47}$$

Here $\kappa = 0, \pm 1$ characterizes as usual the curvature (flat, closed, open) of the 3 dimensional Universe, but we notice the appearance of a "dark radiation" term μ/a^4. The $1/a^4$ behaviour is one of radiation, but the constant μ comes as an integration constant, when integrating the constraint (27.44), a fact that is left as an exercise to prove.

As we mentioned, the bulk formalism gives the same result, even though the methods seem different. There, we consider a moving brane in a static bulk ("mirage cosmology"), and we solve for the bulk geometry (solve the 5 dimensional Einstein equations). After that, we solve for the brane motion, with energy-momentum tensor on the brane, in this fixed geometry and impose the Israel junction conditions. Together with conditions from the bulk this will again lead to the same modified Friedmann equations.

Important Concepts to Remember

- Braneworld cosmology appears as the motion of a brane in a spacetime, and is described either in the brane formalism or the bulk formalism. It primarily describes "late time" cosmology, close to the RD and MD regimes.
- In the bulk formalism, it is called "mirage cosmology"; we use the brane formalism.
- The embedding of a brane in a bulk is defined by the Gauss–Codazzi equations (the Gauss equation and the Codazzi equation), depending on the Riemann curvature tensor and the extrinsic curvaturre.
- For braneworld cosmology, we must assume a negative cosmological constant in the bulk Λ and on the brane λ (as a part of the brane energy-momentum tensor $\tau_{\mu\nu}$), related by $\Lambda \propto -\lambda^2$, like in the RS model.
- Then, instead of $H^2 \propto \rho^2 + \Lambda$, we get a linear term, $H^2 \propto \rho(1+\rho)$, with a small correction to the usual cosmology.
- A static fifth dimension can be achieved by stabilizing the radion via the Goldberger-Wise mechanism, with a massive bulk scalar field with a Higgs potential on the brane.
- The Israel junction conditions are equivalent to the Einstein's equations on the brane, and related the jump in bulk quantities with at most one derivative with the delta function source on the brane, via integration of the (second order) Einstein equations.
- The Israel junction conditions are $[g^{(4)}_{\mu\nu}] = 0$ and $[K_\nu{}^\mu - K\delta^\mu_\nu] = -\kappa_5^2 S_\nu{}^\mu$, with $S_\nu{}^\mu$ the energy-momentum tensor on the brane, including its cosmological constant λ.
- One obtains a corrected FLRW cosmology, with dark radiation appearing as an integration constant.

Further reading: The original braneworld cosmology, with $H^2 \propto \rho^2$, appeared in [36]. Almost immediately, the correction introduced by the compensating brane cosmological constant, obtaining $H^2 \propto \rho + \rho^2$, was found in [37, 38]. The Goldberger-Wise radion stabilization mechanism was presented in [39]. Taking this into account, the cosmology with $\dot{b} = 0$ was considered in [40] and especially [41], which is now the standard treatment, which I mostly followed. Reviews including braneworld cosmology are [42], which is actually an early review on string cosmology, and mentions braneworld cosmology in this context, and the general modified gravity and cosmology review [43], which contains a comparison of the brane and bulk formalisms, and cosmological perturbations in them. The ADD scenario was defined in [44], extended (via strings) in [45], and the two Randall-Sundrum scenarios were defined in [46,

47]. The validity of the Israel junction conditions, as coming from the Einstein's equations, was studied in [48].

Exercises

(1) Show that the Gauss–Codazzi equations imply

$$
\begin{aligned}
{}^{(4)}R_{\mu\nu} - \frac{1}{2} g^{(4)}_{\mu\nu} {}^{(4)}R = & \left({}^{(5)}R_{\rho\sigma} - \frac{1}{2} g^{(5)}_{\rho\sigma} {}^{(5)}R \right) g^{(4)}_{\mu}{}^{\rho} g^{(4)}_{\nu}{}^{\sigma} + {}^{(5)}R_{\rho\sigma} n^{\rho} n^{\sigma} g^{(4)}_{\mu\nu} \\
& + K K_{\mu\nu} - K_{\mu}{}^{\rho} K_{\nu\rho} - \frac{1}{2} g^{(4)}_{\mu\nu} (K^2 - K^{\alpha\beta} K_{\alpha\beta}) - \tilde{E}_{\mu\nu} \,,
\end{aligned}
\tag{27.48}
$$

where

$$
\tilde{E}_{\mu\nu} = {}^{(5)}R^{\alpha}{}_{\beta\rho\sigma} n_{\alpha} n^{\rho} g^{(4)}_{\mu}{}^{\beta} g^{(4)}_{\nu}{}^{\sigma} \,. \tag{27.49}
$$

(2) Derive (argue for) the Israel junction conditions from integrating the Einstein's equations around the brane.
(3) Find the modified Friedman equations with μ/a^4 integration constant term from the integration of the equation for $D^{\mu} E_{\mu\nu}$.
(4) Find the solution for $a(t)$ solving the modified Friedmann equations (27.47) for a flat Universe ($\kappa = 0$), given a scale factor a_0 at time t_0, with density ρ_0 and pressure P_0.

Chapter 28
The KKLT Scenario for de Sitter Backgrounds in String Theory

In this chapter we will learn how to obtain de Sitter backgrounds in string theory. This is a difficult problem. In particular, there was a "no-go theorem" about the non-existence of de Sitter backgrounds in the supergravity approximation of string theory, by Juan Maldacena and Carlos Núñez. Kachru, Kallosh, Linde and Trivedi (KKLT) have shown for the first time that it is in fact possible, with the introduction of some nonperturbative string information.

However, it is worth noting that recently there was a lot of controversy on whether this is truly possible, in particular articles by I. Bena and collaborators claming the approximations used in the calculations are not valid, and there is no true de Sitter vacuum and, even more recently, the "swampland conjecture" of Obied, Ooguri, Spodyneiko and Vafa, claiming that there is a string theory version of the no-go theorem that precludes de Sitter vacua, and we can have only a certain type of metastable situations, leading to quintessence models.

28.1 Supersymmetric Vacua Are AdS

Let us prove that in supergravity, the constraint of having supersymmetry for the vacuum implies it is of AdS type ($\Lambda < 0$), not dS type.

The scalar potential coming out of supergravity coupled to an $\mathcal{N} = 1$ chiral superfield and an $\mathcal{N} = 1$ vector superfield is

$$\begin{aligned} V = e^{K/M_{\rm Pl}^2}\left(g^{\alpha\bar\beta}D_\alpha W\overline{(D_\beta W)} - \frac{3}{M_{\rm Pl}^2}W\right) \\ +\frac{1}{2}({\rm Re} f)^{-1\,AB}D_A D_B\ , \end{aligned} \tag{28.1}$$

where on the first line we have the F-term from the chiral superfield, and on the second we have the D-term from the vector superfield, with f the gauge kinetic function.

H. Năstase, *Cosmology and String Theory*, Fundamental Theories of Physics 197, https://doi.org/10.1007/978-3-030-15077-8_28

The condition for a supersymmetric vacuum is, as we said, $Q|0\rangle = 0$, which in terms of fields means that the variation of the fields in the background must be zero, $\delta_{\rm SUSY}$ fields$=0$. But the variation of a boson is a fermion, which has zero VEV due to Lorentz invariance, so we only need to impose the vanishing of the variation of the fermions (which is a bosonic field). The variation of the spin 1/2 fields is as follows. The variation of the fermion in the chiral multiplet is

$$\delta_{\rm SUSY}\chi^\alpha = -\frac{1}{\sqrt{2}}e^{\frac{K}{2M_{\rm Pl}^2}}g^{\alpha\bar\beta}\overline{D_\beta W}. \tag{28.2}$$

On the other hand, the variation of the gaugino (the fermion in the vector multiplet) is

$$\delta_{\rm SUSY}P_L\lambda^A = \frac{i}{2}({\rm Re}f)^{-1AB}D_B\ , \tag{28.3}$$

where A, B is an index in the adjoint of the gauge group. We therefore see that the conditions for supersymmetry are

$$D_\alpha W = 0;\qquad D_A = 0. \tag{28.4}$$

The condition on the chiral superfield in rigid susy (when $1/M_{\rm Pl}^2 \to 0$) is indeed $\partial_\alpha W = 0$, and the condition for the vector is $D_A = 0$.

But what about the gravitino? The variation of the gravitino is

$$\delta_{\rm SUSY}P_L\psi_\mu = \frac{1}{2M_{\rm Pl}^2}e^{\frac{K}{2M_{\rm Pl}^2}}W\gamma_\mu\ , \tag{28.5}$$

and we note that it is not strictly speaking linear in the fields: it has a constant piece. In fact, when putting the fields in their vacuum, the right hand side will become a constant (W is constant), which signifies the presence of the AdS space, as we will see. In the vacuum, we obtain

$$V = -3\frac{|W|^2}{M_{\rm Pl}^2}e^{\frac{K}{M_{\rm Pl}^2}} < 0. \tag{28.6}$$

So indeed we have an AdS background, associated with a constant W.

In order to obtain a de Sitter background, we therefore need to break supersymmetry. But the breaking must be done in a *controlled* way, so that we can calculate. We will be doing it by adding antibranes in a background generated by branes (with charges opposite to the ones of the background).

28.2 General Set-up for KKLT

The general set-up that we will use for the KKLT scenario is of type IIB string theory flux compactification, with some corrections.

More precisely, we will treat type IIB string theory compactified on a CY_3 space, *with varying* $\tau = a + ie^{-\phi}$. Note that for this case there is a fancy name: it is called "F-theory on CY_4". Unlike M-theory, which is string theory at strong coupling, but is genuinely 11 dimensional: there is dynamics in the 11th dimension, as well as Lorentz invariance, in the F-theory case, this is just a mathematical trick. We write the theory as a 12 dimensional theory with signature (10, 2), but we have no dynamics of the extra dimensions (fields don't depend on the 2 extra coordinates), nor Lorentz invariance. Then τ acts as the modulus of a torus (if τ is constant) on which we compactify, or (if τ varies) as the torus fiber of a fibration of a CY_4 space with a CY_3 base. We also have a G-flux, with

$$G_3 = F_3 - \tau H_3 \tag{28.7}$$

being nonzero on the compact space. As we said, we have the GVW superpotential

$$W = \int_M G \wedge \Omega. \tag{28.8}$$

The volume modulus on the CY_3 is

$$\rho = \frac{b}{\sqrt{2}} + ie^{4u}, \tag{28.9}$$

where b comes from a B-field, and the metric on the compact space is $g_{mn} \propto e^{2u}$.

The Kähler potential for the moduli ρ, τ, Ω, where Ω is the (unique) holomorphic 3-form on the CY_3 is

$$\frac{K}{M_{\rm Pl}^2} = -3\ln[-i(\rho - \bar\rho)] - \ln[-i(\tau - \bar\tau)] - \ln\left[-i\int_M \Omega \wedge \bar\Omega\right]. \tag{28.10}$$

If W is independent of ρ (like in the GVW superpotential above), then

$$\begin{aligned} D_\rho W &= -\frac{3M_{\rm Pl}^2}{(\rho - \bar\rho)} W \\ g_{\rho\bar\rho} &= -\frac{3M_{\rm Pl}^2}{(\rho - \bar\rho)^2} \\ g^{\rho\bar\rho} D_\rho W \overline{D_\rho W} &= \frac{3}{M_{\rm Pl}^2}|W|^2. \end{aligned} \tag{28.11}$$

That means that in the potential, the constant piece $|W|^2$ cancels against the ρ contribution, so

$$V = e^{\frac{K}{M_{\rm Pl}^2}}\left[\sum_{a,b} g^{a\bar{b}} D_a W \overline{D_b W} - \frac{3}{M_{\rm Pl}^2}|W|^2\right]$$
$$= e^{\frac{K}{M_{\rm Pl}^2}}\left[\sum_{i,j} g^{i\bar{j}} D_i W \overline{D_j W}\right] \geq 0\,, \tag{28.12}$$

where a, b are all the moduli, and i, j are a, b less ρ. Note that now the potential is positive definite! So it admits a zero energy supersymmetric vacuum (Minkowski), though of course we still cannot have a de Sitter vacuum.

But in reality, while there are no *perturbative* ρ contributions to the superpotential (since the volume is perturbatively a modulus), there are *nonperturbative* contributions to it, coming from two sources:

(1) Euclidean D3-brane "instantons". These are extensions of the usual field theoretic instantons, which are Euclidean solutions localized both in space and time in 3+1 dimensions. Now we consider Euclidean signature D3-branes wrapping 4-cycles in the geometry, i.e., all of their coordinates (even the Euclidean "worldvolume time") compact, leaving an instanton (localized in space and time) in our 3+1 dimensions.

Then, as we saw, $g_{mn} \propto e^{2u}$, where the volume of the 4-cycle is $\sqrt{g} \propto e^{4u} = \text{Im}\,\rho$. Then the contribution $e^{iS} \sim e^{iVol}$ (the action for the wrapped D3-brane is proportional to the volume of the 4-cycle) becomes a contribution to the superpotential,

$$W_{\rm inst.} = T(z_i) e^{2\pi i \rho}\,, \tag{28.13}$$

where z_i are the complex structure moduli.

(2) Gluino condensation in some $\mathcal{N} = 1$ supersymmetric $SU(N_c)$ gauge theory on D7-branes wrapping 4-cycles.

If we have D7-branes, wrapped on a 4-cycle on the internal manifold, they will still be extended in the 3+1 noncompact directions, and have a $SU(N_c)$ gauge theory on them. But in a supersymmetric gauge theory we can have spontaneous supersymmetry breaking, appearing through the condensation of the gluinos λ, namely a nonzero VEV for a gluino bilinear $\langle\bar{\lambda}\lambda\rangle$. The superpotential responsible for this is $W = \Lambda_{N_c}^3$, where Λ_{N_c} is the dynamically generated scale, of the type $e^{\frac{iS_i}{N_c}}$. Here the S_i is the instanton-type action, which, since $g_{YM}^2 = 4\pi g_s$ on the D3-brane (relation between the open and closed string couplings) and $g_{YM}^2 = \frac{4\pi g_s}{R^4}$ on D7-brane, gives

$$\frac{8\pi^2}{g_{YM}^2} = 2\pi\frac{R^4}{g_s} = 2\pi \text{Im}\rho. \tag{28.14}$$

Here we have used the exact relation $\text{Im}\,\rho = e^{4u} = R^4/g_s$ (since e^{2u} plays the role of R^2, as common factor of the compact metric). Finally, we have

$$W_{\text{gluino}} = Ae^{\frac{2\pi i\rho}{N_c}}. \tag{28.15}$$

Together, the two nonperturbative sources above are generically written as

$$W = Ae^{ia\rho}. \tag{28.16}$$

On the other hand, the perturbative superpotential was independent of ρ, so we will describe it as the constant W_0. In total, we have the superpotential

$$W = W_0 + Ae^{ia\rho}. \tag{28.17}$$

28.3 Moduli Stabilization in KKLT

In general, flux compactifications have the property that the string coupling g_s, and by extension τ, as well as the complex structure (shape) moduli are stabilized (fixed) at a scale $m \sim \frac{\alpha'}{R^3}$ (it is proportional to α', and by dimensional analysis, must also be proportional to $1/R^3$).

To understand this, we take the example of a flux of the usual electromagnetic field $F_{\mu\nu}$ on S^2, which is quantized,

$$\int_{S^2} F_{\mu\nu} dx^\mu \wedge dx^\nu = N. \tag{28.18}$$

The reason for this is the *Dirac quantization condition* in 4 dimensions. The electric charge e and magnetic charge g can be defined as

$$\int_{S^2} *F = e; \quad \int_{S^2} F = g. \tag{28.19}$$

Then the Dirac quantization condition, obtained by considering the quantum mechanics invariance under gauge transformations (we will not explain this further here) is

$$\frac{1}{2\pi} eg \in \mathbb{Z}. \tag{28.20}$$

The result of the quantization condition is that, if *a single magnetic monopole exists in the Universe*, then electric charge is quantized. Since we know experimentally that electric charge seems quantized, it seems very likely that there is a magnetic monopole somewhere.

Reversely, the existence of a single electric charge (which we know to be true) implies that $\int_{S^2} F$ is quantized, as we said. Then, it follows that the solution for the electromagnetic field is

$$F_{\mu\nu} = \frac{N}{4\pi R^2} \epsilon_{\mu\nu}. \tag{28.21}$$

But in turn that means that, if we now *deform* the 2-sphere, it will cost energy (since it will deform the relation above), which means that the complex structure (shape) moduli will be stabilized.

The quantization condition was generalized to p-forms by Nepomechie and Teitelboim. Defining

$$\mu_p = \int *F_{p+2} \,, \tag{28.22}$$

in 10 dimensions the generalized quantization condition is

$$\frac{1}{2\pi}\mu_p \mu_{6-p} \in \mathbb{Z}. \tag{28.23}$$

That means that all the moduli are stabilized, except ρ (in fact, also the Kähler moduli different than ρ are stabilized). We will not describe the details here. We can therefore ignore all other moduli, and concentrate on K and W for ρ. The Kähler potential is, as we saw

$$K = -3\ln[-i(\rho - \bar{\rho})] \,, \tag{28.24}$$

and the superpotential is

$$W = W_0 + Ae^{ia\rho}. \tag{28.25}$$

Here $a \in \mathbb{R}_+$, and W_0, $A \in \mathbb{R}$, and we can also take $W_0 < 0$. We will also consider the case when there is no nonzero axion, so $\rho = i\sigma$.

In a supersymmetric vacuum, as we saw, we need to put $D_\rho W = 0$. But now

$$D_\rho W = iAae^{ia\rho} - \frac{3}{\rho - \bar{\rho}}(W_0 + Ae^{ia\rho}) \,, \tag{28.26}$$

and on $\rho = i\sigma$, we obtain

$$D_\rho W|_{\rho=i\sigma} = i\left[Aae^{-a\sigma} + \frac{3}{2\sigma}(W_0 + Ae^{-a\sigma})\right]. \tag{28.27}$$

Moreover, on $\rho = i\sigma$, we also have

$$\begin{aligned} K|_{\rho=i\sigma} &= -2\ln 2\sigma \\ W|_{\rho=i\sigma} &= W_0 + Ae^{-a\sigma} \\ g_{\rho\bar{\rho}}|_{\rho=i\sigma} &= -\left.\frac{3}{(\rho-\bar{\rho})^2}\right|_{\rho=i\sigma} = \frac{3}{4\sigma^2}. \end{aligned} \tag{28.28}$$

Then the supersymmetric vacuum condition $D_\rho W|_{\rho=i\sigma_0}$ becomes

$$W_0 = -Ae^{-a\sigma_0}a\left(1 + \frac{2a\sigma_0}{3}\right). \tag{28.29}$$

The potential at the minimum is then

$$V(\rho = i\sigma_0) = e^{K/M_{\rm Pl}^2}\left(|D_\rho W|^2 - \frac{3}{M_{\rm Pl}^2}|W|^2\right)\Big|_{\rho=i\sigma_0} = -\frac{3}{M_{\rm Pl}^2}|W|^2 e^{K/M_{\rm Pl}^2}|_{\rho=i\sigma_0}$$
$$= -\frac{A^2a^2}{6\sigma_0}e^{-2a\sigma_0} < 0\,, \tag{28.30}$$

where in the last equality we have used the condition at the minimum (28.29). We see that again, the minimum is of AdS type.

In order to trust the calculation above, we must have small worldsheet string α' corrections, which means that the compactification volume must be large in α' units, or $\sigma \gg 1$. Moreover, we have used a single exponential $Ae^{ia\rho}$, but at least in the instanton calculation, there could be an infinite series of instantons of the type $A_n e^{ian\rho}$. In order for the leading instanton to be a good approximation, we need $a\sigma > 1$. Together, these two conditions are satisfied by

$$a < 1, \quad \sigma \gg 1, \quad a\sigma_0 > 1. \tag{28.31}$$

Moreover, we can then *choose* $W_0 \ll 1$, since this is now consistent with (28.29), independently of A. We will see that we need this condition for the de Sitter vacua.

28.4 Compactification

We now give a few details on the compactification. A bit more will be given in the next chapter. The background metric (in Einstein frame) will be of the type

$$ds^2_{10,E} = e^{2A(y)}\eta_{\mu\nu}dx^\mu dx^\nu + e^{-2A(y)}\tilde{g}_{mn}(y)dy^m dy^n\,, \tag{28.32}$$

and it will be a solution of the type IIB equations of motion known as "Klebanov–Strassler" (KS), after its discoverers.

The solution has a radial coordinate $y = \sqrt{\sum_n (y^n)^2}$ and as a function of it, the volume of the extra coordinates changes from a minimum value to $+\infty$, forming what is known as a (semi-infinite) "cigar" geometry. But instead of this infinite (decompactified) geometry, we cut the KS geometry at some $y_{\rm max}$, and glue onto it a half of a CY_3 geometry, in order to make it compact. This can be done smoothly, so that we still have a solution of the Einstein's equations. Moreover, then we add fluxes in order to stabilize the moduli.

In this construction then, there is a minimum value for the volume, i.e., a minimum value for $A(y)$, giving

$$a_{\rm min} = e^{A_{\rm min}} \sim e^{-\frac{2\pi K}{3g_s M}} \ll 1. \tag{28.33}$$

Here M, K are integers defining the fluxes through the geometry, and we can choose them such as to have $a_{\min} \ll 1$.

28.5 de Sitter Vacua

We finally come to the construction of de Sitter vacua. Since the background has D-brane charges, a way to break supersymmetry is to introduce anti-D-branes. Specifically, we will introduce an anti-D3-brane in the geometry, at position y_0. Then, if

$$a_0 \equiv e^{A(y_0)} , \tag{28.34}$$

the anti-D3-brane adds the extra energy density to the system of

$$\delta V = \frac{2a_0^4 T_3}{g_s^4} \frac{1}{(\mathrm{Im}\rho)^3}. \tag{28.35}$$

We will not derive this formula, only argue about the factor $a_0^4 T_3$: here T_3 is the 3-brane tension, so the energy should be proportional to it, and a_0 is the warp factor, so for a 4-dimensional world-volume, we should indeed have this factor (the volume of the anti-D3-brane worldvolume is a_0^4 times the volume of the Minkowski coordinates).

Then, because of the a_0^4 factor, the anti-D3-brane will run off to the minimum of a_0, $a_{\min}$, the end of the KS throat (or cigar). For more general validity, we write

$$\delta V = \frac{D}{(\mathrm{Im}\rho)^3}. \tag{28.36}$$

This potential is to be added to the potential coming from the supersymmetric case, obtaining

$$\begin{aligned} V &= \frac{1}{(2\sigma)^3}\left[\frac{4\sigma^2}{3}\frac{9}{4\sigma^2}\left(W_0 + Ae^{-a\sigma}\left(1+\frac{2a\sigma}{3}\right)\right)^2 - 3(W_0 + Ae^{-a\sigma})^2\right] + \frac{D}{\sigma^3} \\ &= \frac{Ae^{-a\sigma}}{(2\sigma)^3}\left[6\left(1+\frac{2a\sigma}{3}\right)W_0 + 3Ae^{-a\sigma}\left(1+\frac{4a\sigma}{3}+\frac{4a^2\sigma^2}{9}\right) - 6W_0 - 3Ae^{-a\sigma}\right] + \frac{D}{\sigma^3} \\ &= \frac{aAe^{-a\sigma}}{2\sigma^2}\left[W_0 + Ae^{-a\sigma}\left(1+\frac{a\sigma}{3}\right)\right] + \frac{D}{\sigma^3}. \end{aligned} \tag{28.37}$$

The shape of the resulting potential, with the parameters chosen as explained above, is: the potential drops from infinity at zero, down to a positive value (de Sitter vacuum), then to a maximum, and then runs off to zero at infinity. One can in fact fine tune the coefficient D of the susy breaking term to obtain a vacuum value V_0 very close to zero, consistent with what we see now (very small cosmological constant).

The fine tuning of D is not as extreme as in the cosmological constant, but is fine nevertheless. Note that D is the parameter responsible for supersymmetry breaking, which lifts a would-be AdS background to a dS background.

Important Concepts to Remember

- In supergravity, supersymmetric vacua are AdS vacua.
- With a volume modulus ρ for KK compactification and W independent of ρ, we obtain a positive definite potential $V \geq 0$, with a Minkowski vacuum.
- But there are nonperturbative contributions to $W(\rho)$ from Euclidean D3-brane instantons and gluino condensation on D7-branes wrapping 4-cycles, giving generically $W = Ae^{ia\rho}$, with $a > 0$, besides the constant W_0 perturbative term.
- The moduli are stabilized by integer fluxes on the compact space.
- The solution is a compactified version of the Klebanov–Strassler geometry, where the "cigar" geometry in y has a minimum volume at one sides, and is cut off and glued onto a half of CY_3 at the other.
- The supersymmetric minimum is AdS, but by adding antibranes sitting at the tip of the cigar, with $V = D/(\text{Im}\rho)^3$, we get a de Sitter minimum.

Further reading: The original KKLT paper is [49].

Exercises

(1) Consider a superpotential

$$W = (A\rho + B)e^{ia\rho} + C \ , \tag{28.38}$$

where $A \neq 0$, $a \in \mathbb{R}_+$. Find the conditions for a supersymmetric minimum.
(2) Add an anti-D3-brane to the previous exercise. Do we obtain a dS vacuum?
(3) Consider the superpotential

$$W = A\eta(\rho)^{-6} e^{\frac{3i\tau}{4b}} + B \ , \tag{28.39}$$

where $\eta(x)$ is the Dedekind eta function. Is the volume stabilized? How about the dilaton?
(4) Find a way to obtain a dS vacuum for the superpotential at exercise 3 (by breaking supersymmetry in a controllable way).

Chapter 29
The KKLMMT Scenario for Inflation and Generalizations

In the previous chapter, we have shown how KKLT have obtained a de Sitter background in string theory, with a geometry made from branes, cut off in the UV and smoothly joined onto a CY_3 half of space, and an antibrane at the tip (minimum of the KS throat). In this chapter, we will present the KKLMMT scenario for string inflation based on it, of the type of brane-antibrane inflation already analyzed before in the book.

Since the potential energy found by KKLT depended on the volume modulus ρ, it would have been a natural assumption to believe that we could construct an inflationary scenario for a potential for ρ. Or perhaps by letting the antibrane slide towards the tip of the KS geometry r_0, with its position being the inflaton. But in fact, the inflaton is neither: the volume modulus ρ becomes stabilized at the de Sitter minimum, as we saw, and the antibrane has too steep a potential, leading to the eta problem ($\eta \sim 1$). In fact, in order to obtain inflation, we must introduce a sliding D3-brane in the KS throat geometry, and then its position r_1, or rather the brane-antibrane separation $(r_1 - r_0)$, acts as the inflaton, just like in usual brane-antibrane inflation.

Note that, since the KS geometry itself is made from D3-branes, if we would put only the sliding D3-brane in the geometry, we would have no potential, since the solution would be supersymmetric (we can always add parallel branes without breaking extra supersymmetry. But because there is an anti-brane, breaking susy, we get a potential: the anti-D3 modifies the background (there is a back-reaction), and the D3-brane feels this modification.

This would give inflation as it stands, but we will see that we need to fix the other moduli, and then this generically will spoil the inflationary potential, and we would get $\eta \sim 1$. So in fact, as emphasized in the KKLMMT paper, the scenario doesn't really describe inflation in string theory, but rather the opposite: that generically, we *don't* obtain inflation, if we want to stabilize all the moduli.

H. Năstase, *Cosmology and String Theory*, Fundamental Theories of Physics 197, https://doi.org/10.1007/978-3-030-15077-8_29

29.1 Simplified Model for KKLMMT

We will start with the analysis of a simplified model, where we replaced the somewhat complicated geometry of the KS throat with a simpler solution, $AdS_5 \times X^5$, coming from N D3-branes. But then we must *phenomenologically* cut off the radial r direction of AdS_5 in the IR at r_0 and the UV at $r_{\max}$, in order to make the space compact, thus creating a Randall–Sundrum (RS) like construction (the RS 2-brane model has an IR and a UV brane cutting off a slice of AdS space). The cut-offs r_0 and $r_{\max}$ can be chosen independently of R (as we will see later in the KS construction). The compact space X^5 is a generalization of the pure D3-brane case, that gives S^5. We can have X^5 be a Sasaki–Einstein space in general, for instance a space obtained by dividing (making an equivalence under) a discrete group like $\mathbb{Z}_2$.

The metric of the space is

$$ds^2 = \frac{r^2}{R^2}(-dt^2 + d\vec{x}^2) + \frac{R^2}{r^2}dr^2 + ds^2_{X^5}\ , \tag{29.1}$$

where the radius is defined by an integer N (the flux through AdS, or the number of D3-branes generating it), as

$$R^4 = 4\pi a g_s N \alpha'^2. \tag{29.2}$$

Here a is a phenomenological parameter that allows us to cover several cases, including the KS case. In the pure D3-brane case, $a = 1$, but for KS we will see that $a \neq 1$.

The sliding D3-brane is put at a large distance from the tip where we have the anti-D3-brane, i.e.,

$$r_0 \ll r_1. \tag{29.3}$$

Calculation of the Potential

In order to calculate the potential, we first rewrite the metric in the way that it was originally described, as coming from N D3-branes, namely in terms of a harmonic function $H_3(r)$,

$$\begin{aligned} ds^2 &= H_3^{-1/2}(r)(-dt^2 + d\vec{x}^2) + H_3^{1/2}(r)\left(dr^2 + \frac{r^2}{R^2}\tilde{g}_{ab}dy^a dy^b\right) \\ ds^2_{X^5} &= \tilde{g}_{ab}dy^a dy^b \\ H_3(r) &= \frac{R^4}{r^4}. \end{aligned} \tag{29.4}$$

Note that the harmonic function is a harmonic function of the transverse space,

$$ds_6^2 = dr^2 + \frac{r^2}{R^2}\tilde{g}_{ab}dy^a dy^b\ , \tag{29.5}$$

which for $X^5 = S^5$ is simply flat 6 dimensional space, and the harmonic function is the usual one, $1 + R^4/r^4$, with the 1 dropped because we are in the near-horizon $(r \to 0)$ limit.

Moreover, we have a 4-form potential coupling to the D3-brane of value

$$(C_4)_{tx^1x^2x^3x^4} = \frac{r^4}{R^4} \equiv H_3^{-1}(r) \Rightarrow$$
$$(F_5)_{rtx^1x^2x^3x^4} = \partial_r H_3^{-1}(r) = \frac{4r^3}{R^4}. \quad (29.6)$$

As we said, in order to calculate the potential, we should calculate the back-reaction effect of the anti-D3-brane on the background, and then calculate the action of the sliding D3-brane in the modified background. But since the force exerted by the D3-brane on the anti-D3-brane equals the force exerted by the anti-D3-brane on the D3-brane, we can calculate the potential due to the D3 and anti-D3 by reversely considering how the sliding D3-brane modifies the background, and calculating the action of the anti-D3-brane in the modified background.

This is easier to do, since the harmonic function of parallel branes with no anti-branes (a supersymmetric configuration) is simply the sum of the harmonic functions. So we must make in the above solution the substitution

$$H_3(r) \to H_3'(r) = \frac{R^4}{r^4} + \delta h(r) , \quad (29.7)$$

where $\delta h(r)$ is again a harmonic function of the transverse space. The coefficient of $\delta h(r)$ is one instead of N, and centered at position r_1 instead of $r = 0$, i.e.,

$$\Delta_6 \delta h(r) = C\delta^{(6)}(\vec{r} - \vec{r}_1) \Rightarrow$$
$$\delta h(r) = \frac{R^4}{N} \frac{1}{|\vec{r} - \vec{r}_1|^4} \simeq \frac{R^4}{N} \frac{1}{r_1^4} , \quad (29.8)$$

where in the last equation we have used that $r \ll r_1$, since $r = r_0$ is the position of the anti-D3-brane (where we will calculate the action). In conclusion, we have

$$H_3'(r) = R^4 \left(\frac{1}{r^4} + \frac{1}{N} \frac{1}{r_1^4} \right). \quad (29.9)$$

In this modified background, we calculate the potential by evaluation the action of the anti-D3-brane (the DBI plus the WZ term) at the position $r = r_0$,

$$S_{\text{DBI+WZ},D3/\overline{D3}-\text{brane}}(r = r_0) = -T_3 \int \sqrt{-g_{(4)}} d^4x H_3^{-\frac{p+1}{4}}(r_0) \sqrt{1 + H_3(r_0)\partial_\mu r_0 \partial^\mu r_0}$$
$$\pm T_3 \int (C_4)_{tx^1x^2x^3x^4} dt dx^1 dx^2 dx^3 dx^4. \quad (29.10)$$

The plus (upper) sign is for the D3-brane, when the leading term in DBI cancels against the leading term in the WZ term, and ignoring $\delta h(r)$, we have only the kinetic term,

$$S \simeq -T_3 \int d^4x \frac{r_0^4}{R^4}\sqrt{1+\frac{R^4}{r_0^4}\partial_\mu r_0 \partial^\mu r_0} \pm T_3 \int (C_4)_{tx^1x^2x^3x^4} dt dx^1 dx^2 dx^3 dx^4$$
$$\simeq -\frac{1}{2}\int d^4x T_3 \partial_\mu r_0 \partial^\mu r_0. \tag{29.11}$$

We have used the fact that the 4 dimensional metric is flat, and we see that we obtain the canonical kinetic term, modulo a trivial rescaling, such that the canonical scalar is

$$\phi = \sqrt{T_3} r_0. \tag{29.12}$$

As we have already shown, the 3-brane tension is

$$T_3 = \frac{1}{(2\pi)^3 g_s \alpha'^2}. \tag{29.13}$$

Then, by introducing the modified harmonic function $H_3'(r)$ and considering the anti-D3-brane case at $r = r_0$, we find the potential

$$V(r_0, r_1) = 2T_3 \frac{r_0^4}{R^4}\left(1 - \frac{1}{N}\frac{r_0^4}{r_1^4}\right), \tag{29.14}$$

which we leave as an exercise to prove.

While this potential is in principle a function of both r_0 and r_1, only r_1 can vary, and in the physical KS case r_0 is fixed.

Then, as a function of the inflaton ϕ, we obtain the slow-roll parameters

$$\epsilon \equiv \frac{M_{Pl}^2}{2}\left(\frac{V'}{V}\right)^2 = \left(\frac{M_{Pl}^2}{T_3 r_1^2}\right)\frac{1}{2N}\left(\frac{4r_0^4}{r_1^4}\right)^2 \ll 1$$
$$\eta \equiv M_{Pl}^2\left(\frac{V''}{V}\right) = -\left(\frac{12 M_{Pl}^2}{T_3 r_1^2}\right)\frac{1}{N}\left(\frac{r_0}{r_1}\right)^4 \ll 1. \tag{29.15}$$

We have obtained $\epsilon, \eta \ll 1$ since all the 3 factors in each are $\ll 1$, including

$$\frac{M_{Pl}^2}{T_3 r_1^2} = \frac{(2\pi)^3 M_{Pl}^2 g_s \alpha'^2}{r_1^2} \ll 1, \tag{29.16}$$

where the ratio is much smaller than one, since if the distance r_1 (brane-antibrane distance) is not much larger than the Planck and string scales, we would need to consider quantum string corrections.

29.2 String Theory Example for KKLMMT Construction

We now move on to the standard example in string theory, where instead of the $AdS_5 \times S^5$ space we have the KS geometry. This is an example in type IIB string theory, with G-flux, i.e., with flux for

$$G_3 = F_3 - \tau H_3 , \tag{29.17}$$

on 3-cycles in the internal space. As described in the previous chapter, the total KS space is written as

$$ds^2 = e^{2A(y)}\eta_{\mu\nu}dx^\mu dx^\nu + e^{-2A(y)}g_{mn}(y)dy^m dy^n. \tag{29.18}$$

At the tip of KS throat, there is an S^3 of finite size, leading to the fact that the warp factor $a = e^A$ has a minimum. Then, at a certain $y = y_{\max}$, we cut off the KS space, and glue half of a CY_3 space, specifically a fibration over the 5 dimensional $T^{1,1}$ space (I will not explain what it is), as in Fig. 29.1.

There are integer G-fluxes defined on 3-cycles A and B,

$$\frac{1}{(2\pi)^2\alpha'}\int_A F_3 = M; \quad \frac{1}{(2\pi)^2\alpha'}\int_B H_3 = -K , \tag{29.19}$$

where A is the S^3 at the tip of the KS geometry, and B is a dual 3-cycle. We can choose $M, K \gg 1$. The KS geometry is not known analytically over the whole range of r, only near the tip, and far away from the tip. Far away from the tip, we have the metric

$$ds^2 = H^{-1/2}(r)\eta_{\mu\nu}dx^\mu dx^\nu + H^{+1/2}(r)(dr^2 + r^2 ds^2_{T^{1,1}}) , \tag{29.20}$$

and the harmonic function is a log-corrected $AdS_5 \times S^5$ with $a \neq 1$,

$$H(r) = \frac{27\pi}{4r^4}\alpha'^2 g_s M\left[K + g_s M\left(\frac{3}{8\pi} + \frac{3}{2\pi}\ln\frac{r}{r_{\max}}\right)\right]. \tag{29.21}$$

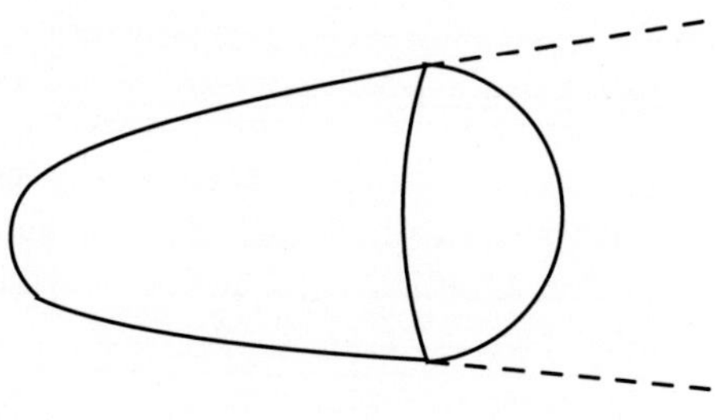

Fig. 29.1 KKLMMT construction: take the KS "cigar geometry", or semi-infinite almost cone with a smooth tip, and cut it off at some position, and glue half a CY space there, to make it compact

As we see, the leading term (without the log correction) is an AdS space defined by a radius of the type already used,

$$R^4 = \frac{27}{16} 4\pi g_s N \alpha'^2 , \tag{29.22}$$

so with $a = 27/16$, and where $N \equiv MK$ is the *product* of the two fluxes. The presence of $T^{1,1}$ in the metric means that we can *smoothly* join a half of a CY_3 space which is a fibration over $T^{1,1}$ onto the KS throat cut off at some finite r. It also means that the volume modulus of the CY will then be proportional to the volume modulus of the whole resulting compact space.

As explained in the previous chapter, the warp factor has a minimum, but the minimum is defined by the *ratio* of fluxes, as

$$\left(\frac{r_0}{R}\right)^4 = a_0^4 \equiv a_{\text{min}}^4 = e^{-\frac{8\pi K}{3 g_s M}} . \tag{29.23}$$

This ends the description of the "model for inflation". We leave as an exercise to calculate the inflaton potential in this physical KS case.

29.3 Volume Stabilization and Problems

We now turn to the problems: when we try to stabilize the moduli, we generically obtain an $\eta \sim \mathcal{O}(1)$. Consider the complex volume modulus ρ, rescaled as

$$\rho_{\text{KKLMMT}} = -i \rho_{\text{KKLT}} , \tag{29.24}$$

and the complexified D3-brane position ϕ that becomes the inflaton. Then the Kähler potential for ρ is, as we already explained, simply from KK compactification, is $K = -3 \log(\rho + \bar{\rho})$. Adding the D3-brane moduli, as we also explained, is done by adding a Kähler potential $\kappa(\phi, \bar{\phi})$ inside the log,

$$K(\rho, \bar{\rho}, \phi, \bar{\phi}) = -\log(\rho + \bar{\rho} - \kappa(\phi, \bar{\phi})) , \tag{29.25}$$

where $\kappa(\phi, \bar{\phi})$ starts with a canonical term,

$$\kappa(\phi, \bar{\phi}) = \bar{\phi}\phi + \cdots \tag{29.26}$$

A partial explanation for the above structure is that the moduli space (the leading kinetic term $g_{ij}\partial\phi^i\partial\phi^j$) of D-branes has metric

$$ds^2 = \frac{3}{2r^2}\left(dr^2 + \left(d\chi + \frac{i}{2}\partial_j \kappa d\phi^i - \frac{i}{2}\partial_{\bar{j}}\kappa d\bar{\phi}^{\bar{j}}\right)^2\right) + \frac{3}{r}\partial_i \partial_{\bar{j}} \kappa d\phi^i d\bar{\phi}^{\bar{j}} . \tag{29.27}$$

Here $r \sim e^{4u}$ is proportional to the volume of the CY space (and thus of the whole compact space). One would be tempted to work with the complex modulus $r + i\chi$, but that is not generated by a Kähler potential. Instead, one must choose the complex parameter ρ, whose imaginary part is the axion (as we said in the previous chapter), while the real part is defined by

$$2r = \rho + \bar{\rho} - \kappa(\phi, \bar{\phi}) \simeq \rho + \bar{\rho} - \phi\bar{\phi} + \cdots . \tag{29.28}$$

If, as we said, the perturbative superpotential equals the tree level contribution, which is constant,

$$W = W_0 , \tag{29.29}$$

then, since (a labels ϕ^i and ρ)

$$g^{a\bar{b}} \partial_a K \partial_{\bar{b}} K = 3 , \tag{29.30}$$

we obtain $V = 0$, so indeed ρ and ϕ are *moduli.*

If however, $W = W(\phi^i)$ only, and stabilizes ϕ^i, then *in a supersymmetric vacuum*, i.e., with $\partial_\phi W = 0$, the potential will be independent of ρ, so ρ is still a modulus.

Note that the Kähler potential K that we wrote does not have good transformation properties in the limit that K is given by κ only, namely at large enough (and stabilized) volume, so that the D-brane will move approximately in flat space, i.e., will have a canonical kinetic term. That however can be fixed by assigning the modified Kähler transformation for κ,

$$\begin{aligned} &\kappa(\phi, \bar{\phi}) \to \kappa + f(\phi) + \overline{f(\phi)} \\ &\rho \to \rho + f, \quad \bar{\rho} \to \bar{\rho} + \bar{f}. \end{aligned} \tag{29.31}$$

Finally, the volume modulus ρ can be stabilized by the nonperturbative contribution of the KKLT model (see the previous chapter),

$$W = W_0 + Ae^{-a\rho_{\text{KKLMMT}}} = W_0 + Ae^{ia\rho_{\text{KKLT}}}. \tag{29.32}$$

But this stabilization has the generic problem that for volume $\propto e^{4u} \propto r$ stabilized at large r, the potential will be generically of the type

$$V(r, \phi) \sim \frac{X(\rho)}{r^\alpha} = \frac{X(\rho)}{(\rho - \bar{\phi}\phi/2 + \cdots)^\alpha}. \tag{29.33}$$

Then if ρ is stabilized at $\rho \simeq \rho_0$, by expanding the above potential around $\phi = 0$ and considering zero axion (imaginary part of ρ is zero, so ρ is real) we obtain

$$V \simeq V_0 \left(1 + \alpha \frac{\phi\bar{\phi}}{2\rho_0} + \cdots \right) , \tag{29.34}$$

where $\alpha \sim \mathcal{O}(1)$, and also $\rho_0 \sim \mathcal{O}(1)$ in Planck units (the volume modulus stabilized close to the Planck scale, as is natural), then $\eta \sim \mathcal{O}(1)$.

Of course, if instead of $X(\rho)$ we have a more general $X(\rho, \phi)$, then we will get an extra contribution

$$V \simeq V_0 \left(1 + \alpha \frac{\phi\bar{\phi}}{2\rho_0} + \frac{\partial_\phi \partial_{\bar{\phi}} X(\rho_0, 0)}{X(\rho_0, 0)} \phi\bar{\phi} + \cdots \right), \tag{29.35}$$

and by fine-tuning we could arrange for the two terms to (almost) cancel, so that we get $\eta \ll 1$. But it means that this doesn't happen generically.

29.4 KKLMMT Generalization with Magnetic Flux

Since volume stabilization is such an important issue, we must look for generalizations. One such generalization is in the paper by Abe, Higaki and Kobayashi [51], which shows that we can have not just the $Ae^{i\rho_{\rm KKLT}}$ superpotential contribution with $a > 0$, but also one with

$$Be^{-ia\rho}, \tag{29.36}$$

again with $a > 0$, and with

$$B = Ce^{-m_9 c \langle S \rangle}. \tag{29.37}$$

Here the dilaton modulus is S,

$$2\pi S = e^{-\phi} - ic_0, \tag{29.38}$$

where c_0 is the axion,

$$c = \frac{8\pi^2}{N_9}. \tag{29.39}$$

This superpotential comes from gluino condensation on magnetized D9-branes added to the KKLT construction, such that the coupling constant of the theory on the D9-branes is given by

$$\frac{1}{g_{D9}^2} = |m_9 {\rm Re} S - w_9 {\rm Re T}|, \tag{29.40}$$

and $T = -i\rho_{\rm KKLT} = \rho_{\rm KKLMMT}$.

Moreover, a term of this type in the superpotential would be required by imposing the condition of *T-duality invariance* (as well as the condition for "modular invariance" about which we will not say anything). Indeed, we saw that T-duality, relating small compact volumes with large compact volumes, is a symmetry of string theory, so the potential, that blows up at very small volumes, should also blow up at large ones. In fact, it was argued that the superpotential arising from gaugino condensation from compactification on tori is

$$W(T, S) \sim \eta(iT)^{-6} e^{-\frac{3S}{8\pi b}} , \tag{29.41}$$

where $\eta(x)$ is the Dedekind eta function. At large volume, Re $T \to \infty$, the superpotential behaves as

$$W \propto e^{\pi T/2} , \tag{29.42}$$

with a proportionality coefficient of the order of

$$\sim e^{-\frac{3S}{8\pi b}} , \tag{29.43}$$

and $b \sim \mathcal{O}(1)$ is an RG flow factor.

We should mention however that there is some controversy about this calculation. Despite T-duality invariance being needed, it was argued that it is unphysical (from the point of view of the string coupling dependence in string theory) to have the potential blowing up as the volume blows up.

29.5 Type IIA Model Corresponding to KKLMMT

Until now we have worked in type IIB string theory, but an alternative model in type IIA string theory has been proposed.

The background that replaces KS in this case is the doubly Wick rotated nonextremal D4-brane background considered before, when discussing brane inflation, namely

$$\begin{aligned} ds^2 &= H_4^{-1/2}(r)(-dt^2 + d\vec{x}_3^2 + f(r)dx_6^2) + H_4^{+1/2}(r)\left(\frac{dr^2}{f(r)} + r^2 d\Omega_4^2\right) \\ C_{(5)} &= \frac{1}{g_s} H_4^{-1}(r) dt \wedge dx^1 \wedge \cdots \wedge dx^4 \\ f(r) &= 1 - \left(\frac{r_H}{r}\right)^3 \\ H_4(r) &= 1 + \alpha_4 \left(\frac{r_4}{r}\right)^3 . \end{aligned} \tag{29.44}$$

The geometry is now:

- a cigar geometry for r over S^4, with a physical cut-off at $r_{\min} = r_H$, and an explicit cut-off at $r_{\max}$.
- a cigar geometry for r over x_6, again cut off at $r_{\max}$, and at r_H.

Then, at $r_{\max}$, we can glue a half of a CY_3 space W over the space (r, x_6, S^4), and finally find a construction that is T-dual to KKLT.

Model 1

We can construct a model by adding a sliding D4-brane in the above construction. Even though it looks different, it is actually the same construction we had before, since the fact that the background is nonextremal means that we have anti-D4-branes in it (instead of the probe anti-D4-brane situated at the tip of the geometry).

Model 2

We can construct an alternative model by putting $r_H = 0$, i.e., considering a supersymmetric background, without any anti-D4-branes in it, but then adding a sliding anti-D4-brane probe to break supersymmetry and generate a potential.

It is left as an exercise to show that the potentials we obtain in the two models, by calculating as before the DBI+WZ action for the probe in the background, are

$$\begin{aligned} V_1(\phi)|_{\text{plateau}} &\simeq \frac{T_4(2\pi R)}{g_s}\left(\frac{\phi_H}{\phi_4}\right)^3\left[N - \frac{1}{2\alpha_4}\left(1 + \frac{\phi_4^3}{4\phi^3} - \frac{\phi^3}{\alpha_4\phi_4^3}\right)\right] \\ V_2(\phi) &= \frac{2T_4(2\pi R)}{g_s}\frac{1}{1+\frac{\phi_4^3}{\phi^3}}. \end{aligned} \tag{29.45}$$

Important Concepts to Remember

- The KKLMMT scenario is a modification of the KKLT scenario, where the volume modulus (in the presence of the anti-D3-brane at the tip) is stabilized at its de Sitter minimum, and the inflaton is a sliding D3-brane, or rather the D3-anti-D3 separation, like in usual brane-antibrane inflation.
- A simplified model has an $AdS_5 \times X^5$ space, cut off at r_0 and $r_{\max}$ (chosen independently of R, the radius of the geometry), coming from N D3-branes.
- To calculate the potential, instead of calculating the backreaction of the anti-D3 on the geometry, and then the potential for the sliding D3, we calculate how the sliding D3 modifies the background, and then the potential for the fixed anti-D3 in the modified geometry.
- The resulting potential has a plateau followed by a drop, $V(\phi) \sim V_0(1 - \alpha/\phi^4)$, and has $\epsilon, \eta \ll 1$.
- In the physical string theory case, with the KS geometry, still r_0 and $r_{\max}$ are chosen independently, as r_0 depends on a ratio of fluxes K/M, whereas the radius R on the product $N = MK$, and $r_{\max}$ is arbitrary.
- When we try to stabilize the volume modulus ρ through a nonperturbative superpotential like the KKLT one, generically we obtain a $\mathcal{O}(1)$ contribution to η, i.e., a quadratic contribution to $V(\phi)$ of order one in Planck units.

- We can also generate a $W = Be^{-ia\rho}$ (opposite sign in the exponent) for the KKLT scenario by adding magnetized D9-branes. Moreover, such a term would be required for T-duality invariance. The resulting potential has a minimum (it blows up in the IR and the UV).
- A type IIA version of KKLT can be obtained by replacing the KS geometry with the doubly Wick rotated nonextremal D4-brane geometry,
- One model has a sliding D4-brane in the nonextremal geometry, another has a sliding anti-D4-brane in the extremal geometry.

Further reading: The original KKLMMT paper is [50]. The KKLT model with magnetic flux is [51], and the type IIA model is [35].

Exercises

(1) Prove that the KKLMMT potential for the $AdS_5 \times S^5$ background is

$$V(r_0, r_1) = \frac{2T_3 r_0^4}{R^4}\left(1 - \frac{1}{N}\frac{r_0^4}{r_1^4}\right). \tag{29.46}$$

(2) Repeat the exercise for the RS background, to find its $V(r_0, r_1)$.
(3) Check that for the case T-dual to KKLMMT, we have

$$\begin{aligned} V_1(\phi)|_{\text{plateau}} &\simeq \frac{T4(2\pi R)}{g_s}\left(\frac{\phi_H}{\phi_4}\right)^3\left[N - \frac{1}{2\alpha}\left(1 + \frac{\phi_H^3}{4\phi^3} - \frac{\phi^3}{\alpha_4\phi_4^3}\right)\right] \\ V_2(\phi) &= \frac{2T_4(2\pi R)}{g_s}\frac{1}{1+\frac{\phi_4^3}{\phi^3}}. \end{aligned} \tag{29.47}$$

(4) Consider the modification of the KKLMMT model with superpotential

$$W(T, S) \sim \eta(iT)^{-6}e^{-\frac{3S}{8\pi b}}, \tag{29.48}$$

and calculate the scalar potential, and check whether all scalars are stabilized.

Chapter 30
The Ekpyrotic Scenario

In this chapter we will describe the ekpyrotic scenario, which is an alternative to inflation, inspired from string theory, defined in a 2001 paper by Khoury, Ovrut, Steinhardt and Turok. The general idea is that we have two "end of the world" branes that move towards each other in a fifth direction, and their collision represents the Big Bang.

In fact, the original idea of the ekpyrosis (coming from the ancient Greek word for the end of the Universe in a conflagration) came from a string theory construction (shown in the original paper, and explained in more detail later): In heterotic M-theory, the strong coupling version of the heterotic string, the "eleventh" direction, of radius $R = g_s^{2/3}$, has two "end of the world" M9-branes at its ends. In fact, the eleventh direction is an interval $S^1/\mathbb{Z}_2$, obtained by identifying ("orbifolding") a circle under reflection, which means that the M9-branes at the endpoint sit on the orbifold fixed points at the end of the interval. Each M9-brane has an E_8 group on them, for a total $E_8 \times E_8$ gauge group. But in this construction, there is also an M5-brane in the bulk, moving towards one of the M9-branes, and finally colliding with it. However, we will see that this mechanism is not favoured in modern ekpyrotic constructions, and instead we must have only two "end of the world" branes moving towards each other and colliding.

So it would seem like it has some similarities with brane-antibrane inflation, but that is only superficial. In the brane-antibrane inflation scenario the Universe is always expanding, as the separation decreases, and the collision corresponds to reheating; moreover, contrary to what will be seen here, the potential was positive, leading to an inflating Universe, with the inflaton being the separation.

By contrast, now almost everything relevant happens *before* the Big Bang, in a *contracting phase* with *negative* potential. The primordial fluctuations are created not by quantum fluctuations in some scalar during the expanding phase, but rather by fluctuations on the branes, *before the collision*.

We now solve in a different way the smoothness problem (the fact that the Universe is uniform and isotropic on the largest scales) and the related horizon problem, which means that when extrapolating in the past the current cosmology we get that there are many patches in the current sky that were causally disconnected (72 patches that were

H. Năstase, *Cosmology and String Theory*, Fundamental Theories of Physics 197, https://doi.org/10.1007/978-3-030-15077-8_30

disconnected on the surface of last scattering), and yet they seem to be correlated. These problems are now solved because the branes colliding are held together by tension as a single object, and the collision happens at every point on the branes in the same time. So the connectedness of various regions is due to the tension, or said another way, due to the fact that the Universe is much older than the Big Bang, so there was some causal connection of the various points *before the Big Bang*. On the other hand, in inflation these puzzles are solved by the exponential expansion due to the inflaton, *after the Big Bang*.

Like in the case of inflation, we can have an *effective field theory description* for most of the evolution, in which the fifth dimension is integrated over (somewhat similar to what happens in dimensional reduction), and the interbrane separation becomes a scalar field in 4 dimensions, $\phi(t, \vec{x})$. We can explain the early stages of the ekpyrotic scenario entirely in terms of evolution of the scalar ϕ coupled to gravity, just like in inflation. However, the later stages must take into consideration the actual higher dimensional set-up.

30.1 The Ekpyrotic Phase

The essential tool of the ekpyrotic model is the existence of a *contracting* ekpyrotic phase. It is due to a potential that has the form

$$V = -V_0 e^{-c\phi} , \tag{30.1}$$

which can be obtained as the potential for the interaction of the bulk brane with the boundary brane in the original heterotic M-theory construction sketched above. However, it was realized that we only need to have the above potential, not necessarily the construction with the bulk brane (though without the bulk brane it becomes unclear how to justify the above potential).

Note that the above potential, *if considered with a positive sign*, gave an exact solution for inflation (as we saw in Chap. 9, of part I). In fact, we can use the same algebra to solve it also now.

The KG equation for the scalars (we consider several scalars for generality) and the two Friedmann equations (the second being the dependent one) are

$$\begin{aligned} &\ddot{\phi} + 3H\dot{\phi} + V_{i,\phi_i} = 0 \\ &H^2 = \frac{1}{3}\left[\sum_i \frac{1}{2}\dot{\phi}_i^2 + \sum_i V_i(\phi_i)\right] \\ &\dot{H} = -\frac{1}{2}\sum_i \dot{\phi}_i^2 . \end{aligned} \tag{30.2}$$

Here the potential is $V = \sum_i V_i(\phi_i)$, where

$$V_i(\phi_i) = -V_i e^{-c_i\phi_i}\ , \tag{30.3}$$

and we consider $c_i \gg 1$ for all i. Then we can show (which we leave as a simple exercise) that there is a scaling solution,

$$\begin{aligned} a(t) &= (-t)^p \\ \phi_i &= \frac{2}{c_i}\log\left(-t\sqrt{\frac{c_i^2 V_i}{2}}\right) \\ p &= \sum_i \frac{2}{c_i^2} \end{aligned} \tag{30.4}$$

Here $t < 0$, so the evolution is from $t = -\infty$ to $t = 0$, i.e., towards a Big Crunch, after which we have the Big Bang, and our current cosmology. Since we must have $c_i \gg 1$ for all i, it follows that $p \ll 1$, so the time evolution is very slow ($a(t) \simeq$ const.). Note that by substituting ϕ_i in the potential, we obtain the time evolution

$$V_i(\phi_i(t)) = -\frac{2}{c_i^2 t^2}. \tag{30.5}$$

On the other hand, comparing $a(t)$ with the evolution due to a single matter component with equation of state $P = w\rho$, which is

$$a(t) \propto t^{\frac{2}{3(1+w)}}\ , \tag{30.6}$$

we obtain

$$w = \frac{2}{3p} - 1 \gg 1. \tag{30.7}$$

Thus we need matter with $w \gg 1$ in order to obtain the ekpyrotic phase.

Note that for a single i,

$$w = \frac{2}{3p} - 1 = \frac{c^2}{3} - 1\ , \tag{30.8}$$

so the condition $w > 1$ (certainly needed if $w \gg 1$, but also will be needed later) implies

$$c > \sqrt{6}. \tag{30.9}$$

Solution to the Flatness Problem

As we said, for $p \ll 1$, we have $a(t) \simeq$ constant, but on the other hand we obtain

$$H = \frac{\dot{a}}{a} \propto \frac{1}{t}\ , \tag{30.10}$$

which means that we can solve the flatness problem (the fact that the Universe is approximately flat near the Big Bang, $\Omega \simeq 1$, independent of what we have now) in a way similar to that done during inflation: In inflation it was solved by having aH grow by e^{60} (60 e-folds), though that was done by H being constant and a increasing by e^{60}. We can solve it now again by having aH increase by 60 e-folds, but rather by having a constant, and instead H increasing by e^{60}, which in a contracting phase means that

$$|t_{\rm beg}| \geq e^{60}|t_{\rm end}|. \tag{30.11}$$

Note that the end time of the ekpyrotic phase can be close enough to the Planck scale, so that the condition on $|t_{\rm beg}|$ is not necessarily very stringent.

We mentioned that $w > 1$ is needed, and in fact in order to avoid chaotic behaviour, the Belinsky–Khalatnikov–Lifshitz (BKL) singularity, an instability towards an anisotropic, homogenous, chaotically oscillating solution (the anisotropies oscillate), we need $w > 1$.

During the ekpyrotic phase, the Universe contracts, but the energy density of a component with equation of state $P = w\rho$ behaves as

$$\rho \propto a^{-3(1+w)} , \tag{30.12}$$

which means that as t evolves, the highest w will dominate (whereas in the usual expanding phase, the lowest w dominates asymptotically: the cosmological constant with $w = -1$ gets to dominate eventually). That justifies the choice of $w \gg 1$ for the ekpyrotic phase: it is the last w to dominate during the contraction.

30.2 Kinetic Phase

Further, as one approaches even more the Big Crunch (and thus the Planck scale), eventually the potential (which is defined by some energy scale) must become irrelevant (the potential cannot increase without bound, but the kinetic energy will increase), so we have

$$S = -\frac{1}{2}\int d^4x\sqrt{-g}(\partial_\mu\phi)^2 , \tag{30.13}$$

leading to the energy-momentum tensor

$$T_{\mu\nu}(\phi) = \partial_\mu\phi\partial_\nu\phi - \frac{1}{2}g_{\mu\nu}(\partial_\rho\phi)^2. \tag{30.14}$$

Considering a uniform scalar, so $\phi = \phi(t)$, we obtain

$$T_{00} = T_{ii} = \frac{1}{2}\dot{\phi}^2 \Rightarrow P = \rho \Rightarrow w = 1. \tag{30.15}$$

This is called (as we explained in part I of the book) stiff matter. The KG and Friedmann equations for this case are

$$\ddot{\phi} + 3H\dot{\phi} = 0$$
$$3H^2 = \frac{1}{2}\dot{\phi}^2 = -\dot{H}. \tag{30.16}$$

From the second equation, we obtain

$$a(t) \propto e^{\frac{\phi(t)}{\sqrt{6}}}, \tag{30.17}$$

whereas from $3H^2 = -3\dot{H}$ we get

$$a(t) = a_0(-t)^{1/3}. \tag{30.18}$$

Comparing the two, we find

$$\phi(t) = \sqrt{\frac{2}{3}} \ln(-t) + \phi_0. \tag{30.19}$$

Finally, we find that indeed, as for stiff matter

$$\rho = \frac{\dot{\phi}^2}{2} \propto a^{-6}. \tag{30.20}$$

At the end of the kinetic phase, we should reach the Planck scale. That means that before that, the fifth dimension will become important in the analysis. In turn, that means that it is important to remember that ϕ represents the distance between the colliding branes in the fifth dimension, which is the length of a line segment $S^1/\mathbb{Z}_2$.

In KK dimensional reduction from dimension D down to dimension d, as we saw in Chap. 13, the nonlinear ansatz for gravity (that gives the correct Einstein action in the lower dimension) is

$$ds_D^2 = \Delta^{-\frac{1}{d-2}} ds_d^2 + g_{mn} dx^m dx^n, \tag{30.21}$$

where $\Delta = \det g_{mn}$. When reducing from $D = 5$ down to $d = 4$, since we have a single extra dimension, $g_{55} = R^2$, so the reduction ansatz is

$$ds_5^2 = R^{-1} ds_4^2 + R^2 dy^2. \tag{30.22}$$

However, using the general formulas for the action for the scalar R, we can define the relation to the canonical scalar ϕ (identified with the same scalar ϕ used until now in the ekpyrotic and kinetic phases), as

$$R \equiv e^{\sqrt{\frac{2}{3}}\phi_{\rm can}}. \tag{30.23}$$

Then substituting into the 5 dimensional metric, we obtain the ansatz (and dropping the index "can" on the canonical scalar)

$$ds_5^2 = e^{-\sqrt{\frac{2}{3}}\phi} ds_4^2 + e^{2\sqrt{\frac{2}{3}}\phi} dy^2 , \tag{30.24}$$

for the canonical 4 dimensional effective action (valid during the kinetic phase, when the scalar potential can be ignored)

$$S_{\text{red}} = \int d^4x \sqrt{-g} \left[\frac{1}{2\kappa^2} \mathcal{R} - \frac{1}{2} (\partial_\mu \phi)^2 \right]. \tag{30.25}$$

Since ϕ is the ekpyrotic scalar, we can substitute in the 5 dimensional ansatz (30.24) the solution (30.18), (30.19), with $a_0 = 1$, $\phi_0 = 0$, and then we find

$$ds_5^2 = (-t)^{-2/3}[-dt^2 + (-t)^{2/3} d\vec{x}_3^2] + (-t)^{4/3} dy^2. \tag{30.26}$$

Making the change of variables

$$T \equiv \frac{3}{2}(-t)^{2/3}; \quad y = \frac{2}{3}\bar{y} , \tag{30.27}$$

we find the metric

$$ds_5^2 = -dT^2 + T^2 d\bar{y}^2 + d\vec{x}_3^2. \tag{30.28}$$

But note that by a Wick rotation, the $(T, \bar{y})$ space is a cone, with radial coordinate T and angular one $\bar{y}$, which means that the space is actually flat, except at the collision point $T = 0$ (so $t = 0$). Indeed, even without the Wick rotation, by making a coordinate transformation

$$u = T \cosh y; \quad v = T \sinh y , \tag{30.29}$$

we find the flat space (except at $T = 0$, which makes the coordinate transformation singular)

$$ds_5^2 = -du^2 + dv^2 + d\vec{x}_3^2. \tag{30.30}$$

If we didn't have $\vec{x}_3$, nor the compactification and orbifolding of the y coordinate, the resulting space would be called the "Milne Universe". As it is, it is called the "compactified Milne mod $\mathbb{Z}_2$", and as we saw, it describes 2 orbifold planes (where we have the branes) approaching, colliding, and then (during the Big Bang phase) receding from each other.

30.3 String Theory Model

Until now we have presented the general 2-brane model that is preferred, as we will see. But now we also present the original string theory motivation for the model, though as we will shortly see, it is by now excluded for theoretical reasons.

As we mentioned, the model is defined in heterotic M-theory, which has two "end of the world" M9-branes, situated on orbifold fixed points, at a relative distance (radion) $\pi R = \pi l_P g_s^{3/2}$. The M9-branes have each a E_8 gauge group on them, for a total $E_8 \times E_8$. We are interested in the large g_s limit (large R), and in this case we can describe the resulting theory from the point of view of the low energy supergravity, compactified to 4 dimensions. The bulk action is then the 11 dimensional supergravity action

$$S_{11} = \frac{1}{2\kappa_{11}^2}\int_{M^{11}}\sqrt{-\hat{g}}\left[\hat{\mathcal{R}} - \hat{F}_{MNPQ}\hat{F}^{MNPQ} - \frac{\sqrt{2}}{1728}\epsilon^{M_1\dots M_{11}}\hat{A}_{M_1M_2M_3}\hat{F}_{M_4\dots M_7}\hat{F}_{M_8\dots M_{11}}\right]. \tag{30.31}$$

But there is also an anomaly action, arising from a one loop quantum worldsheet string correction (so it is of order α'), that lives only on the boundary, i.e., on the 2 M9-branes (defined by $i = 1, 2$), so that

$$S_{\rm bdy} = \frac{1}{8\pi\kappa_{11}^2}\left(\frac{\kappa_{11}}{4\pi}\right)^{2/3}\sum_{i=1,2}\int_{\mathcal{M}_{10}^{(i)}}\sqrt{-g}\left[{\rm Tr}(F^{(i)})^2 - \frac{1}{2}{\rm Tr}R^2\right]. \tag{30.32}$$

Moreover, the 6 extra coordinates (except the "eleventh one" and our 4 dimensions) must be compactified on a CY_3 space, as usual. From the general formula for nonlinear KK compactification (30.21), we obtain

$$ds_{11}^2 = \Delta^{-1/3}ds_5^2 + g_{mn}dy^m dy^n\ , \tag{30.33}$$

with $\Delta = \det g_{mn} = V^2$, where V is the volume of the CY_3, so

$$ds_{11}^2 = V^{-2/3}ds_5^2 + g_{mn}dy^m dy^n. \tag{30.34}$$

To cancel the anomaly coming from the two boundary terms, not locally but in the 4 dimensional effective field theory, we can choose the standard embedding (under KK compactification of string theory), of the gauge field on one of the branes (in one E_8 group) into the spin connection, while putting to zero the other. This means that, under dimensional reduction, the same lower dimensional field appears both in the gauge field and in part of the spin connection, and it results in having

$${\rm Tr}[F^{(1)} \wedge F^{(1)}] = {\rm Tr}[R \wedge R], \quad F^{(2)} = 0. \tag{30.35}$$

Lastly, we must introduce the M5-brane moving in the bulk, which interacts with the boundary branes via an exchange of M2-branes. One can calculate in M-theory the resulting 4 dimensional superpotential (in effective field theory),

$$W \sim e^{-cY}\,, \tag{30.36}$$

where Y is the chiral superfield corresponding to the distance between the M5-brane and the boundary M9-brane. But using the general formula for the $\mathcal{N}=1$ supergravity coupled to chiral superfields,

$$V = e^{K/M_{\rm Pl}^2}\left[g^{i\bar{j}}D_i W \overline{D_j W} - \frac{3}{M_{\rm Pl}^2}|W|^2\right]\,, \tag{30.37}$$

we find

$$V = -V_0 e^{-\tilde{c}\phi}\,, \tag{30.38}$$

as advertised.

Under dimensional reduction on CY_3, we obtain the effective 5 dimensional action

$$\begin{aligned} S_5 = \frac{M_5^3}{2}\int_{\mathcal{M}_5} d^5x\sqrt{-g}\left[\mathcal{R} - \frac{1}{2}(\partial\phi)^2 - \frac{3}{2}e^{2\phi}\frac{\mathcal{F}^2}{5!}\right] - 3\sum_{i=1,2,3}\alpha_i M_5^3\int_{\mathcal{M}_4} d^4\xi_i\left(\sqrt{-h_{(i)}}e^{-\phi}\right. \\ \left. -\frac{1}{4!}\epsilon^{\mu\nu\rho\sigma}\mathcal{A}_{\alpha\beta\gamma\delta}\partial_\mu X^\alpha\partial_\nu X^\beta\partial_\rho X^\gamma\partial_\sigma X^\delta\right)\,, \end{aligned} \tag{30.39}$$

where the 3-brane terms come from the dimensional reduction of the two M9-branes on CY_3, and of the M5-brane on a 2-cycle, both leading to 3-branes in 5 dimensions.

We can then show that, for $\alpha_1 = -\alpha$, $\alpha_2 = \alpha - \beta$, $\alpha_3 = \beta$, we have a BPS solution defined by a one dimensional harmonic function (in the coordinate transverse to the branes) $D(y)$,

$$\begin{aligned} ds^2 &= D(y)(-N^2 d\tau^2 + A^2 d\vec{x}_3^2) + B^2 D^2(y) dy^2 \\ e^\phi &= BD^3(y) \\ \mathcal{F}_{0123y} &= -\alpha A^3 N B^{-1} D^{-2}(y), \quad \text{for} \quad y < Y \\ &= -(\alpha - \beta) A^3 N B^{-1} D^{-2}(y), \quad \text{for} \quad y > Y \\ D(y) &= \alpha y + C, \quad \text{for} \quad y < Y \\ &= (\alpha - \beta) y + C + \beta Y, \quad \text{for} \quad y > Y \end{aligned} \tag{30.40}$$

Here $y = Y$ is the position of the bulk brane, and the boundary branes are situated at $y = 0$ and $y = R$ (so $0 \le y \le R$). One can use a moduli space approximation, which means to promote the constant parameters like the position Y to fields = functions of time τ, in particular $Y(t)$, resolve the equations of motion to linear order in the perturbation, and find the effective action on the moduli space, i.e., on the space of $Y(t)$ and other fields. One will find that the effective description that we used during the kinetic phase, i.e., with a canonical scalar, is a good one.

30.4 Solving an Issue Related to the Null Energy Condition

After the initial ekpyrotic paper, it was observed in a paper by the same authors plus Seiberg that the fact that the null energy condition is satisfied presents a problem with the original ekpyrotic scenario described above, that can be solved if we consider only the end of the world branes to be the colliding ones.

The null energy condition is the most general of the energy conditions, believed to always hold. The energy conditions are conditions that replace the condition of positivity of the energy density in general relativity. Generally, the null energy condition is described as $T_{\mu\nu}n^\mu n^\nu \geq 0$, for all null n^μ vectors, i.e., $n^\mu n_\mu = 0$, but in the case of a perfect fluid, it reduces simply to

$$\rho + P \geq 0. \tag{30.41}$$

For gravity coupled to scalars (the various moduli of a compactification) and with a potential V,

$$S = \int d^4x\sqrt{-g}\left[\frac{1}{2\kappa_4^2}\mathcal{R} - \frac{1}{2}g^{\mu\nu}G_{IJ}(\phi^K)\partial_\mu\phi^I\partial_\nu\phi^J - V(\phi^K)\right], \tag{30.42}$$

which has the energy-momentum tensor

$$T_{\mu\nu} = G_{IJ}\partial_\mu\phi^I\partial_\nu\phi^J - g_{\mu\nu}\left[\frac{1}{2}g^{\alpha\beta}G_{IJ}\partial_\alpha\phi^I\partial_\beta\phi^J + V\right], \tag{30.43}$$

the null energy condition becomes

$$\rho + P \equiv T_{00} - \frac{1}{3}g^{ij}T_{ij} = G_{IJ}\dot{\phi}^I\dot{\phi}^J \geq 0. \tag{30.44}$$

The null energy condition is indeed satisfied, just because of the positivity of the kinetic term for the scalars, i.e., because of the absence of ghosts in a normal theory.

But by substituting this result in the second Friedmann equation, we obtain

$$\dot{H} = -4\pi G_N(\rho + P) = -4\pi G_N G_{IJ}\dot{\phi}^I\dot{\phi}^J \leq 0. \tag{30.45}$$

However, that means that we cannot have a reversal from a contracting phase, during which $H < 0$, to an expanding phase, during which $H > 0$, via matter satisfying the null energy condition, since it gives $\dot{H} < 0$.

In order to violate this simple theorem, we must either:

- have terms in the action that are higher order in derivatives and violate the null energy condition, or
- pass through a *singularity*, where we can jump discontinuously from $H < 0$ to $H > 0$, so where $\dot{H}$ is not well defined.

Clearly we are not in the first case, so we must be in the second. But the problem with it is that we then cannot calculate things. However, it turns out that there are two descriptions of the physics, one in which there is a singularity, which is the usual cosmology that we described, but there is another one where there are no singularities, so we can calculate things.

Consider the kinetic dominated phase, with $V \simeq 0$, near the singularity, and the FLRW metric in conformal time η, so

$$ds^2 = a^2(\eta)[-d\eta^2 + d\vec{x}_3^2]. \tag{30.46}$$

The solution for $(a(t), \phi(t))$ is rewritten as

$$\begin{aligned} a &= a(1)\eta^{1/2}; \\ \phi &= \phi(0) + \sqrt{3}\epsilon \log|\eta| , \end{aligned} \tag{30.47}$$

where $\epsilon = \pm 1$.

But we can now remember that the scalar ϕ parametrizes the radion, which is related to the dilaton of string theory (being the "eleventh dimension") as

$$R = R_0 e^{\sqrt{\frac{2}{3}}\phi_{\rm can}} = l_P g_s^{2/3} = l_P e^{\frac{2}{3}\phi_{\rm dilaton}}. \tag{30.48}$$

But that means that we can define a *string frame*, as the natural metric that appears in string theory (from string theory calculations), related to the Einstein metric by

$$g_{\mu\nu,s} = e^{\phi_{\rm dilaton}/c} g_{\mu\nu,E} , \tag{30.49}$$

where

$$c = \sqrt{\frac{d-2}{2}}. \tag{30.50}$$

Indeed, for $d = 10$, we have $c = 2$, as we said in part II, whereas for $d = 4$ we have $c = 1$. The string frame action in d dimensions is

$$S_{\rm string} = \int d^d x \sqrt{-g} e^{-c\phi_{\rm dilaton}} [\mathcal{R}[g_s] + c^2 g^{\mu\nu} \partial_\mu \phi_{\rm dilaton} \partial_\nu \phi_{\rm dilaton}]. \tag{30.51}$$

Defining the string time τ_s by

$$d\tau_s = a_s d\eta , \tag{30.52}$$

we find the string frame solution (we leave it as an exercise to check it is satisfied)

$$\begin{aligned} a_s &= a_s(1)|\tau_s|^{\frac{\epsilon}{\sqrt{3}}} \\ \phi_{\rm dilaton} &= \phi_{\rm dilaton}(0) + (\epsilon\sqrt{3} - 1)\log|\tau_s| \\ \epsilon &= \pm 1. \end{aligned} \tag{30.53}$$

Here $\epsilon = \pm 1$ is used to compare the ekpyrotic case with a previously considered case.

$\epsilon = -1$ corresponds to the "pre-Big Bang" scenario previously considered by Veneziano in string theory. However, as we can easily see, it corresponds to strong coupling near the Big Crunch/Big Bang,

$$g_s = e^{\phi_{\text{dilaton}}} \to \infty \quad \text{for} \quad t \to 0- \,. \tag{30.54}$$

That means that in order to describe what happens at the cosmological singularity, we must make assumptions about strongly coupled string theory, which are of course not reliable.

$\epsilon = +1$ on the other hand corresponds to the ekpyrotic case considered in this chapter, and as we can easily see has instead weak coupling at the cosmological singularity,

$$g_s = e^{\phi_{\text{dilaton}}} \to 0 \quad \text{for} \quad t \to 0- \tag{30.55}$$

That means that on the contrary, now we can trust the physics at the cosmological singularity, since it is classical (zero coupling).

But if there is a singularity, how can we trust anything that goes through it not to be modified? This is needed, since we need to consider perturbations, which are generated during the ekpyrotic phase.

The answer lies in the fact that there is another frame for gravity, besides the Einstein and string frames, in which physics is non-singular. Indeed, consider the metric frame

$$\bar{g}_{\mu\nu} \equiv e^{-\frac{\phi_{\text{dilaton}}}{\sqrt{3}}} g_{\mu\nu} \,, \tag{30.56}$$

which leads to the solution

$$\begin{aligned} \bar{a}(t) &= A = \text{constant} \\ e^{\frac{\phi_{\text{dilaton}}}{\sqrt{12}}} &= B|t|^{1/2}. \end{aligned} \tag{30.57}$$

Therefore as we can see, while again we have the string coupling $g_s = e^{\phi_{\text{dilaton}}}$ vanishing at the cosmological singularity, there is in fact no singularity in the metric, since the spatial metric is constant.

The existence of these coordinates should have been guessed from the brane picture. Indeed, the original description of the cosmology was in terms of a collision of branes, and clearly in that picture there is nothing singular happening with the 3 dimensional spatial volume at the collision point: it still looks finite; in fact, the whole 3 dimensional space on the branes doesn't look like it changes at all near the transition, just like in the $\bar{a}(t)$ coordinates above.

The important observation is that the mechanism described here *only* works if the ekpyrotic scalar ϕ is the radion, which is identified in heterotic M-theory with the dilaton ϕ_{dilaton}. That means that we need to only have the boundary branes colliding. If we would have instead the bulk brane colliding with the boundary brane, at fixed interval (radion) distance R, so fixed dilaton, the mechanism described here would not work. That eliminates the calculable scenario which was the original motivation for ekpyrosis.

30.5 Fluctuations in Ekpyrotic Models

The most important issue nowadays, given that we can measure very precisely the CMBR spectrum, is to find the spectrum of fluctuations.

- One possible assumption is that the full metric, including fluctuations, goes unchanged through the bounce from Big Crunch to Big Bang. There have been many papers claiming this, with various variants for it.
- But there have been also papers claiming that there are dynamical issues happening at the cosmological singularity. I believe that the issue is not yet settled.

If the first assumption is made, then, defining the quantity

$$\epsilon \equiv \frac{3}{2}(1+w) \gg 1 , \quad (30.58)$$

after a long calculation that will not be reproduced here, one finds the tilt of the power spectrum coming in the Newtonian potential Φ as

$$n_\Phi - 1 = 1 - \left|\frac{\epsilon+1}{\epsilon-1}\right| \ll 1. \quad (30.59)$$

But we need to calculate the spectrum of the curvature perturbations ζ, as we saw in part I. It turns out that this needs a mixing of modes that is model dependent, so the result above is modified in general. Note that we get $n - 1 \ll 1$, like for inflation (though we still need to check that the same happens for ζ fluctuations).

But the thing that most distinguishes the ekpyrotic scenario from the inflationary one is the spectrum of tensor fluctuations. One finds that this spectrum has the tilt

$$n_T = 3 - \left|\frac{\epsilon-3}{\epsilon-1}\right| \simeq 2 , \quad (30.60)$$

with contrast to inflation, where we find also $n_T \simeq 1$ (a scale invariant spectrum). Thus we find a very blue spectrum. But since we haven't even seen tensor perturbations, much less measure the tilt of their spectrum, this distinguishing feature is unlikely to be tested in the very near future.

Important Concepts to Remember

- The ekpyrotic scenario is based on two end-of-the-world branes, moving towards each other and colliding.
- Unlike inflation, everything relevant happens before the Big Bang, in a contracting phase with a negative potential.
- The original ekpyrotic model was in heterotic M-theory, with a bulk M5-brane colliding with one of the M9-branes carrying E_8 gauge groups.
- Fluctuations are on the branes, before the Big Bang (collision), and the horizon problem is fixed because of the cohesion of the branes before the collision, and their pre-Big Bang causal connection.
- There is an effective field theory description in terms of $\phi(\vec{x}, t)$, the interbrane separation.
- For the contracting ekpyrotic phase, we need $V = -V_0 e^{-c\phi}$, with $c \gg 1$, which leads to $a(t) \propto (-t)^p$, $p \ll 1$, corresponding to an effective equation of state with $w \gg 1$.
- After the ekpyrotic phase, we have a kinetic phase, dominated by a kinetic term, with $w = 1$ (stiff matter), and after it we reach the Planck scale.
- In this late stage the 5 dimensional picture prevails, and we obtain a "compactified Milne Universe mod $\mathbb{Z}_2$", for 2 orbifold planes colliding and receding.
- The null energy condition (NEC) $\rho + P \geq 0$, satisfied by a scalar field with a potential, implies via the Friedmann equation that $\dot{H} = -4\pi G_N(\rho + P) \leq 0$, i.e., we cannot go from a contracting ($H < 0$) Universe to an expanding ($H > 0$) Universe, unless we pass through a singularity or violate the NEC.
- While we pass through a singularity, there is a gravitational frame in which things are smooth (constant scale factor as a function of time). With a choice $\epsilon = +1$, we even have $g_s \to 0$ at the singularity, so we can pass through it.
- Like for inflation, scalar fluctuations have $n_\Phi - 1 < 0$ and $\ll 1$, however tensor fluctuations have $n_T - 1 \simeq 1$, unlike inflation, where it is also $\ll 1$.

Further reading: The original paper on the ekpyrotic scenario is [52]. The analysis of the transition to the Big Bang and the necessity of 2-brane model was done in [53]. A good review of the ekpyrotic (and cyclic) scenarios is [54].

Exercises

(1) Check that

$$\begin{aligned}
a(t) &= (-t)^p \\
\phi_i &= \frac{2}{c_i} \log\left(-t\sqrt{\frac{c_i^2 V_i}{2}}\right) \\
p &= \sum_i \frac{2}{c_i^2}
\end{aligned} \tag{30.61}$$

solves the equations of motion of the cosmology with potential $V = \sum_i V_i(\phi_i)$, where

$$V_i(\phi_i) = -V_i e^{-c_i \phi_i}. \tag{30.62}$$

(2) Check that the solution

$$\begin{aligned} ds^2 &= D(y)(-N^2 d\tau^2 + A^2 d\vec{x}_3^2) + B^2 D^2(y) dy^2 \\ e^\phi &= B D^3(y) \\ \mathcal{F}_{0123y} &= -\alpha A^3 N B^{-1} D^{-2}(y), \quad \text{for} \quad y < Y \\ &= -(\alpha - \beta) A^3 N B^{-1} D^{-2}(y), \quad \text{for} \quad y > Y \\ D(y) &= \alpha y + C, \quad \text{for} \quad y < Y \\ &= (\alpha - \beta) y + C + \beta Y, \quad \text{for} \quad y > Y \end{aligned} \tag{30.63}$$

satisfies the ϕ and $\mathcal{F}$ equations of motion for the action

$$\begin{aligned} S_5 = \frac{M_5^3}{2} \int_{\mathcal{M}_5} d^5 x \sqrt{-g} \left[\mathcal{R} - \frac{1}{2} (\partial\phi)^2 - \frac{3}{2} e^{2\phi} \frac{\mathcal{F}^2}{5!} \right] - 3 \sum_{i=1,2,3} \alpha_i M_5^3 \int_{\mathcal{M}_4} d^4 \xi_i \left(\sqrt{-h_{(i)}} e^{-\phi} \right. \\ \left. - \frac{1}{4!} \epsilon^{\mu\nu\rho\sigma} \mathcal{A}_{\alpha\beta\gamma\delta} \partial_\mu X^\alpha \partial_\nu X^\beta \partial_\rho X^\gamma \partial_\sigma X^\delta \right) , \end{aligned} \tag{30.64}$$

for $\alpha_1 = -\alpha$, $\alpha_2 = \alpha - \beta$, $\alpha_3 = \beta$.
(3) Check that the string frame solution

$$\begin{aligned} a_s &= a_s(1) |\tau_s|^{\frac{\epsilon}{\sqrt{3}}} \\ \phi &= \phi(0) + (\epsilon\sqrt{3} - 1) \log |\tau_s| \\ \epsilon &= \pm 1 \end{aligned} \tag{30.65}$$

satisfies the equations of motion of the action

$$S = \int d^d x \sqrt{-g_s} e^{-c\phi} (R[g_s] + c^2 g^{\mu\nu} \partial_\mu \phi \partial_\nu \phi) , \tag{30.66}$$

for

$$c = \sqrt{\frac{d-2}{2}}. \tag{30.67}$$

(4) Show that the 5 dimensional action (30.39) is obtained by dimensional reduction on CY_3 from the 11 dimensional supergravity action (30.31).

Chapter 31
The Cyclic and New Ekpyrotic Scenarios

In this chapter we will study variants of the original ekpyrotic scenario, which are introduced in order to solve some problems with the original model. These are the cyclic model of Steinhardt and Turok, and the new ekpyrotic model of Buchbinder, Khoury and Ovrut.

31.1 The Cyclic Model

The cyclic model is described within 4 dimensional effective field theory, by a potential $V(\phi)$ that has the following properties:

- it must go asymptotically, for $\phi \to +\infty$, to a constant V_0, giving the "cosmological constant", or rather dark energy, today. Only for small ϕ close to zero, the potential bends down and goes through zero. Thus for $\phi > 0$ we are in the standard cosmology, dominated asymptotically by dark energy, with equation of state $w \simeq -1$.
- for negative ϕ, until a certain value, we must have an ekpyrotic phase, with $w \gg 1$, so the potential must have a steep (exponential) drop, until a minimum value of $-V_{\rm end}$.

Combining these two features, we have the following potential:

$$V = V_0(1 - e^{-c\phi}) \,, \tag{31.1}$$

where from the ekpyrotic condition of $w \gg 1$, as we saw in the previous chapter, we need $c \gg 1$. In fact, from making contact with experiments we will see that we need $c \gtrsim 10$.

- Finally, as we said last time, the potential cannot increase indefinitely, and near the Planck scale we should have kinetic energy domination. That means that the potential should reverse, forming a well shape with minimum at $-V_{\rm end}$, and go

H. Năstase, *Cosmology and String Theory*, Fundamental Theories of Physics 197, https://doi.org/10.1007/978-3-030-15077-8_31

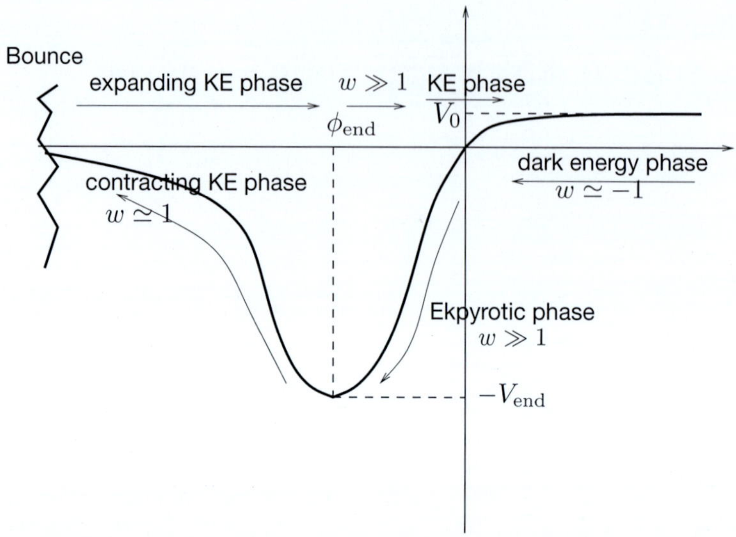

Fig. 31.1 The potential for the cyclic model, with a sketch for the various phases of evolution in it

back to zero. The kinetic domination phase applies for this upward branch of the potential, with equation of state $w \simeq 1$ (stiff matter). At the end, when one reaches zero, the bounce from Big Crunch to Big Bang should happen.

The potential is sketched in Fig. 31.1.

So the picture of the cyclic cosmology is as follows. Cosmology happens in cycles, expansion to a maximum, followed by contraction to zero (Big Crunch), followed by Big Bang and expansion. We start in a dark energy phase, with kinetic energy that is just shy of the one needed to reach the asymptotic value V_0, and after a maximum, we start to contract and go back to smaller ϕ. For $\phi < 0$ we enter a contracting ekpyrotic phase, followed by the contracting kinetic phase, followed by a bounce. Then we start expanding in a kinetic phase, followed by a quick ekpyrotic phase (that can be neglected), and then we are back to standard cosmology: radiation domination, followed by dark energy domination. And the cycle repeats.

We will see that the entropy increases in each cycle, but so does the volume, which increases exponentially over a cycle, so in effect, we are left with the same entropy density.

The scalar ϕ corresponds in the original string theory construction to the distance between the colliding brane, and the bounce to the collision between branes (that generates entropy).

In the brane picture, we have 2 branes, for a bulk dimensionally reduced metric

$$ds_{11}^2 = V^{-2/3} ds_5^2 + V^{1/3} \tilde{g}_{mn} dy^m dy^n . \tag{31.2}$$

Here note that $V = \sqrt{\det g} = \Delta^{1/2}$ is the volume of the internal 6 dimensional space, hence the $\Delta^{-1/(d-2)} = \Delta^{-1/3} = V^{-2/3}$ prefactor in front of the 5 dimensional metric. We have also taken out of the metric a factor of $V^{1/3} = \det g^{1/6}$, so that the remaining metric has unit determinant, $\det \tilde{g}_{mn} = 1$. Moreover, $V = e^{\phi}$ is the 4 dimensional dilaton. Then the effective action in 5 dimensions is

$$S_5 = \frac{1}{2\kappa_5^2}\int d^5x\sqrt{-g}\left[\mathcal{R} - \frac{\partial_\mu V\partial^\mu V}{V^2} - \frac{6\alpha^2}{V^2}\right] + \frac{12\alpha}{2\kappa_5^2}\left[-\int_{y=+1} d^4x\sqrt{-g}\frac{1}{V} + \int_{y=-1} d^4x\sqrt{-g}\frac{1}{V}\right]. \quad (31.3)$$

Note that the potential term is now, once we canonically normalize V to ϕ, $\sim e^{-\sqrt{2}\phi}$, and that the brane sources are at $y = \pm 1$. We obtain a domain wall solution,

$$\begin{aligned} ds_5^2 &= h^{2/5}(y)[A^2(-d\tau^2 + d\vec{x}^2) + B^2 dy^2] \\ V &= Bh^{6/5}(y) \\ h(y) &= 5\alpha y + C\ , \end{aligned} \quad (31.4)$$

and symmetric with respect to $y = \pm 1$, so that it actually satisfies the Israel junction conditions at these branes.

In this way we have obtained two static branes, but we can use the usual moduli space approximation for the brane solution, where we make the parameters in the (solitonic) solution, including the position of the (solitonic) brane, as functions of time, $A, B, C \to A(\tau), B(\tau), C(\tau)$. Then as usual, we don't have anymore a solution, but we can re-solve the equations of motion to linear order in the velocities (in the time dependence), and find the *motion on moduli space*, and its associated effective action on moduli space. In this way, we find the colliding brane solution and its effective field theory.

Above we have obtained the exponential term in the potential (like we also had in the previous chapter). But the positive constant part of the potential V_0, which would mean a de Sitter background, is harder to obtain.

During the ekpyrotic phase, we have a very steep drop in potential, and we generate perturbations, but unlike the case of inflation, we do so in a "fast-roll" approximation, the opposite to the "slow-roll" of inflation. For a single field ekpyrosis, the spectral tilt of generated fluctuations can be found to be

$$n_s - 1 \simeq -4(\epsilon + \eta)\ , \quad (31.5)$$

where however now ϵ and η are "fast-roll" parameters

$$\epsilon = \frac{1}{M_{\rm Pl}^2}\left(\frac{V}{V'}\right)^2$$

$$\eta = 1 - \frac{V''V}{V'^2} . \quad (31.6)$$

The “fast-roll” approximation means that $\epsilon, \eta \ll 1$, so indeed we obtain $n_s - 1 \ll 1$, as it observed experimentally. Moreover, using the potential above, we find

$$n_s - 1 = -\frac{4}{c^2} , \quad (31.7)$$

and from the experimental value $n_s - 1 \simeq -0.04$, we find indeed

$$c \gtrsim 10. \quad (31.8)$$

Note that for *two-field ekpyrosis*, one obtains instead

$$n_s - 1 \simeq +4\epsilon - 4\eta. \quad (31.9)$$

We now turn to the evolution of the space, i.e., the evolution of $a(t)$ and $H(t)$.

- During the *dark energy phase*, $a(t)$ will grow a certain number of e-folds, by definition $N_{\rm DE}$, so $a(t)$ increases by $e^{N_{\rm DE}}$, whereas $H(t)$ remains approximately constant.
- During the *ekpyrotic phase*, since, as we saw last time,

$$V(\phi(t)) \propto \frac{1}{t^2} , \quad (31.10)$$

and the time evolution is

$$a(t) \propto (-t)^p , \quad H(t) \propto \frac{1}{t} , \quad (31.11)$$

we obtain

$$\begin{aligned} a(t) &\propto V(\phi)^{-\frac{p}{2}} \\ H(t) &\propto V(\phi)^{1/2} , \end{aligned} \quad (31.12)$$

so $a(t)$ shrinks by

$$\left| \frac{V_{\rm end}}{V_{\rm beg}} \right|^{-p/2} \sim \mathcal{O}(1) , \quad (31.13)$$

where this factor is not large, since $p \ll 1$. On the other hand, $H(t)$ grows by

$$\left| \frac{V_{\rm end}}{V_{\rm beg}} \right|^{1/2} \equiv e^{N_{ek}} , \quad (31.14)$$

where in the last equality we have defined the number of e-folds of ekpyrosis.

- After the ekpyrotic phase, we have a period of kinetic domination, starting at $V_{\rm end}$ (the bottom of the potential well), followed by the bounce, that creates energy from collisions (*reheats* the Universe), followed by kinetic phase again, and a short (and irrelevant, since it will not generate more perturbations during expansion) ekpyrotic phase, until *thermalizing* at a reheat temperature T_r (so that $\rho_{\rm reheat} \sim T_r^4$), once we reach positive potential. This then transitions into the standard radiation dominated cosmology. In effect then, we start with $\rho \sim |V_{\rm end}| \sim T^4$, and we end with $\rho \sim T_r^4$. Since during this (mostly) kinetic dominated phase, we have $w \gg 1$, so $\rho \propto a^{-6} \propto t^{-2}$ (in general $\rho \propto a^{-3(1+w)}$), and therefore $a(t)$ increases by

$$\left(\frac{|V_{\rm end}|^{1/4}}{T_r}\right)^{2/3} \equiv e^{\frac{2}{3}\gamma_{\rm KE}}. \tag{31.15}$$

On the other hand, $H(t) \propto 1/t \propto \rho^{1/2} \propto T^2$, so shrinks by

$$\left(\frac{|V_{\rm end}|^{1/4}}{T_r}\right)^{-2} = e^{-2\gamma_{\rm KE}}. \tag{31.16}$$

- During the standard radiation dominated cosmology, $\rho \propto a^{-4} \propto 1/t^2 \propto T^4$, so $a(t)$ increases by

$$\frac{T_r}{T_0} \equiv e^{N_{\rm rad}}\ , \tag{31.17}$$

where T_0 is the CMBR temperature today. Also, $H(t) \propto 1/t \propto \sqrt{\rho} \propto T^2$, so shrinks by

$$e^{-2N_{\rm rad}}. \tag{31.18}$$

Putting everything together, we find that $a(t)$ grows in a cycle by

$$\exp\left(N_{\rm DE} + \frac{2}{3}\gamma_{\rm KE} + N_{\rm rad}\right). \tag{31.19}$$

On the other hand, $H(t)$ is modified by

$$\frac{\sqrt{|V_{\rm end}|}}{\sqrt{V_{\rm beg}}}\frac{T_r^2}{\sqrt{|V_{\rm end}|}}\frac{T_0^2}{T_r^2} = \frac{T_0^2}{V_{\rm beg}^{1/2}}. \tag{31.20}$$

But this ratio is approximately one, as indeed it should be, since over a cycle, $H(t)$ must return to the same value, the one given by the dark energy V_0. In another way, this gives a constraint on N_{ek}, since the total number of e-folds in $H(t)$ is $N_{ek} - 2(\gamma_{\rm KE} + N_{\rm rad})$:

$$N_{ek} \simeq 2(\gamma_{\rm KE} + N_{\rm rad}). \tag{31.21}$$

We have some constraints on the reheat temperature T_r. First, like in the case of inflation, T_r must be larger than the Big Bang Nucleosynthesis temperature of about 1MeV, which is the first time that we know precisely what happens in cosmology (we know nuclear reactions, and we can find that changing things would mess up the primordial abundances). Second, we know it must be smaller than the GUT scale of about 10^{16} GeV, to avoid the monopole problem (the appearance of undiluted solitons coming from the GUT phase transition). All in all,

$$1\text{MeV} < T_r < 10^{16}\,\text{GeV}. \tag{31.22}$$

But another constraint appears purely from the condition to have radiation domination start only after the crossing of the potential well. Since during the kinetic dominated phase (assuming that roughly the same exponential potential is valid), as we saw last time,

$$\phi(t) \propto \sqrt{\frac{2}{3}}\ln(t) \Rightarrow V(\phi(t)) \propto t^{-c\sqrt{\frac{2}{3}}}, \tag{31.23}$$

and assuming that the potential at the beginning of the phase is of the order of $|V_{\text{end}}|$ (we work in order of magnitude only anyway) and after the crossing is of about V_0, we obtain

$$t_{\text{cross}} = t_{\text{end}}\left(\frac{|V_{\text{end}}|}{V_0}\right)^{\sqrt{\frac{3}{2}}\frac{1}{c}}. \tag{31.24}$$

Then, from the condition that reheating happens (just) after crossing, $t_r \gtrsim t_{\text{cross}}$, and from the fact that $t_r \sim H_r^{-1}$ ($H \propto 1/t$), and $t_r \sim 1/\sqrt{\rho} \sim T_r^{-2}$, we find

$$T_r \lesssim |V_{\text{end}}|^{1/4}\left|\frac{V_0}{V_{\text{end}}}\right|^{\sqrt{\frac{3p}{16}}}. \tag{31.25}$$

As an example, considering a reasonable $|V_{\text{end}}| \sim (10^{15}\,\text{GeV})^4$, and a $p \sim 10^{-3}$, we find

$$T_r \lesssim 10^{12}\,\text{GeV}. \tag{31.26}$$

31.2 The New Ekpyrotic Cosmology

New ekpyrotic cosmology does not attempt to modify the basic scenario of the ekpyrotic case (unlike the cyclic cosmology), but rather attempts to deal with the fact that during the kinetic dominated phase we must switch from a contracting ($H < 0$) Universe to an expanding one ($H > 0$), and as we saw, there is a simple theorem saying we can either have a singularity, or violate the null energy condition.

In the ekpyrotic and cyclic cosmologies, the assumption is that it can happen since we have a singularity, but on the other hand we can calculate what happens to the spectrum of fluctuations since it can go untouched through it, due to the existence of a frame where there is no singularity, and where the string coupling goes to zero at the transition. But various points in that reasoning have been contested, hence the need for an alternative, and the logical alternative is to consider a violation of the null energy condition.

Now a random violation of the null energy condition is bad, since the NEC is assumed to hold in some generality. But one example of a better violation is in the system known as *ghost condensate* (since there is the potential to have a ghost, which of course would also be bad; but that is avoided).

The action one takes is a general function of the kinetic variable

$$X \equiv -\frac{1}{2m^4}(\partial\phi)^2 > 0\,, \tag{31.27}$$

so

$$S = \int d^4x\sqrt{-g}P(X)\,, \tag{31.28}$$

where $P(X)$ is an arbitrary function. Note that a case where $P(X) \simeq A + BX$ at small X, with various higher derivative corrections, has been considered before, going under the name of k-essence if it tries to define dark energy, or k-inflation (or k-flation) for the application to inflation. This would be the case of the DBI action, which is actually

$$P(X) = \sqrt{1+X} \simeq 1 + X + \cdots \tag{31.29}$$

But here we will not be interested in that case. For a cosmology with uniform field ϕ and scale factor, so $a = a(t)$, $\phi = \phi(t)$, the equation of motion coming from the above action is

$$\frac{d}{dt}(a^3 P_{,X}\dot{\phi}) = 0. \tag{31.30}$$

But we want to consider the case that $P(X)$ has *minimum* at some finite X, which can be taken to be (by a redefinition of fields and of m) $X_0 = 1/2$. Then $X_0 = 1/2$ is an *exact* solution of the above equation of motion (since $P_{,X}(X_0) = 0$), and it corresponds to

$$\phi = -m^2 t. \tag{31.31}$$

Note that the sign is arbitrary, and we have chosen the minus for convenience.

The energy-momentum tensor coming from the above action, if $M = m$, is

$$T_{\mu\nu} = g_{\mu\nu}M^4 P(X) + P_{,X}\partial_\mu\phi\partial_\nu\phi\,, \tag{31.32}$$

and it leads to the energy density and pressure (in Minkowski space)

$$\begin{aligned} \rho &= T_{00} = M^4(2P_{,X} - P) \\ P &= T_{ii} = M^4 P(X). \end{aligned} \tag{31.33}$$

At the extremum X_0, where $P_{,X}(X_0) = 0$, we have $P = -\rho$, so $w = -1$, just like a cosmological constant. This is not surprising, since then we can just approximate the action as $S = \int d^4x\sqrt{-g}M^4 P(X_0)$, which is indeed a cosmological constant.

We now consider fluctuations around this minimum. Define

$$\phi(t, \vec{x}) = -m^2 t + \pi(t, \vec{x}) \,, \tag{31.34}$$

where $\pi(t, \vec{x})$ is the fluctuation. Then

$$X \simeq X_0 - \frac{\dot{\pi}}{m^2} + \frac{\dot{\pi}^2 - (\vec{\nabla}\pi)^2}{2m^4} + \mathcal{O}(\pi^3) \,, \tag{31.35}$$

where we have used the fact that $X_0 = 1/2$. Substituting in the action, we find

$$\begin{aligned} S \simeq \int d^4x\sqrt{-g}\Big[& M^4 P(X_0) - P_{,X}(X_0)M^4\frac{\dot{\pi}}{m^2} \\ & + \left(P_{,X}(X_0) + 2X_0 P_{,XX}(X_0)\right)\dot{\pi}^2 - P_{,X}(X_0)(\vec{\nabla}\pi)^2\Big] + \cdots \end{aligned} \tag{31.36}$$

But at the minimum $P_{,X}(X_0) = 0$, we have no ghosts: the leading kinetic term $\dot{\pi}^2$ has the usual sign (the sign would be opposite, and we would have a ghost, at a maximum of $P(X)$). We must then consider a higher order action for the gradient of π (one not encoded in $P(X)$), namely

$$\mathcal{L}_{\text{gradient}} = -\frac{1}{M'^2}(\vec{\nabla}^2\pi)^2 \,, \tag{31.37}$$

besides the kinetic term

$$\mathcal{L}_{\text{kinetic}} = P_{,XX}(1/2)\dot{\pi}^2 \,, \tag{31.38}$$

leading to a dispersion relation for plane waves

$$\omega^2 \sim \frac{k^4}{M'^2}. \tag{31.39}$$

Also at the minimum, expanding

$$P_{,X} \simeq P_{,X}(X_0) - P_{,XX}(X_0)\frac{\dot{\pi}}{m^2} + \cdots \,, \tag{31.40}$$

and adding a potential V (that can absorb the constant $-P(X_0 = 1/2)$), we obtain near the minimum the energy density and pressure

$$\rho \simeq -\frac{KM^4\dot{\pi}}{m^2} + V$$
$$P \simeq -V\,, \tag{31.41}$$

where

$$K \equiv P_{,XX}(1/2). \tag{31.42}$$

Then the second Friedmann equation, which was $M_{\rm Pl}^2\dot{H} = -1/2(\rho + P)$, and which by the NEC should have been negative, is now

$$M_{\rm Pl}^2\dot{H} = \frac{1}{2}\frac{KM^4\dot{\pi}}{m^2}\,, \tag{31.43}$$

and can have either sign, depending on $\dot{\pi}$. In particular, if $\dot{\pi} > 0$, we have $\dot{H} > 0$, unlike standard cosmology.

Then the KG equation for the scalar (in the presence of a potential) becomes

$$\frac{M^4}{m^4}\frac{1}{a^3}\partial_t(a^3 P_{,X}\dot{\phi}) = -V_{,\phi}\,, \tag{31.44}$$

which becomes further

$$\ddot{\pi} + 3H\dot{\phi} = -\frac{V_{,\phi}}{K}\frac{m^4}{M^4}. \tag{31.45}$$

We consider the same field ϕ to be the ghost condensate field, and the ekpyrotic field. Then, during the ekpyrotic phase, we want to have the usual situation, so

$$M^4 P(X) \simeq m^4 X = -\frac{1}{2}(\partial_\mu\phi)^2\,, \tag{31.46}$$

whereas during the kinetic dominated phase we need to have a ghost condensate phase, so

$$P(X) \simeq \frac{K}{2}\left(X - \frac{1}{2}\right)^2. \tag{31.47}$$

Of course, it could be that $P(X) \simeq X$ (linear) at small X, and then we have the minimum at 1/2, but that would mean that there would be a maximum in between 0 and 1/2, and a maximum would mean we do have a perturbative ghost (as we already argued), which is bad.

So instead, it must be that *after* the minimum at $X_0 = 1/2$, $P(X)$ starts growing almost linearly, which is the branch that is associated with ekpyrosis. In fact, in order to have a violation of the NEC as we want, we must have a potential that is approximately

$$V(\phi) \simeq \alpha\Lambda^4\left(1 - \beta\frac{\Lambda^2}{m^2}\frac{\phi}{M_{\rm Pl}}\right)\,, \tag{31.48}$$

with $\alpha, \beta > 0$. The constant part is as it should be from just $P(X_0)$.

This ghost condensation seems then to solve most of our problems with the ekpyrotic phase, however it is unclear how one would obtain it in string theory, or some UV consistent theory. So we have traded a solution for the cosmology to not being able to obtain the model as an effective field theory, at least not in an obvious way.

There are conditions on the model one could impose. Noting that $H(t) \propto V^{1/2}$ (from the Friedmann equation), during ekpyrosis we find

$$e^{\mathcal{N}} \equiv \frac{H_{\rm min}}{H_{\rm ekp}} = \sqrt{\frac{V_{\rm min}}{V_{\rm beg}}}. \tag{31.49}$$

One can in fact show that we need

$$e^{-2\mathcal{N}} |V_{\rm min}| = V_{\rm beg} \gg m^4 , \tag{31.50}$$

and also $|V_{\rm min}| \ll M^4 K/p$, leading to a total interval for the potential at the minimum,

$$m^4 e^{2\mathcal{N}} \ll |V_{\rm min}| \ll \frac{M^4 K}{p}. \tag{31.51}$$

Important Concepts to Remember

- The cyclic model is based on a potential that goes asymptotically to a small constant $V_0 > 0$ as $\phi \to \infty$, and for negative ϕ has a steep exponential drop until a minimum value of $-V_{\rm end}$, and then goes back to zero at even more negative ϕ. For the first two stages, a good choice is $V = V_0(1 - e^{-c\phi})$.
- Cosmology happens in cycles, with one cycle being a dark energy phase, with expansion to a maximum, then contraction, then contracting ekpyrotic phase, then kinetic phase, then bounce. Then expanding kinetic phase, quick ekpyrotic phase, and back to standard cosmology: RD, MD, dark energy phase.
- Entropy and volume increase exponentially over a cycle, with a steady entropy density.
- During the ekpyrotic phase, we have fluctuations being generated in a "fast-roll" approximation, with parameters ϵ, η, such that $n_s - 1 = -4(\epsilon + \eta)$ for single-field ekpyrosis.
- At the end of the short expanding ekpyrotic phase, and entering into the RD era, we have a thermalization at a reheating temperature T_r, which must be > 1MeV, and has an upper bound as well.
- New ekpyrotic cosmology breaks the NEC through a ghost condensate.
- A ghost condensate has an action $S = \int \sqrt{-g} P(X)$, where $X = -(\partial\phi)^2/2m^4$, that has a minimum at $X_0 = 1/2$, i.e., $\phi = -m^2 t$. After the minimum, it starts growing almost linearly, for an usual kinetic term.

Further reading: The cyclic scenario was defined in [55]. The new ekpyrotic scenario was defined in [56]. A good review of the ekpyrotic and cyclic scenarios is [54].

Exercises

(1) Calculate the evolution of the "fast-roll" parameters ϵ, η in the full potential $V = V_0(1 - e^{-c\phi})$, and evaluate the limits for the ekpyrotic and dark energy phases. Repeat for the "slow-roll" inflation parameters ϵ, η.
(2) Consider a ghost condensate with $P(X) = (X - 1/2)^2$. Calculate the equation of state $w(X)$ at arbitrary X.
(3) Show that the null energy condition can be violated if

$$V(\phi) \simeq \alpha\Lambda^4 \left(1 - \beta \frac{\Lambda^2}{m^2} \frac{\phi}{M_{\rm Pl}}\right) , \tag{31.52}$$

with $\alpha, \beta > 0$.
(4) For the case at exercise 2, calculate the time evolution of the scale factor, $a(t)$.

Chapter 32
String Gas Cosmology: Basics and Brandenberger-Vafa Scenario

In this chapter we address the first truly stringy cosmological model, the string gas model, started by Brandenberger and Vafa. The reason for the model starts with a very simple observation: If inflation lasts for more than 70 e-folds (which admitedly is not necessary, we need more than 50–60 e-folds, but restricting $N_e < 70$ would not be needed for experimental reasons, only for calculational reasons, which seems unreasonable), then current cosmological scales, $\sim H_0^{-1}$, that leave an imprint in the CMBR, were less than Planck size $l_{\rm Pl}$ at the initial time. But then we *cannot* trust the effective field theory of inflation, and we would need a full quantum gravity theory, like string theory.

The problem is that we don't have a nonperturbative quantum gravity theory (string theory), whereas for the sub-Planckian regime, this is exactly what we would need. The string gas model, and its generalization, the brane gas model, to be described in the next chapter, is an attempt to formulate a simple string theory model that still preserves some nonperturbative information, specifically T-duality.

Indeed, T-duality says that small scales are the same as large scales in string theory. For a compactified dimension, physics with radius R is the same as physics with radius α'/R, valid even non-perturbatively. If therefore we introduce this T-duality information into the model, we have a model that has some chance of working. The T-duality invariance is in fact related to the existence of a *Hagedorn temperature*, that acts like a phase transition temperature.

32.1 Hagedorn Temperature

The existence of the Hagedorn temperature is due to the exponential increase in the degeneracy of string states (the number of string states with a given energy) at large energy, which is found to be

$$d(m) \simeq e^{4\pi\sqrt{\alpha'}m}\,, \tag{32.1}$$

H. Năstase, *Cosmology and String Theory*, Fundamental Theories of Physics 197, https://doi.org/10.1007/978-3-030-15077-8_32

where the energy is

$$E_h = m = \frac{4}{\sqrt{\alpha'}}(h-1) , \tag{32.2}$$

and h is the conformal weight of the state.

But the thermal partition function at temperature $T = 1/\beta$ is

$$Z(\beta) = \sum_h e^{-\beta E_h} d(m) , \tag{32.3}$$

which in the large m limit becomes the integral

$$Z(\beta) = \int_0^\infty dm e^{4\pi\sqrt{\alpha'}m} e^{-\beta m} , \tag{32.4}$$

which diverges if

$$T > T_H = \frac{1}{4\pi\sqrt{\alpha'}}. \tag{32.5}$$

Here T_H is the *Hagedorn temperature*, where the diverging partition function signals a phase transition, to a system described in terms of different degrees of freedom. Note that at $T = T_H$,

$$\beta = \beta_H = 4\pi\sqrt{\alpha'} \Rightarrow d(m) = e^{b_H m}. \tag{32.6}$$

The exponential increase in the number of states, signaling a phase transition, is also observed in QCD, close to the deconfinement phase transition (and also related to the degeneracy of states of the QCD string).

But the existence of the Hagedorn temperature is related to T-duality invariance, since we need to consider compact directions in order to find β_H. If we consider some directions uncompactified, we find that the specific heat of the system is negative, $C < 0$, whereas to find $C > 0$, we need to consider all directions compactified.

If we consider a fully compactified (9-dimensional in the case of the superstring) space, the density of states at large energy is found to be of the type

$$d(E) = c\frac{e^{\beta_H E}}{E}. \tag{32.7}$$

A (long) string calculation finds indeed this form, with $c = 1$, density of states at level n of

$$d_n = (2n)^{-\frac{11}{4}} e^{2\pi\sqrt{2n}} , \tag{32.8}$$

where the energy of a string state at large level n is $M^2 = E^2 \simeq n/\alpha'$, so we obtain the previous result. We also can obtain positive specific heat, $C > 0$.

The assumptions for the calculation are:

- homogenous fields, i.e., functions of time t only.
- adiabatic approximation, or small time t dependence, which means that we can ignore higher derivatives, i.e., higher α' corrections (quantum worldsheet string corrections).
- weak coupling $g_s \ll 1$, which means that we can ignore quantum string corrections as well.
- all dimensions to be compactified, as we saw.

32.2 3 Large Spatial Dimensions

Given that we need to have all dimensions compactified for $C > 0$, but we know there are 3 very large, apparently uncompactified, dimensions, the question is, how this came to be? It seems natural to start at the Planck scale, with all dimensions compactified with the same scale. Through the cosmological time evolution, it seems natural to have the dimensions start expanding, but we have to explain why only 3 expand, while the other ones remain at the Planck scale. Since we have strings with winding, whose energy is proportional to the size of the compact dimensions ($E = R/\alpha'$), when the dimension grows classical, so do the strings, becoming cosmic strings, and pressing back against the expansion of the dimension. The only way to prevent that is to have annihilation of two closed strings with opposite winding into an unwound string, $w + \bar{w} \leftrightarrow 0$, through a "pair of pants" interaction.

But since the dimension is large and classical, so is the string, so we need to answer whether there is an interaction of *classical* strings, determined solely by geometry. As time evolves, two classical (one dimensional) strings extending in the whole dimension(s) (infinite, if the dimension(s) is/are infinite) oriented arbitrarily and moving in an arbitrary direction, will eventually intersect if the motion is in 3 dimensions, but not if there are 4, since two arbitrary infinite lines form a 3 dimensional space. Of course, there is the case, when the strings are parallel, and never intersect, but that is irrelevant since it has zero measure. Therefore 3 large spatial dimensions is the maximum possible number that allows for annihilation of wound strings of opposite windings, thus avoiding the energy cost that can slow their expansion. In more than 3 spatial dimensions, the windings never annihilate, and the dimensions would stop expanding.

This is then the explanation of 3 large spatial dimensions, even when starting with an arbitrary state of small compact dimensions.

32.3 String Equations of Motion

The idea of the string gas model is very simple: we consider a gas of free (non-interacting) strings, each with Polyakov action. Of course, this is not a very good model: the strings are actually interacting, and non-perturbative, so it is not clear why such a simple model can work so well. On the other hand, in many theories there is some effective description in terms of some free modes, not necessarily the fundamental ones. Perhaps in this case the string gas is made up of effective modes, not the fundamental ones. Or maybe it simply is too simple a model.

For a generic NS-NS background $(g_{\mu\nu}, B_{\mu\nu}, \phi)$, the Polyakov action is

$$S_P = -\frac{1}{4\pi\alpha'}\int d^2\sigma\left[\sqrt{-\gamma}\gamma^{ab}g_{\mu\nu}(X)\partial_a X^\mu\partial_b X^\nu + \epsilon^{ab}B_{\mu\nu}(X)\partial_a X^\mu\partial_b X^\nu \right. \\ \left. +\alpha'\sqrt{-\gamma}\phi(X)\mathcal{R}^{(2)}\right]. \tag{32.9}$$

Its equations of motion are:

- the X^μ equation of motion, $\delta S/\delta X^\mu = 0$, giving

$$\partial_a(\sqrt{-\gamma}\gamma^{ab}\partial_b X^\mu) + \Gamma^\mu{}_{\lambda\nu}\sqrt{-\gamma}\gamma^{ab}\partial_a X^\lambda\partial_b X^\nu + \frac{1}{2}H^\mu{}_{\lambda\nu}\epsilon^{ab}\partial_a X^\lambda\partial_b X^\nu = 0\,, \tag{32.10}$$

- the γ^{ab} equation of motion, $\delta S/\delta\gamma^{ab} = 0$,

$$g_{\mu\nu}(X)\left[\partial_a X^\mu\partial_b X^\nu - \frac{1}{2}\gamma_{ab}\gamma^{cd}\partial_c X^\mu\partial_d X^\nu\right] = 0. \tag{32.11}$$

The Polyakov action is classically Weyl invariant, which means that

$$-\frac{2}{\sqrt{-\gamma}}\gamma^{ab}\frac{\delta S_P}{\delta\gamma_{ab}} = T^a{}_a = 0. \tag{32.12}$$

But for local symmetries like Weyl invariance, we need to not have quantum anomalies, i.e., the symmetry must be preserved at the quantum level. Considering a one-loop quantum action on the string worldsheet, i.e., an α' correction, the quantum Weyl (or conformal) invariance condition becomes

$$T^a{}_a = \beta^G_{\mu\nu}\sqrt{-\gamma}\gamma^{ab}\partial_a X^\mu\partial_b X^\nu + \beta^B_{\mu\nu}\epsilon^{ab}\partial_a X^\mu\partial_b X^\nu + \beta^\phi\sqrt{-\gamma}\mathcal{R}^{(2)}\,, \tag{32.13}$$

where

$$\begin{aligned}
\beta^G_{\mu\nu} &= R_{\mu\nu} + 2\nabla_\mu\nabla_\nu\phi - \frac{1}{4}H_{\mu\rho\sigma}H_\nu{}^{\rho\sigma} + \mathcal{O}(\alpha')\\
\beta^B_{\mu\nu} &= \nabla^\rho H_{\rho\mu\nu} - 2\nabla^\rho\phi H_{\rho\mu\nu} + \mathcal{O}(\alpha')\\
\beta^\phi &= \frac{1}{\alpha'}\left(\frac{D-26}{48\pi^2}\right) + 4\nabla_\rho\phi\nabla^\rho\phi - 4\nabla^2\phi - \mathcal{R}^{(2)} + \frac{1}{12}H_{\mu\nu\rho}H^{\mu\nu\rho} + \mathcal{O}(\alpha').
\end{aligned} \tag{32.14}$$

In the last equation, the constant is written for the bosonic string in a noncritical dimension; for the superstring we have a different one.

The equations of motion can be obtained the spacetime effective action

$$S_{\rm ST} = \frac{1}{2\kappa_D^2}\int d^D x\sqrt{-G}e^{-2\phi}\left[R + c + 4(\nabla\phi)^2 - \frac{1}{12}H_{\mu\nu\rho}\right], \tag{32.15}$$

where

$$2\kappa_D^2 = (2\pi\sqrt{\alpha'})^{D-2}\frac{g_s^2}{2\pi}. \tag{32.16}$$

The string background that solves the spacetime equations of motion can be thought of self-consistently as being a condensate of the string modes, but we can still consider other quantum mechanical strings moving in this background.

32.4 String Gas Cosmological Model

The string gas model has T-duality invariance built in, and in the context of the cosmological model, it becomes scale factor duality.

From homogeneity (meaning that the metric components are only functions of time t) and the cosmological ansatz, the solution is

$$\begin{aligned} ds^2 &= -dt^2 + \sum_{i=1}^{d} a_i^2(t)dx_i^2 \\ B_{\mu\nu} &= 0; \quad \phi = \phi(t). \end{aligned} \tag{32.17}$$

Here $d = D - 1$ is the number of space dimensions. We define the log of the scale factors by

$$a_i(t) = e^{\lambda_i(t)}. \tag{32.18}$$

In the case of FLRW cosmology, the scale factors in the 3 noncompact (or rather, large) spatial dimensions are the same, but here we assume a more general ansatz. We also assume that the compact space is a square torus geometry, defined only by the radii (scale factors) of the extra dimensions. The total volume of space (all the dimensions) is

$$V = \prod_{i=1}^{d} e^{\lambda_i}. \tag{32.19}$$

The T-duality invariant dilaton is

$$\tilde{\phi} = \phi - \frac{1}{2}\ln V = \phi - \frac{1}{2}\sum_{i=1}^{d}\lambda_i. \tag{32.20}$$

Then T-duality leaves $\tilde{\phi}$ invariant, while inverting the scale factors,

$$a_i \to \frac{1}{a_i} \Rightarrow \lambda_i \to -\lambda_i. \tag{32.21}$$

To construct the cosmological model, we add a string source S_P to the spacetime action $S_{\rm ST}$, obtaining the modified equations of motion for $g_{\mu\nu}$ and $B_{\mu\nu}$,

$$\begin{aligned}
&\frac{\delta(S_{\rm ST}+S_P)}{\delta g_{\mu\nu}} = 0 \Rightarrow \\
&R_{\mu\nu} + 2\nabla_\mu\nabla_\nu\phi - \frac{1}{4}H_{\mu\rho\sigma}H_\nu{}^{\rho\sigma} = -\frac{\kappa_D^2 e^{2\phi}}{2\pi\alpha'\sqrt{-g}}\int d^2\sigma\sqrt{-\gamma}\gamma^{ab}\partial_a X^\mu \partial_b X^\nu \delta^{(D)}(X - X(\sigma,\tau)) \\
&\frac{\delta(S_{\rm ST}+S_P)}{\delta B_{\mu\nu}} = 0 \Rightarrow \\
&\nabla_\mu(e^{-2\phi}H^{\mu\nu\rho}) = \frac{\kappa_D^2}{\pi\alpha'\sqrt{-g}}\int d^2\sigma\epsilon^{ab}\partial_a X^\mu\partial_b X^\nu\delta^{(D)}(X - X(\sigma,\tau)) \,,
\end{aligned} \tag{32.22}$$

whereas the ϕ equation is unmodified, since the ϕ term in S_P has the topological $\sqrt{-g}\mathcal{R}^{(2)}$ dependence,

$$4\nabla_\rho\phi\nabla^\rho\phi - 4\nabla^2\phi - \mathcal{R}^{(2)} + \frac{1}{12}H_{\mu\nu\rho}H^{\mu\nu\rho} = 0. \tag{32.23}$$

The energy-momentum tensor of the string source is

$$T^{\mu\nu} = \frac{2}{\sqrt{-g}}\frac{\delta S_P}{\delta g_{\mu\nu}} = -\frac{1}{2\pi\alpha'\sqrt{-g}}\int d^2\sigma\partial_a X^\mu\partial_b X^\nu\delta^{(D)}(X - X(\sigma,\tau)). \tag{32.24}$$

We will use an *adiabatic approximation*, which means that the time variation of the scale factors is slow enough so that we can consider them approximately constant, for the purpose of using flat space formulas for the equations of motion,

$$\partial_+\partial_- X^\mu = 0 \,, \tag{32.25}$$

and for the spectrum. In particular, the left-moving and right-moving momenta for the compact dimensions are as usual

$$\begin{aligned}
p_{R,m} &= \frac{\sqrt{\alpha'}}{R}n_m - \frac{R}{\sqrt{\alpha'}}w_m \\
p_{L,m} &= \frac{\sqrt{\alpha'}}{R}n_m + \frac{R}{\sqrt{\alpha'}}w_m.
\end{aligned} \tag{32.26}$$

In conformal gauge ($\gamma_{ab} = \delta_{ab}$) and using also the light-cone gauge $X^0 = \tau\sqrt{\alpha'}E$ for the residual symmetry (conformal invariance), where $E = p^+$ stands for the energy of the string state, we find

$$\alpha' E^2 = \alpha' \vec{p}^2 + \frac{p_L^2 + p_R^2}{2} + 2(N_L + N_R + a_L + a_R). \tag{32.27}$$

We leave the proof (the same as in flat space) as an exercise. Here $\vec{p}$ is the momentum in the 3 noncompact dimensions, N_L and N_R are occupation numbers for left- and right-movers, and a_L, a_R are the normal ordering constants. In the bosonic string case, $(a_L, a_R) = (-1, -1)$. We also obtain the usual level matching condition

$$p_L^2 - p_R^2 = 4n_i w_i = 4(N_R - N_L - a_R - a_L). \tag{32.28}$$

Eliminating $N_R + a_R$ between the two equations, we find

$$E = \sqrt{\vec{p}^2 + G^{mn}\left(n_m + \frac{w_m}{\alpha'}\right)\left(n_n + \frac{w_n}{\alpha'}\right) + \frac{4}{\alpha'}(N_L + a_L)}\ , \tag{32.29}$$

where G^{mn} is the inverse metric in the compact dimensions.

Replacing this result in the conformal (and lightcone) gauge into the form of the energy-momentum tensor $T_{\mu\nu}$, as well as the solution for X^μ, we obtain

$$T^{00} = \ldots = \frac{E}{\sqrt{-G_{D-1}}}\delta^{(D-1)}(X^i - X^i(\sigma, \tau)). \tag{32.30}$$

We next generalize to N string sources, i.e., a gas made up of N strings. Consider that the gas is split into species s, for instance in the above gauge, parametrized by n_m, w_m, N_L, N_R. Then the energy density is

$$\rho = \sum_s \tilde{n}_s E_s\ , \tag{32.31}$$

where $\tilde{n}_s = n_s/V$ is the number density of species s, and E_s is the energy of a single state of species s. The pressures in directions i are

$$P_i = -\frac{1}{V}\frac{\partial(\rho V)}{\partial \lambda_i}. \tag{32.32}$$

Here V is, as before, the total $d = D - 1$ dimensional volume (in the noncompact *and* the compact directions),

$$V = \prod_{l=1}^{d} e^{\lambda_l}. \tag{32.33}$$

Example

Consider the simplest case, of a gas of bosonic strings, with $a_L = a_R = -1$. Moreover, consider that it is made up only of:

- winding modes, with $w_m \neq 0$, but $n_m = 0$, $N_L = N_R = 1$ (so that $N_L + a_L = N_R + a_R = 0$)
- momentum modes, with $n_m \neq 0$, but with $w_m = 0$, $N_L = N_R = 1$ (so that $N_L + a_L = N_R + a_R = 0$).

Since the radius in direction l is $R_l = e^{\lambda_l}$, and winding modes have energy $n_w^{(l)} R_l = n_w^{(l)} e^{+\lambda_l}$, it follows that the energy density in the case of compactification of all the $d = D - 1$ spatial directions is (now the species s are simply the modes in various compact directions)

$$\rho_w = \sum_{l=1}^{d} \tilde{n}_w^{(l)} e^{\lambda_l(t)} = \frac{1}{V} \sum_{l=1}^{d} n_w^{(l)} e^{\lambda_l(t)} , \tag{32.34}$$

and from the general formula for the pressures, we find

$$P_w^{(l)} = -\tilde{n}_w e^{\lambda_l(t)}. \tag{32.35}$$

In the case that all the d directions are identical, we find

$$P_w = -\frac{1}{d} \rho_w. \tag{32.36}$$

For the momentum modes, the energy is $n_m^{(l)}/R_l = n_w^{(l)} e^{-\lambda_l}$, so that the energy density is

$$\rho_m = \sum_{l=1}^{d} \tilde{n}_m^{(l)} e^{-\lambda_l(t)} = \frac{1}{V} \sum_{l=1}^{d} n_m^{(l)} e^{-\lambda_l(t)} , \tag{32.37}$$

the pressures are now

$$P_m^{(l)} = \tilde{n}_m e^{-\lambda_l(t)} , \tag{32.38}$$

and in the case that all d directions are identical, we find

$$P_m = +\frac{1}{d} \rho_m. \tag{32.39}$$

The results are consistent with the fact that winding modes provide a kind of zero point energy, leading to negative pressure, whereas momentum modes give the usual positive pressure.

32.5 Cosmological Early Time Evolution

We can now consider a simplified model of *cosmological* early time evolution, in which the noncompact dimensions take the FLRW form, i.e., with equal scale factors, $\lambda_1 = \lambda_2 = \lambda_3 \equiv \lambda$, whereas for simplicity we consider all the compact scale factors to be identical, $\lambda_4 = \cdots = \lambda_d = \nu$. We consider also a solution with a constant NS-NS field strength H in the noncompact spatial directions H_{123}, i.e.,

$$\begin{aligned} ds^2 &= -dt^2 + e^{2\lambda(t)}d\vec{x}^2 + e^{2\nu(t)}d\vec{y}^2 \\ H_3 &= h dx^1 \wedge dx^2 \wedge dx^3 \\ \phi &= \phi(t). \end{aligned} \tag{32.40}$$

Though it might seem that a constant field with indices, that picks up a certain direction or directions as special breaks Lorentz invariance, in this case it doesn't break more than what the FLRW solution already does. Indeed, as we already mentioned when talking about the CMBR and the fact that we can measure the velocity of the Earth with respect to it, since we are in general relativity with a non-Minkowski metric, we don't have a (global) Lorentz invariance anymore, only a local Lorentz invariance in an infinitesimal neighbourhood of a point. The (global) cosmological FLRW solution picks out a special (global) frame, that defines the cosmological time. This is equivalent to picking out a special vector A_0, or equivalently, the Poincaré dual spatial 3-tensor H_{123}.

Adding the string sources, as we have seen, adds on the right hand side of the $g^{\mu\nu}$ and ϕ equations of motion terms involving $T_{\mu\nu}$, as

$$\begin{aligned} & R_{\mu\nu} + 2\nabla_\mu\nabla_\nu\phi - \frac{1}{4}H_{\mu\rho\sigma}H_\nu{}^{\rho\sigma} = 2\kappa^2 e^{2\phi}T_{\mu\nu} = 16\pi G_{10}e^{2\phi}T_{\mu\nu} \\ & 4\nabla_\rho\phi\nabla^\rho\phi - 4\nabla^2\phi - \mathcal{R}^{(2)} + \frac{1}{12}H_{\mu\nu\rho}H^{\mu\nu\rho} = 2\kappa^2 e^{2\phi}T^\mu{}_\mu = 16\pi G_{10}e^{2\phi}T^\mu{}_\mu . \end{aligned} \tag{32.41}$$

The $H_{\mu\nu\rho}$ equation of motion is automatically satisfied by the ansatz, as one can check.

With this simplified ansatz, the total spacetime volume $V = e^{3\lambda}e^{6\nu}$, so the energy of the string gas is

$$E = \rho V = \rho e^{3\lambda+6\nu} , \tag{32.42}$$

and the pressures times volume are

$$\bar{P}_i \equiv P_i V = P_i e^{3\lambda+6\nu} , \tag{32.43}$$

and they split into P_λ (usual pressures, times volume, in the noncompact directions), and P_ν (in the compact dimensions).

The Einstein ($g_{\mu\nu}$) and dilaton (ϕ) equations of motion become on our ansatz

$$\begin{aligned}
& c - 3\dot{\lambda} - 6\dot{\nu}^2 + 4\dot{\tilde{\phi}}^2 - \frac{h^2}{2}e^{-6\lambda} = e^{2\tilde{\phi}}E \\
& \ddot{\lambda} - 2\dot{\tilde{\phi}}\dot{\lambda} - \frac{h^2}{2}e^{-6\lambda} = \frac{1}{2}e^{2\tilde{\phi}}\bar{P}_\lambda \\
& \ddot{\nu} - 2\dot{\tilde{\phi}}\dot{\nu} = \frac{1}{2}e^{2\tilde{\phi}}\bar{P}_\nu \\
& 2\ddot{\tilde{\phi}} - 3\dot{\lambda}^2 - 6\dot{\nu}^2 = \frac{1}{2}e^{2\tilde{\phi}}E .
\end{aligned} \tag{32.44}$$

We also need to impose the conservation equation for the energy-momentum tensor, $\partial^\mu T_{\mu\nu} = 0$, which becomes

$$\dot{E} + 3\dot{\lambda}\bar{P}_\lambda + 6\dot{\nu}\bar{P}_\nu = 0. \tag{32.45}$$

These equations, with ρ, P_λ, P_ν given by the gas of string momentum and winding modes, cannot be solved analytically, but we can make simulations of the time evolution.

One finds a period of "cosmological loitering", during which $\dot{\lambda} = 0$, $\dot{\tilde{\phi}} = 0$, i.e., our 3 spatial dimensions don't evolve. (in fact, in the Einstein frame, they contract) This lasts until all winding modes annihilate. After its end, our 3 spatial dimensions can grow large. The existence of this period solves the horizon problem in a different way than in inflation. Rather than having a small patch blow exponentially way outside the horizon, the size remains constant as the Hubble size evolves, but the effect is the same. Also the isotropy (if not the homogeneity) problem is solved, as initial differences between the various directions get smoothed over.

32.6 Moduli Stabilization

All the "radions", the radii (or sizes) of the extra dimensions, will be stabilized by the string gas model. Going back to the general ansatz for the compact dimensions, i.e.,

$$e^{2\nu}d\vec{y}^2 \to \sum_{\alpha=1}^{6} b_\alpha^2(t)dy_\alpha^2 , \tag{32.46}$$

one finds the equation of motion for them

$$\ddot{b}_\alpha + \left(3H + \sum_{\beta=1,\beta\neq\alpha}^{6} \frac{\dot{b}_\beta}{b_\beta}\right)\dot{b}_\alpha = \sum_{m,n} 8\pi G_N \frac{\mu_{m,n}}{\epsilon_{m,n}} S , \tag{32.47}$$

where $\mu_{m,n}$ is the number density of string states with momentum n nad winding m, $\epsilon_{m,n}$ is the energy of this state, and

$$S = \sum_\alpha \left(\frac{m_\alpha}{b_\alpha}\right)^2 - \sum_\alpha n_\alpha^2 b_\alpha^2 + \frac{2}{D-1}\left[\vec{n}\cdot\vec{n} + \vec{n}\cdot\vec{m} + 2(N-1)\right]. \qquad (32.48)$$

The last term will vanish at the self-dual (T-duality invariant) radius, i.e., it will be stabilized (since it minimizes the right hand side of the equation of motion, the "potential").

Important Concepts to Remember

- For an inflation that lasts more than 70 e-folds, we find that current Hubble scales were initially sub-Planckian, hence it makes no sense to treat them in effective field theory (inflation).
- The Hagedorn temperature $T_H = 1/(4\pi\sqrt{\alpha'})$ is the maximum temperature of a gas of strings, after which the partition function diverges due to the exponential number density of states.
- The existence of the Hagedorn temperature is related to T-duality invariance. The calculation of the density of states for a fully compactified theory (the only case that gives positive specific heat $C > 0$) gives the exponential behaviour.
- From the initially all compactified dimensions, only 3 expand and become large, since only for those can strings find each other, interact, and annihilate their winding (which otherwise keeps the dimensions small).
- The string equations of motion are obtained from quantum conformal invariance of the string worldsheet, and can be derived from a spacetime effective action.
- Considering a gas of strings, with momentum modes and winding modes, the winding modes have equation of state $P = -\rho/d$ and the momentum modes have equation of state $P = +\rho/d$.
- Early time cosmology has a "cosmological loitering" phase, in which the 3 spatial dimensions don't evolve (in fact, contract in the Einstein frame), and winding modes annihilate. It solves the horizon and isotropy problem differently than in inflation.
- Moduli get stabilized during the time evolution.

Further reading: The original paper of Brandenberger and Vafa is [57]. Good reviews for string gas cosmology are [58, 59] and the recent one [60].

Exercises

(1) Prove that for $N_e > 70$-folds of inflation current cosmological scales ($\sim H_0^{-1}$) were $l_{\rm Pl}$ at initial time.

(2) Show that in conformal gauge, with $X^\rho = \tau\sqrt{\alpha'}E$, we obtain

$$\alpha' E^2 = \alpha' \vec{p}^2 + \frac{p_L^2 + p_R^2}{2} + 2(N_L + N_R + a_L + a_R). \qquad (32.49)$$

(3) Show that the conservation equation $\partial^{\mu} T_{\mu\nu} = 0$ becomes

$$\dot{E} + 3\dot{\lambda}\bar{P}_{\lambda} + 6\dot{\nu}\bar{P}_{\nu} = 0. \tag{32.50}$$

(4) Show that the equations of motion of the spacetime action plus the string source, $S_{\mathrm{ST}} + S_P$, are given by (32.22).

Chapter 33
String Gas and Brane Gas Developments

In this chapter we will first describe a simple generalization of the previously defined string gas cosmology, namely brane gas; and then we will calculate the spectrum of scalar and tensor perturbations in string gas cosmology.

33.1 Brane Gas

The idea of the brane gas is obvious from its name: replace the gas of strings from the previous chapter with a gas of p-branes.

The p-brane action is taken to be a DBI action for interaction with a NS-NS background, and with an electromagnetic field on its worldvolume, so

$$S_{\rm DBI,p} = -T_p \int d^{p+1}\xi e^{-\phi}\sqrt{-\det(\tilde{g}_{ab} + \tilde{B}_{ab} + 2\pi\alpha' F_{ab})}\;, \tag{33.1}$$

where $\tilde{g}_{ab}$ and $\tilde{B}_{ab}$ are induced NS-NS metric and B-fields on the $p+1$ dimensional worldvolume, and F_{ab} is the field strength of the electromagnetic field on the worldvolume,

$$\begin{aligned}\tilde{g}_{ab} &= g_{\mu\nu}\partial_a X^\mu \partial_b X^\nu \\ \tilde{B}_{ab} &= B_{\mu\nu}\partial_a X^\mu \partial_b X^\nu .\end{aligned} \tag{33.2}$$

In a static gauge, $X^a = \xi^a$, $X^i = \phi^i$, we find

$$\begin{aligned}\tilde{g}_{ab} &= g_{ab} + g_{ij}\partial_a\phi^i\partial_b\phi^j + g_{ib}\partial_a\phi^i + g_{ia}\partial_b\phi^i \\ \tilde{B}_{ab} &= B_{ab} + B_{ij}\partial_a\phi^i\partial_b\phi^j + B_{i[a}\partial_{b]}\phi^i .\end{aligned} \tag{33.3}$$

H. Năstase, *Cosmology and String Theory*, Fundamental Theories of Physics 197, https://doi.org/10.1007/978-3-030-15077-8_33

The tension of the p-brane is

$$T_p = \frac{1}{(2\pi\sqrt{\alpha'})^{p+1}} \frac{\pi}{g_s} \tag{33.4}$$

We also consider a cosmological background in conformal time, where for simplicity we consider the same scale factor for all directions,

$$\begin{aligned} ds^2 &= a^2(\eta)[-d\eta^2 + d\vec{x}^2] \\ B_{\mu\nu} &= 0. \end{aligned} \tag{33.5}$$

Substituting it into the DBI action, and expanding in the number of derivatives, we find

$$S_{\rm DBI,p} \simeq -T_p \int d^{p+1}\xi a(\eta)^{p+1} e^{-\phi} \left[1 + \frac{1}{2}(\partial_a \phi)^2 + \pi^2 \alpha'^2 a^{-4} F_{ab}^2 + \cdots \right]. \tag{33.6}$$

We can now calculate the energy-momentum tensor T^{ab} for the analog of winding modes, which are now "wrapped" modes of the brane (all p directions wrap some of the -compact- spatial directions). From it, we can calculate the energy density ρ and the pressure P, obtaining

$$P_i = w_i \rho; \quad w_p = -\frac{p}{d}. \tag{33.7}$$

Here d is the total number of space dimensions. For a 3-brane in $3+1$ dimensional space, we would obtain $w = -1$, since this would then truly be a cosmological constant. On the other hand, for momentum modes, we have as usual $w = +1/d$, whereas in general normal matter has $w \in [0, 1]$.

The equations of motion for the cosmology are obtain from the equations of the previous chapter (32.44), considered for $c = 0$ (i.e., in the critical dimension, $D = 26$ for the bosonic string and $D = 10$ for the superstring), $h = 0$ (since we now consider only $B_{\mu\nu} = 0$), and $\lambda = \nu$ since we now consider the same scale factor for all dimensions, which is certainly true during the first, *Hagedorn*, or cosmological loitering, phase of the cosmological evolution, before our 3 spatial dimensions start expanding. Indeed, in the following we will only consider this Hagedorn phase. The equations are then

$$\begin{aligned} -d\dot{\lambda} + 4\dot{\tilde{\phi}}^2 &= e^{2\tilde{\phi}} E \\ \ddot{\lambda} - 2\dot{\tilde{\phi}}\dot{\lambda} &= \frac{1}{2} e^{2\tilde{\phi}} \bar{P} \\ 2\ddot{\tilde{\phi}} - d\dot{\lambda}^2 &= \frac{1}{2} e^{2\tilde{\phi}} E. \end{aligned} \tag{33.8}$$

Here as usual $\tilde{\phi} = \phi - \frac{d}{2}\lambda$ is the T-duality invariant dilaton. The energy and pressure follow from the previous brane action. Indeed, we can say that, since the energy is proportional to the brane tension times the spatial volume,

$$E \propto T_p[a(\eta)]^p \,, \tag{33.9}$$

the effective potential in 10 dimensional string frame is

$$V_{\text{eff}}(\lambda) = \beta_p e^{2\tilde{\phi}} e^{p\lambda}. \tag{33.10}$$

Here $e^{p\lambda} = a(\eta)^p$ is the spatial volume of the p-brane, $\beta_p \propto T_p$. In more generality, if our 3 dimensions have scale factor a and the other d have scale factor b (valid after the Hagedorn phase), the effective potential in string frame would be

$$V_s^{(4+d)} = \frac{N}{(2\pi\sqrt{\alpha'})^4} \frac{b^k}{a^3 b^d} \,, \tag{33.11}$$

where the $1/(2\pi\sqrt{\alpha'})^4 = T_3$ is a 3-brane tension, N is the number of branes, $a^3 b^d$ is the total volume of space, and $k = 1$ for the p-branes and $k = -1$ for momentum modes is the dependence coming either from the p-dimensional volume of the branes wrapped over the compact space (so that $p \leq d$) or from the $1/R = 1/b$ momentum of the momentum modes.

The importance of the p-brane gas model comes from a generalization of the argument about the string gas selecting 3 large dimensions. For two generic classical p-branes to intersect as time evolves, we need a large $2p + 1$ dimensional space, since two generic p-branes form a $2p + 1$ dimensional space. That means that, through the interaction, we can eventually annihilate all p-branes of opposite wrapping number only if the dimension of space is $2p + 1$ or smaller. But higher p branes annihilate slower than lower p ones (there's a larger phase space to be covered), which means that we create a hierarchy of dimensions.

Then, for instance, in the presence of strings and 2-branes (which annihilate in 5+1 dimensions), we will create 3 very large (almost infinite) dimensions (because of the strings), but also two larger dimensions (because of the 2-branes), ending up with a *large extra dimensions* scenario. This is the main interest of the brane gas: the possibility of hierarchies for dimensions, allowing for a large extra dimensions scenario.

33.2 Cosmological Perturbations Generalities

The cosmological evolution we will be interested in happens in a *Hagedorn phase*. Like we said in the previous chapter, during this phase (also called cosmological loitering phase), in Einstein frame the horizon scale $H^{-1}(t)$ *shrinks*, solving the horizon problem in the opposite way to inflation: there the scales blow up exponentially, and H^{-1} is fixed, here scales are evolving slowly, but the horizon shrinks. It in fact shrinks until it reaches Hagedorn time t_R, where it is minimum,

$$H^{-1}(t_R) \sim \frac{M_{\rm Pl}}{T_H^2}\,, \tag{33.12}$$

and then, for $t > t_R$, when many scales have gone already outside the horizon, we start standard cosmology, and H^{-1} starts increasing again. Perturbations observed in the CMBR will be created close to this minimum, so this is what we will try to calculate.

We therefore move on to the calculation of cosmological perturbations, which will leave an imprint in the CMBR. We want to calculate the scalar and tensor perturbations, using a starting point similar to that for inflation: The most general perturbation of the metric with scalar and tensor components, in a Newtonian gauge with $\Phi = \Psi$ and in conformal time, is

$$ds^2 = a^2(\eta)\left\{-(1+2\Phi(x,\eta))d\eta^2 + [(1-2\Phi(x,\eta))\delta_{ij} + h_{ij}]dx^i dx^j\right\}. \tag{33.13}$$

Here Φ will generate scalar perturbations, and h_{ij} tensor perturbations.

For the Newton potential Φ, we have the Newton equation relating it to a small density fluctuation acting as a source,

$$\nabla^2\Phi = 4\pi G_N \delta\rho. \tag{33.14}$$

In momentum space, we have

$$-k^2\Phi(k) = 4\pi G_N\delta\rho(k) = 4\pi G_N \delta T^0{}_0(k). \tag{33.15}$$

For the tensor perturbation h_{ij}, for $i \neq j$, we have similarly

$$\nabla^2 h_{ij} = 4\pi G_N \delta T^i{}_j\,, \tag{33.16}$$

or in momentum space

$$-k^2 h_{ij}(k) = 4\pi G_N \delta T^i{}_j. \tag{33.17}$$

Substituting in the correlators that we want to calculate, we obtain (for $i \neq j$)

$$\begin{aligned}\langle|\Phi(k)|^2\rangle &= \frac{(4\pi G_N)^2}{k^4}\langle\delta T^0{}_0(k)\delta T^0{}_0(k)\rangle\\ \langle|h(k)|^2\rangle &= \frac{(4\pi G_N)^2}{k^4}\langle\delta T^i{}_j(k)\delta T^i{}_j(k)\rangle.\end{aligned} \tag{33.18}$$

33.3 Correlators from Thermodynamics

We now express the energy-momentum tensor correlators in terms of thermodynamical quantities, which will be later calculated for the string gas.

For a general system with a metric component g_{00}, the partition function Z, free energy F and Hamiltonian H, for states s, are related by

$$e^{-\beta F} = Z = \sum_s e^{-\beta\sqrt{-g_{00}}H(s)}. \tag{33.19}$$

For the strings, with a nontrivial g_{00}, but in a static configuration, so the Lagrangian equals minus the Hamiltonian, $\mathcal{L} = -H$, we have the action (written in terms of the proper time $d\tau = dt\sqrt{-g_{00}}$)

$$S = \int dt\sqrt{-g_{00}}H. \tag{33.20}$$

In a path integral formulation in Euclidean space, $\int dt = \beta = 1/T$.

But since the energy-momentum tensor is

$$T_{\mu\nu} = \frac{2}{\sqrt{-g}}\frac{\delta S}{\delta g^{\mu\nu}} \Rightarrow T^\mu{}_\nu = \frac{2g^{\mu\lambda}}{\sqrt{-g}}\frac{\delta S}{\delta g^{\lambda\nu}}, \tag{33.21}$$

and the variational derivative δ means normal derivative d, when we drop the time integral $\int dt$, we obtain in a state s

$$T^\mu{}_\nu(s) = \frac{2g^{\mu\lambda}}{\sqrt{-g}}\frac{d}{dg^{\lambda\nu}}[-\sqrt{-g_{00}}H(s)]. \tag{33.22}$$

But a general ensemble average gives

$$\langle M\rangle \equiv \frac{1}{Z}\sum_s e^{-\beta\sqrt{-g_{00}}H(s)}M, \tag{33.23}$$

so applying it for $T^\mu{}_\nu$ gives

$$\langle T^\mu{}_\nu\rangle = \frac{2g^{\mu\lambda}}{\sqrt{-g}}\frac{\delta}{\delta g^{\lambda\nu}}\ln Z, \tag{33.24}$$

since

$$\delta e^{-\beta\sqrt{-g_{00}}H(s)} = e^{-\beta\sqrt{-g_{00}}H(s)}\delta[-\beta\sqrt{-g_{00}}H(s)] = e^{-\beta\sqrt{-g_{00}}H(s)}d[-\sqrt{-g_{00}}H(s)]. \tag{33.25}$$

Here we have used that $\beta = \int dt$ and the system is time-independent.

Now considering a second derivative will give the correlator of the fluctuations of the energy-momentum tensor,

$$\begin{aligned}\langle \delta T^{\mu}{}_{\nu}\delta T^{\sigma}{}_{\rho}\rangle &= \langle T^{\mu}{}_{\nu}T^{\sigma}{}_{\rho}\rangle - \langle T^{\mu}{}_{\nu}\rangle\langle T^{\sigma}{}_{\rho}\rangle \\ &= \frac{2g^{\mu\alpha}}{\sqrt{-g}}\frac{\delta}{\delta g^{\alpha\nu}}\left(\frac{g^{\sigma\lambda}}{\sqrt{-g}}\frac{\delta}{\delta g^{\lambda\rho}}\ln Z\right) + \frac{2g^{\sigma\alpha}}{\sqrt{-g}}\frac{\delta}{\delta g^{\alpha\rho}}\left(\frac{g^{\mu\lambda}}{\sqrt{-g}}\frac{\delta}{\delta g^{\lambda\nu}}\ln Z\right).\end{aligned} \tag{33.26}$$

Applying this for $\mu = \nu = \rho = \sigma = 0$, we obtain

$$\begin{aligned}\langle \delta T^{0}{}_{0}\delta T^{0}{}_{0}\rangle &= \langle(\delta\rho)^2\rangle \\ &= \frac{4g^{00}}{\sqrt{-g}}\frac{\delta}{\delta g^{00}}\left(\frac{g^{00}}{\sqrt{-g}}\frac{\delta}{\delta g^{00}}\ln Z\right).\end{aligned} \tag{33.27}$$

Considering a 3 dimensional spatial metric $g_{ij} \simeq R^2\delta_{ij}$, we obtain

$$\langle(\delta\rho)^2\rangle = \frac{1}{R^6}\left(\frac{2g_{00}}{\sqrt{-g_{00}}}\frac{\delta}{\delta g_{00}}\right)^2 \ln Z = \frac{1}{R^6}\left(\frac{\delta}{\delta\sqrt{-g_{00}}}\right)^2 \ln Z. \tag{33.28}$$

But since $\ln Z = -\beta F$, and, when acting on Z, $\delta/\delta\sqrt{-g_{00}} = d/(\sqrt{-g_{00}}d\beta)$, we obtain

$$\langle(\delta\rho)^2\rangle = -\frac{1}{R^6(-g_{00})}\frac{d^2}{d\beta^2}(\beta F) = -\frac{1}{R^6(-g_{00})}\frac{\partial}{\partial\beta}\left(F + \beta\frac{\partial}{\partial\beta}F\right). \tag{33.29}$$

From the thermodynamic relations $F = E - TS$ and $dF = -SdT - PdV$, with $V = R^3$, we find

$$E = F + TS = F + \beta\left(\frac{\partial F}{\partial\beta}\right), \tag{33.30}$$

which means that

$$T^2 C_V = T^2\left(\frac{\partial E}{\partial T}\right)_V = -\left(\frac{\partial E}{\partial\beta}\right)_V = -\frac{\partial}{\partial\beta}\left(F + \beta\left(\frac{\partial F}{\partial\beta}\right)_V\right)_V. \tag{33.31}$$

Finally, we obtain

$$\langle(\delta\rho)^2\rangle = \frac{T^2}{R^6}C_V. \tag{33.32}$$

Next we apply the same methods for the spatial components,

$$\begin{aligned}\langle \delta T^{i}{}_{j}\delta T^{i}{}_{j}\rangle &= \langle T^{i}{}_{j}T^{i}{}_{j}\rangle - \langle T^{i}{}_{j}\rangle\langle T^{i}{}_{j}\rangle \\ &= \frac{4g^{ik}}{\sqrt{-g}}\frac{\delta}{\delta g^{kj}}\left(\frac{g^{il}}{\sqrt{-g}}\frac{\delta}{\delta g^{lj}}\ln Z\right),\end{aligned} \tag{33.33}$$

with $i \neq j$. With $g_{ij} = R^2(\delta_{ij} + h_{ij}) \simeq R^2\delta_{ij}$, we obtain

$$\langle \delta T^i{}_j \delta T^i{}_j \rangle = \frac{4\delta R^2}{\sqrt{-g_{00}}R^3} \frac{\delta}{\delta R^2} \left(\frac{R^2}{\sqrt{-g_{00}}R^3} \frac{\delta}{\delta R^2} \right) \ln Z = \left(\frac{1}{\sqrt{-g_{00}}R^3} \frac{\delta}{\delta \ln R} \right)^2 \ln Z. \tag{33.34}$$

But since $\delta/\delta R$, when acting on Z, removes also an $\int dt = \beta$, so $\delta/\delta R = (1/\beta)d/dR$, and $\ln Z = -\beta F$, we obtain

$$\langle \delta T^i{}_j \delta T^i{}_j \rangle = \frac{1}{-g_{00}\beta} \frac{1}{R^3} \frac{\partial}{\partial \ln R} \left(-\frac{1}{R^3} \frac{\partial F}{\partial \ln R} \right). \tag{33.35}$$

Then, since $dE = TdS - PdV$ and $dF = -SdT - PdV$ and $V = R^3$, we have

$$p = T\left(\frac{\partial S}{\partial V}\right)_E = -\left(\frac{\partial F}{\partial V}\right)_T = -\frac{1}{3R^3}\left(\frac{\partial F}{\partial \ln R}\right)_T \tag{33.36}$$

so that finally

$$\langle \delta T^i{}_j \delta T^i{}_j \rangle = \frac{3}{(-g_{00})\beta R^2} \left(\frac{\partial p}{\partial R}\right)_T. \tag{33.37}$$

33.4 Thermodynamics for String Gas

With the general thermodynamics formulas, we could calculate the fluctuation correlator for any quantum gas. But in particular, we want to apply for the gas of strings. We need to calculate the total number of states of given energy E and volume R^3, $\Omega(E, R)$, and from it the entropy

$$S(E, R) = \ln \Omega(E, R). \tag{33.38}$$

A string calculation finds

$$\Omega(E, R) \simeq \beta_H e^{\beta_H E + n_H V}[1 + \delta\Omega_{(1)}], \tag{33.39}$$

where $\beta_H = 1/T_H$ with T_H the Hagedorn temperature, $n_H \sim l_s^{-3}$ is the Hagedorn number density and $\delta\Omega_{(1)}$, with $-\delta\Omega_{(1)} \ll 1$, comes from a density of states, and is found to be

$$\delta\Omega_{(1)} \simeq -\frac{(\beta_H E)^5}{5!} e^{-(\beta_H - \beta_1)(E - \rho_H V)}, \tag{33.40}$$

where $\rho_H \sim l_s^{-4}$ is a Hagedorn density, and β_1 is the closest singularity to β_H, which obeys

$$\beta_H - \beta_1 \sim \frac{l_s^3}{R^2}; \qquad R \gg l_s$$
$$\sim \frac{R^2}{l_s}; \qquad R \ll l_s. \tag{33.41}$$

We then obtain

$$S(E,R) \simeq \beta_H E + n_H V + \ln[1 + \delta\Omega_{(1)}] \,, \tag{33.42}$$

and, since $\frac{1}{T} = \left(\frac{\partial S}{\partial E}\right)_V$, we find

$$T(E,R) = \left[\left(\frac{\partial S}{\partial E}\right)_V\right]^{-1} \simeq \left[\beta_H + \left(\frac{\partial \delta\Omega_{(1)}}{\partial E}\right)_V\right]^{-1} \simeq T_H\left[1 - \frac{1}{\beta_H}\left(\frac{\partial \Omega_{(1)}}{\partial E}\right)_V\right]$$
$$\simeq T_H\left[1 + \frac{\beta_H - \beta_1}{\beta_H}\delta\Omega_{(1)}\right]. \tag{33.43}$$

Inverting this relation, and using $\beta_H - \beta_1 \simeq l_s^3/R^2$, since we are interested in the regime $R \gg l_s$, we obtain

$$l_s^3 \delta\Omega_{(1)} \simeq -\frac{R^2}{T_H}\left(1 - \frac{T}{T_H}\right). \tag{33.44}$$

Then inverting the relation (33.40) to find E as a function of $\Omega_{(1)}$, we obtain

$$E \simeq \frac{1}{\beta_H - \beta_1} \ln \frac{1}{-\delta\Omega_{(1)}} \simeq \frac{R^2}{l_s^3} \ln\left[\frac{l_s^3 T}{R^2\left(1 - \frac{T}{T_H}\right)}\right]. \tag{33.45}$$

Then finally

$$C_V = \left(\frac{\partial E}{\partial T}\right)_V \simeq \frac{R^2}{l_s^3}\frac{\partial}{\partial T} \ln \frac{T}{1 - \frac{T}{T_H}} = \frac{R^2/l_s^3}{T\left(1 - \frac{T}{T_H}\right)}. \tag{33.46}$$

For the pressure, the formula for the entropy (33.42) gives

$$p = T\left(\frac{\partial S}{\partial V}\right)_E \simeq n_H T + T\left(\frac{\partial}{\partial V}\delta\Omega_{(1)}\right)_E. \tag{33.47}$$

But

$$\left(\frac{\partial \delta\Omega_{(1)}}{\partial V}\right)_E = \left[-\frac{\partial(\beta_H - \beta_1)}{\partial V}(E - \rho_H V) + \rho_H(\beta_H - \beta_1)\right]\delta\Omega_{(1)}. \tag{33.48}$$

Using $\beta_H - \beta_1 = l_s^3/R^2$, $\rho_H = l_s^{-4}$, $V = R^3$, E from (33.45) and $\delta\Omega_{(1)}$ from (33.44), we find after some algebra

$$p \simeq n_H T - \frac{T}{T_H}\frac{1-\frac{T}{T_H}}{l_s^3}\left\{\frac{5}{3l_s} + \frac{2}{3R}\ln\left[\frac{l_s^3 T}{R^2\left(1-\frac{T}{T_H}\right)}\right]\right\}. \tag{33.49}$$

If we are very close to the Hagedorn temperature, $T \simeq T_H$, we can approximate further

$$p \simeq n_H T_H - \frac{2}{3}\frac{1-\frac{T}{T_H}}{l_s^3 R}\ln\left[\frac{l_s^3 T}{R^2\left(1-\frac{T}{T_H}\right)}\right] \tag{33.50}$$

Moreover, we can calculate the derivative

$$\left(\frac{\partial p}{\partial R}\right)_T \simeq \frac{T}{T_H}\frac{1-\frac{T}{T_H}}{l_s^3 R^2}\left(\frac{4}{3} + \frac{2}{3}\ln\left[\frac{l_s^3 T}{R^2\left(1-\frac{T}{T_H}\right)}\right]\right). \tag{33.51}$$

Finally then, we can calculate the energy-momentum tensor correlators,

$$\langle(\delta\rho)^2\rangle = \frac{T^2}{R^6}C_V \simeq \frac{l_s^{-3}}{R^4}\frac{T}{\left(1-\frac{T}{T_H}\right)} \tag{33.52}$$

and, for $i \neq j$,

$$\langle\delta T^i{}_j \delta T^i{}_j\rangle = \frac{3T}{R^2}\left(\frac{\partial p}{\partial R}\right)_T \simeq \frac{3T^2}{T_H}\frac{1-\frac{T}{T_H}}{l_s^3 R^4}\left(\frac{4}{3} + \frac{2}{3}\ln\left[\frac{l_s^3 T}{R^2\left(1-\frac{T}{T_H}\right)}\right]\right). \tag{33.53}$$

For $T \simeq T_H \simeq l_s^{-1}$, we can approximate further,

$$\langle\delta T^i{}_j \delta T^i{}_j\rangle \simeq \frac{2T\left(1-\frac{T}{T_H}\right)}{l_s^3 R^4}\ln\left[\frac{l_s^2}{R^2\left(1-\frac{T}{T_H}\right)}\right]. \tag{33.54}$$

33.5 String Gas Power Spectra and Tilts

We can finally relate to CMBR physics. The correlators of $\Phi(k)$ and $|h(k)|$ give rise to scalar and tensor power spectra in the CMBR. It is assumed that the fluctuations in the CMBR are close to the end of the Hagedorn phase, and the beginning of standard cosmology, like in the case of inflation (where the fluctuations are computed towards the end of inflation).

Scalar Power Spectrum

As we saw in part I of the book, the scalar power spectrum in the CMBR and its tilt n_s is defined by

$$P_\Phi(k) = \frac{k^3}{2\pi^2}|\Phi(k)|^2 \propto k^{n_s-1}\,, \tag{33.55}$$

so

$$n_s - 1 = \frac{d}{d\ln k}\ln P_\Phi(k). \tag{33.56}$$

But $\langle|\Phi(k)|^2\rangle$ was related to $\langle(\delta\rho)^2\rangle$ in (33.18), which gives

$$P_\Phi(k) = \frac{8G_N^2}{k}\langle(\delta\rho(k))^2\rangle. \tag{33.57}$$

Note however that we need a density fluctuation in momentum space, $\delta\rho(k)$, whereas the correlators we have calculated were in position space. We can relate them by considering a sphere of radius $R = 1/k$. Then $\delta\rho = \delta M/R^3 = k^3\delta M$, whereas in momentum space (because of the Fourier transform)

$$\langle(\delta\rho(k))^2\rangle = \frac{1}{V}\langle(\delta M)^2\rangle = k^3\langle(\delta M)^2\rangle = \frac{1}{k^3}\langle(\delta\rho)^2\rangle. \tag{33.58}$$

Then, from the density correlator (33.52), we find

$$P_\Phi(k) = \frac{8G_N}{l_s^3}\frac{T(k)}{1-\frac{T(k)}{T_H}}. \tag{33.59}$$

We can rewrite this, using $T_H \sim l_s^{-1}$, as

$$P_\Phi(k) \sim \left(\frac{l_{\rm Pl}}{l_s}\right)^4\frac{T(k)}{T_H}\frac{1}{1-\frac{T(k)}{T_H}} \tag{33.60}$$

The first observation ia that the spectrum is almost scale invariant, $n_s - 1 \simeq 0$, since there is no explicit k dependence, and only a very small scale implicit dependence in the temperature $T(k)$. That is due to the fact that, as in inflation, we calculate the fluctuation when the scale exits the horizon (since it is frozen in there, and then goes back inside the horizon only now), at $t_i(k)$, just before the minimum horizon size time t_R.

But moreover, we have a slight red tilt, as observed experimentally, since, like in inflation, larger scales (larger wavelengths λ, thus smaller k) exit the horizon sooner (at smaller $t_i(k)$), when the temperature $T(k)$ is higher: the temperature slightly decreases with time at the end of the Hagedorn phase, while remaining smaller than the maximum one, T_H. That means that $dT(k)/dk < 0$.

Specifically, we obtain

$$n_s - 1 = \frac{d\ln P_\Phi}{d\ln k} = \left(\frac{1}{1-\frac{T(k)}{T_H}}+1\right)k\frac{d}{dk}\frac{T(k)}{T_H} \simeq \frac{1}{1-\frac{T(k)}{T_H}}k\frac{d}{dk}\frac{T(k)}{T_H} < 0. \tag{33.61}$$

It is however impossible to calculate analytically dT/dk, so that is the most we can say: $n_s - 1$ can be small and negative, as observed experimentally.

Tensor Power Spectrum

Similarly, we define for the tensor perturbation

$$P_h(k) = \frac{k^3}{2\pi^2}|h(k)|^2 \propto k^{n_T-1}. \tag{33.62}$$

We have defined the power with a -1, even though sometimes it is defined without it, in order to have symmetry with the scalar spectrum.

Then in the same way as for the scalar case, we find

$$P_h(k) = \frac{8G_N^2}{k}\langle\delta T^i{}_j(k)\delta T^i{}_j(k)\rangle\,, \tag{33.63}$$

but we need to relate to the previously calculated one by considering a sphere of radius $R = 1/k$, so

$$P_h(k) = \frac{8G_N^2}{k^4}\langle\delta T^i{}_j(R)\delta T^i{}_j(R)\rangle\,, \tag{33.64}$$

and, using (33.53), we find

$$P_h(k) \simeq \frac{16G_N^2}{l_s^3}\frac{T^2\left(1-\frac{T}{T_H}\right)}{T_H}\ln\left[\frac{l_s^2}{R^2\left(1-\frac{T}{T_H}\right)}\right], \tag{33.65}$$

that becomes

$$P_h(k) \simeq \left(\frac{l_{\rm Pl}}{l_s}\right)^4\frac{T^2}{T_H^2}\left(1-\frac{T}{T_H}\right)\ln\left[\frac{l_s^2}{R^2\left(1-\frac{T}{T_H}\right)}\right]. \tag{33.66}$$

Again we see that the tensor spectrum is approximately scale invariant, $n_T - 1 \simeq 0$, but now we find the tilt

$$n_T - 1 = \frac{d \ln P_h(k)}{d \ln k} = \left(-\frac{1}{1 - \frac{T}{T_H}} + 2 \right) k \frac{d}{dk} \frac{T(k)}{T_H} = -(n_s - 1) \frac{2\frac{T(k)}{T_H} - 1}{2 - \frac{T(k)}{T_H}}$$
$$\simeq -(n_s - 1) > 0 \quad (33.67)$$

So the characteristic property of string gas cosmology is this blue tilt for the tensor spectrum, with the same deviation from scale invariance as the red tilt of the scalar spectrum. Inputting the experimentally observed scalar tilt from the Planck data, we expect that

$$n_T - 1 \simeq 0.03. \quad (33.68)$$

Moreover, the tensor to scalar ratio is now

$$r = \frac{P_h(k)}{P_\Phi(k)} \simeq \left(1 - \frac{T(k)}{T_H}\right)^2 \ln \left[\frac{1}{l_s^2 k^2} \left(1 - \frac{T(k)}{T_H}\right) \right] \ll 1 \,, \quad (33.69)$$

so a small ratio, consistent with observations, as well as being similar to string inflation models in that respect.

Important Concepts to Remember

- For a gas of Dp-branes, the wrapping modes have equation of state $w_p = -p/d$, generalizing the one for the strings.
- The addition of p-branes to strings means that we can have a hierarchy of large dimensions: besides the 3 very large ones, we can have two smaller but still large ones for 2-branes, etc.
- Two-point correlators of the energy-momentum tensor can be expressed in terms of thermodynamic quantities: for $\delta\rho$ in terms of C_V, and for spatial components in terms of $\partial P / \partial V|_T$.
- The thermodynamics of a gas of strings allows us to calculate the two-point correlators of energy-momentum tensor, and from them the power spectra of scalar (for $\delta\rho$) and tensor (for the spatial components) modes.
- Besides a small tensor-to-scalar ratio $r \ll 1$, which is also possible in inflation (especially appears in string inflation), we obtain the characteristic $n_T - 1 = -(n_s - 1)$.

Further reading: The spectrum of perturbations in string gas cosmology was found in [61], and the brane has generalization was defined in [62]. Good reviews for string gas and brane gas cosmology are [58, 59] and the recent one [60].

Exercises

(1) What order is the Hagedorn phase transition?
(2) What is the equation of state near the Hagedorn transition?
(3) Use experimental data to constrain the function $T(k)$.

(4) Show that for the metric

$$ds^2 = a^2(\eta)\left\{-(1+2\Phi(x,\eta))d\eta^2 + [(1-2\Phi(x,\eta))\delta_{ij} + h_{ij}]dx^i dx^j\right\} , \tag{33.70}$$

the Einstein equations give

$$\nabla^2\Phi = 4\pi G_N \delta\rho. \tag{33.71}$$

and

$$\nabla^2 h_{ij} = 4\pi G_N \delta T^i{}_j , \tag{33.72}$$

Chapter 34
Chameleon Scalars and String Theory

In this chapter we will study chameleon scalars and their embedding in string theory. It was usually thought the fifth force experiments conducted on Earth, or in the Solar System, would prevent the existence of light scalars.

However, the important observation of Khoury and Weltman, who developed the chameleon idea, was that we can have a mass for the scalar that depends on the local mass density. In a dense environment like the Earth, the chameleon will be very massive, so will elude the usual fifth force constraints. Moreover, we have a "thin-shell" effect, where for a very massive body only a thin shell surrounding it effectively interacts via the scalar force, which allows the chameleon to avoid the constraints on the Newtonian motion.

But on large cosmological scales, the scalar can be extremely light, so that it can have important consequences. As a plus, in string theory in principle we wouldn't need to "stabilize the moduli" (have large masses for them nonperturbatively), but rather they could be chameleons.

34.1 Chameleon (and Symmetron) Models

The action relevant for the chameleon models (as well as for the symmetron variant) is

$$S = \int d^4x\sqrt{-g}\left[\frac{M_{\rm Pl}^2}{2}R[g] - \frac{1}{2}g^{\mu\nu}\partial_\mu\phi\partial_\nu\phi - V(\phi)\right] + \int d^4x\mathcal{L}_m[\tilde{g}, \psi] , \tag{34.1}$$

where $\mathcal{L}_m$ is a matter action, depending on matter fields ψ, but which couples to the conformally-related metric

$$\tilde{g}_{\mu\nu} = A^2(\phi)g_{\mu\nu}. \tag{34.2}$$

Here $g_{\mu\nu}$ is the Einstein frame metric. The equation of motion for the scalar field (KG) is

H. Năstase, *Cosmology and String Theory*, Fundamental Theories of Physics 197, https://doi.org/10.1007/978-3-030-15077-8_34

$$\Box_g \phi - V_{,\phi} + A^3(\phi) A_{,\phi} \tilde{T} = 0 \,, \tag{34.3}$$

where

$$\tilde{T} = \tilde{T}_{\mu\nu} \tilde{g}^{\mu\nu} \tag{34.4}$$

is the trace of the matter (or Jordan) frame energy-momentum tensor

$$\tilde{T}_{\mu\nu} \equiv -\frac{2}{\sqrt{-\tilde{g}}} \frac{\delta \mathcal{L}_m}{\delta \tilde{g}^{\mu\nu}}. \tag{34.5}$$

This tensor is conserved in the $\tilde{g}_{\mu\nu}$ (Jordan) frame as usual, i.e.,

$$\tilde{\nabla}_\mu \tilde{T}^\mu{}_\nu = 0. \tag{34.6}$$

We define the density in the Jordan frame as usual, by

$$\tilde{T}^\mu{}_\nu = \mathrm{diag}(-\tilde{\rho}, \tilde{P}, \tilde{P}, \tilde{P}) \,, \tag{34.7}$$

and for dust matter ($P \simeq 0$) we have

$$\tilde{T}^\mu{}_\mu \simeq -\tilde{\rho}. \tag{34.8}$$

The energy-momentum tensor in the Jordan frame can be related to the one in Einstein frame by

$$\begin{aligned} \tilde{T}_{\mu\nu} &= \frac{1}{A^4} \frac{-2}{\sqrt{-g}} A^2 \frac{\delta \mathcal{L}_m}{\delta g^{\mu\nu}} \Rightarrow \\ \tilde{T}^\mu{}_\nu &= \frac{1}{A^4} \left(-\frac{2}{\sqrt{-g}} g^{\mu\rho} \frac{\delta \mathcal{L}_m}{\delta g^{\rho\nu}} \right) \Rightarrow \\ -\tilde{\rho} &= \frac{1}{A^4} T^\mu{}_\mu. \end{aligned} \tag{34.9}$$

On the other hand, the zero component of the conservation equation becomes

$$\begin{aligned} \tilde{\nabla}_\mu \tilde{T}^\mu{}_0 &= \tilde{\nabla}_0 \tilde{T}^0{}_0 + \tilde{\nabla}_i \tilde{T}^i{}_0 \\ &= \nabla_0 \tilde{T}^0{}_0 + \tilde{\Gamma}^i_{i0} \tilde{T}^i{}_0 = \nabla_0 \tilde{T}^0{}_0 + \left(3\frac{\dot{A}}{A} + \Gamma^i_{i0} \right) \tilde{T}^0{}_0 \\ &= \nabla_0 \tilde{T}^0{}_0 + \left(3\frac{\dot{A}}{A} + \nabla_i \right) \tilde{T}^i{}_0. \end{aligned} \tag{34.10}$$

Then the conservation equation becomes

$$\left(\nabla_0 + 3\frac{\dot{A}}{A}\right)\tilde{\rho} = 0. \tag{34.11}$$

That means that if we define the density ρ by

$$\rho \equiv A^3 \tilde{\rho}\ , \tag{34.12}$$

then it will be conserved (time independent in a non-expanding Universe) in Einstein frame. In particular, it will be independent of ϕ. Moreover we obtain

$$T^{\mu}{}_{\mu} = A\rho\ , \tag{34.13}$$

which means that the equation of motion for ϕ is

$$\Box_g \phi = V_{,\phi} + A_{,\phi}\rho. \tag{34.14}$$

Finally, that means that we have an *effective potential*

$$V_{\text{eff}}(\phi) = V(\phi) + \rho A(\phi). \tag{34.15}$$

For a spherically symmetric solution, the equation of motion for the chameleon is

$$\frac{d^2\phi}{dr^2} + \frac{2}{r}\frac{d\phi}{dr} = V_{,\phi} + A_{,\phi}\rho. \tag{34.16}$$

We have now the interesting possibility that even though $V(\phi)$ might not have a minimum, $V_{\text{eff}}(\phi)$ can, and the effective mass at the minimum of the effective potential is

$$m_{\text{eff}}^2 = V_{,\phi\phi}(\phi_{\min}) + A_{,\phi\phi}(\phi_{\min})\rho. \tag{34.17}$$

Since the ρ term increases generically with ϕ, we need $V_{,\phi} < 0$ over the period of interest in order to have a minimum for the effective potential (the sum of the two terms). For stability, i.e., in order to have a positive mass squared, we need $V_{,\phi\phi} > 0$. Moreover, for increasing of the effective mass with ρ (so that the mass on Earth is bigger than outside it), we need $V_{,\phi\phi\phi} < 0$.

Examples of potentials that obey these constraints are the inverse power law,

$$V \sim \frac{1}{\phi^n}\ ,\quad n > 0\ , \tag{34.18}$$

the exponential

$$V \sim e^{-a\phi}\ , \tag{34.19}$$

or more complicated exponentials.

A typical coupling function $A(\phi)$, of the type that also is natural in string theory, is

$$A(\phi) \simeq e^{g\frac{\phi}{M_{\rm Pl}}} \simeq 1 + g\frac{\phi}{M_{\rm Pl}}. \tag{34.20}$$

The main difference (with respect to generic chameleons) in the case of the symmetron model is an $A(\phi)$ that doesn't have a linear term, only a quadratic one, i.e., $A(\phi) \simeq (\phi - \phi_0)^2$.

34.2 Thin-Shell Effect

Consider a spherically symmetric object of radius $\mathcal{R}$, density $\rho_{\rm in}$, associated with a field $\phi_{\rm min-in}$, embedded in a medium with density $\rho_{\rm out}$, associated with a field $\phi_{\rm min-out}$.

Then for a sufficiently large object, ϕ will start in the core at $\phi_{\rm min-in}$, and it will start to deviate from this value at $\mathcal{R}_{\rm roll}$. The deviation can be approximated to be quadratic, so we find

$$\begin{aligned}\phi &\simeq \phi_{\rm min-in}; \quad 0 < r < \mathcal{R}_{\rm roll}\\ &\simeq \phi_{\rm min-in} + \frac{g}{2M_{\rm Pl}}\rho_{\rm min}(r - \mathcal{R}_{\rm roll})^2; \qquad \mathcal{R}_{\rm roll} < r < \mathcal{R}.\end{aligned} \tag{34.21}$$

By matching with the outside value for the scalar, we find

$$\frac{\Delta\mathcal{R}}{\mathcal{R}} = \frac{\mathcal{R} - \mathcal{R}_{\rm roll}}{\mathcal{R}} = \frac{\phi_{\rm min-out} - \phi_{\rm min-in}}{6gM_{\rm Pl}\Phi}. \tag{34.22}$$

Here

$$\Phi \equiv \frac{\rho_{\rm min}\mathcal{R}^2}{6M_{\rm Pl}^2} \tag{34.23}$$

is the Newton potential at the surface (compared to infinity, outside).

Finally then, the scalar field outside the sphere is a *screened* one, by a factor of $\Delta\mathcal{R}/\mathcal{R}$,

$$\phi_{\rm screened} \simeq -\frac{g}{4\pi M_{\rm Pl}}\frac{\Delta\mathcal{R}}{\mathcal{R}}\frac{Me^{-m_{\rm min-out}r}}{r} + \phi_{\rm min-out}. \tag{34.24}$$

We leave this as an exercise to prove.

In turn, that means that the effective coupling between two large bodies is

$$g_{\rm eff}^2 = g^2\left(\frac{\Delta\mathcal{R}}{\mathcal{R}}\right)_1\left(\frac{\Delta\mathcal{R}}{\mathcal{R}}\right)_2. \tag{34.25}$$

Concluding this subsection and experimental constraints, generically we have

$$m_{\rm eff}^2(\text{Earth}) \gg V''(\phi_{\rm out}) , \tag{34.26}$$

which means that the chameleon is quite massive on Earth, and avoids fifth force detection.

Moreover, in the Solar System, we have

$$\left(\frac{\Delta\mathcal{R}}{\mathcal{R}}\right)_{\rm planet} \ll 1 , \tag{34.27}$$

which means that we don't see the scalar force in the Solar System because of the thin-shell effect: the effective coupling is very small.

On the other hand, the chameleon is left to be light on cosmological scales, which means that we can have cosmological effects.

34.3 String Theory Embedding

The chameleon idea is very welcome in string theory. Indeed, we saw that we have many moduli that need stabilization, and when we try to do stabilize all, we usually spoil inflation. It would be of great help therefore if some of these moduli would not need to be stabilized, but rather would be chameleons. Then they could be hidden on Earth and in the Solar System, but would still be detectable cosmologically.

One way to embed the chameleons in string, or rather string-inspired, models, takes ϕ as the volume modulus (related to the volume of the compact dimensions) in a KK dimensional reduction.

Under KK dimensional reduction, we generically have

$$ds_D^2 = R^2 ds_d^2 + g_{mn}dx^m dx^n + \cdots , \tag{34.28}$$

where

$$ds_d^2 = g_{\mu\nu}dx^\mu dx^\nu \tag{34.29}$$

is the d-dimensional Einstein frame.

Here

$$R = \Delta^{-\frac{1}{d-2}}; \quad \Delta = \sqrt{\det g_{mn}}. \tag{34.30}$$

Under the important assumption that the matter action is simple (becomes the usual matter) in the ds_D^2 Jordan frame rather than the Einstein frame (which fact is not obvious a priori in KK compactifications), we have the identification

$$A^2(\phi) = R^2 = \Delta^{-\frac{2}{d-2}} = (\det g_{mn})^{-\frac{2}{d-2}}. \tag{34.31}$$

Choosing the physical case of $D = 10$, $d = 4$, we have

$$A^2(\phi) = R^2 = \Delta^{-1} = \frac{1}{V_6 M_{\rm Pl}^6} = \frac{1}{r^6 M_{\rm Pl}^6}. \tag{34.32}$$

In general, in KK compactifications in string theory, the canonical scalar is related to R by

$$\phi = \frac{M_{\rm Pl}}{g} \ln \frac{R}{R_*}, \tag{34.33}$$

for some dimensionless coupling g and some constant R_*.

But for the volume modulus there must be some nonperturbative potential stabilizing it, translating into a potential for R.

Around the minimum (stabilization point) of this potential, we expand:

- for $R \gtrsim R_*$,

$$V(R) = M_{\rm Pl}^4[-\alpha(R - R_*) + \beta(R - R_*)^2]. \tag{34.34}$$

The minimum of the potential is at

$$R_{\rm min} = R_* + \frac{\alpha}{2\beta}. \tag{34.35}$$

- for $R \lesssim R_*$, the potential is a steep exponential,

$$V(R) = M_{\rm Pl}^4 v[e^{\gamma(R^{-k} - R_*^{-k})} - 1], \tag{34.36}$$

where $\alpha, \beta, \gamma, k, v > 0$ are dimensionless. By matching the two behaviour at $R = R_*$, we find the relation

$$\alpha = v \frac{k\gamma}{R_*^{k-1}}. \tag{34.37}$$

Example: KKLT

As an example of this postulated behaviour, we show that the KKLT model is of this type.

Consider a complex volume modulus ρ, then

$$\sigma = {\rm Im}\rho = \frac{M_s^4 r^4}{g_s} = \frac{M_{10}^4 r^4}{2\sqrt{\phi} g_s}, \tag{34.38}$$

where M_{10} is the 10 dimensional Planck scale and $M_s = 1/(2\pi\sqrt{\alpha'})$ is the string scale.

But if instead, we would have the relation

$$\sigma = \frac{M_{\rm Pl}^4 r^4}{g_s} = \frac{R^{-4/3}}{g_s}\,, \tag{34.39}$$

which differs from the above by a large factor a that can be absorbed in the exponent of the KKLT superpotential (though now a doesn't appear naturally anymore), then we obtain the picture advocated above.

As we already saw, the Kähler potential for the volume modulus is

$$K = -3\ln[-i(\rho - \bar{\rho})]\,, \tag{34.40}$$

and the KKLT superpotential is

$$W = W_0 + Ae^{ia\rho}. \tag{34.41}$$

Moreover, we can obtain also $a < 0$, which is what we will need, by modifying the standard KKLT scenario, as we saw.

We can prove (we leave this as an exercise) that the supersymmetric KKLT potential near its minimum can be approximated by

$$V_{\rm KKLT} - V_0 \simeq \frac{A^2 a^2}{6M_{\rm Pl}^2 \sigma_*}\left[e^{2|a|\sigma} - e^{2|a|\sigma_0}\right]. \tag{34.42}$$

34.4 Experimental Constraints

We can now put the experimental constraints on this model. Fifth force experiments on Earth give

$$g_{\rm eff}^2 \lesssim 10^{-3}\,, \tag{34.43}$$

which means

$$\frac{3\Delta\mathcal{R}}{\mathcal{R}} = \frac{\phi_{\rm vac} - \phi_{\rm test-mass}}{2gM_{\rm Pl}\Phi_{\rm test-mass}} \lesssim 10^{-3/2}\,, \tag{34.44}$$

that finally implies

$$\phi_{\rm vac} - \phi_{\rm test} \lesssim 10^{-29} M_{\rm Pl}. \tag{34.45}$$

After some algebra, this is shown to imply

$$R_*^k \lesssim 10^{-30}\gamma k. \tag{34.46}$$

Taking the values close to the constraint, we obtain the range of the chameleon in various environments as

$$m_{\rm eff}^{-1} \lesssim \frac{0.2\,{\rm mm}}{\sqrt{\rho[{\rm g/cm^3}]}}\,, \tag{34.47}$$

which means that in the atmosphere

$$m_{\text{atm}}^{-1} \lesssim 1\,\text{cm}\,, \tag{34.48}$$

and in the Solar System

$$m_{\text{Solar system}}^{-1} \lesssim 10^5\,\text{Km}. \tag{34.49}$$

But moreover, in the Solar System, the thin-shell factors are extremely small. Even for the Moon we find

$$\left(\frac{3\Delta\mathcal{R}}{\mathcal{R}}\right)_{\text{Moon}} \lesssim 10^{-10}. \tag{34.50}$$

We need to impose that the galaxy itself is screened, otherwise its local field will not be given by its density. Then

$$\left(\frac{3\Delta\mathcal{R}}{\mathcal{R}}\right)_{\text{galaxy}} < 1\,, \tag{34.51}$$

which in the end imposes the constraint

$$\frac{\alpha}{\beta} \lesssim 10^{-6} R_*. \tag{34.52}$$

On the other hand, the cosmological density must be low, so that ϕ doesn't lie on the steep exponential part. This, after some algebra, implies

$$\alpha \geq 10^{-120} R_*^{-1}. \tag{34.53}$$

Finally, we find that the mass of the chameleon on cosmological scales obeys

$$m_{\text{cosmo}} = \sqrt{2}gM_{\text{Pl}}\sqrt{\beta}R_* \gtrsim 10^3 H_0. \tag{34.54}$$

Applying this for KKLT, we find

$$|V_0| \simeq 10^{-174} M_{\text{Pl}}^4; \quad W_0 \simeq 10^{-42} M_{\text{Pl}}^3; \quad A \sim M_{\text{Pl}}^3 e^{-10^{30}}. \tag{34.55}$$

However, the last constraint doesn't look so absurd when it is rewritten as

$$Ae^{-a\sigma} = M_{\text{Pl}}^3 e^{|a|(\sigma-\sigma_0)}\,, \tag{34.56}$$

with $\sigma_0 \sim 10^{30}$.

However, it turns out that we need to apply the constraints directly to KKLT, the general model not being a good enough approximation. Then we find instead

$$\begin{aligned}
&|a|\sigma_{\rm min} > 10^6\\
&|V_0| \geq \frac{10^{-120}}{|a|\sigma_{\rm min}} M_{\rm Pl}^4 \gtrsim 10^{-159} M_{\rm Pl}^4\\
&W_0 \gtrsim 10^{-30} M_{\rm Pl}^3.
\end{aligned} \tag{34.57}$$

Those are not that important modifications. However, we also find that now

$$m_{\rm cosmo} \gtrsim 10^{15} H_0\ , \tag{34.58}$$

which is an important difference, since now the chameleon is not relevant for cosmological distances anymore. But in fact it turns out that with two KKLT exponentials in the superpotential instead of one (very close in value to each other), we can achieve the general formula of $m_{\rm cosmo} \geq 10^3 H_0$.

Important Concepts to Remember

- Chameleon scalars are scalars with mass that depends on the local density, that are massive on Earth and avoid detection.
- They also have a thin-shell effect: only a thin shell of a massive body interacts via the scalar, avoiding detection for planets in the Solar System.
- The chameleon is obtained when the matter couples to $\tilde{g}_{\mu\nu} = A^2(\phi) g_{\mu\nu}$, resulting in an effective potential $V_{\rm eff} = V + \rho A(\phi)$.
- The density $\rho = A^3 \tilde{\rho}$ is conserved in Einstein frame.
- The thin-shell effect gives a screening coefficient of $3\Delta\mathcal{R}/\mathcal{R} = \Delta\phi/(6g M_{\rm Pl}\Phi)$, with Φ the Newton potential at the surface.
- We can embed in string theory with ϕ being the volume modulus.
- To obtain a chameleon, consider a potential for stabilizing R that has a minimum, and a steep exponential on one side, $e^{\gamma R^{-k}}$.
- The KKLT model gives such an example, with $k = 4/3$.
- Generically, we obtain $m_{\rm cosmo} \gtrsim 10^3 H_0$. In simple KKLT, we obtain instead $m_{\rm cosmo} \gtrsim 10^{15} H_0$, but with two exponentials we can get the general formula.

Further reading: The chameleon scalars were defined in [63, 64]. The embedding in string theory described here was done in [65].

Exercises

(1) Check that the screened solution outside the spherical mass source is

$$\phi_{\rm screened} \simeq -\frac{g}{4\pi M_{\rm Pl}} \frac{\Delta\mathcal{R}}{\mathcal{R}} \frac{M e^{-m_{\rm min-out} r}}{r} + \phi_{\rm min-out}. \tag{34.59}$$

(2) Prove that we can approximate the KKLT potential around the minimum by

$$V_{\rm KKLT} - V_0 \simeq \frac{A^2 a^2}{6 M_{\rm Pl}^2 \sigma_*} \left[e^{2|a|\sigma} - e^{2|a|\sigma_0}\right]. \tag{34.60}$$

(3) Check and explain the origin of the difference in chameleon mass on cosmological scales for the KKLT model and for the generic model with a minimum and a steep exponential on one side.
(4) Is the thin-shell effect, encapsulated by the solution at exercise 1, still valid for the symmetron case, with $A(\phi) = (\phi - \phi_0)^2$?

Chapter 35
Axion Inflation and Axion Monodromy from String Theory

In this chapter, we will learn about axion inflation in field theory, how axions appear in string theory, and a specific set-up for string inflation that has been very popular recently, axion monodromy.

35.1 Axions in Field Theory

In a nonabelian gauge theory, we can have instantons, self-dual solutions of the Yang–Mills equations in Euclidean space (satisfying $F^a_{\mu\nu} = \pm * F^a_{\mu\nu} \equiv \frac{1}{2}\epsilon_{\mu\nu\rho\sigma}F^{a\,\rho\sigma}$), carrying the topological number called the second Chern number (or Pontryagin number, or instanton number)

$$c_2 = \int_{M_4} \frac{1}{32\pi^2}\,\mathrm{Tr}[F_{\mu\nu} * F^{\mu\nu}] \in \mathbb{Z}. \tag{35.1}$$

The instanton action is $S = 8\pi^2/g^2$, and at the quantum level it gives transitions between states with different *winding numbers* n for the gauge configuration on the spatial $\mathbb{R}^3$. Therefore gauge fields with a given winding number n are in general not a good vacuum of the theory by itself. A good vacuum is a superposition of the states $|n\rangle$ with winding number n, parametrized by a constant θ as

$$|\theta\rangle = \sum_{n\geq 0} e^{-in\theta}|n\rangle. \tag{35.2}$$

This is called the θ vacuum, and here θ is an a priori free parameter. The effects of this parameter can be repackaged as a parameter multiplying the topological term giving the instanton in the Lagrangian,

$$\mathcal{L}_{\text{top}} = \frac{\tilde{\theta}g^2}{32\pi^2}F^a_{\mu\nu} * F^{a\,\mu\nu}. \tag{35.3}$$

H. Năstase, *Cosmology and String Theory*, Fundamental Theories of Physics 197, https://doi.org/10.1007/978-3-030-15077-8_35

Actually, the parameter in the Lagrangian contains, besides θ, a quantum contribution from fermions coupled to it,

$$\tilde{\theta} = \theta + arg(\det M) \ , \tag{35.4}$$

where M is the fermion mass matrix.

But in the real world, in QCD, it turns out that θ is measured (bound) to be extremely small, less than 10^{-10}. To explain why this parameter is so small, the solution proposed by Peccei and Quinn is to first promore θ to a pseudoscalar field $\theta(x)$ and then assume there is a global chiral symmetry (called Peccei-Quinn symmetry) that acts on it as a shift symmetry, $\theta \to \theta$+const., and on the fermions, for instance by multiplying the mass term by a γ_5 phase, effectively $M \to e^{i\delta\theta\gamma_5} M$, and which is spontaneously broken at some scale f_{PQ}. The Nambu-Goldstone boson for its spontaneous breaking is the pseudoscalar field called the axion $\theta(x) = a(x)$, and the breaking is due to instanton effects.

This chiral symmetry is anomalous since we have fermions charged under it, and in the presence of an instanton it generates the topological term. Indeed, the anomaly of a global symmetry in a gauge theory is

$$\partial_\mu j_\mu = \frac{C_a g^2}{16\pi^2} \mathrm{Tr}[F_{\mu\nu} * F^{\mu\nu}] \ , \tag{35.5}$$

where C_a depends on the number and charges of the fermions. Therefore in the presence of an instanton it generates a contribution to the action given by the above, times the axion $\theta(x) = a(x)$.

This in turn means a potential term for the axion, fixing it to have the total VEV equal to zero, the potential coming from the instanton contributions, and breaking the symmetry at some scale, in particular giving a mass Λ_{QCD}^2/f_{PQ}. Here Λ_{QCD} is the dynamically generated scale of the gauge theory, and $f_{PQ} = f$ is also called *axion decay constant*.

The Lagrangian for the axion can thus be taken to be

$$\mathcal{L}(a) = -\frac{1}{2} f^2 (\partial_\mu a)^2 - \Lambda^4 [1 - \cos(a)] + \cdots \ , \tag{35.6}$$

where the dots represent higher derivative corrections and multi-instanton contributions, or, by defining a canonically normalized field $\phi = fa$,

$$\mathcal{L}(\phi) = -\frac{1}{2}(\partial_\mu \phi)^2 - \Lambda^4 \left[1 - \cos\frac{\phi}{f}\right] + \cdots \simeq -\frac{1}{2}(\partial_\mu \phi)^2 - \frac{\Lambda^4}{2f^2}\phi^2 + \cdots \tag{35.7}$$

Then the axion has now just a periodicitiy $a \to a + 2\pi$, or $\phi \to \phi + 2\pi f$, instead of the general shift symmetry.

35.2 Axions in String Theory

In string theory, we look for fields $a(x)$ that have shift symmetries $a \to a+$const., and which then can be broken by nonperturbative effects to constant shifts.

There are many such fields in string compactifications, arising from the integration of p-form potentials over p-cycles in the compact space.

Specifically:

- in type IIA string theory we would have the NS-NS 2-form $B_{\mu\nu}$ integrated over 2-cycles Σ_2^I, the RR one-form C_μ integrated over one-cycles Σ_1^I, and the RR 3-form $C_{\mu\nu\rho}$ integrated over 3-cycles Σ_3^I.
- in type IIB, where we will concentrate, we have again the NS-NS 2-form $B_{\mu\nu}$, but now also the R-R 2-form $C_{\mu\nu}$ integrated over 2-cycles Σ_2^I, the R-R 4-form $C_{\mu\nu\rho\sigma}$ integrated over 4-cycles Σ_4^I. We also have 3 universal axions, 4 dimensional pseudoscalars (not integrated over cycles), the IIB axion C, as well as the *4-dimensional* duals b and c of $B_{\mu\nu}$ and $C_{\mu\nu}$ ($dB_2 = H_3 = *_4 H_1,\ H_1 = db$, $dC_2 = F_3 = *_4 F_1$, $F_1 = dc$).

We then have, in type IIB,

$$b_I = \frac{1}{2\pi\alpha'}\int_{\Sigma_2^I} B_2\,,\quad c_I = \frac{1}{2\pi\alpha'}\int_{\Sigma_2^I} C_2\,,\quad \tilde{c}_I = \frac{1}{2\pi\alpha'}\int_{\Sigma_4^I} C_4\,, \tag{35.8}$$

together with C, b and c. Together, these are called a (the dimensionless axion defined before, with period 2π).

Generically, the p-forms have gauge invariances, $\delta C_p = d\Lambda_{p-1}$, which leads, under integration on a cycle, to $\delta\int_{\Sigma_p} C_p = \int_{\Sigma_p} d\Lambda_{p-1}$, which can be taken to be an arbitrary constant, under a suitable choice of gauge parameter $(p-1)$-form Λ_{p-1}. For instance, for $C_1 = d\lambda$, we have

$$\delta c = \delta\oint_{S^1} d\lambda = \int_{S^1} dx^\mu \partial_\mu\lambda = \lambda(x + 2\pi R) - \lambda(x)\,, \tag{35.9}$$

which can be chosen to be what we want.

Thus classically, we have $\delta a = const.$ as a symmetry. But at the quantum level, we can have *worldsheet, or D-brane, instantons*. A D-p-brane is a source for a C_{p+1}, so if an *Euclidean* (worldsheet signature) D-brane wraps that same cycle, there is a priori a contribution to the axion action, that breaks the shift symmetry. The same thing happens for a fundamental string Euclidean worldsheet wrapping a 2-cycle. Because of the Euclidean signature, these are both instantonic solutions.

Let us describe this last example in a bit more detail. The action for the string worldsheet contains the term

$$S_B = -\frac{1}{4\pi\alpha'}\int_{\Sigma_2^I} d^2\sigma\epsilon^{ab}\partial_a X^M \partial_b X^N B_{MN}(X) = -\frac{1}{2\pi\alpha'}\int_{\Sigma_2^I} B_2 \equiv -b(x^\mu)^I\,, \tag{35.10}$$

where $b(x^\mu)$ is the dimensionless axion ($a(x^\mu)$ in the general notation), and we have denoted $M=(\mu,i)$, with μ being 4 dimensional (noncompact) index and i compact index. This term contributes to the axion action. We can expand the coordinates X^M around some point $X^M_{(0)}$.

Considering only $X^M_{(0)}$, so that the B-field is a constant, both in the worldsheet and in spacetime, we get that the contribution is

$$\delta S_B=-\frac{1}{4\pi\alpha'}\int_{\Sigma_2^I}\partial^2\sigma\partial_a\left(\epsilon^{ab}X^M\partial_b X^N B_{MN}(X^K_{(0)})\right)\,, \tag{35.11}$$

and this vanishes unless there is a boundary or a topologically nontrivial cycle for Σ_2^I. The terms involving the fluctuations in X^M will come with derivatives, $\partial_P B_{MN}(X_{(0)})$, translating in $\partial_\mu b$ terms in spacetime, which respect the shift symmetry.

But the worldsheet instantons, worldsheets wrapping nontrivial cycles Σ_2^I, give instantonic contributions to the action. B_2 couples to the Kähler form J for the 2-cycle into a complex field, as we saw before, so $(J+iB_2)$ is integrated over Σ_2, for an instanton action

$$S_{\rm instanton}=\exp\left(-\frac{1}{2\pi\alpha'}\int_{\Sigma_2^I}(J+iB_2)\right)=e^{-ib^I}\,. \tag{35.12}$$

This is a contribution to the superpotential for the b axion, which has therefore a periodicity $b\to b+2\pi$, as expected.

To determine the axion decay constant f, we must calculate the kinetic term for the axion. In the case of b, the 10 dimensional kinetic term, coming from the type II supergravity action with coefficient $-1/(4\kappa^2)$, is[1]

$$S_{\rm B,sugra}=-\frac{1}{2(2\pi)^7 g_s^2\alpha'^4}\int_{M_{10}}d^{10}x\sqrt{-g}|dB_2|^2. \tag{35.13}$$

We expand B_2 in a basis Ω^I of 2-forms dual to the cycles Σ_2^I,

$$B_2=\sum_I b_I(x)\Omega^I\,, \tag{35.14}$$

where $b_I(x)$ are the 4-dimensional fields, and the 2-forms Ω^I are normalized by (note that B_2 is dimensionless, and b_I are also, so from this, and the instanton action below, the normalization must be)

[1] If we call, as usual, the coefficient of the Einstein action as $1/(2\kappa^2)$, then one can find that the D-string tension is $T_{D1}=4\pi^{5/2}\alpha'/\kappa$, from matching the D-brane action against a string theory calculation. The string tension is $T_{F1}=1/(2\pi\alpha')$, so their ratio is $T_{F1}/T_{D1}=\kappa/(8\pi^{7/2}\alpha'^2)$, and is usually defined to be equal to g_s, leading to $2\kappa^2=(2\pi)^7g_s^2\alpha'^4$.

$$\int_{\Sigma_2^I} \Omega^J = 2\pi\alpha' \delta_I^J . \tag{35.15}$$

Then, substituting in the supergravity action (note that $|F_p|^2 \equiv \frac{1}{p!} g^{\mu_1\nu_1} \dots g^{\mu_p\nu_p} F_{\mu_1\dots\mu_p} F_{\nu_1\dots\nu_p}$), we obtain

$$S_{\text{B,sugra}} = -\frac{1}{2} \int d^4x \sqrt{-g} g^{IJ} (\partial_\mu b_I)(\partial^\mu b_J) , \tag{35.16}$$

where the sigma model metric on the axion space is (from the above, with $1/3! = 1/6$ coefficient)

$$g^{IJ} \equiv \frac{1}{6(2\pi)^7 g_s^2 \alpha'^4} \int_{K_6} \Omega^I \wedge *_6 \Omega^J . \tag{35.17}$$

But, from the Einstein action, which in string theory is found to be ($2\kappa^2$ was considered before)

$$S = \frac{1}{2\kappa^2} \int_{M_{10}} R = \frac{1}{(2\pi)^7 g_s^2 \alpha'^4} \int_{M_{10}} R = \frac{M_{\text{Pl}}^2}{2} \int_{M_4} R_{(4)} + \cdots , \tag{35.18}$$

we get that (M_{Pl} is the *4-dimensional* Planck scale)

$$M_{\text{Pl}}^2 = \frac{2V}{(2\pi)^7 g_s^2 \alpha'^4} , \tag{35.19}$$

where V is the volume of the compact space K_6. For an isotropic compactification with scale L, so that $V = L^6$, and because of our normalization of Ω^I ($\int_{\Sigma_2^I} \Omega^J = 2\pi\alpha' \delta_I^J$), we find $\int_{K_6} \Omega^I \wedge *_6 \Omega^J \simeq L^2 \alpha'^2 \delta^{IJ}$. Finally, diagonalizing the metric g^{IJ} to $f^2 \delta^{IJ}$, we obtain

$$g^{IJ} \simeq \frac{M_{\text{Pl}}^2 \alpha'^2}{12 L^4} \delta^{IJ} \Rightarrow \frac{f^2}{M_{\text{Pl}}^2} \simeq \frac{\alpha'^2}{12 L^4} . \tag{35.20}$$

Since in order to trust computations we need $L \gg \sqrt{\alpha'}$, it means we have $f \ll M_{\text{Pl}}$.

Axionic Inflation from String Theory

But for n_s to be compatible with "natural inflation" coming from the axion action, the Planck experiment (see [76], Fig. 11) finds that one needs $f \gtrsim 10 M_{\text{Pl}}$. Note that the result depends very drastically on priors; the cited formula is for a uniform prior for $\log f$. But as we saw, super-Planckian values for f are problematic.

So it seems that the simplest models with the string theory axions don't work, and we need to do something else. There are 3 main solutions to this: a fine tuning in the case of two axions similar to "racetrack" inflation fine tuning, a version of N-flation (inflation with a radial field in an N-dimensional space of inflatons), and the best candidate from string theory, axion monodromy inflation.

The first two cases amount to having f's that are sub-Planckian for each axion, but the effective axion has a trans-Planckian decay constant, and therefore field range.

35.3 "Racetrack" with Two Axions

It is a known fact that if we fine-tune the sum of two exponentials or trigonometric functions (which are exponentials of imaginary argument), we can obtain a function that has a longer period, so it looks "flatter". This is basically the mechanism for "racetrack" inflation, which is an inflationary potential made up of two close by exponentials.

A similar mechanism can be constructed for axion potentials, where the some of two cosine axionic potentials will lead to a potential with a much longer period, i.e., range or axion decay constant f.

One considers two axionic scalars, each of which couples to a linear combination of two confining nonabelian gauge groups, that lead to a cosine potential. That is, the coupling term in the action is (from the form (35.3) for a rescaled dimensionless axion $\tilde{\theta} = \varphi/f$, where φ has a canonical kinetic term, and for two different group factors a, b)

$$S_{\text{coupling}} = \int d^4x \sum_{i=1,2} \frac{\varphi_i}{f_i} \left(\frac{c_{ia}}{32\pi^2} \operatorname{Tr}\left[F^{(a)} \wedge F^{(a)}\right] + \frac{c_{ib}}{32\pi^2} \operatorname{Tr}\left[F^{(b)} \wedge F^{(b)}\right] \right), \tag{35.21}$$

and it leads at low energies to the axion potential

$$V = \Lambda_a^4 \left[1 - \cos\left(c_{1a}\frac{\varphi_1}{f_1} + c_{2a}\frac{\varphi_2}{f_2}\right)\right] + \Lambda_b^4 \left[1 - \cos\left(c_{1b}\frac{\varphi_1}{f_1} + c_{2b}\frac{\varphi_2}{f_2}\right)\right]. \tag{35.22}$$

Consider now that we have approximately (but to a high precision)

$$C \equiv \frac{c_{2a}}{c_{1a}} \simeq \frac{c_{2b}}{c_{1b}}. \tag{35.23}$$

Then the potential becomes

$$V \simeq \Lambda_a^4 + \Lambda_b^4 - \Lambda_a^4 \cos\frac{c_{1a}}{f_1}\left(\varphi_1 + C\frac{f_1}{f_2}\varphi_2\right) - \Lambda_b^4 \cos\frac{c_{1b}}{f_1}\left(\varphi_1 + C\frac{f_1}{f_2}\varphi_2\right). \tag{35.24}$$

As we can see, the potential is now a function of only a linear combination of the fields, $\varphi_1 + C\frac{f_1}{f_2}\varphi_2$, and is independent of the transverse linear combination, which is canonically normalized as

$$\xi \equiv \frac{c_{1a}\varphi_2 f_2 - c_{2a}\varphi_1 f_1}{\sqrt{c_{2a}^2 f_1^2 + c_{1a}^2 f_2}} \propto \varphi_2 - C\varphi_1 \frac{f_1}{f_2}. \tag{35.25}$$

Note that we have

$$(\partial\varphi_1)^2 + (\partial\varphi_2)^2 = \left(\frac{\varphi_1 + C\frac{f_2}{f_1}\varphi_2}{\sqrt{1 + C^2\frac{f_1^2}{f_2^2}}}\right)^2 + \left(\frac{\varphi_2 - C\frac{f_2}{f_1}\varphi_1}{\sqrt{1 + C^2\frac{f_1^2}{f_2^2}}}\right)^2, \tag{35.26}$$

and the second term is $(\partial\xi)^2$, whereas the first is the canonically normalized $\varphi_1 + C\frac{f_1}{f_2}\varphi_2$.

That means that ξ has no potential, therefore infinite range (infinite periodicity, or axion decay constant f). If we remember that the ratios of constants c_{ia} are not exactly equal, we can rewrite the potential in terms of the two modes defined above (we define now $C \equiv c_{2a}/c_{1s}$), as

$$\begin{aligned}
V &= \Lambda_a^4\left[1 - \cos\left(\frac{c_{1a}\sqrt{1 + C^2\frac{f_1^2}{f_2^2}}}{f_1}\frac{\varphi_1 + C\frac{f_2}{f_1}\varphi_2}{\sqrt{1 + C^2\frac{f_1^2}{f_2^2}}}\right)\right] + \Lambda_b^4 [1 - \\
&\quad - \cos\left(\frac{c_{1a}}{f_1}\frac{c_{1a}c_{1b}f_2^2 + c_{2a}c_{2b}f_1^2}{c_{1a}^2 f_2^2 + c_{2a}^2 f_1^2}\left(\varphi_1 + \frac{c_{2a}}{c_{1a}}\frac{\varphi}{f_2}\right) + \frac{(c_{1a}f_2\varphi_2 - c_{2a}f_1\varphi_1)(c_{2b}c_{1a} - c_{2a}c_{1b})}{c_{2a}^2 f_1^2 + c_{1a}^2 f_2^2}\right)\Bigg] \\
&\equiv \Lambda_a^4\left[1 - \cos\left(\frac{c_{1a}\sqrt{1 + C^2\frac{f_1^2}{f_2^2}}}{f_1}\frac{\varphi_1 + C\frac{f_2}{f_1}\varphi_2}{\sqrt{1 + C^2\frac{f_1^2}{f_2^2}}}\right)\right] \\
&\quad + \Lambda_b^4\left[1 - \cos\left(\alpha\frac{\varphi_1 + C\frac{f_2}{f_1}\varphi_2}{\sqrt{1 + C^2\frac{f_1^2}{f_2^2}}} + \beta c_{1b}\frac{\xi}{f_\xi}\right)\right],
\end{aligned} \tag{35.27}$$

where $\beta = 1$. We can find the above relation by putting arbitrary coefficients α, β and fixing them in order to get the same thing on both sides. Then, the axion decay constant for the mode ξ becomes

$$f_\xi = \frac{c_{1b}\sqrt{c_{2a}^2 f_1^2 + c_{1a}^2 f_2^2}}{|c_{2b}c_{1a} - c_{2a}c_{1b}|}, \tag{35.28}$$

so it can easily be made to be transplanckian, in particular $f_\xi > 10M_{\rm Pl}$.

35.4 N-Flation

Another known mechanism to extend the range of the inflaton so that it can become transplanckian is to consider N copies of the same scalar field, which leads to an effective field that is the radius in N dimensional field space, effectively

$$\Phi^2 \equiv \sum_{i=1}^{N} (\phi_i)^2 = N|\phi|^2 \Rightarrow \Phi = \sqrt{N}|\phi|. \tag{35.29}$$

This enhancing of the range by $\sqrt{N}$ due to the presence of N fields is called N-flation.

To be more precise, for an action

$$S = \sum_{i=1}^{N} \int d^4x \left[-\frac{1}{2}(\partial_\mu \phi_i)^2 - V_i(\phi_i) \right] , \tag{35.30}$$

calculating the KG equations, we see that each scalar field experiences its potential V_i,

$$\ddot{\phi}_i + 3H\dot{\phi}_i = -\partial_i \phi_i , \tag{35.31}$$

but Hubble friction coming from all of them, since the Friedmann equation in the inflationary regime (negligible kinetic energy) is

$$3H^2 M_{\rm Pl}^2 \simeq \sum_{i=1}^{N} V_i. \tag{35.32}$$

The simplest case is when all the axions have the same mass, $m_i = \Lambda_i^2/f_i \equiv m$, and for field displacements small with respect to their relevant scale, $\phi_i \ll f_i$, so we can approximate the potential by a quadratic one. Then

$$V = \sum_i V_i \simeq \frac{1}{2} m^2 \sum_{i=1}^{N} \phi_i^2 \equiv \frac{1}{2} m^2 \Phi^2 , \tag{35.33}$$

so indeed the effective field is the radius in field space, Φ, enhanced by a $\sqrt{N}$ with respect to the individual displacements.

In order to have a Φ with a range larger than $10 M_{\rm Pl}$, and still have calculability and reasonable individual field displacements, we need a number N of axions larger than about 1000.

Note that the model says that the radial coordinate is different than the Cartesian system of coordinates. That is so, since the individual fields have periodicities $\phi_i \sim \phi + 2\pi f_i$, restricting their range, but the radial coordinate doesn't.

35.5 Axion Monodromy Inflation

The most popular type of model, originally proposed by McAllister, Silverstein and Westphal, is called axion monodromy inflation, developed from a model of inflation from general string monodromy by Silverstein and Westphal [67]. In the following I will follow mostly the arguments in [67].

Monodromy refers to the fact that when going around a closed loop in some parameter space, a system doesn't return to its original state, but to a modified one. An example relevant for the present case would be a helix or spiral, where upon performing a circular motion, the system also evolves in a third direction.

Indeed, in certain special string compactifications, while going around what looks like a loop in moduli space, specifically varying the axion across its "periodicity" range, in reality the potential energy V is modified, thus the system is not, after all, periodic. That leads to a substantial enhancement of the period, or field range, in effect becoming infinite, which as we saw before, was needed in order to make axionic inflation compatible with data.

All the relevant string monodromies refer to D-branes moving around in a (naive) configuration space, and returning to a modified state after a loop. The generic D-brane action is, as we saw,

$$S = S_{\text{DBI}} + S_{\text{WZ}} = -\frac{1}{(2\pi)^p \alpha'^{\frac{p+1}{2}} g_s} \int d^{p+1}\sigma \Big[e^{-(\Phi-\Phi_0)} \sqrt{-\det(G_{MN} + B_{MN})\partial_a X^M \partial_b X^N} + \left(\sum_p C^{(p)}\right) e^{-B} e^{2\pi\alpha' F} \Big]. \tag{35.34}$$

The simplest case to consider is of a D5-brane in type IIB string theory, wrapped on a 2-cycle Σ_2, and the axion is the B-field wrapped on it, $b = \frac{1}{2\pi\alpha'} \int_{\Sigma_2} B_2$. Then the relevant term in the wrapped D5-brane action is

$$S = \frac{1}{(2\pi)^5 g_s (\alpha')^3} \int_{M_4 \times \Sigma_2} d^6\sigma \sqrt{-\det G_{ab} + B_{ab}}\,, \tag{35.35}$$

where G_{ab} and B_{ab} are the induced metric and B-field. Since we have, for a compact space with $g_{11} = g_{22} = l^2$,

$$\det(G_{ab} + B_{ab}) = \det_{M_4} g \cdot \det \begin{pmatrix} g_{11} & B_{12} \\ -B_{12} & g_{22} \end{pmatrix} = \det_{M_4} g \cdot ((g_{11})^2 - (B_{12})^2) = \det_{M_4} g \cdot (l^4 - b^2\alpha'^2)\,, \tag{35.36}$$

the term in the action becomes a potential for the axion b,

$$S = \int d^4x \sqrt{-g} \frac{1}{(2\pi)^5 g_s (\alpha')^3} \sqrt{l^4 - b^2\alpha'^2} \equiv \int d^4x \sqrt{-g} V(b). \tag{35.37}$$

At large b, this becomes

$$V(b) \simeq \frac{1}{(2\pi)^6 g_s (\alpha')^2} b. \tag{35.38}$$

For the axion b, we have seen that the kinetic term is $-\frac{1}{2} f^2 (\partial_\mu b)^2$, and then in terms of the canonically normalized scalar $\phi = fb$, we have the action

$$S = \int d^4 x \sqrt{-g} \left[-\frac{1}{2}(\partial_\mu \phi)^2 - \mu^3 \phi + V_{\rm per}(\phi) \right], \quad \mu^3 \equiv \frac{1}{f} \frac{1}{(2\pi)^6 g_s (\alpha')^2}, \tag{35.39}$$

where $V_{\rm per}(\phi)$ is the $1 - \cos(\phi/f)$ potential for the axion, which can be neglected for the needed large b.

This action is the action for a chaotic inflation model with linear potential, but protected against corrections for super-Planckian displacements by the shift symmetry, which is just mildly broken by the new, linear, potential term.

Another possibility for inflaton monodromy from string theory can appear for D-branes wrapped on cycles in some special manifolds, where the position of the D-brane on the cycle is just the inflaton itself.

Such a manifold is the Nil 3-manifold with metric

$$\frac{ds^2}{\alpha'} = L_{u_1}^2 du_1^2 + L_{u_2}^2 du_2^2 + L_x^2 (dx' + M u_1 du_2)^2 , \tag{35.40}$$

compactified by identification under the shifts

$$\begin{aligned}
(x, u_1, u_2) &\to (x + 1, u_1, u_2) \\
(x, u_1, u_2) &\to \left(x - \frac{M}{2} u_2, u_1 + 1, u_2 \right) \\
(x, u_1, u_2) &\to \left(x + \frac{M}{2} u_1, u_1, u_2 + 1 \right).
\end{aligned} \tag{35.41}$$

For a D4-brane in type IIA string theory wrapped on u_2 and moving in u_1 in the above manifold, the DBI term in the D4-brane action gives

$$S = -\frac{1}{(2\pi)^4 g_s \alpha'^{5/2}} \int_{M_4 \times u_2} d^5 \sigma \sqrt{-g_4} \sqrt{g_{u_2 u_2} (1 - \alpha' g_{u_1 u_1} \dot{u}_1^2)}. \tag{35.42}$$

Doing trivially the integral over u_1 (with period 1), since nothing depends on it, and defining $L^3 \equiv L_{u_1} L_{u_2} L_x$ and L_u by $L^3 \equiv L_u^2 L_x$ and then $\beta = L_{u_2}/L_{u_1}$, we obtain a 4 dimensional action

$$S = -\frac{1}{(2\pi)^4 g_s \alpha'^{5/2}} \int_{M_4} d^4 x \sqrt{-g} \sqrt{\beta L_u^2 + L_x^2 M^2 u_1^2} \sqrt{1 - \alpha' \frac{L_u^2}{\beta} \dot{u}_1^2}. \tag{35.43}$$

We see that again we obtain (for $\dot{u}_1 = 0$) a potential term $V(u_1)$ of the form $\sqrt{K^2 + u_1^2}$ as before, also going to the linear term u_1 at infinity, but now the presence of the nontrivial kinetic term for u_1 means that in terms of the canonical scalar ϕ we obtain

$$S = \int d^4x \sqrt{-g_4} \left(\frac{1}{2}\dot{\phi}_1^2 - \frac{\sqrt{\beta} L_u}{(2\pi)^4 g_s \alpha'^2} \sqrt{1 + \frac{M^2}{\beta} \frac{L_x^2}{L_u^2} u_1^2(\phi)} \right). \tag{35.44}$$

At very large u_1, we see from the above action that the first square root becomes $\propto u_1$, which means that both the potential becomes linear in u_1, and the kinetic term becomes $\propto u_1 \dot{u}_1^2$, leading to a canonical scalar $\phi \propto u_1^{3/2}$, and a potential

$$V(\phi) \propto u_1 \propto \phi^{2/3}. \tag{35.45}$$

Conditions on Axion Monodromy Models

As with all string inflation models, axion monodromy has plenty of constraints to be satisfied, and it is very difficult to satisfy them all.

For one, we need as always when we have D-branes, to consider the consistency of the compactification.

Indeed, first *"tadpole cancellation,"* which amounts to having a vanishing total D-brane charge on the compact manifold (otherwise the flux has nowhere to go on the compact manifold) imposes some restrictions on possible models. As usual, this can be fixed by having an anti-D-brane as some other point in the compactification, or by adding orbifold fixed points.

Second, we need to *stabilize the moduli,* which as we saw in general, can introduce terms that spoil the flatness of the inflation potential. If the shift symmetry is not badly broken, it is now easier to satisfy this constraint than in general string inflation models, but we still have generically a problem.

For instance, if we consider the same b modulus as above, and a single Kähler modulus T, with superpotential and Kähler potential

$$\begin{aligned} W &= W_0 + Ae^{-T} \\ K &= -3\ln(T + \bar{T} + \gamma b^2)\,, \end{aligned} \tag{35.46}$$

then the holomorphic volume T is stabilized, but we obtain terms in the scalar potential that come from wrapped branes, and therefore depend on the "physical volume" $V \propto (T + \bar{T} + \gamma b^2)^{3/2}$. That means that the potential depends on both T and b, so any flat direction is lifted.

However, for axions coming from R-R fields, like the S-dual of the B field, the C_2 of type IIB string theory leading to a c axion, the problems are aleviated, as there is no tree level c term in the Kähler potential K.

There are nonperturbative corrections to the superpotential, coming from Euclidean strings and branes wrapping cycles, as well as nonperturbative gauge theory corrections on branes (like gaugino condensation on D7-branes). These need to be taken into consideration, and to make sure nothing spoils the potential we want.

We also need to consider the *backreaction* of the branes on the geometry: in most we cannot consistently treat them in the probe approximation.

35.6 Dante's Inferno

A model that fixes some of the problems above is a variant of axion monodromy with two axions, called "Dante's Inferno". In effective field theory, we consider two axions r and θ, with axion decay constants that satisfy $f_r < f_\theta$ and

$$S = \int d^4x\sqrt{-g}\left[-\frac{1}{2}(\partial_\mu r)^2 - \frac{1}{2}(\partial_\mu \theta)^2 - V(r,\theta)\right]$$
$$V(r,\theta) = W(r) + \Lambda^4\left[1 - \cos\left(\frac{r}{f_r} - \frac{\theta}{f_\theta}\right)\right]. \tag{35.47}$$

We see that the names r and θ are related to the fact that only r has a term that breaks the shift symmetry of the potential, $W(r)$, though the metric in field space is $ds^2 = dr^2 + d\theta^2$, so it is not really a "radius". The specific case studied is $W(r) = \frac{1}{2}m^2r^2$. However, if we represent (r,θ) as cylindrical coordinates, then $V(r,\theta)$ looks like a spiral with edges for each spire, which brings to mind the description of the Inferno put forward by Dante, hence the name.

If we choose parameters that satisfy

$$f_r \ll f_\theta \ll M_{\rm Pl}\,, \quad \Lambda^4 \gg f m^2 r_{\rm in}\,, \tag{35.48}$$

where $f \ll r_\in < M_{\rm Pl}$ for the initial condition in r, then the excitation in the $\tilde{r}$ direction, appearing in the cosine potential, is very massive, and $\tilde{r}$ is effetively fixed. The remaining coordinate $\tilde{\theta}$ is the inflaton. The field redefinition (rotation) is

$$\begin{pmatrix}\tilde{r}\\ \tilde{\theta}\end{pmatrix} = \begin{pmatrix}\cos\xi & \sin\xi\\ -\sin\xi & \cos\xi\end{pmatrix}\begin{pmatrix}r\\ \theta\end{pmatrix}\,, \tag{35.49}$$

where

$$\sin\xi = \frac{f_r}{\sqrt{f_r^2+f_\theta^2}}\,, \quad \cos\xi = \frac{f_\theta}{\sqrt{f_r^2+f_\theta^2}}\,, \quad f = \frac{f_r f_\theta}{\sqrt{f_r^2+f_\theta^2}}. \tag{35.50}$$

In our limit on the parameters, we obtain that, since $\tilde{r}$ is almost fixed, the effective field is

$$\phi_{\rm eff} = \tilde{\theta} \simeq \theta, \tag{35.51}$$

its potential is

$$V_{\text{eff}}(\phi_{\text{eff}}) = \frac{1}{2} m_{\text{eff}}^2 \phi_{\text{eff}}^2 , \tag{35.52}$$

and the effective mass of the field is given by

$$m_{\text{eff}}^2 = m^2 \sin^2 \xi \simeq m^2 \frac{f_r^2}{f_\theta^2}. \tag{35.53}$$

That means that we obtain a classic chaotic inflation model (other power laws are also possible), but the advantages are:

- the effective mass is suppressed compared with the mass in the Lagrangian: $m_{\text{eff}} \simeq m f_r / f_\theta \ll m$.
- the Lyth bound is modified: indeed, the Lyth bound about transplanckianity needed for nontrivial r ratio of tensors to scalars refers to the effective field $\phi_{\text{eff}} \simeq \theta$, which needs to be $\gg M_{\text{Pl}}$. But the θ direction is periodic, which helps it avoid large corrections. On the other hand the range in r, which is the field with a mass term, is

$$\Delta r \simeq \frac{f_r}{\sqrt{f_r^2 + f_\theta^2}} \Delta\phi_{\text{eff}} \simeq \frac{f_r}{f_\theta} \Delta\phi_{\text{eff}} \ll \Delta\phi_{\text{eff}} , \tag{35.54}$$

and can be made subplanckian.

More importantly, the model can be embedded into string theory as follows. Consider an Euclidean D1-brane instanton, wrapping a 2-cycle Σ that is a linear combination of the cycles Σ_r and Σ_θ that correspond to the axions r and θ (the axions are fields integrated on them). Then we obtain a term in the potential of the required form,

$$\cos\left(\alpha \frac{r}{f_r} - \beta \frac{\theta}{f_\theta}\right). \tag{35.55}$$

One also finds that it is not hard to obtain the needed hierarchy $f_r / f_\theta \ll 1$.

35.7 Phenomenology of Axion Monodromy Inflation

The axion monodromy potential is

$$V = V_0(\phi) + \Lambda^4 \cos \frac{\phi}{f} , \tag{35.56}$$

and one needs the second term to be subdominant, but not small,

$$b_* \equiv \frac{\Lambda^4}{V_0'(\phi_*) f} < 1\,, \tag{35.57}$$

where ϕ_* is the value of the inflaton when the pivot scale k_* exits the horizon.

There are many models of axion inflation from string theory, but there are a few common signatures for them. Among those, we have:

- because of the modulated potential for the axion, one finds a modulation in the amplitude of the power spectrum $\Delta_R^2(k)$, to leading order in b_*,

$$\Delta_R^2(k) = \Delta_R^2(k_*) \left(\frac{k}{k_*}\right)^{n_s-1} \left[1 + A\cos\left(\frac{\phi_k}{f}\right)\right]\,, \tag{35.58}$$

  where $A \propto b_*$.
- the non-Gaussianity is oscillating, also because of the oscillating potential, and its possible resonance means that the constraints on the parameters from observations are looser than for usual models.
- the inflation is coupled to gauge fields, which results in gauge field production.
- the produced gauge fields have a stress tensor that can source gravitational waves, which are *chiral*, i.e., of only one chirality: the one that is unstable for the gauge field.

Important Concepts to Remember

- The vacuum state in the presence of instantons is parametrized by a phase θ, mixing winding numbers, and which can be represented as a parameter in the Lagrangian, multiplying the instanton action $F * F$.
- Axions can explain the smallness of θ: the anomaly equation in the presence of instantons means a cosine potential for θ, minimized at $\theta = 0$; $\theta(x)$ is called the axion field.
- Axions in string theory can appear as p-form fields integrated over cycles, or 4-dimensional duals of p-form fields, and are classically shift symmetric.
- At the quantum level, the string theory axions can get nonperturbative corrections from worldsheet, or D-brane, instantons.
- For a good axionic string inflation model, we need $f > 10 M_{\rm Pl}$, which is not possible for a generic string axion.
- We can have a fine-tuned "racetrack" with two axions, where an effective remaining axion has a very large decay constant f.
- We can have N-flation for the axions, where the effective field is the radial coordinate of N identical fields, and $\Phi \simeq \sqrt{N}|\phi|$.
- Axion monodromy means that the axion gets an extra $V_0(\phi)$, besides the axionic potential.
- For the axion monodromy model of a D5-brane in type IIB, with a B-field on Σ_2 as the axion, we have at large b a linear potential, leading to a chaotic inflation model.

- For a simple string monodromy model, of D4-brane moving on a cycle in a Nil 3-manifold, the inflaton potential is $V \propto \phi^{2/3}$.
- Dante's Inferno is a 2-axion monodromy model, with one field having a steep potential like a mass term, and a linear combination having the consine potential, and a hierarchy on the decay constants $f_r \ll f_\theta$.
- Axion monodromy produces an oscillating power spectrum, oscillating non-Gaussianity, gauge fields, and chiral gravitational waves.

Further reading: For more about axions, see [1], Chap. 10. For more about axions in string theory, and axion monodromy models, see [31], Sects. 3.2 and 5.4. The original axion monodromy scenario was defined in [66], following the model of [67] for inflation from monodromy in string theory, and the "Dante's Inferno" scenario was defined in [68].

Exercises

(1) Show the details for the calculation of the DBI action of the D4-brane on the Nil 3-manifold (35.40) leading to the action (35.44).
(2) Show explicitly that for

$$\begin{aligned} W &= W_0 + Ae^{-T} \\ K &= -3\ln(T + \bar{T} + \gamma b^2) , \end{aligned} \tag{35.59}$$

the holomorphic volume T is stabilized.
(3) Calculate the CMBR parameters r and n_s for the Dante's Inferno model, as a function of its parameters.
(4) Consider the D-brane action for $p = 5$. Does the WZ action affect the potential for the axion ϕ for any constant $C^{(p)}$?

Chapter 36
Fuzzy Dark Matter from String Theory

In this chapter, we will examine an alternative to the usual ΛCDM model, where the cold dark matter is replaced by a "fuzzy dark matter", and we will see that it appears more naturally in string theory. In fact, it is based on the axionic models from the previous chapter.

36.1 Fuzzy Dark Matter from String Theory

We have seen that a priori, the two most important relevant cases of dark matter models were cold dark matter, the relatively heavy, nonrelativistic matter that can only interact with itself but not others, and hot dark matter, which meant thermalized dark matter, in practice neutrinos, which are still relativistic (with a temperature comparable to the one of the CMBR), and are thermalized (because of their early times interaction with the other particles). Neutrinos are pretty much the only possibilities for hot dark matter, and have masses of less than 10^{-2} eV. As we saw in Chap. 3, the hot dark matter scenario is actually ruled out, though a small hot component to CDM ("warm dark matter") is still possible.

But there is a third possibility, one that is intriguing, and relates to the axions of the previous chapter. We can have dark matter that is extremely light, with masses of the order of $m \sim 10^{-22}$ eV, specifically a boson with de Broglie wavelength of $\lambda \sim 1kpc$, yet it is not thermal (extremely cold). That means that the boson, despite being much lighter than the neutrinos, is nonrelativistic, and is quantum on scales smaller than $1kpc$. The reason this scenario is possible is that the boson is not thermalized like the neutrinos.

The boson one will consider is an axion a, with action

$$S = \int d^4x \sqrt{-g} \left[-\frac{1}{2} f^2 (\partial_\mu a)^2 - \Lambda^4 (1 - \cos a) \right]. \tag{36.1}$$

H. Năstase, *Cosmology and String Theory*, Fundamental Theories of Physics 197, https://doi.org/10.1007/978-3-030-15077-8_36

We want the mass m to be

$$m = \frac{\Lambda^2}{f} \sim 10^{-22} - 10^{-21}\,\text{eV}. \tag{36.2}$$

From string theory, the generic axion decay constant is below the Planck scale, and above the grand unification (GUT) scale, so

$$10^{18}\,\text{GeV} \gtrsim f \gtrsim 10^{16}\,\text{GeV}\,, \tag{36.3}$$

though it is worth noting that most models from the previous chapter obtained a larger f for the effective axion. The confining scale Λ is obtained from instanton effects, and can in principle be suppressed by susy effects, with scale Λ_S, and we have

$$\Lambda^4 \sim M_{\text{Pl}}^2 \Lambda_S^2 e^{-S_{\text{inst}}}\,, \tag{36.4}$$

where S_{inst} is the instanton action, and Λ_S has, in susy breaking scenarios (or without susy at all), the value

$$10^{18}\,\text{GeV} \gtrsim \Lambda \gtrsim 10^4\,\text{GeV}. \tag{36.5}$$

To obtain a mass of $m \sim 10^{-22}$ eV, if $\Lambda = 10^4$ GeV we need $S_{\text{inst}} = 165$, and if $\Lambda = 10^{18}$ GeV, we need $S_{\text{inst}} = 230$. But from the Standard Model, the instanton action is $S_{\text{inst}} = 8\pi^2/g^2 \equiv 2\pi/\alpha_G$, where $\alpha_G = g^2/(4\pi)$ is the Standard Model gauge coupling at the GUT scale $\sim 10^{16}$ GeV. For $\alpha_G = 1/20$, we get $S_{\text{inst}} = 126$, and for $\alpha_G = 1/30$, we get $S_{\text{inst}} = 188$, so there is overlap with the desired region.

36.2 Axion Dynamics in the Expanding Universe

Having seen what the basic set-up is, and that it can be derived from string theory, we turn to describing its consequences.

The first thing to do is to set up how the axion evolves from the beginning of the Universe until today.

One assumes that the axion starts off as a constant field with some random initial value a_0. Then we evolve it in FLRW Universe $ds^2 = -c^2dt^2 + R^2(t)d\vec{x}^2$ as a time dependent field, $a = a(t)$. The KG equation of motion for the axion is then

$$\ddot{a} + 3H\dot{a} + m^2 \sin a = 0. \tag{36.6}$$

The solution is approximately constant for $H \gtrsim m$, and for $H \lesssim m$, it oscillates with angular frequency m, damped as $R(t)^{-3/2}$ due to the usual Hubble friction term.

The transition between the two regimes happens at a temperature T_0 such that (since we assume that the Universe is radiation dominated at early times) $\rho \simeq \rho_{\text{rad}} \sim T_0^4 \sim H^2 M_{\text{Pl}}^2 \sim m^2 M_{\text{Pl}}^2$, so

$$T_0 \sim \sqrt{m M_{\rm Pl}} \sim 500\,\text{eV}\ , \tag{36.7}$$

or a redshift of about $z \sim 2 \times 10^6$, which is after nucleosynthesis. Note that matter domination starts at about $T_1 \sim 1$ eV, so this is still after T_0.

Since the axion is about constant until $T_0 \sim 500$ eV, its potential energy (there is no kinetic energy since $\vec{\nabla} a = 0$ and $\dot{a} \simeq 0$ on the solution) $V \sim \Lambda^4$ acts as a component of dark energy ("cosmological constant"), being constant.

But after $T_0 \sim 500$ eV, the axion starts to oscillate (damped by Hubble friction), and it thus behaves like a condensate of bosons of zero spatial momentum, with energy density dropping like $R(t)^{-3}$, as for any nonrelativistic matter. It thus behaves like an extremely cold nonrelativistic dark matter (of negligible temperature).

One can also show that the axion self-interaction (due to the cosine potential) becomes extremely quickly negligible, in fact already at $T = 10^{-5/3} T_0$.

We can fix the the axion decay constant by noting that after T_0 (when radiation has $\rho_{\rm rad} \sim T_0^4$ and (fuzzy dark) matter has $\rho_m \sim \Lambda^4 = f^2 m^2$) and until T_1, when $\rho_m \sim \rho_{\rm rad}$, we have $\rho_m / \rho_{\rm rad} \propto T$, so

$$\left.\frac{\rho_m}{\rho_{\rm rad}}\right|_{eq.} \sim \frac{\Lambda^4}{T_0^4} \times \frac{T_0}{T_1} \simeq 1 \Rightarrow f = \sqrt{\frac{\Lambda^2}{m}} \sim \frac{M_{\rm Pl}^{3/2} T_1^{1/2}}{m^{1/4}} \sim 0.5 \times 10^{17}\,\text{GeV}\ , \tag{36.8}$$

which is consistent with the range required earlier.

36.3 Fuzzy Dark Matter as a Nonrelativistic Quantum System

The main applications of fuzzy dark matter are for galaxies and structures present in the Hubble volume, as in the case of cold dark matter, so we need to find out its description today. The first thing to note is that we can describe the fuzzy dark matter as a nonrelativistic quantum system.

The axion must be nonrelativistic, by our assumption. We thus write its action by reintroducing $\hbar$ and c, with the purpose of taking the nonrelativistic limit:

$$S = \int \frac{d^4x}{\hbar c^2} \sqrt{-g} \left[-\frac{1}{2} g^{\mu\nu} \partial_\mu \phi \partial_\nu \phi - \frac{1}{2} \frac{m^2 c^2}{\hbar^2} \phi^2 \right]\ , \tag{36.9}$$

where the field ϕ has dimensions of energy and $x^0 = ct$. The de Broglie wavelength is given by

$$\frac{\lambda}{2\pi} = \frac{\hbar}{mv} = 1.92 kpc \left(\frac{10^{-22}\,\text{eV}}{m} \right) \left(\frac{10\,\text{km/s}}{v} \right)\ , \tag{36.10}$$

which means that in order to get a de Broglie wavelength of about kpc, we need $m \sim 10^{-22}$ and $v \sim 10$ km/s, which indeed makes it very non-relativistic.

In the nonrelativistic limit, we must take out a phase containing the rest energy of the particle, $e^{-i\frac{E_0 t}{\hbar}} = e^{-i\frac{mc^2}{\hbar^2}}$, as well as a constant that rescales the action to the nonrelativistic action. Moreover, then the field multiplying it is complex, and we must add the complex conjugate, so

$$\phi = \sqrt{\frac{\hbar^2 c}{2m}} \left(\psi e^{-i\frac{mc^2}{\hbar}} + \psi^* e^{+i\frac{mc^2}{\hbar}} \right). \tag{36.11}$$

Substituting in the action, we obtain

$$S = \int d^4x \sqrt{-g} \left[-|g^{00}| \psi^* i\hbar \partial_t \psi + g^{rr} \frac{\hbar^2}{2m} (\vec{\nabla}\psi^*)(\vec{\nabla}\psi) + (|g^{00}| - 1) \frac{mc^2}{2} \psi^* \psi \right]. \tag{36.12}$$

Considering a background of the type of a FLRW metric with scale $R(t)$, perturbed by a Newton potential $\Phi(\vec{r}, t)$,

$$ds^2 = -\left(1 + \frac{2\Phi}{c^2}\right) c^2 dt^2 + R^2(t) \left(1 - \frac{2\Phi}{c^2}\right) d\vec{r}^2 , \tag{36.13}$$

and defining as usual the Hubble constant as $H = \dot{R}/R$, we find the equation of motion (varying with respect to ψ^*)

$$i\hbar \left(\dot{\psi} + \frac{3}{2} H \psi \right) = \left(-\frac{\hbar^2}{2mR^2(t)} \vec{\nabla}^2 + m\Phi \right) \psi. \tag{36.14}$$

This is just a modified Schrödinger equation, with terms coming from the Hubble expansion and the Newton potential, as expected for the nonrelativistic limit of a scalar field theory.

36.4 Superfluid Picture

However, we are in a classical situation, with small quantum fluctuations, so we must be able to view ψ as some kind of classical field. That is only possible if we view the dark matter as a superfluid. A regular fluid is a description of a quantum mechanical system at large distances (ignoring terms of higher orders in derivatives), whereas a superfluid is a quantum fluid with phase coherence: the wavefunction of the superfluid defines both the density ρ of the fluid, through the interpretation of $|\psi|^2$ as probability density, i.e., number density, or ρ/m, whereas the $U(1)$ current $\vec{j}$ defined from the phase is understood as $\rho\vec{v}$. That means that

$$\psi = \sqrt{\frac{\rho}{m}} e^{i\theta} \tag{36.15}$$

and the velocity is

$$\vec{v} = \frac{\vec{j}}{\rho} = \frac{1}{m\psi^*\psi}\frac{-i\hbar}{2R(t)}(\psi^*\vec{\nabla}\psi - \psi\vec{\nabla}\psi^*) = \frac{\hbar}{mR(t)}\vec{\nabla}\theta. \tag{36.16}$$

Because of $\vec{v} \propto \vec{\nabla}\theta$, we have zero vorticity, $\vec{\omega} \equiv \vec{\nabla} \times \vec{v} = 0$.

Replacing the above ψ in the modified Schrödinger equation, separating the equation into one for the modulus and one for the phase, then taking $\vec{\nabla}$ of the phase and replacing $\vec{\nabla}\theta$ by $\vec{v}$, we obtain two equations that look like the ones of the fluid in comoving coordinates,

$$\begin{aligned} &\dot{\rho} + 3H\rho + \frac{1}{R}\vec{\nabla}\cdot(\rho\vec{v}) = 0 \\ &\dot{\vec{v}} + H\vec{v} + \frac{1}{R}(\vec{v}\cdot\vec{\nabla})\vec{v} = -\frac{1}{R}\vec{\nabla}\Phi + \frac{\hbar^2}{2R^3m^2}\vec{\nabla}\left(\frac{\vec{\nabla}^2\sqrt{\rho}}{\sqrt{\rho}}\right). \end{aligned} \tag{36.17}$$

The first is like the continuity equation, and the second like the Euler equation, generalized to the expanding Universe. These are known as the Madelung equations, and we note that the only new feature is the "quantum pressure term," the last term in the Euler equation.

One can numerically simulate the Madelung equations to find specific fuzzy dark matter solutions.

36.5 Schrödinger–Poisson Equation and Spherical Soliton Solution

However, there is another way to approach solutions of the equations of motion (36.14). For many applications for galaxies, it is enough to ignore the Hubble expansion, so put $H = 0$ and $R(t) = 1$. Then we have a usual Schrödinger equation, turned into a time independent one in the usual way, by $\psi(\vec{r}, t) = e^{-iEt}\psi(\vec{r})$, so

$$-\frac{\hbar^2}{2m}\vec{\nabla}^2\psi(\vec{r}) + m\Phi\psi(\vec{r}) = E\psi(\vec{r}). \tag{36.18}$$

The new feature is that now, as before, $|\psi|^2$ is a probability density, so $m|\psi|^2$ is a mass density, that sources the Newton potential, so we also have the Poisson equation

$$\vec{\nabla}^2\Phi = 4\pi G_N m|\psi|^2. \tag{36.19}$$

For the solution defining a galaxy, we consider an isolated system, assuming that as $|\vec{r}| \to \infty$, the variables $\psi(\vec{r})$ and $\Phi(\vec{r})$ go to zero.

Considering a spherical solution, which can be chosen to be real, we obtain the equations

$$-\frac{\hbar^2}{2mr^2}\frac{d}{dr}r^2\frac{d\psi}{dr} + m\Phi\psi = E\psi$$
$$\frac{1}{r^2}\frac{d}{dr}r^2\frac{d\Phi}{dr} = 4\pi G_N m\psi^2. \quad (36.20)$$

We obtain soliton solutions ordered by an energy level n (for eigenstates of the Schrödinger equation), ψ_n and Φ_n. It turns out that we can scale out dimensions from all the parameters, given an overall mass $M = \int \rho = m \int |\psi|^2$ for the soliton, and express them in terms of numerically calculable dimensionless numbers. We find that the central density, central potential, and half-mass radius are

$$\rho_c = \left(\frac{Gm^2}{\hbar^2}\right)^3 M^4 \rho_n\,, \quad \Phi_c = \left(\frac{GMm}{\hbar}\right)^2 \phi_n\,, \quad r_{1/2} = \frac{\hbar^2}{GMm^2} f_n. \quad (36.21)$$

Numerically, we see that $f_0 \simeq 3.9251$ and f_n increases for $n > 0$, $\rho_0 = 0.00440$ and ρ_n decreases for $n > 0$.

36.6 Observational Consequences

First, the soliton formulas above mean that

$$r_{1/2} \geq 3.925\frac{\hbar^2}{GMm^2}; \quad \rho_c \leq 0.0044\left(\frac{Gm^2}{\hbar^2}\right)^3 M^4\,, \quad (36.22)$$

so there is a minimum half-mass radius, of about $0.33kpc$ for $M = 10^9 M_{\rm Sun}$ and $m = 10^{-22}$ eV, and a maximum central density, of about $7.05 M_{\rm Sun}/pc^3$ for the same values.

There is also a maximum mass for the soliton, since the central Newton potential Φ of the soliton cannot exceed $\sim c^2$, which from (36.21) gives about $\hbar c/(G_N m)$. More precisely, one finds

$$M_{\rm max} = 0.633\frac{\hbar c}{G_N m}\,, \quad (36.23)$$

which is about $8.46 \times 10^{11} M_{\rm Sun}$ for $m = 10^{-22}$ eV.

For distances large compared to the de Broglie wavelength, which as we saw, was about kpc order, fuzzy dark matter behaves like ordinary dark matter, since it looks classical. So for any scale larger than galactic, we cannot see a difference.

Differences occur on smaller scales. The minimum half-mass radius and maximum central density are examples.

The soliton core is expected to be surrounded by a virialized halo with the usual (CDM) Navarro-Frenk-White (NFW) profile, since those distances are larger than the de Broglie wavelength. The maximum mass above is only the mass of the soliton core, there are FDM halos with much larger mass.

From simulations, one finds a minimum mass for dark matter halos of the order of $10^7 M_{\text{Sun}}(m/10^{-22}\,\text{eV})^{-3/2}$.

Dynamical friction for clusters moving in dwarf galaxies is reduced considerably in FDM, due to various effects, including the quantum effects on small scales (smaller than the de Broglie wavelength) of the superfluid. That means that the timescale for relaxation is greatly increased, as observed experimentally in the globular clusters of the Fornax dwarf galaxy, a satellite galaxy of our own Milky Way galaxy.

The only possible contradiction with observations of the fuzzy dark matter scenario is from the Lyman-α absorbtion spectrum in galaxies, but there are many unknown effects, and it is not clear if the constraint, which seems to favor a mass m of about $10 - 20 \times 10^{-22}$ eV rather than the $1 - 10 \times 10^{-22}$ eV range favoured by the rest of observations, is correct.

Important Concepts to Remember

- Fuzzy dark matter is an alternative to cold dark matter, with an ultra-light boson with a mass $m \sim 10^{-22}$ eV, and a de Broglie wavelength of about $1kpc$, which means it is classical on larger scales.
- The boson is ultra-cold, so is nonrelativistic despite its extremely small mass.
- The best candidate is an axion, specifically a string theory axion with a decay constant f in between the Planck and GUT scales, and an energy density scale larger than 10^4 GeV.
- The axion behaves like a dark energy component, being a constant, until $T_0 \sim \sqrt{m M_{\text{Pl}}} \sim 500\,\text{eV}$, and as regular dark matter afterwards.
- FDM behaves like a nonrelativistic quantum system, satisfying a modified Schrödinger equation for the wavefunction, with a Hubble friction and a Newton potential term.
- FDM can be also recast as a superfluid satisfying Madelung equations, with a quantum pressure term.
- On galaxy scales, the Schrödinger–Poisson system admits spherical scaling soliton solutions.
- The soliton picture results in a maximum central core density, a minimum half-mass radius, a maximum core mass, and an NFW halo outside.

Further reading: The fuzzy dark matter scenario with string theory applications was defined in [69].

Exercises

(1) Show that the free massive scalar action (36.9) gives in the nonrelativistic limit (36.12), and that its equations of motion in the FLRW metric perturbed by the Newton potential, (36.13), are given by the modified Schrödinger equation (36.14).
(2) Show that the modified Schrödinger equation (36.14) leads to the Madelung equations (36.17).

(3) Prove the scaling relations (36.21) for the radial soliton solutions of the Schrödinger–Poisson equations.
(4) Find the solution for the equation of motion of the axion field (KG) in the expanding Universe, (36.6).

Chapter 37
Holographic Cosmology

In this chapter we will consider holographic approaches to cosmology, that is, to take advantage of the duality to a weakly coupled (therefore calculable) field theory provided by the AdS/CFT correspondence and its generalizations, gauge/gravity duality, in order to calculate the behaviour near the strongly coupled gravity near the cosmological singularity.

There have been various attempts at using holography in this way, with various degrees of success. Here I will focus on two examples, which provide more concrete observational consequences, though they both come with caveats.

In one approach, started by McFadden and Skenderis, one starts with a "domain wall" spacetime, which is an Euclidean version of the FLRW cosmology in the spatially flat ($k = 0$) case. One then constructs a *holographic phenomenological approach*, by calculating the results of a generic super-renormalizable quantum field theory (at the 3-dimensional boundary of the domain wall spacetime) momentum space 2-point function and, after analytical continuation to Lorentzian signature, compare the resulting power spectra against observations.

In another approach, based on some exact (*top down holography*) type IIB supergravity cosmological solutions, one calculates the transition of cosmological perturbations through a (strongly coupled) cosmological singularity, by doing a calculation in the (weakly coupled) gauge theory, assumed to be $\mathcal{N} = 4$ SYM with a time-dependent coupling constant g_{YM}, and then mapping to the dual cosmology.

37.1 Generic Holographic Cosmology Map

The generic idea about using holography to calculate correlation functions in cosmology is due to the seminal paper of Maldacena from 2002, which calculated for the first time inflationary non-Gaussianity in the 3-point functions. Maldacena proposed that holography in an asymptotically de Sitter background should use the

H. Năstase, *Cosmology and String Theory*, Fundamental Theories of Physics 197, https://doi.org/10.1007/978-3-030-15077-8_37

correspondence between the wavefunction of the Universe $\Psi[h_{ij}, \phi]$, with a given spatial 3-metric h_{ij} and scalar ϕ, and the dual field theory partition function, with the partition function $Z_{\rm QFT}[h_{ij}, \phi]$ depending on sources h_{ij} and ϕ for the operators,

$$\Psi[h_{ij}, \phi] = Z_{\rm QFT}[h_{ij}, \phi]. \tag{37.1}$$

The correspondence is analytically continued from (asymptotically) AdS space. The fields in the cosmology side start in the Bunch–Davies vacuum, corresponding to the analytical continuation from Euclidean signature.

The the stress tensor T_{ij} on the boundary is expressed in terms of the wavefunction Ψ in the bulk as

$$T_{ij}(x) = \frac{\delta Z_{\rm QFT}[h_{ij}, \phi]}{\sqrt{h}\delta h^{ij}(x)}\bigg|_{\delta h_{ij}=0,\phi=0} = \frac{\delta \Psi[h_{ij}, \phi]}{\sqrt{h}\delta h^{ij}(x)}\bigg|_{\delta h_{ij}=0,\phi=0}. \tag{37.2}$$

This is understood as an operator statement, and valid inside correlation functions. Analogously, we have for the boundary operator $\mathcal{O}$ coupling with ϕ a variation of the usual AdS/CFT relation,

$$\langle \mathcal{O}(x_1)\dots\mathcal{O}(x_n)\rangle = \frac{\delta^n Z_{\rm QFT}[h_{ij}, \phi]}{\delta\phi(x_1)\dots\delta\phi(x_n)}\bigg|_{\delta h_{ij}=\phi=0} = \frac{\delta^n \Psi[h_{ij}, \phi]}{\delta\phi(x_1)\dots\delta\phi(x_n)}\bigg|_{\delta h_{ij}=\phi=0}. \tag{37.3}$$

This then integrates to the wavefunction of the Universe

$$\Psi[\phi] = \exp\left[\sum_{n=2}^{\infty}\frac{1}{n!}\int dx_1\dots\int dx_n\langle\mathcal{O}(x_1)\dots\mathcal{O}(x_n)\rangle\right]. \tag{37.4}$$

and similarly when we include h_{ij}, dual to T_{ij} correlators.

The wavefunction $\Psi[h_{ij}, \phi]$ can be computed in the cosmology as $\Psi \sim e^{iS_{\rm cl}}$, where $S_{\rm cl}$ is the classical action on the on-shell fields with the given boundary values. But then correlation functions of h_{ij} and ϕ are evaluated as usual in quantum mechanics by integrating with the probability $|\Psi|^2$, e.g.

$$\langle h_{ij}(x_1)h_{kl}(x_2)\rangle = \int Dh_{mn}|\Psi[h_{mn}, \phi]|^2 h_{ij}(x_1)h_{kl}(x_2). \tag{37.5}$$

Since both sides of the equality are real, only the real part of $iS_{\rm cl}$ contributes, but we obtain a crucial factor of 2 from the square of the wavefunction:

$$\langle h_{ij}(x_1)h_{kl}(x_2)\rangle = \int Dh_{mn}e^{-2{\rm Im}S_{\rm cl}}h_{ij}(x_1)h_{kl}(x_2). \tag{37.6}$$

If the gravity theory is strongly coupled, this correlation function might be difficult to calculate. Otherwise, for a theory approximately in de Sitter space, these correlators for superhorizon distances, which are what is needed for inflationary CMB calculations, can be evaluated (and were evaluated by Maldacena). But in a different background, it might be hard to calculate.

Note that this is the correlation function that we want, which is a bit different than the correlation functions in usual AdS/CFT, which are correlation functions of the operators $\mathcal{O}(x)$ on the boundary field theory, and not of the sources $\phi(x)$ and $h_{ij}(x)$ associated with them. If we want to use holography to calculate the correlators of ϕ and h, we need to find a way to relate them to the correlators of $\mathcal{O}$ and T_{ij} in the quantum field theory.

37.2 Domain Wall/Cosmology Correspondence

The first holographic approach is based on the "domain wall/cosmology correspondence," which means a certain Wick rotation in the context of holographic models.

The "domain wall" spacetime is really a FLRW cosmological model, analytically continued to Euclidean signature. Together, the two signature solutions are written as

$$\begin{aligned} ds^2 &= \eta dz^2 + a^2(z)[\delta_{ij} + h_{ij}(z, \vec{x})]dx^i dx^j \\ \Phi &= \phi(z) + \delta\phi(z, \vec{x}). \end{aligned} \tag{37.7}$$

Here $\eta = -1$ is FLRW cosmology, and $\eta = +1$ is the domain wall, obtained by the Wick rotation $t = -iz$. h_{ij} is a gravitational fluctuation, $\delta\phi$ a dilaton fluctuation, and $a(z)$ the scale factor. The Euclidean vacuum becomes the standard Bunch-Davies vacuum when Wick rotating to the cosmology. Both signature solutions come from the same action, just with a different overall sign,

$$S = \frac{\eta}{2\kappa^2}\int d^4x\sqrt{|g|}[-R + (\partial_\mu\phi)^2 + 2\kappa^2 V(\Phi)]. \tag{37.8}$$

If $\phi(z)$ is monotonic, we can invert it to $z(\phi)$, and thus define a "fake superpotential" $W(\phi)$ by

$$H(z) = -\frac{1}{2}W(\phi(z)). \tag{37.9}$$

Then the equations of motion of the gravity plus scalar system are

$$\frac{\dot{a}}{a} = -\frac{1}{2}W(\phi(z))\,, \quad \dot{\phi} = W'\,, \quad 2\eta\kappa^2 V = W'^2 - \frac{3}{2}W^2. \tag{37.10}$$

We can add perturbations on top of the background satisfying the above conditions, and as usual consider

$$h_{ij} = -2\psi(z, \vec{x})\delta_{ij}\,, \tag{37.11}$$

go to comoving momentum space $\vec{q}$, and construct the usual gauge invariant variable

$$\zeta = \psi + \frac{H}{\dot{\phi}}\delta\phi \tag{37.12}$$

and transverse traceless perturbation γ_{ij}, and the slow roll parameter

$$\epsilon_H = -\frac{\dot{H}}{H^2} = 2\frac{W'^2}{W^2}. \tag{37.13}$$

Then the equations of motion become as usual

$$\begin{aligned} &\ddot{\zeta} + \left(3H + \frac{\dot{\epsilon}_H}{\epsilon_H}\right)\dot{\zeta} - \eta\frac{q^2}{a^2}\zeta = 0 \\ &\ddot{\gamma}_{ij} + 3H\dot{\gamma}_{ij} - \eta\frac{q^2}{a^2}\gamma_{ij}. \end{aligned} \tag{37.14}$$

The analytical continuation from Euclidean to Lorentzian signature is

$$\bar{\kappa}^2 = -\kappa^2 , \quad \bar{q} = -iq. \tag{37.15}$$

The scalar and tensor (superhorizon) power spectra are momentum space two-point functions of the perturbations ζ and γ_{ij} that go to a constant at late time values,

$$\begin{aligned} \Delta_S^2(q) &= \frac{q^3}{2\pi^2}\langle\zeta(q)\zeta(-q)\rangle = \frac{q^3}{2\pi^2}|\zeta_{q(0)}|^2 \\ \Delta_T^2(q) &= \frac{q^3}{2\pi^2}\langle\gamma_{ij}(q)\gamma_{ij}(-q)\rangle = \frac{2q^3}{\pi^2}|\gamma_{q(0)}|^2 , \end{aligned} \tag{37.16}$$

where $\zeta_{q(0)}$ and $\gamma_{q(0)}$ are the late time values of the perturbations, which have at early times the Bunch-Davies vacuum condition $\zeta_q, \gamma_q \sim e^{-iq\tau}$ for conformal time $\tau \to -\infty$.

Define the canonical conjugate momenta to ζ_q and γ_q,

$$\Pi_q^{(\zeta)} = \frac{2}{\kappa^2}\epsilon_H a^3 \dot{\zeta}_q , \quad \Pi_q^{(\gamma)} = \frac{1}{4\kappa^2}a^3\dot{\gamma}_q , \tag{37.17}$$

and impose (operatorial) canonical commutation conditions for them,

$$\begin{aligned} &\zeta_q \Pi_q^{(\zeta)*} - \Pi_q^{(\zeta)}\zeta_q^* = i \\ &2(\gamma_q \Pi_q^{(\gamma)*} - \Pi_q^{(\gamma)}\gamma_q^*) = i. \end{aligned} \tag{37.18}$$

The specific functions for ζ_q and γ_q imply a given $\dot{\zeta}_q, \dot{\gamma}_q$, therefore a given $\Pi_q^{(\zeta)}$ and $\Pi_q^{(\gamma)}$, which can be described by saying that there is a response to ζ_q and γ_q, defined by response functions Ω and E,

$$\Pi_q^{(\zeta)} = \Omega\zeta_q \ , \quad \Pi_q^{(\gamma)} = E\gamma_q . \tag{37.19}$$

Substituting this functional form in the canonical commutation relations above, the real parts cancel, and we obtain the conditions

$$|\zeta_q|^2 = -\frac{1}{2\mathrm{Im}\Omega} \ , \quad |\gamma_q|^2 = -\frac{1}{4\mathrm{Im}E} . \tag{37.20}$$

Substituting in the superhorizon power spectra, we obtain

$$\begin{aligned}\Delta_S^2(q) &= \frac{-q^3}{4\pi^2 \mathrm{Im}\Omega_{(0)}(q)} \ , \\ \Delta_T^2(q) &= \frac{-q^3}{2\pi^2 \mathrm{Im}E_{(0)}(q)} .\end{aligned} \tag{37.21}$$

In the domain wall spacetime, i.e., in Euclidean signature, we can define the same variables, via the analytical continuation

$$\bar{\Omega}(-iq) = \Omega(q) \ , \quad \bar{E}(-iq) = E(q) \ , \tag{37.22}$$

leading to the definitions

$$\bar{\Pi}_{\bar{q}}^{(\zeta)} = -\bar{\Omega}\zeta_{\bar{q}} \ , \quad \bar{\Pi}_q^{(\gamma)} = -\bar{E}\gamma_{\bar{q}} . \tag{37.23}$$

37.3 Holographic Calculation

To calculate holographically the power spectra, we must calculate $\bar{\Omega}_{(0)}$ and $\bar{E}_{(0)}$ holographically, and then make the analytical continuation to cosmology.

For asymptotically AdS or dS domain walls, the holographic prescription is understood. In particular, for the dS case relevant for inflation, this was done by Maldacena, as reviewed at the beginning of the chapter.

We can then adopt the same logic as there for the relevant case here, which is of a power law corresponding to FLRW, at least asymptotically

$$a(z) \sim (z/z_0)^n \ , \quad \phi \sim \sqrt{2n}\log(z/z_0) . \tag{37.24}$$

One calculates holographically, as described at the begining of the chapter, the two-point function of energy-momentum tensors on the boundary, decomposed in general as

$$\langle T_{ij}(\bar{q})T_{kl}(-\bar{q})\rangle = A(\bar{q})\Pi_{ijkl} + B(\bar{q})\pi_{ij}\pi_{kl} \ , \tag{37.25}$$

where

$$\Pi_{ijkl} = \pi_{i(k}\pi_{l)j} - \frac{1}{2}\pi_{ij}\pi_{kl}\ , \quad \pi_{ij} = \delta_{ij} - \frac{\bar{q}_i\bar{q}_j}{\bar{q}^2} \tag{37.26}$$

are the 4-index transverse traceless projection operator, and the 2-index transverse projection operator (a 4-index trace), respectively.

But a holographic calculation based on the holographic renormalization approach of Skenderis and Papadimitriou (that will not be reproduced here) can be done proving that the coefficients $A(\bar{q})$ and $B(\bar{q})$ are related to the response functions Ω an E at large times as

$$A(\bar{q}) = 4\bar{E}_{(0)}(\bar{q})\ , \quad B(\bar{q}) = \frac{1}{4}\bar{\Omega}_{(0)}(\bar{q})\ , \tag{37.27}$$

leading finally to

$$\Delta_S^2(\bar{q}) = \frac{-(i\bar{q})^3}{16\pi^2 \mathrm{Im} B(\bar{q})}\ , \quad \Delta_T^2(\bar{q}) = \frac{-2(i\bar{q})^3}{\pi^2 \mathrm{Im} A(\bar{q})}. \tag{37.28}$$

Note that in principle, the same holographic calculation can be made using the Maldacena map (37.1) to calculate the h_{ij} two-point function from the T_{ij} two-point function on the boundary. The result is the same as above.

The last step is going back to Lorentzian signature, by Wick rotating the momenta, as well as the rank N of the QFT gauge group,

$$\bar{q} = -iq\ , \quad \bar{N} = -iN. \tag{37.29}$$

37.4 Phenomenological QFT Calculation

The quantum field theory in which a calculation is done is the most general super-renormalizable gauge theory on the Euclidean 3 dimensional boundary. In 3 dimensions, Yukawa couplings and ϕ^4 couplings are dimensional, therefore superrenormalizable, and there are no higher powers of fields with dimensional couplings. The fields are gauge field $A_i = A_i^a T_a$, scalars $\phi^M = \phi^{aM} T_a$ and fermions $\psi^L = \psi^{aL} T_a$ which are in the adjoint of $SU(N)$ and have flavor indices M, L.

The Euclidean space action is then

$$\begin{aligned} S_{\mathrm{QFT}} &= \int d^3x\, \mathrm{Tr}\left[\frac{1}{2}F_{ij}F^{ij} + \delta_{M_1M_2}D_i\Phi^{M_1}\Phi^{M_2} + 2\delta_{L_1L_2}\bar{\psi}^{L_1}\gamma^i D_i\psi^{L_2}\right.\\ &\quad \left.+\sqrt{2}g_{YM}\mu_{ML_1L_2}\Phi^M\bar{\psi}^{L_1}\psi^{L_2} + \frac{1}{6}g_{YM}^2\lambda_{M_1\dots M_4}\Phi^{M_1}\dots\Phi^{M_4}\right]\\ &= \frac{1}{g_{YM}^2}\int d^3x\, \mathrm{Tr}\left[\frac{1}{2}F_{ij}F^{ij} + \delta_{M_1M_2}D_i\Phi^{M_1}\Phi^{M_2} + 2\delta_{L_1L_2}\bar{\psi}^{L_1}\gamma^i D_i\psi^{L_2}\right.\\ &\quad \left.+\sqrt{2}\mu_{ML_1L_2}\Phi^M\bar{\psi}^{L_1}\psi^{L_2} + \frac{1}{6}\lambda_{M_1\dots M_4}\Phi^{M_1}\dots\Phi^{M_4}\right]\ , \end{aligned} \tag{37.30}$$

where in going from the first form to the second we have rescaled the fields by g_{YM}. Here $\lambda_{M_1 \ldots M_4}$ and $\mu_{ML_1L_2}$ are dimensionless couplings, $\mathrm{Tr}[T_aT_b] = \frac{1}{2}\delta_{ab}$, and in the first line $[A_i] = 1/2, [\Phi^M] = 1/2, [\psi^L] = 1$, whereas in the second $[A_i] = 1, [\Phi^M] = 1, [\psi^L] = 3/2$, as in 4 dimensions.

The fact that in the second form the dimensions are in g^2_{YM}, outside the action, but inside there are no dimensional parameters, is called "generalized conformal structure," and it implies a scaling form for 2-point functions, which in turn implies a scaling form for the coefficients A and B: the dimensions are contained in powers of momenta only, and these come in quantum corrections via unknown functions of the effective coupling

$$g^2_{\mathrm{eff}} = \frac{g^2 N}{q}. \qquad (37.31)$$

Since also A and B (the whole 2-point function) scales as N^2 in the large N limit, as we saw in general AdS/CFT, we have the general scalings

$$A(q, N) = q^3 N^2 f_T(g^2_{\mathrm{eff}}) \; , \quad B(q, N) = \frac{1}{4} q^3 N^2 f(g^2_{\mathrm{eff}}) \; , \qquad (37.32)$$

where the 1/4 factor is conventional.

Finally, we need to make the Wick rotation to Lorentzian signature, which takes

$$\bar{q}^3 \bar{N}^2 = -i q^3 N^2 \; , \quad g^2_{\mathrm{eff}}(\bar{q}, \bar{N}) = g^2_{\mathrm{eff}}(q, N) \; , \qquad (37.33)$$

and thus the measured superhorizon power spectra in terms of the Euclidean QFT are

$$\Delta^2_S(q) = \frac{q^3}{4\pi^2 N^2 f(g^2_{\mathrm{eff}})} \; , \quad \Delta^2_T(q) = \frac{2q^3}{\pi^2 N^2 f_T(g^2_{\mathrm{eff}})}. \qquad (37.34)$$

The interesting regime is when the gravitational description is strongly coupled, which results in a weakly coupled dual QFT, meaning we can calculate f and f_T in perturbation theory. Under general considerations, we find in perturbation theory

$$\begin{aligned} f(g^2_{\mathrm{eff}}) &= f_0 \left[1 - f_1 g^2_{\mathrm{eff}} \ln g^2_{\mathrm{eff}} + f_2 g^2_{\mathrm{eff}} + \mathcal{O}(g^4_{\mathrm{eff}})\right] \\ f_T(g^2_{\mathrm{eff}}) &= f_{T0} \left[1 - f_{T1} g^2_{\mathrm{eff}} \ln g^2_{\mathrm{eff}} + f_{T2} g^2_{\mathrm{eff}} + \mathcal{O}(g^4_{\mathrm{eff}})\right]. \end{aligned} \qquad (37.35)$$

Here f_0 and f_{T0} are obtained from a 1-loop computation, and f_1, f_{T1} and f_2, f_{T2} are from a two-loop computation. In it, we set the RG scale μ equal to the pivot scale q_* of the observational CMBR spectrum.

We can define dimensionless variables g, β and g_T, β_T via

$$\begin{aligned} f_1 g^2_{YM} N &= g q_* \; , \quad \ln \beta = -\frac{f_2}{f_1} - \ln |f_1| \; , \\ f_{T1} g^2_{YM} N &= g q_* \; , \quad \ln \beta_T = -\frac{f_{T2}}{f_{T1}} - \ln |f_{T1}| \; , \end{aligned} \qquad (37.36)$$

which parametrizes the power spectrum as

$$\Delta_S^2(q) = \frac{\Delta_0^2}{1 + (gq_*/q)\ln|q/\beta g q_*| + \mathcal{O}(gq_*/q)^2}$$
$$\Delta_T^2(q) = \frac{\Delta_{0T}^2}{1 + (g_T q_*/q)\ln|q/\beta_T g q_*| + \mathcal{O}(g_T q_*/q)^2}, \quad (37.37)$$

where

$$\Delta_0^2 = \frac{1}{4\pi^2 N^2 f_0}, \quad \Delta_{T0}^2 = \frac{2}{\pi^2 N^2 f_{T0}}. \quad (37.38)$$

The parameters f_0, f_1, f_2 (or g, β, Δ_0^2) have been calculated at 2-loops in terms of the number of scalars and fermions, and the dimensionless couplings μ and λ, but the formulas are long and not very illuminating. An important observation is that, in order to be able to fit against CMBR data, we need to introduce a nonminimal coupling between scalars and gravity, in the form of a term in the action of

$$\frac{1}{2g_{YM}^2}\int d^4x\sqrt{-g}\sum_M R(\Phi^M)^2\xi_M. \quad (37.39)$$

As far as the tensor perturbations, the most relevant number is the tensor to scalar ratio, r. In terms of the number of scalars N_S and number of fermions N_F, the formula one finds is

$$r = 32\frac{1+\sum_{M=1}^{N_S}(1-8\xi_M)^2}{1+2N_F+N_S}. \quad (37.40)$$

In the limit $N_S \gg N_F$ and $N_S \gg 1$, and for $\xi_M = \xi$ independent of the scalar, we find

$$r \simeq 32(1-8\xi)^2, \quad (37.41)$$

The holographic spectrum (37.37) has been compared with the data, as a parametrization, different than the one inspired by inflation with ΛCDM, with an n_s that runs, so n_s and $\alpha_s = dn_s/d\ln q$,

$$\Delta_S^2(q) = \Delta_0^2(q_*)\left(\frac{q}{q_*}\right)^{(n_s-1)+\frac{\alpha_s}{2}\ln\frac{q}{q_*}}. \quad (37.42)$$

The idea in using α_s as a parameter as well is that now we have the same number of parameters: besides the pivot scale q_*, and the normalization $\Delta_0(q_*)$, we have on one side g and β, and on the other n_s and α_s. The result is that both fits (the inflationary one with n_s and α_s, and the phenomenologic holographic cosmology one, with β and g) are compatible with CMBR data.

In the case of the phenomenological holographic cosmology, we find that theories with only gauge fields and fermions are excluded, but if we introduce scalars *and* nonminimal couplings to gravity for them, we can fit the CMBR. To give one example of a good fit, if we choose a common nonminimality parameter $\xi = 0.133$, giving $r = 0.12$, and a common $\lambda = 1$, we find $N = 2995$ and $N_S = 23255$ (so $N^2 \gg N_S$, as needed).

From the fits, one finds that generically, $f_1 g_{\rm eff}^2 \ln g_{\rm eff}^2 \simeq 1$ at $l \simeq 35$. Thus at very low multipoles ($l < 30$), perturbation theory breaks down, and we would need instead to use lattice quantum field theory to give predictions. Restricting to the reliable part of the data $l > 30$ (known reliably from a theoretical point of view), the holographic cosmology fits the CMBR data just as well as the ΛCDM model with inflation.

37.5 Time Dependent Field Theory Model: Top Down Holography

Another interesting approach to holographic cosmology starts with the original AdS/CFT set-up, of the duality between $\mathcal{N} = 4$ SYM theory in 3+1 dimensions, and string theory (in the supergravity limit) on $AdS_5 \times S^5$, with metric

$$ds_{10}^2 = \frac{R^2}{z^2}[dz^2 + \eta_{\mu\nu}dx^\mu dx^\nu] + R^2 d\Omega_5^2\,, \tag{37.43}$$

and 5-form

$$F_{(5)} = \omega_5 + *_{10}\omega_5\,, \tag{37.44}$$

where ω_5 is the volume form on the 5-sphere.

Instead of considering a 4 dimensional bulk cosmology like before, one considers a 5 dimensional bulk, with a spatially flat ($k = 0$) 3+1 dimensional cosmology *near the boundary*, i.e., consider instead a 10 dimensional metric

$$ds_{10}^2 = \frac{R^2}{z^2}[dz^2 + \tilde{g}_{\mu\nu}dx^\mu dx^\nu] + R^2 d\Omega_5^2\,, \tag{37.45}$$

where the boundary metric is

$$ds_4^2 = \tilde{g}_{\mu\nu}dx^\mu dx^\nu = -dT^2 + a^2(T)\delta_{ij}dx^i dx^j. \tag{37.46}$$

Yet, this is still considered as a *bulk* geometry, with a dual field theory still living in *flat* spacetime, as the boundary is conformally flat:

$$ds_4^2 = a^2(T)\left[-\frac{dT^2}{a^2(T)} + \delta_{ij}dx^i dx^j\right] \equiv a^2(t)[-dt^2 + \delta_{ij}dx^i dx^j]\,, \tag{37.47}$$

where t is conformal time (which we called η in the context of cosmology), which is regular time in the boundary field theory.

In order to be a type of "top-down" model, we need to find a gravitational solution of type IIB that can still be viewed as a dual to some sort of modification of $\mathcal{N} = 4$ SYM. The way this happens is that one introduces a nontrivial dilaton field, specifically time dependent, $\phi = \phi(T)$.

The type IIB supergravity equations of motion with metric and dilaton (Einstein and Klein-Gordon) for our ansatz,

$$\tilde{R}_{\mu\nu} = \frac{1}{2}\partial_\mu\phi\partial_\nu\phi$$
$$\partial_A(\sqrt{-g}G^{AB}\partial_B\phi) = 0 , \quad (37.48)$$

where g_{AB} is the full 10-dimensional metric and $\tilde{R}_{\mu\nu}$ is the Ricci tensor of $\tilde{g}_{\mu\nu}$, have the solution

$$a(T) = \left(\frac{3|T|}{2R}\right)^{1/3} , \quad \phi(T) = \frac{2}{\sqrt{3}}\ln\frac{|T|}{R} + \phi_0 , \quad (37.49)$$

valid both for $T > 0$ and for $T < 0$. Thus the cosmology presents a Big Crunch, followed by Big Bang, singularity. In conformal time t ($T = \frac{2R}{3}(t/R)^{3/2}$), we have

$$ds_4^2 = \tilde{g}_{\mu\nu}dx^\mu dx^\nu = \frac{|t|}{R}[-dt^2 + \delta_{ij}dx^i dx^j] , \quad \phi(t) = \sqrt{3}\ln\frac{|t|}{R} + \phi_0. \quad (37.50)$$

This implies

$$e^\phi = g_s\left(\frac{|t|}{R}\right)^{\sqrt{3}} \quad (37.51)$$

On the boundary theory, since in string theory we have the map

$$\frac{g_{YM}^2}{4\pi} = g_s = e^{\langle\phi\rangle} , \quad (37.52)$$

having a time dependent dilaton $\phi = \phi(t)$ amounts to having a time dependent Yang-Mills coupling $g_{YM} = g_{YM}(t)$, specifically

$$g_{YM}^2(t) = g_{YM,0}^2\left(\frac{|t|}{R}\right)^{\sqrt{3}} . \quad (37.53)$$

We thus consider a gauge theory action for the gauge fields

$$S_{YM} = -\frac{1}{4}\int d^4y(g_{YM}^2(t))^{-1}\,\mathrm{Tr}[F_{\mu\nu}F^{\mu\nu}]. \quad (37.54)$$

37.6 Spectrum of Fluctuations Going Through Singularity, from Holography

The specific question to be answered relates to the cosmological evolution near the singularity, where gravity is strongly coupled, therefore not calculable, but the dual field theory is weakly coupled, therefore quite calculable. The question, of relevance for models with Big Crunch followed by Big Bang, is of the propagation of fluctuations through the singularity.

The general strategy is: assume perturbations in the cosmological gravity theory, at $t \to -\infty$, which can be calculated in the gravitational background, or assumed as scale invariant. Translate them to gauge theory, then evolve the gauge theory perturbations, and finally reconstruct the perturbations at $t \to +\infty$.

In the gauge theory, one works in the gauge $A_0 = 0$ and $\partial^i A_i = 0$, so A_i are the only variables. Moreover, one is interested in the IR limit, in which $\vec{k}^2$ is small, so the operator corresponding to the dilaton perturbation $\delta\phi_k$ (in momentum space) is

$$\mathcal{O}_k = \text{Tr}[F_{\mu\nu}^2(k,t)] \simeq 2\dot{A}_{i;k}^2. \tag{37.55}$$

To relate the scalar perturbation to the field theory, we note first that a boundary fluctuation $\delta\phi_k$ must equal a boundary fluctuation of the operator $\text{Tr}[F_{\mu\nu}^2] \simeq 2\dot{A}_i^2$ in momentum space. But

$$\begin{aligned}\dot{A}_i^2 &= \int d^3k_1 d^3k_2 \dot{A}_i(k_1)\dot{A}_i(k_2)e^{i(k_1+k_2)x}V \\ &= \int d^3k e^{ikx}\sqrt{V}\int d^3k'\frac{1}{4}\dot{A}_i\left(\frac{k+k'}{2}\right)\dot{A}_i\left(\frac{k-k'}{2}\right)\sqrt{V},\end{aligned} \tag{37.56}$$

where we have defined $k_{1,2} = (k \pm k')/2$. Doing the integral over k, and identifying the Fourier space coefficient with the one of $\delta\phi_k/2$, we obtain

$$M_{\text{Pl}}^4\delta\phi_k = \frac{1}{4}\int d^3k'\dot{A}_i\left(\frac{k+k'}{2}\right)\dot{A}_i\left(\frac{k-k'}{2}\right)\sqrt{V}. \tag{37.57}$$

Then doing the integral in the region $k' < k$, where we can approximately put $k' = 0$ in the arguments of A_i, and obtain $\sim k^3$ from d^3k, we get

$$m_{\text{Pl}}^4\delta\phi_k \sim k^3\dot{A}_{i;k}^2\sqrt{V}. \tag{37.58}$$

If the spectrum of $\delta\phi_k$ is scale invariant, so $\propto k^{-3/2}$ (so $\mathcal{P}_k \sim k^3\delta\phi_k^2$ is k independent), then we obtain the initial gauge perturbation

$$\dot{A}_{i;k}^{\text{in}} \sim k^{-9/4}. \tag{37.59}$$

After performing the gauge theory calculation, we should propagate the operator $\mathcal{O}$ back to the bulk, via a "smearing function", a sort of causal bulk to boundary propagator, that is restricted to tbe nonzero only on a causal wedge relating the point z in the bulk with a slice of time $|t'-t| < z$ on the boundary, by

$$\delta\phi(z,\vec{x},t) = \int dt' d^3x' K(\vec{x}',t|z,\vec{x},t)\mathcal{O}(\vec{x}',t'). \tag{37.60}$$

Taking the Fourier transform, we find (defining $\vec{x}'-\vec{x}=\vec{y}'$ and $t'-t=s$)

$$\begin{aligned}\delta\phi(z,k,t) &= V^{-1/2}\int d^3x\delta\phi(z,x,t)e^{ik\cdot x}\\ &= \int dsd^3y'K(\vec{y}',t+s|z,0,t)e^{-\vec{k}\cdot\vec{y}'}\mathcal{O}(k,t+s)\\ &\simeq \int ds\left[\int d^3y'K(s+t,\vec{y}'|z,0,t)\right]\mathcal{O}(k,t+s).\end{aligned} \tag{37.61}$$

In this, we must substitute $\mathcal{O}(k,t+s)=2\dot{A}_i^2$ resulting from the $\dot{A}_i$ in the gauge theory calculation.

This strategy is quite broadly applicable, for various holographic duals.

Results

In the case of the $\mathcal{N}=4$ SYM with time-dependent coupling, the calculation of the passage through the singularity is as follows. One can redefine fields to

$$\tilde{A}_i = e^{-\phi/2}A_i\ , \tag{37.62}$$

which when substituted back in the gauge theory action gives an equation of motion for a KG field with a time dependent mass,

$$-\partial_\mu\partial^\mu\tilde{A}_i + M^2_{\rm SYM}(t)\tilde{A}_i = 0\ , \tag{37.63}$$

where

$$M^2_{\rm SYM}(t) = \frac{\ddot{\phi}}{2} - \frac{\dot{\phi}^2}{4} = -\frac{\sqrt{3}(\sqrt{3}+2)}{4t^2}\ , \tag{37.64}$$

where we can replace $\sqrt{3}$ by whatever power α we have in $g^2_{YM}\propto e^{\phi(t)}\propto |t|^\alpha$. It turns out that, because the mass blows up at the singularity at $t=0$, there is a singularity in the propagation of the modes through $t=0$.

To get a sensible result, one needs to regularize, for instance capping off $M_{\rm SYM}(t)$ at $t=-\xi$, maintaining it constant until $t=+\xi$. Then one finds that the spectrum is unchanged as it passes through the singularity (perhaps expected due to the perturbative nature of SYM, but one encounters subtleties along the way, as one has branch cuts). Only the amplitude is changed, between a time $-t$ and a time $+t$ as

$$\tilde{A}_k(+t) \simeq \left(\frac{t}{\xi}\right)^{2\nu_g} \tilde{A}_k(-t) , \tag{37.65}$$

where the exponent is

$$2\nu_g = 1 + \sqrt{3} \to 1 + \alpha. \tag{37.66}$$

Then one can go back to the cosmological spectrum, and find that the spectrum as it passes through the perturbation is unchanged, and remains scale invariant. Only the amplitude gets enhanced.

But the calculation is not reliable: we needed to regularize the equations by hand, and even so, we obtain some singular behaviour.

Nevertheless, the method explained here could be more broadly used.

Important Concepts to Remember

- In the holographic cosmology approach, one can equate the wavefunction of the Universe, as a function of spatial metric h_{ij} with the boundary partition function for the same source, $\Psi[h_{ij}] = Z[h_{ij}]$.
- We are interested in correlation functions of h_{ij} in the gravity side, but we can calculate instead correlation functions of the operators dual to them (T_{ij} here) in the dual QFT. We must learn how to relate the two.
- In the domain wall/cosmology correspondence, one can make an Euclidean space continuation (Wick rotation) of a 4 dimensional cosmology with an $a(t)$ to an Euclidean domain wall with $a(z)$, which can have a holographic dual.
- A phenomenological approach to holographic cosmology can be obtained by relating the scalar and tensor power spectra in cosmology, $\Delta^2_{0,S/T}(k)$, with response functions Ω and E that take fluctuations into their canonical conjugates.
- Holographically, Ω and E are calculated from the coefficients A and B of the 2-point functions of T_{ij}, and an analytical continuation to Lorentzian signature.
- As a dual QFT, one considers a general gauge+fermions+scalars+nonminimal coupling to R theory, and matches against CMBR data.
- The matching, considered only over the region $l > 30$ in which the perturbative calculation is valid (as opposed to some lattice QFT one), is as good as ΛCDM with inflation.
- For top-down holography, one can consider a 5 dimensional bulk, deformed $AdS_5 \times S^5$, made up of a 3+1 dimensional flat FLRW metric, which is the metric near the boundary, and is conformal to flat space, and a 4 dimensional $\mathcal{N} = 4$ SYM in flat space.
- One uses a type IIB supergravity solution with a time dependent dilaton and, since $e^{\phi(t)} \sim g_s \sim g^2_{YM}$, a time dependent coupling in $\mathcal{N} = 4$ SYM.
- Fluctuations can be mapped from cosmology to field theory, then evolved through the singularity (which is strongly coupled on the gravity side), then mapped back to cosmology.

Further reading: Maldacena's map holographic map on correlation functions, and from wavefunctions to partition functions, was defined in [70]. The phenomenological holographic cosmology was defined in [71], and its effects for Planck data were studied in [72, 73]. The time dependent field theory model was studied in [74, 75].

Exercises

(1) Write down the 1-loop Feynman diagram(s) for the 2-point function of the energy-momentum tensor in (37.30) and the integral expression using the Feynman rules.
(2) Show that the holographic cosmology ansatz for the scalar spectrum Δ_S^2 in (37.37) and the inflationary ansatz in (37.42) are not that different in terms of their CMBR values, so they have a chance of being alternative fits to the data.
(3) Show that the equations of motion of type IIB supergravity (37.48) have the solutions (37.49).
(4) Show that the equation of motion of the gauge field, under the rescaling $\tilde{A}_i = e^{-\phi/2} A_i$, under the solution (37.49), gives the KG equation with mass (37.64).

References

1. E.W. Kolb and M.S. Turner, "The early Universe," Westview Press, 1994.
2. V. Mukhanov, "Physical foundations of cosmology," Cambridge University Press, 2005.
3. Scott Dodelson, "Modern cosmology," Academic Press, 2003.
4. P.J.E. Peebles, "Principles of Physical Cosmology", Princeton University Press, 1993.
5. F. Couchot, S. Henrot-Versillé, O. Perdereau, S. Plaszczynski, B. Rouillé d'Orfeuil, M. Spinelli and M. Tristram, "Cosmological constraints on the neutrino mass including systematic uncertainties," Astron. Astrophys. **606**, A104 (2017) [arXiv:1703.10829 [astro-ph.CO]].
6. K. Freese, "Review of Observational Evidence for Dark Matter in the Universe and in upcoming searches for Dark Stars," EAS Publ. Ser. **36**, 113 (2009) [arXiv:0812.4005 [astro-ph]].
7. G. D'Amico, M. Kamionkowski and K. Sigurdson, "Dark matter astrophysics," arXiv:0907.1912.
8. D. Clowe, M. Bradac, A.H. Gonzalez, M. Markevitch, S.W. Randall, C. Jones and D. Zaritsky, "A direct empirical proof of the existence of dark matter," Astrophys. J. **648**, L109 (2006) [arXiv:astro-ph/0608407].
9. D.H. Lyth and A.R. Liddle, "The primordial density perturbation: Cosmology, inflation, and the origin of structure," Cambridge University Press, 2009.
10. A.R. Liddle and D.H. Lyth, "Cosmological inflation and large-scale structure," Cambridge University Press, 2000 (for inflation and the CMBR).
11. Daniel Baumann, "TASI lectures on inflation," arXiv:0907.5424 [hep-th].
12. S. Weinberg, "Cosmology", Oxford University Press, 2008.
13. M.J. Duff, B.E.W. Nilsson and C.N. Pope, "Kaluza-Klein Supergravity," Phys. Rept. **130**, 1 (1986).
14. H. Nastase, D. Vaman and P. van Nieuwenhuizen, "Consistency of the AdS(7) x S(4) reduction and the origin of selfduality in odd dimensions," Nucl. Phys. B **581**, 179 (2000) [arXiv:hep-th/9911238].
15. P. West, "Introduction to supersymmetry and supergravity", Wspc, 1990.
16. J. Wess and J. Bagger, "Supersymmetry and supergravity", Princeton University Press, 1992.
17. D. Freedman and A. van Proeyen, "Supergravity", Cambridge University Press, 2012.
18. P. Van Nieuwenhuizen, "Supergravity," Phys. Rept. **68**, 189 (1981).
19. S. Weiberg, "The quantum theory of fields, vol. 3: Supersymmetry", Cambridge University Press, 2005.
20. Barton Zwiebach, "A first course in string theory", Cambridge University Press, 2009.
21. J. Polchinski, "String theory", vols. 1 and 2, Cambridge University Press, 2005.
22. M. Green, J. Schwarz and E. Witten, "Superstring theory", Cambridge University Press, 2012.
23. Clifford V. Johnson, "D-branes", Cambridge University Press, 2003.

H. Năstase, *Cosmology and String Theory*, Fundamental Theories of Physics 197, https://doi.org/10.1007/978-3-030-15077-8

24. J.M. Maldacena, "The Large N limit of superconformal field theories and supergravity," Int. J. Theor. Phys. **38**, 1113 (1999) [Adv. Theor. Math. Phys. **2**, 231 (1998)] [arXiv:hep-th/9711200].
25. S.S. Gubser, I.R. Klebanov and A.M. Polyakov, "Gauge theory correlators from noncritical string theory," Phys. Lett. B **428**, 105 (1998) [arXiv:hep-th/9802109].
26. E. Witten, "Anti-de Sitter space and holography," Adv. Theor. Math. Phys. **2**, 253 (1998) [arXiv:hep-th/9802150].
27. Horatiu Nastase, "Introduction to the ADS/CFT Correspondence", Cambridge University Press, 2015.
28. M. Ammon and J. Erdmenger, "Gauge/gravity duality and applications," Cambridge University Press, 2015.
29. D. Baumann and L. McAllister, "Advances in Inflation in String Theory," Ann. Rev. Nucl. Part. Sci. **59**, 67 (2009) [arXiv:0901.0265 [hep-th]].
30. L. McAllister and E. Silverstein, "String Cosmology: A Review," Gen. Rel. Grav. **40**, 565 (2008) [arXiv:0710.2951 [hep-th]].
31. D. Baumann and L. McAllister, "Inflation and String Theory," arXiv:1404.2601 [hep-th].
32. D. Baumann and L. McAllister, "Inflation and String Theory," Cambridge University Press, 2015.
33. H. Bernardo and H. Nastase, "Small field inflation in $\mathcal{N} = 1$ supergravity with a single chiral superfield," JHEP **1609**, 071 (2016) [arXiv:1605.01934 [hep-th]].
34. C.P. Burgess, M. Majumdar, D. Nolte, F. Quevedo, G. Rajesh and R.J. Zhang, "The Inflationary brane anti-brane universe," JHEP **0107**, 047 (2001) [arXiv:hep-th/0105204].
35. H. Nastase and J. Sonnenschein, "Towards brane-antibrane inflation in type II_A: The Holographic MQCD model," JHEP **1202**, 040 (2012) [arXiv:1109.6813 [hep-th]].
36. P. Binetruy, C. Deffayet and D. Langlois, "Nonconventional cosmology from a brane universe," Nucl. Phys. B **565**, 269 (2000) [arXiv:hep-th/9905012].
37. J.M. Cline, C. Grojean and G. Servant, "Cosmological expansion in the presence of extra dimensions," Phys. Rev. Lett. **83**, 4245 (1999) [arXiv:hep-ph/9906523].
38. C. Csaki, M. Graesser, L. Randall and J. Terning, "Cosmology of brane models with radion stabilization," Phys. Rev. D **62**, 045015 (2000) [arXiv:hep-ph/9911406].
39. W.D. Goldberger and M.B. Wise, "Modulus stabilization with bulk fields," Phys. Rev. Lett. **83**, 4922 (1999) [arXiv:hep-ph/9907447].
40. P. Binetruy, C. Deffayet, U. Ellwanger and D. Langlois, "Brane cosmological evolution in a bulk with cosmological constant," Phys. Lett. B **477**, 285 (2000) [arXiv:hep-th/9910219].
41. T. Shiromizu, K.i. Maeda and M. Sasaki, "The Einstein equation on the 3-brane world," Phys. Rev. D **62**, 024012 (2000) [arXiv:gr-qc/9910076].
42. F. Quevedo, "Lectures on string/brane cosmology," Class. Quant. Grav. **19**, 5721 (2002) [arXiv:hep-th/0210292].
43. T. Clifton, P.G. Ferreira, A. Padilla and C. Skordis, "Modified Gravity and Cosmology," Phys. Rept. **513**, 1 (2012) [arXiv:1106.2476 [astro-ph.CO]].
44. N. Arkani-Hamed, S. Dimopoulos and G.R. Dvali, "The Hierarchy problem and new dimensions at a millimeter," Phys. Lett. B **429**, 263 (1998) [arXiv:hep-ph/9803315].
45. I. Antoniadis, N. Arkani-Hamed, S. Dimopoulos and G.R. Dvali, "New dimensions at a millimeter to a Fermi and superstrings at a TeV," Phys. Lett. B **436**, 257 (1998) [arXiv:hep-ph/9804398].
46. L. Randall and R. Sundrum, "A Large mass hierarchy from a small extra dimension," Phys. Rev. Lett. **83**, 3370 (1999) [arXiv:hep-ph/9905221].
47. L. Randall and R. Sundrum, "An Alternative to compactification," Phys. Rev. Lett. **83**, 4690 (1999) [arXiv:hep-th/9906064].
48. D. Marolf and S. Yaida, "Energy conditions and junction conditions," Phys. Rev. D **72**, 044016 (2005) [arXiv:gr-qc/0505048].
49. S. Kachru, R. Kallosh, A.D. Linde and S.P. Trivedi, "De Sitter vacua in string theory," Phys. Rev. D **68**, 046005 (2003) [arXiv:hep-th/0301240].
50. S. Kachru, R. Kallosh, A.D. Linde, J.M. Maldacena, L.P. McAllister and S.P. Trivedi, "Towards inflation in string theory," JCAP **0310**, 013 (2003) [arXiv:hep-th/0308055].

51. H. Abe, T. Higaki and T. Kobayashi, Phys. Rev. D **73**, 046005 (2006) https://doi.org/10.1103/PhysRevD.73.046005 [hep-th/0511160].
52. J. Khoury, B.A. Ovrut, P.J. Steinhardt and N. Turok, "The Ekpyrotic universe: Colliding branes and the origin of the hot big bang," Phys. Rev. D **64**, 123522 (2001) [arXiv:hep-th/0103239].
53. J. Khoury, B.A. Ovrut, N. Seiberg, P.J. Steinhardt and N. Turok, "From big crunch to big bang," Phys. Rev. D **65**, 086007 (2002) [arXiv:hep-th/0108187].
54. J.L. Lehners, "Ekpyrotic and Cyclic Cosmology," Phys. Rept. **465**, 223 (2008) [arXiv:0806.1245 [astro-ph]].
55. P.J. Steinhardt and N. Turok, "Cosmic evolution in a cyclic universe," Phys. Rev. D **65**, 126003 (2002) [arXiv:hep-th/0111098].
56. E.I. Buchbinder, J. Khoury and B.A. Ovrut, "New Ekpyrotic cosmology," Phys. Rev. D **76**, 123503 (2007) [arXiv:hep-th/0702154].
57. R.H. Brandenberger and C. Vafa, "Superstrings in the Early Universe," Nucl. Phys. B **316**, 391 (1989).
58. R.H. Brandenberger, "String Gas Cosmology," String Cosmology, J. Erdmenger (Editor). Wiley, 2009. pp. 193–230 [arXiv:0808.0746 [hep-th]].
59. T. Battefeld and S. Watson, "String gas cosmology," Rev. Mod. Phys. **78**, 435 (2006) [arXiv:hep-th/0510022].
60. R.H. Brandenberger, "String Gas Cosmology after Planck," Class. Quant. Grav. **32**, no. 23, 234002 (2015) [arXiv:1505.02381 [hep-th]].
61. A. Nayeri, R.H. Brandenberger and C. Vafa, "Producing a scale-invariant spectrum of perturbations in a Hagedorn phase of string cosmology," Phys. Rev. Lett. **97**, 021302 (2006) [arXiv:hep-th/0511140].
62. S. Alexander, R.H. Brandenberger and D.A. Easson, "Brane gases in the early universe," Phys. Rev. D **62**, 103509 (2000) [arXiv:hep-th/0005212].
63. J. Khoury and A. Weltman, "Chameleon fields: Awaiting surprises for tests of gravity in space," Phys. Rev. Lett. **93**, 171104 (2004) [arXiv:astro-ph/0309300].
64. J. Khoury and A. Weltman, "Chameleon cosmology," Phys. Rev. D **69**, 044026 (2004) [arXiv:astro-ph/0309411].
65. K. Hinterbichler, J. Khoury and H. Nastase, "Towards a UV Completion for Chameleon Scalar Theories," JHEP **1103**, 061 (2011) Erratum: [JHEP **1106**, 072 (2011)] [arXiv:1012.4462 [hep-th]].
66. L. McAllister, E. Silverstein and A. Westphal, "Gravity Waves and Linear Inflation from Axion Monodromy," Phys. Rev. D **82**, 046003 (2010) [arXiv:0808.0706 [hep-th]].
67. E. Silverstein and A. Westphal, "Monodromy in the CMB: Gravity Waves and String Inflation," Phys. Rev. D **78**, 106003 (2008) [arXiv:0803.3085 [hep-th]].
68. M. Berg, E. Pajer and S. Sjors, "Dante's Inferno," Phys. Rev. D **81**, 103535 (2010) [arXiv:0912.1341 [hep-th]].
69. L. Hui, J.P. Ostriker, S. Tremaine and E. Witten, "Ultralight scalars as cosmological dark matter," Phys. Rev. D **95**, no. 4, 043541 (2017) [arXiv:1610.08297 [astro-ph.CO]].
70. J.M. Maldacena, "Non-Gaussian features of primordial fluctuations in single field inflationary models," JHEP **0305**, 013 (2003) [arXiv:astro-ph/0210603].
71. P. McFadden and K. Skenderis, "Holography for Cosmology," Phys. Rev. D **81**, 021301 (2010) [arXiv:0907.5542 [hep-th]].
72. N. Afshordi, C. Coriano, L. Delle Rose, E. Gould and K. Skenderis, "From Planck data to Planck era: Observational tests of Holographic Cosmology," Phys. Rev. Lett. **118**, no. 4, 041301 (2017) [arXiv:1607.04878 [astro-ph.CO]].
73. N. Afshordi, E. Gould and K. Skenderis, "Constraining holographic cosmology using Planck data," Phys. Rev. D **95**, no. 12, 123505 (2017) [arXiv:1703.05385 [astro-ph.CO]].
74. R.H. Brandenberger, E.G.M. Ferreira, I.A. Morrison, Y.F. Cai, S.R. Das and Y. Wang, "Fluctuations in a cosmology with a spacelike singularity and their gauge theory dual description," Phys. Rev. D **94**, no. 8, 083508 (2016) [arXiv:1601.00231 [hep-th]].

75. E.G.M. Ferreira and R. Brandenberger, “Holographic Curvature Perturbations in a Cosmology with a Space-Like Singularity,” JCAP **1607**, no. 07, 030 (2016) [arXiv:1602.08152 [hep-th]].
76. P.A.R. Ade *et al.* [Planck Collaboration], “Planck 2013 results. XXII. Constraints on inflation,” Astron. Astrophys. **571**, A22 (2014) [arXiv:1303.5082 [astro-ph.CO]].
77. N. Aghanim *et al.* [Planck Collaboration], “Planck 2018 results. VI. Cosmological parameters,” arXiv:1807.06209 [astro-ph.CO].

Index

H. Năstase, *Cosmology and String Theory*, Fundamental Theories of Physics 197, https://doi.org/10.1007/978-3-030-15077-8

The manufacturer's authorised representative in the EU is Springer
Nature Customer Service Centre GmbH, Europaplatz 3, 69115 Heidelberg,
Germany. If you have any concerns regarding our products, please
contact ProductSafety@springernature.com

Printed and bound by CPI Group (UK) Ltd, Croydon, CR0 4YY
30/09/2026
02224059-0002